# ISSUES IN AGRICULTURAL DEVELOPMENT

IAAE OCCASIONAL PAPER SERIES

# Issues in Agricultural Development

## Sustainability and Cooperation

Edited by
Margot Bellamy and
Bruce Greenshields

International Association of Agricultural Economists

Published by
Dartmouth Publishing Company Limited
Gower House
Croft Road
Aldershot
Hants GU11 3HR
UK

Dartmouth Publishing Company
Old Post Road
Brookfield
Vermont 05036
USA


**British Library Cataloguing in Publication Data**
Issues in agricultural development: sustainability and cooperation
(IAAE Occasional Paper No. 6)
    1. Agricultural economics.
    I. Bellamy, Margot A.   II. Greenshields, Bruce L.   III.
International Association of Agricultural Economists.
    IV. Series.
    338.1

    ISBN 1 85521 302 8

Printed and Bound in Great Britain by
Athenaeum Press Ltd, Newcastle upon Tyne.

# Contents

# Foreword

This book is the sixth in the IAAE *Occasional Paper* series. It contains the 45 contributed papers—together with the discussion opening and summary of the general discussion—presented at the Twenty-First International Conference of Agricultural Economists, Tokyo, Japan, 22–29 August 1991. A companion publication—*Sustainable Agricultural Development: The Role of International Cooperation*—contains the invited papers presented at the conference, and also appears under the imprint of Dartmouth Publishing Company.

The contributed papers were refereed using a double-blind review process, and the acceptance rate for presentation in contributed paper sessions and publication in full in this series was 12 percent. The 27 referees, representing 20 countries and 6 continents, were Jean-Marc Boussard (Institut National de la Recherche Agronomique, France), Daniel Bromley (University of Wisconsin), Colin Brown (University of Queensland), Yang Boo Choe (Korea Rural Economics Institute), Elmar da Cruz (Empresa Brasileira de Pesquisa Agropecuária), José Luis Fernández-Cavada (Universidad Politécnica de Madrid), Murray Hawkins (Curtin University), Naraomi Imamura (Tokyo University), Odin Knudsen (World Bank), Ulrich Koester (Universität Kiel), Robert Lindner (University of Western Australia), Laurent Martens (State University Gent), Richard Meyer (Ohio State University), Wilfred Mwangi (Centro Internacional de Mejoramiento de Maiz y Trigo, Ethiopia), Hiroyuki Nishimura (Kyoto University), Peter Nuthall (Lincoln University), George Peters (University of Oxford), Ewa Rabinowicz (Swedish University of Agricultural Sciences), Mária Sebestyén-Kostyál (Budapest University of Economic Sciences), Secondo Tarditi (Università di Siena), Dušan Tomić (Economic Institute, Yugoslavia), C.L.F. van der Meer (National Council for Agricultural Research, Netherlands), Michele Veeman (University of Alberta), T.K. Warley (University of Guelph), Kelley White (Food and Agriculture Organization of the United Nations), Yan Ruizhen (People's University of China), and Antonio Yúnez-Naude (Colegio de México).

The International Association of Agricultural Economists is grateful to these referees for their efforts in reviewing the papers, to Bruce Greenshields of the Economic Research Service of the US Department of Agriculture for organizing the contributed paper competition and sessions, to Margot Bellamy of CAB International and Bruce Greenshields for editing this volume, to Gwen Matlock and Verla Rape of the Economic Research Service for their production and copy editing assistance, and to John Lee, the Administrator of the Economic Research Service, for his support of this part of the Association's activities. A particular debt is owed to Bruce Greenshields, who is not only the coeditor of this volume but who has also simultaneously assumed responsibility for the Association's journal, *Agricultural Economics*. We also extend our thanks to our publishers and to members of the Japanese Organizing Committee for the Tokyo Conference.

The views expressed herein are not necessarily those of the IAAE or of the institutions with which the contributors or editors are connected.

G.H. Peters
Proceedings Editor, IAAE

# Labour Supply of Aged Farmers in Japan

*Ryuichi Shigeno*[1]

**Abstract:** There is a rapid trend towards an ageing society in Japanese rural areas; aged farmers thus play an important part in agricultural production. The purpose of this paper is to analyse labour supply behaviour of aged farmers. After clarifying the main features of their labour supply, a labour supply function is estimated using data from the 1985 Survey of the Farm Household Economy.

Major findings are that: (1) aged farmers sometimes gain pleasure from working in agriculture, but their estimated wage and income affect whether they decide to be permanent or supplementary workers; (2) the participation rate in agriculture of aged females tends to increase if their husbands work in agriculture; (3) regional differences in probabilities of aged farmer employment can be explained by the income and wage effect and type of farming; and (4) pension payments for farmers have a negative effect on labour supply of the elderly; i.e., the participation rate in the agricultural labour force decreases about 7 percent because of pension payments.

## Introduction

The ratio of aged persons to total family members has been increasing rapidly in Japan in recent years. According to the 1985 Census of Agriculture, the percentage of farm members over 65 increased by 17.3 percent in a 10-year period, and by a significantly greater percentage than the 10.3 percent of elderly among the general population. The percentage of aged persons who work in agriculture has nevertheless remained high. The result is that the employment rate for farmers over 65 has reached 29.1 percent, and the role that the aged play in agricultural production is becoming extremely important.

The main purpose of this paper is to estimate the labour supply function of aged farmers, to analyse the factors affecting the labour supply of aged farmers, and to clarify their quantitative relationships. Based on the estimated labour supply function, the causes of regional differences in the labour supply of aged farmers and in farm size are also examined.

## The Special Nature of the Aged Farmer Workforce

The following three points may be singled out as being special features of the aged farmer workforce in Japan:

### Work that Provides a Meaningful Existence

The percentage of aged farmers who work is fairly high when compared with employment of the aged in nonagricultural sectors. One reason for this is that there are many different kinds of agricultural work and diversity among family farms, allowing for flexibility in letting the aged work only when they are able to and so desire. For this reason, many aged persons work in agriculture to obtain a meaningful existence.

The point may be confirmed through estimation of the production function and shadow wage rate, using data from the 1985 Survey of the Farm Household Economy.[2] The results indicate that the shadow wage rate becomes significantly different due to the unique employment status of the aged. Accordingly, the shadow wage rate for farms in which the aged work in agriculture for a relatively long time is fairly low compared to the rate for farms without an aged labour force.

The low shadow wage rate in farms with aged members implies that labour is surplus on these farms. At the same time, it means that the elderly are able to decide their own work schedule, and is one proof of agriculture's reputation as meaningful or healthy employment.

1

## The Aged as a Marginal Labour Force

There are regional disparities in the employment rate of aged members of farming families in Japan. For instance, the rate for males over 70 years old ranges from 76 percent in the Chugoku region to 36 percent in the Tohoku region, and for females over 70 years old from 34 percent in the Chugoku region to 8 percent in the Tohoku region. Such regional disparities exist because aged farmers are regarded as a marginal labour force. The marginal labour force operates in the boundary between the labour force and the nonworking sector, and whether it is employed or not depends on the financial state of individual households and also on employment status. Therefore, if the elderly are regarded as a marginal labour force, then the regional variance in their employment rates can be shown to be influenced by such economic factors as regional differences in agricultural structure, crop combinations, and incomes of other family members. The employment rate of aged farmers also varies widely depending on the size of the farm.

The decision on the amount of hours worked will in many cases not be explainable by economic analysis alone; it will be influenced by such noneconomic considerations as quality of life and health management. On the other hand, in the decision on whether or not to engage in agriculture as work, economic conditions are likely to have a substantial effect on participation of the aged in agriculture. This paper, therefore, is limited to the topic of whether aged farmers choose to work or not, and analyses of the factors that affect their work activities.

## Isolation from the Labour Market

In general, it is difficult for aged persons to work in the nonagricultural sector in Japan. Many companies have age limits of 55 or 60 years of age, and there is not much demand for temporary employment of the elderly. Even so, quite a few males aged 60–70 work in the nonagricultural sector, but they are largely self-employed. It is therefore probably correct to assume that aged members of farming families are isolated from the labour market. In construction of labour supply models, this point is important. If farms are faced with a competitive labour market, then resource allocation may be set independently of consumption behaviour. Many current farm labour supply models are recursive; i.e., they fix the derived demand of labour from a balanced production sector, and based on this decide the allocation between self-employment and hired labour. However, if the aged members of farms are isolated from the labour market, it becomes difficult to apply a recursive model, and it is desirable to construct a model that takes into consideration both the production and consumption sectors.

# Labour Supply Model for Aged Farmers

First, we must assume the consumption-leisure preference of aged farmers as (1):

(1) $\quad U = U \, (H–L, \, C_1, \, \dots \, C_i)$

subject to the restrictions facing the elderly:

(2) $\quad \Sigma P_i C_i = \pi(q; \, L) + Y_o + Y_p$, and $L \leq H$

where $H$ is available time of the elderly, $L$ the agricultural working hours of the elderly, $C_i$ the amount of consumption goods, $P_i$ the price of consumption goods, $\pi$ the profit function for farms, $q$ the input price vector, $Y_o$ assistance from other family members, and $Y_p$ the pension. Since there is no labour market for the elderly, it is impossible to estimate their market wage rate. The income from agriculture will therefore be calculated using a profit function such as (2).

The labour supply function of the elderly will be derived by maximizing (1) subject to (2). Since (2) is nonlinear, it is not easy to solve this problem. If linear homogeneity of the production function is assumed, then it is possible to solve it simply as described below.

If the production function is both linear and homogeneous, the profit function may be written as follows:

(3) $\pi(q; L) = L\pi(q)$

Equation (3) implies that the profit does not depend on hours of work.

$W^*$ is defined as profit per hour of work or shadow wage rate. $W^*$ can be derived from the farm production function:

(4) $W^* = \dfrac{L\pi(q)}{L} = \pi(q)$

subject to:

(2)′ $\Sigma P_i C_i = W^* L + Y$

where $Y = Y_o + Y_p$. From a solution of (1) subject to (2), the labour supply function can be derived.[3] Therefore:

(5) $L = \Omega(p, W^*, Y)$

is specified in hours of work, but this model can be converted to the logit model. Consider $W^{*\prime}$, such that $W^* = W^{*\prime}$ where $T = \Omega(p, W^{*\prime}, Y)$. $T$ is the hours of work that the aged farmer would be fully engaged in agriculture. $W^{*\prime}$ indicates the boundary between permanent agricultural work and part-time work or unemployment, such that:

(6) $L \geq T$ if $W^* \geq W^{*\prime}$, and $L \leq T$ if $W^* \leq W^{*\prime}$

If $R$ is defined as the probability of participation by the elderly in agriculture, then:

(7) $R = prob(W^* \geq W^{*\prime})$

Equation (7) is the probability density function, and $p$, $W^*$, and $Y$ from (5) can be regarded as its parameters. Therefore,

(8) $R = \Phi(p, W^*, Y)$

Equation (8) is the labour supply function of aged farmers.

To estimate Equation (8), it can be approximated to a linear function of its independent variables. Then, the equation to be estimated assumes the following form:

(9) $R = \dfrac{1}{1 + e^{(a + bW^* + cY + dZ)}}$

where $a$, $b$, $c$, and $d$ are parameters to be estimated and $Z$ is a vector of constraints that arise from the characteristics of each farm, which are:

ME: Husband's employment status
KID: Number of children under 6 years old
Type-of-farm dummies: vegetable, fruit, and dairy

3

Table 1—Empirical Results of the Logit Analysis

| Variable | Males | | Females (1) | | Females (2) | |
|---|---|---|---|---|---|---|
| | 60–70 | 70+ | 60–70 | 70+ | 60–70 | 70+ |
| $Y$ | -0.1750E-4*** (-7.173) | -0.1108E-4*** (-4.420) | -0.8063E-5*** (-3.391) | -0.2649E-5* (-1.779) | -0.8924E-5*** (-3.477) | -0.7879E-6 (-0.833) |
| $W^*$ | 0.0193*** (5.209) | 0.0106*** (4.009) | 0.0039 (0.838) | 0.0082*** (2.772) | 0.0123** (2.385) | 0.0056*** (2.881) |
| $ME$ | | | | | 0.4110*** (11.548) | 0.0989*** (7.515) |
| $KID$ | | | | | -0.0274 (-1.059) | -0.0227* (-1.816) |
| Type-of-farm dummies: | | | | | | |
| $Vegetable$ | 0.1659*** (3.095) | 0.0714 (1.596) | 0.3843*** (8.149) | 0.0804*** (3.200) | 0.2727*** (5.393) | 0.0479*** (2.775) |
| $Fruit$ | 0.0928*** (1.562) | 0.0670 (1.363) | 0.1872*** (3.810) | 0.0464 (1.572) | 0.0770 (1.439) | 0.0334* (1.662) |
| $Dairy$ | 0.4782*** (5.534) | 0.1123* (1.900) | 0.2252*** (3.703) | 0.1005*** (3.044) | 0.1286** (1.971) | 0.0663*** (2.985) |
| Intercepts | 0.4845 | -1.0307 | -0.6778 | -2.3738 | -1.3695 | -3.6294 |
| Sample size | 1,009 | 744 | 959 | 1,028 | 959 | 1,028 |

Note: The numbers in parentheses are $t$-statistics. ***Statistically significant at the 1-percent level. **Statistically significant at the 5-percent level. *Statistically significant at the 10-percent level.

# Empirical Results

## Estimation Results

The maximum likelihood estimation procedure is adopted to estimate Equation (9) using data from the 1985 Survey of the Farm Household Economy (Table 1). The logit coefficients are shown having been changed to partial derivatives at each average value. The coefficients indicate the change in the employment probability of the elderly associated with a one-unit change in the variable. They also indicate elasticities at average values (Table 2).

The sign condition of $Y$ is negative, and it is satisfied for males and females. The sign of the wage coefficient cannot be determined theoretically; however, the estimation results indicate it to be significantly positive.

Wage elasticity has a relatively low value of less than 0.5, which becomes larger as age increases. The low wage elasticity is consistent with regional disparities in the participation rate of the elderly, which become larger in proportion to age. One can also calculate the elasticity of substitution from income and wage elasticities using Slutsky's equation. The estimated elasticities of substitution were all positive, satisfying the theoretical condition.

Table 2—Elasticities Evaluated at the Means

| Sex and age | Income | Wage | Compensated Wage |
|---|---|---|---|
| Males: | | | |
| 60–70 | −0.403 | 0.266 | 0.522 |
| 70+ | −0.730 | 0.339 | 0.505 |
| Females: | | | |
| 60–70 | −0.466 | 0.161 | 0.239 |
| 70+ | −0.246 | 0.423 | 0.433 |

The coefficient of *ME* indicates a strong relationship between the female elderly and their husbands; i.e., whether the husbands are consistently employed in agriculture or not strongly affects their wives' participation rate in agriculture. The coefficient of *KID* is not statistically significant, but the sign is negative as assumed.

For type-of-farm dummies, the coefficients of vegetable and dairy farming have high positive values, possibly because these types of farming need a large amount of labour input as well as cooperation among family members.

## An Explanation of Differences of Participation Rates in Region and Farm Scale

The factors affecting regional disparity in the participation rate in agriculture of the elderly farmers were decomposed into the explanatory variables in Equation (9). The participation rate for 60–70 year-olds in Hokkaido is higher than the average for both male and female, which can be explained by the wage effect and differences in crop combinations. The high ratio of dairy farmers is one factor that raises the participation rate in agriculture.

In the Tohoku region, the high ratio of rice farmers lowers the participation rate. Since all coefficients of the variables for Hokuriku region males are negative, participation rates in agriculture are less than the national average. Due to the prevalence of part-time farming and also the extremely small scale of farming, other family members can receive a higher income in the nonagricultural sector. The participation rate in the Kinki region is also low due to the large incomes of other family members. All effects are positive for males in the Shikoku region, raising the participation rate. In the Kyushu region, low incomes of other family members raise the participation rate for males.

The factors that contribute to scale differences in participation rates of the elderly were also analysed. The results indicate that the low participation rate among small-scale farms is due to the wage effect, and the low participation rate among large-scale farms is due to the income effect.

## The Effects of Pensions on the Participation Rate

Assumptions to compute the effects of pensions are as follows:

(a) The participation rate is affected only by the negative income effect due to pensions.

(b) The amount of the pension is ¥800,000 per year.

(c) The pension is only available to elderly couples.

Based on the above assumptions, a shift in the labour supply curve for 60–70 year-old males was calculated. It was found that pension payments depressed the participation rate of the elderly by approximately 7 percent.

## Summary and Conclusions

There is a rapid trend towards an ageing society in Japanese rural areas, so that aged farmers play an important part in agricultural production. The purpose of this paper was to analyse the labour supply behaviour of aged farmers. After clarifying the main features of their labour supply, a labour supply function was estimated using data from the 1985 Survey of the Farm Household Economy.

Major findings are:

(1) Aged farmers sometimes gain pleasure from working in agriculture. However, their estimated wage and income affect whether they decide to be permanent or supplementary workers.

(2) The participation rate in agriculture of aged females tends to increase if their husbands work in agriculture.

(3) Regional differences in probabilities that an aged farmer works can be explained by the income and wage effect and type of farming.

(4) Pension payments for farmers have a negative effect on labour supply of the elderly; i.e., the participation rate in the agricultural labour force decreases about 7 percent because of pension payments.

## Notes

[1]National Research Institute of Agricultural Economics, Japan.

[2]The functional form and estimation procedures are described in Egaitsu and Shigeno (1983).

[3]The procedure to derive the labour supply function is analogous to that used by Lopetz (1986).

## References

Egaitsu, F., and Shigeno, R., "Rice Production Function and Equilibrium Level of Wage and Land Rent in Postwar Japan," *Journal of Rural Economics*, Vol. 54, 1983, pp. 167–174.
Lopetz, R.L., "Structural Models of the Farm Household that Allow for Independent Utility and Profit-Maximization Decisions," in Singh, Inderjit, Squire, L., and Strauss, J. (Eds.), *Agricultural Household Models*, Johns Hopkins University Press, Baltimore, Md., USA, 1986, Chap. 11, pp. 306–325.

## Discussion Opening—*Shinnosuke Tama* (Hirosaki University)

In Japanese agriculture, the proportion of aged persons who work in agriculture is high and their role is becoming more important. This paper is useful not only in terms of the future of Japanese agriculture but as an analysis of trends in the country's aged society.

The paper identifies the labour of aged farmers as noneconomic activities (i.e., quality of life) and assumes that noneconomic factors influence only the decision on working hours, and that the decision on whether or not to engage in agriculture as work can be explained by economic factors. But the conception of farming as a meaningful existence also influences the decision of aged people to work or not, and economic factors also influence their decision on the amount of working hours. If it is necessary for mathematical modelling to distinguish the influence of noneconomic factors from that of economic ones, the paper should have made the distinction as a procedure for modelling.

Regional disparities of employment rates for aged farmers are related more to the reason why the proportion of aged farmers is so high in Japanese agriculture than to the economic conditions identified in the paper.

The explanation of why the proportion of aged farmers is so high lies in the *Ie* (family) system of Japan. The *Ie* system consists of three elements: family property, family business, and the family name itself. The highest priority in *Ie* is given to the sustained succession of all three elements by primogeniture. To guarantee this succession, a custom of retirement is retained in Japanese rural areas. In other words, the representative rights of *Ie* are bequeathed from a householder to one of his children, usually the oldest son, when the householder becomes a certain age, usually 60. Many farmers inherited from their fathers in this way, and they believe that passing on *Ie* to their sons is their *raison d'être*.

The high proportion of aged farmers means a low proportion of retirement from farming. There are many aged farmers who have a desire to retire from farming, but cannot retire because all their children have left home. Therefore, regional disparities have a strong relation to how many successors of farm households remain in each rural area. There are comparatively many successors in the Tohoku region because of the strong custom of primogeniture there, and few successors in the Chugoku region because of the weakness of that custom there. That implies that employment rates in agriculture of aged farmers are low in Tohoku and high in Chugoku.

The number of farm successors nevertheless depends on economic factors; e.g., farm scale and type of farming. But the behaviour of aged farmers is influenced by the spirit of *Ie*.

*[Other discussion of this paper and the author's reply appear on page 21.]*

# Limitations and Potential of Family Farms: Can Japanese Family Farms Survive?

*Nobuhiro Tsuboi*[1]

**Abstract:** Japanese agriculture is a family-farm-dominated industry. Farms have been transferred from generation to generation by inheritance. Recently, Japanese farms have faced difficulty in transfer of management to succeeding generations, resulting in an increase in the number of ageing farmers and in idle farm resources. The problem is mainly characterized by such factors as underdevelopment of farmland markets, a shift of the sense of commitment to and identification with groups from families to corporations, and a predominance of large organizations in the technical innovation field in highly industrialized societies.

Even if family farms face the difficulties caused by highly industrialized societies and have consequently declined gradually, they still have potential for gentler treatment of the environment by family farming activities, relatively large employment capacity, and assisting in reorientation of modern societies towards family values. This potential is important worldwide in an era faced with the crisis of the collapse of modern societies.

The study of the family farm will thus contribute to the development of sustainable farming and the construction of postmodern societies based on interdisciplinary and international cooperation among researchers worldwide.

## Introduction

Family farms are the predominant type of farming in Japan. Farms have been transferred from generation to generation in inheritance of all kinds of property, including farm resources. However, the number of farmers who are faced with difficulty in transferring the management of the farm to succeeding generations is increasing rapidly, resulting in ageing of the farm operator and an increase in underutilization of farm resources, including the expansion of the amount of degraded farmland. The difficulty of transferring farm management and the expansion of idle farm resources will become more serious in the future.

This paper discusses the limitations of family farms: that full-time family farms with a reasonable income from farming are not transferred to the next generation. The limitations of family farms also include the difficulty in transferring farm resources from farm to farm. Family farms are here defined as farms in which most of the labour required is supplied by family members, and which derive a reasonable income from farming.

The main objective of this paper is to analyse the background of the limitations of family farms and to discuss the future of family farms in Japan. Family farms elsewhere in Eastern Asia may be faced with problems similar to those in Japan. In the USA and Western Europe, attention may be focused on family farms from the viewpoint of environmental problems or revitalization of rural areas. Agricultural reforms in China, Eastern Europe, and the Soviet Union will be influenced by the development of family farms. I hope that this study will contribute to the promotion of structural reforms in agriculture worldwide through international and interdisciplinary research cooperation.

## Background of the Problem

The limitations of family farms in Japan are mainly caused by recent industrialization.

### Liberation of Family Members from Compulsory Succession to Farming Activities

Until the end of the 1950s, farms were transferred from generation to generation without free choice. However, since the 1960s, farm family members have changed their behaviour and have become reluctant to accept compulsory succession. Members of the next generation, even the eldest son, are able to choose jobs according to their desires and abilities. As a result, they may abandon farming to take up other jobs, even those involving commuting. That leads to the underutilization of farm resources.

As fewer members of the succeeding generation become engaged in farming, the farm population is gradually becoming older (Figure 1).

Figure 1—Number of Members of the Next Generation Becoming Engaged in Farming (a), Ratio of Farmers Over 65 Years Old (b), and Average Farm Size (c)

Source: *Pocketbook of Agricultural Statistics*, Ministry of Agriculture, Forestry and Fisheries.

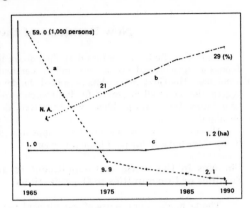

## Underdevelopment of Farmland Markets

The main farm resources in Japan include human capital, farmland, irrigation water, and machinery. Among these resources, farmland and irrigation water have little mobility due to underdevelopment of their markets. Lack of suitable markets is a constraint on farmer introduction of free organizations of labour, machinery, and technology.

If the members of the next generation want to become engaged in farming, they can exploit most of the farm resources in spite of the lack of markets, because they inherit them. However, if they plan to enlarge their farms after succession, it is very difficult for them to buy or rent additional land due to underdevelopment of farmland markets. As a result, the next generation is likely to abandon farming for an easier job.

## Traditional, Small-Scale, and Dispersed Field Pattern

In Japan, the average size of farmland per farm is about 1.2 ha. Each small farm is divided into 13.4 parcels, on average, which are not aggregated but dispersed (Table 1). This farmland system is well known as the small dispersed field pattern in Japan. Although Japan has no suitable farmland markets, occasionally farmers must sell or lease their farmland. However, due to the small dispersed field pattern, farmers who want to buy or rent land to enlarge their farm size cannot obtain additional land in suitable areas for farming.

Table 1—Status of the Small Dispersed Field Pattern of Farms in Japan, 1981

|  | Paddy Field | Upland Field | Whole Farm |
|---|---|---|---|
| Size of parcel (100) | 6.8 | 5.6 | 6.5 |
| Number of parcels per farm | 8.4 | 4.9 | 13.4 |

Notes: Hokkaido is not included in the table. These parcels of paddy fields are aggregated into three units on average.

Source: *New Structural Policy in Agriculture*, Ministry of Agriculture, Forestry and Fisheries, 1985.

When members of farm families select jobs other than farming, the amount of surplus farmland gradually increases. But active farmers cannot use such land efficiently due to the difficulty of enlarging the farm size. These factors are disincentives for the successors of farm families to continue the farm operation or for newcomers to become engaged in farming.

# New Problems of the Family Farm

Since the 1960s, the factors described above have influenced farm structure, and many Japanese agricultural economists and researchers have focused their attention on these problems. The Government has promoted the development of the farmland market and the aggregation of small dispersed fields through various policies. It is anticipated that in the near future these two problems will be solved, due to the surplus and the development of farmland.

However, two new factors have appeared recently, to which the Government and researchers have so far paid little attention.

## Changes in the Sense of Commitment to and Identification with Groups from Families or Communities to Organizations

The farmer's strong sense of commitment to and identification with the family and the community has shifted to organizations such as corporations. This trend is obvious among the younger generation.

In this paper, the sense of commitment to and identification with groups is defined as identity with the target, norm, and value of groups, willingness to work for the sake of groups, and strong desire to remain a member of a group. Corporations are defined as groups that have a specific purpose in common with members.

One of the main factors determining the shift of the sense of commitment to and identification with families and communities to corporations is the change in the character of society from property dominated to corporation dominated. In highly industrialized societies, including Japan, the main factor that determines the social status, social function, and income of individuals is the corporation to which the younger generation especially wants to belong.

The younger generation develops an inferiority complex when employed in family-based businesses such as family farms and family retailing. This feeling is growing gradually in Japan due to the welfare policy and system in which family workers are treated disadvantageously compared with corporation workers. This is also one of the reasons why the next generation of farm families and others do not select farming as a main full-time occupation.

## Predominance of Organizations Promoting Technical Innovation in a Highly Industrialized Society

A modern, highly industrialized society is characterized by rapid development of technological innovations by corporations that can afford the time and capital for research and development technology. The control of time is realized by internal specialization and integration of work in an organization.

In an era of rapid technological innovation, the family-based business cannot contribute to innovations due to its small research and development capacity. Technological innovations include the development of skills and devices in management, new products, marketing, and organizing resources for production.

Until the 1960s, family farms played an important role in technological innovations in specific fields; e.g., plant breeding in Japan. However, they have little opportunity to play this role in a highly industrialized society.

The above factors (i.e., the shift of the sense of commitment to and identification with groups and the predominant role of corporations in technological innovations) are discouraging future generations of farm families from carrying on and nonfarm families from becoming engaged in farming business.

# Arguments Against the Concept of Limitations of Family Farms

There are some arguments against the concept of limitations of family farms.

### The Organization of Farming Units or Group Farming May Enable the Limitations of the Family Farm to be Overcome

Many researchers in Japan emphasize that the organization of farming units or group farming based on individual operations may alleviate the limitations of the family farm. Group farming may overcome to some extent the inefficiency of farming operations caused by small-scale farming.

For the promotion of group farming, farmers must accurately understand the situation and characteristics of family farms in a highly industrialized society. To combine individual operators or managers of family farms into a group is difficult, due to their independence. The formation and the survival of groups of these individual farm operators need more skilful and sensitive principles than those for corporation workers.

These principles can be defined by analysing carefully the limitations of the family farm. However, since many researchers pay little attention to the situation of family farms in highly industrialized societies, they do not understand and cannot solve the problems of the limitations on family farms.

### Family Farms Will Survive with the Support of Business Activities Undertaken by Agricultural Cooperatives

As agricultural cooperatives are organized to supplement the activities of individual farms, the business activities of the cooperatives are expected to solve the problems of family farms. This is also an accepted concept.

However, this concept is inconsistent with the characteristics of farming in Japan. Many members of cooperatives are farmers who are not active enough due to their age. If they transfer farms to younger generations, the members of cooperatives become younger. While they want to transfer farms, they cannot do so due to the problems mentioned above.

It is often thought that the present agricultural cooperatives cannot alleviate the problems of individual farms due to their structural defects. Agricultural cooperatives in Japan control full-time farmers as well as smallholders and part-time farmers who derive little income from farming, pay little attention to the development of farming, and account for the majority of the members of agricultural cooperatives in Japan. It is, therefore, difficult to implement effective measures to develop and support full-time farmers who are in a minority.

The reorganization of the agricultural cooperatives requires that the cooperative managers and members themselves accurately understand the problems of family farms and the shortcomings of agricultural cooperatives. However, most of them have a limited understanding of and little interest in these matters, making it difficult to reorganize agricultural cooperatives. Meanwhile, many family farmers may go out of farming before the cooperatives become reorganized.

# Hypotheses on the Potential of the Family Farm

Even if family farms are facing difficulties in highly industrialized societies and the number of family farms is decreasing, family farms may survive because they have some important favourable characteristics.

The activities of family farms are compatible with the preservation of the natural environment. Human activities that do not depend on high technology and large organizations are generally compatible with the preservation of the natural environment. Family-based farming meets such requirements due to the relatively low level of technology applied and small-scale organization.

Family farms show a relatively large labour absorptive capacity. In the 21st century, employment will become a serious problem worldwide due to population increase. In the Third World, Eastern Europe, and the Soviet Union, this problem is already critical. In these regions, the governments are promoting agricultural reform. If agricultural reform leads to an additional flow of unemployed people from rural areas, the social and economic development programmes will be further delayed.

Highly industrialized societies are aware of the need to recover human values lost as a result of estrangement from their traditional organizations and activities. In the highly industrialized societies, the sense of commitment to and identification with groups is shifting from families or communities to organizations such as corporations. However, these corporations, which are characterized by a strong bureaucratic and hierarchical structure, may alienate people from human values and may make people reconsider the advantages of family-based groups and activities.

At the beginning of industrialization, it was thought that family-based economic activities would decline anyway. However, even in highly industrialized societies, there is a change of values, and some family-based activities may persist, not only in Japan but in countries such as those of Western Europe and in the USA, in which industrial development started earlier. Family farms may provide a new approach to the postmodern society.

Human activities not restricted to commitment to and identification with groups will become important. Philosophers predicted that before the 21st century, human beings may reject the concept of the sense of commitment to and identification with groups, one of the main concepts underlying the analysis of the limitations of family farms in this paper.

Family activities appear to have the effect of creating social stability. In the 21st century, the concept of stability of societies may become as important a concept as liberty and democracy. The family as a fundamental unit may secure social stability. Family-based activities may play similar important roles, and their large labour-absorptive capacity may preserve and promote the stability of society.

## Conclusions

Problems like those of the limitations of family farms appear in other sectors in Japan, such as retailing, fisheries, and forestry. Furthermore, similar problems may appear in the farming sector in other East Asian countries, and in other sectors too.

Family farms in Japan face the limitations mentioned above. Although there are fewer family farms, they may offer some potential for the future. Trends in family farming in Japan should, therefore, be carefully followed. To achieve this objective, new methods of studying the problems of family farms and other family-based activities will be developed.

Economic vitality is necessary, but not enough for the survival of family farms. Other factors, such as the social ones mentioned above, are keys to survival. As the problems of limitations and potential of family farms involve various fields of science, interdisciplinary research is essential. Furthermore, as the problems may appear worldwide, international research cooperation is also needed.

## Note

[1]National Agricultural Research Centre, Japan.

# Discussion Opening—*Geetha Nagarajan* (Ohio State University)

With rapid industrialization and agricultural structural adjustments, family farms in the 1990s are faced with problems of technological development, capital, parity, replacement labour, and succession. The paper by Nobuhiro Tsuboi examines the limitations and potential of family farms in Japan. The author is concerned about the declining number of potential successors and their preparedness to operate family farms, leading to ageing of farmers and increasing the number of idle farms. While the family farms are limited by farmland market imperfection, disintegration of traditional community relationships and dominance of organizations in technical innovations, the ability of family farms to reduce the problems of environmental stress and unemployment in addition to preserving the traditional family values are pointed out as features indicating the potential of family farms.

The problem addressed in the paper is a very interesting one, and stimulates further discussion on the future of family farms. Indeed, family farms are valuable assets, and farming is a way of life to the farm families. However, the inefficiency of family farms due to lack of size economies cannot be denied. Evidence suggests that farm size structure is a major constraint for efficient use of capital, and that increased farm size would lower the production costs and improve accessibility to credit. One needs to analyse whether small and fragmented family farms can earn economic profits from full-time farming without heavy subsidization. Furthermore, with the ageing of farm operators, one needs to be concerned with the possibility of productivity loss due to reduced adoption of new technology and inefficient use of land, capital, and labour resources. One needs to ask whether consolidation of family farms or group farming that allows secured individual land ownership and flexible land use will help in increasing the returns to farming and thereby provide economic incentives for successors.

The high opportunity cost of farming due to the high income gradient and flexible and short work time in nonfarm occupations is also a reason for lack of successors to family farms. Since the majority of farmers provide their children with a nonfarm education and thus eliminate a succeeding generation trained in farming, the successors lose their comparative advantage in farming. The question is whether the emigration of successors can be reduced by creating rural-based nonfarm employment opportunities leading to part-time farming. However, an increase in the number of part-time farms will reduce the chances of consolidation.

Evidence from the USA shows that environmental concerns are better served by larger farms than smaller family farms. Is it possible to find significant differences in the use of environmentally degrading practices among the family and nonfamily farms in Japan, especially when less than 2 percent of family farms are under organic farming? With the increasing urban environmental pollution and increase in capital gains due to rising land prices, it is economically efficient and profitable for small and fragmented family farms to sell or convert their farmland to real estate.

Social values attached to land ownership tend to dominate the economic gains from consolidation and sale of family farms. However, the loss in economic efficiency due to small, fragmented and part-time family farms far exceeds the gain in equity from protecting the family farms. It will be interesting to examine the efficiency gains and equity and welfare losses from consolidation of family farms without subsidization against preservation of family farms by more government incentives.

*[Other discussion of this paper and the author's reply appear on page 21.]*

# Assessing Returns to R&D Expenditures on Post-War Japanese Agricultural Production

*Jyunichi Ito*[1]

**Abstract:** The objective of this paper is to disentangle the question of the influence of R&D investment on resource allocation efficiency and income distribution in Japanese agriculture during 1960–87. The average marginal product of R&D stock as estimated by the cost function approach is calculated to be 4.47 at 1985 prices and 1.84 at current prices, and the internal rates of returns are 45.6 percent and 33.9 percent. In spite of their high profit-earning efficiency, they have drawn near to the profitability criterion, the current interest rate. Although technical progress induced by R&D activities increases social welfare without fail, its distribution among consumers and producers depends, to a large extent, on market circumstances. The empirical results suggest that consumer and producer economic surpluses have collided with respect to the allocation of R&D investment.

## Introduction

Since R.M. Solow shed some light on the contribution of technical progress to economic growth, two kinds of economic themes have been assigned to economists. One is to test the validity of the induced innovation hypothesis and another is to highlight the source of technical progress. Even in the area of agriculture, numerous empirical studies have been undertaken on the former subject, and it seems to be generally accepted that they uphold the validity of the induced innovation hypothesis. For the latter, Griliches (1988) and Akino (1973), in compliance with Schultz's hypothesis, proposed that R&D and rural education are major contributors to technical progress, and they assessed R&D profitability by means of production function analysis.

Attempts to pin down the efficiency of R&D investment, however, have not always succeeded, and much still remains open for further empirical studies. The significance of economic analysis of this subject is in the following two directions. First, as large parts of agricultural research activities are assigned to the public sector, the incentives for the pursuit of profit are likely to be weakened. As a result, the efficiency of R&D activities is liable to be inhibited. The economic implications are thus likely to arouse the concern of not only economists but also investment decision makers. Second, since technical knowledge, apart from ordinary private goods, has a public good attribute, it is hard to grasp its imputed price as evaluated in the market, which finally gives rise to market failure. These two characteristics of agricultural research will probably impede efficient R&D activities.

This paper has two main objectives. The first is to calculate the internal rate of return (*IRR*) of R&D investment in Japanese agriculture during 1960–87. The second is to elucidate how the economic surplus yielded by technical progress is distributed among consumers and producers.

The analytical framework is in line with previous studies, developed in terms of: (1) the source of technical progress is regarded not so much as R&D investment but rather as R&D knowledge stock; (2) in calculating the marginal product of R&D stock, the cost function approach is employed instead of the conventional production function (for the cost function approach to R&D profitability, see Stranahan and Shonkwiler, 1986, and Ito, 1991); and (3) the marginal product of R&D stock evaluated in the market is estimated, taking explicit account of an attribute of technical knowledge as a public good.

## R&D and Extension Activities

Before proceeding with the empirical study, some brief background on research and extension activities adopted in Japan after the second World War is in order. In view of price differences between substantial products both within and outside the country, development of technology that makes it possible to lower production costs is an urgent priority for Japanese agriculture. In this situation, the national research institutes have set out to promote basic research by investing half the amount of R&D expenditure in it and 300 local research institutes have been striving to meet location-specific technical demands. Local institutes' research expenditures currently amount to 2.3 times those of the national institutes.

Figure 1 shows the change in the ratio of R&D investment to the agricultural budget and output after 1950. In spite of a fairly steady improvement in the ratio of R&D expenditures to agricultural output, it is now no more than 0.7–0.8 percent. Given that this ratio ranges from 2–3 percent in other industries, it may be said that investment in agricultural R&D is inadequate.

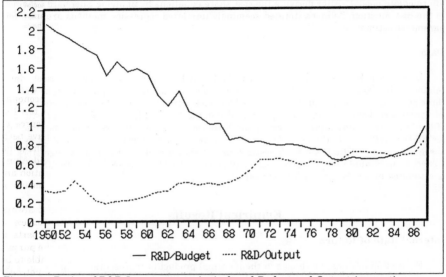

Figure 1—Ratios of R&D Investment to Agricultural Budget and Output (percent)

## Empirical Model and Data

### Model

Previous empirical studies that estimated the marginal product of the R&D knowledge stock are based on total factor productivity (TFP) analysis, assuming a linear homogeneous production function. For all its simplicity, this method fails to measure the contribution of R&D to productivity increase, on the grounds that TFP includes the effect of technical progress as well as that of nonconstant returns to scale (Capalbo, 1988). On the other hand, the cost function approach paves the way for relaxing the restriction of linear homogeneous production technology.

The dual expression of the marginal product of R&D knowledge stock assessed in the market $(\partial F/\partial R)$ can be written as:

$$(1) \quad \frac{\partial F}{\partial R} = \Sigma\Sigma_{ij}n_{ij}\frac{\partial f^{ij}}{\partial R} = \Sigma\Sigma_{ij}n_{ij}\frac{\partial C^{ij}/\partial R}{\partial C^{ij}/\partial Y},$$

where $R$, $Y$, and $n_{ij}$ stand for R&D stock, total output, and the number of farm households belonging to the $i$th region and $j$th operational size (for calculation of the shadow price of fixed inputs from the cost function, see Diewert, 1974). Since technical knowledge has a

15

characteristic of a public good, its imputed price will be evaluated by aggregating the individual ones.

The cost function $C^{ij}=C^{ij}(P, Y; R; Y)$, which has R&D stock as a factor input, is specified in the translog form as follows:

$$
\begin{aligned}
(2) \quad lnC^{ij} &= \alpha_0 + \Sigma\alpha_i lnP_i + \beta_y lnY + \beta_r lnR + \beta_s lnS + \Sigma_i\alpha_{iy} lnP_i lnY + \Sigma_i\alpha_{ir} lnP_i lnR + \Sigma_i\alpha_{is} lnP_i lnS \\
&+ \beta_{yr} lnY lnR + \beta_{rs} lnR lnS + \frac{1}{2}\{\Sigma\Sigma_{ij}\alpha_{ij} lnP_i lnP_j + \beta_{yy}(lnY)^2 + \beta_{rr}(lnR)^2 + \beta_{ss}(lnS)^2\}
\end{aligned}
$$

where $P_j S$ denote the input price and farmland ($j = l$ for labour, $m$ for machinery, and $i$ for intermediate goods). After deriving the cost share equations by applying Shephard's lemma to the cost function, Zellner's iterated seemingly unrelated regression methods are used for parameter estimation.[2]

## Data

If technical knowledge is permitted to be treated as an ordinary tangible asset, data related to the gestation period of investment and depreciation rate of capital stock, as well as those on investment expenditure, are needed for stock estimation. For the former, the average gestation period of research, 6 years, as recorded in the "Annual Report of Research Institutes of the Ministry of Agriculture, Forestry, and Fisheries," is assumed to represent the time lag for R&D payoffs, while the efficient period for a research outcome can be identified as the obsolescence rate of R&D stock. Investigation has shown that the R&D knowledge stock is subject to an annual rate of depreciation of 10 percent. On the basis of these data, R&D stock is estimated by the bench mark year method.

# Empirical Results

## Internal Rate of Return

Figure 2 illustrates the change in the marginal product of representative farms for each operational size in Tokyo and other prefectural areas at 1985 prices. They increased until the end of the 1960s, and declined thereafter, while maintaining a positive correlation with farm size. Assuming technical knowledge as a divisible factor input, the marginal product of each farm is regulated by its operational size.

The legitimacy of the cost function approach can be ascertained by comparing the change in the marginal product with technical progress. In concrete terms, applying Diewert's quadratic lemma to the translog cost function, technical progress can be calculated as a residual (Diewert, 1978).[3] Results indicate that the rate of technical progress has a consistent trend with the change in the marginal product of R&D stock, which attests the validity of the cost function approach.

Figure 3 depicts the change in the aggregated imputed price of R&D stock; its average value is computed to be 4.47 at constant prices and 1.84 at current prices. And the corresponding *IRR(r)* are 45.6 percent and 33.9 percent, given by:

$$
(3) \quad \exp(r\theta) = \int_0^\infty (\partial F/\partial R)(-rt)dt,
$$

ion,where $\theta$ represents the diffusion lag (assumed to be 5 years). After the 1980s, during which the marginal product drastically declined, the marginal product of R&D stock has come close to the profitability criterion of the interest rate.

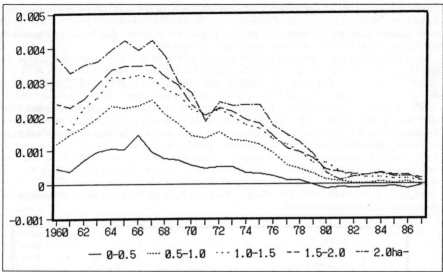

Figure 2—Marginal Product of R&D Stock (1/1000 ¥, 1985 constant prices)

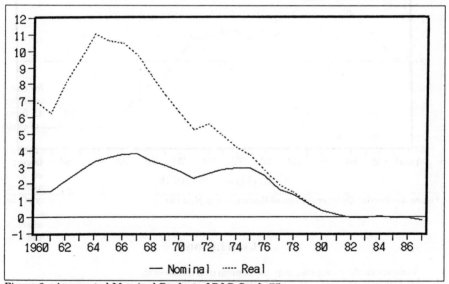

Figure 3—Aggregated Marginal Product of R&D Stock (¥)

## The Distribution of Economic Surplus

The paper now moves on to the second aspect of interest, investigation into the impact of R&D on consumer and producer welfare. As is intuitively understood, a marginal increase in R&D stock brings about a rightward shift of the supply curve of agricultural product; as a result, social welfare certainly increases. However, its distribution among consumers and

17

producers depends to a large extent on the circumstances of the agricultural product market. To be more concrete, some parts of economic surplus yielded by technical progress will revert to consumers through a decrease in product prices, while others due to producers will be contingent on the balance of cost reduction and price decline.

To clarify this, de Gorter and Zilberman's theoretical model (1990) was used. Since the equilibrium price and output in the agricultural market are a function of R&D stock ($R$), consumer utility ($V$) and producer profit ($\pi$) are also functions of $R$, and if R&D stock is determined by government in such a way as to maximize consumer utility (or producer profit), $\partial V/\partial R$ (or $\partial\pi/\partial R$) is equal to zero.[4] Figure 4 discloses the change in the partial differentiation of $V$ and $\pi$ with R&D stock. To the extent that they are positive, R&D investment is insufficient for the respective economic agent, and vice versa. This figure indicates that an additional R&D stock would raise consumer welfare for the period under question, and before 1976 it would also raise producer surplus. Therefore, at least for 1960–76, a marginal provision of R&D input would constitute Pareto improvement. However, from then on there is no feasible allocation of R&D investment where everyone is at least as well off and at least one agent is strictly better off. Accordingly, these results imply that benefits to consumers and producers have recently collided.

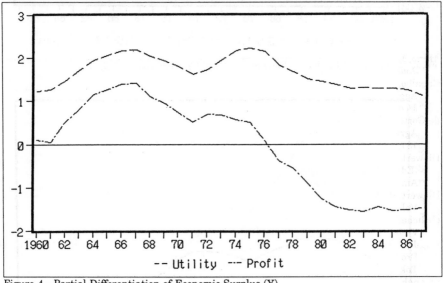

Figure 4—Partial Differentiation of Economic Surplus (¥)

## Conclusion

Some concluding remarks may be summarized as follows:

(1) Although the internal rate of return to R&D investment in Japanese agriculture has been sustaining a high level since the 1960s, its has been trending downwards after peaking in the mid-1960s.

(2) The allocation of R&D investment has attained Pareto efficiency since 1976.

(3) Generally speaking, consumers (producers) cannot become better off without a welfare loss to producers (consumers), which implies that the government is under heavy popular pressure to draw up an appropriate agricultural research policy.

## Notes

[1]National Research Institute of Agricultural Economics, Japan.

[2]Every regularity condition of the cost function is satisfied at each observation.

[3]Technical progress ($-\partial \ln C/\partial \ln t$) captured as a residual is given by:

$$\frac{1}{2}\left(\frac{\partial lnC}{\partial lnt}+\frac{\partial lnC}{\partial lns}\right)[t-s]=\ln\frac{C_t}{C_s}-\Sigma_j\frac{B_{jt}+B_{js}}{2}\ln\frac{P_{jt}}{P_{js}}-\frac{1}{2}\left(\frac{\partial lnC}{\partial lnY_t}+\frac{\partial lnC}{\partial lnY_s}\right)\frac{lnY_t}{lnY_s}-\frac{1}{2}\left(\frac{\partial lnC}{\partial lnS_t}+\frac{\partial lnC}{\partial lnS_s}\right)\frac{lnS_t}{lnS_s},$$

where subscripts $t$, $s$ and $B$ symbolize the time and cost share of the $j$th factor respectively.

[4]The partial differentiation of consumer and producer surplus with R&D stock is expressed as follows:

$$\frac{\partial V}{\partial R}=-\frac{YC_{Y_R}}{1-\eta^D/\eta^S}-i, \text{ and } \frac{\partial \pi}{\partial R}=\frac{YC_{Y_R}}{1-\eta^D/\eta^S}-C_R,$$

where $\eta^D,\eta^S$ are price elasticities of demand and supply (de Gorter and Zilberman, 1990), and sufficient conditions of consumer utility and producer profit maximization are satisfied at each observation.

## References

Akino, M., "The Contribution of Research and Education to Agricultural Production," *Quarterly Journal of Agricultural Economy*, Vol. 27, No.1, 1973, pp. 42–78.

Capalbo, S.M., "A Comparison of Econometric Models of U.S. Agricultural Productivity and Aggregate Technology," in Capalbo, S.M., and Antle, J.M., *Agricultural Productivity, Measurement and Explanation*, Resources for the Future, Washington, D.C., USA, 1988.

de Gorter, H., and Zilberman, D., "On the Political Economy of Public Good Inputs in Agriculture," *American Journal of Agricultural Economics*, Vol. 72, No. 1, 1990, pp. 131–137.

Diewert, W.E., "Application of Duality Theory," in Intriligator, M.D., and Kendrick, D.A. (Eds.), *Frontier of Quantitative Economics, Vol. 2*, North-Holland Publishing Co., Amsterdam, Netherlands, 1974.

Diewert, W.E., "Superlative Index Numbers and Consistency in Aggregation," *Econometrica*, Vol. 46, No. 4, 1978, pp. 883–900.

Griliches, Z., "Research Expenditures and Growth Accounting," in Griliches, Z. (Ed.), *Technology, Education, and Productivity*, Basil Blackwell, Oxford, UK, 1988, pp. 244–267.

Ito, J., "Assessing the Returns of R&D Expenditures on Post-War Japanese Agricultural Production," Research Paper No. 7, National Research Institute of Agricultural Economics, Tokyo, Japan, 1991.

Stranahan, H.A., and Shonkwiler, J.S., "Evaluating the Returns to Post-Harvest Research in Florida Citrus-Processing Subsector," *American Journal of Agricultural Economics*, Vol. 68, No. 1, 1986, pp. 88–94.

---

## Discussion Opening—*Yasuhiko Yuize* (Chiba University)

To produce data on the R&D knowledge stock of Japanese agriculture and to measure its returns on agricultural production can be regarded as pioneering, even heroic. While being highly appreciative of the work, I should also like to point out a few problems in application of the theory to the data.

In order to estimate the effects of R&D knowledge stock on production, this study makes use of the cost function instead of the production function. The production function represents the pure technical input-output relationship. On the other hand, the cost function is a derived function, derived from maximizing profit; i.e., determining the equilibrium relationship between the production function and the market prices of products and of inputs as fixed factors. The study needs to be modified by distinguishing three phases of the agricultural situation in the postwar period in Japan: part-time farming, the government acreage allotment, and land improvement capital.

## Part-Time Farming

Part-time farming is managed as a balance between farm income and nonfarm income, but not only within farm jobs. For instance, farm machinery is often overinvested to save farm labour and to increase nonfarm labour in a farm household. Also, farm households generally use marginal or fringe labour, particularly aged labour, which is not supplied to the labour market to determine the wage rate. Therefore, the marginal productivities of the machinery and labour inputs may lose their equilibrium with the price of machinery and the market wage rate.

## The Government Acreage Allotments after 1970

This administrative power may have skewed input-output relations in farm production. The farm price of rice has not always been consistent with the cropping area of rice because of government policy, and not always with current inputs because of subsidies. Therefore, the marginal productivities of current inputs may not have balanced with their prices.

## Land Improvement Capital

The land variable in the cost function is adopted as a fixed factor, but it does not include land improvement capital. This is a kind of social capital, which is important to farming, especially to rice production. Capital formation is largely dependent on central and local government expenditures, like those on R&D. So, a little of the price of capital is paid by the farmers. This variable should be considered as a fixed factor in the same way as the R&D knowledge stock. Data on the capital stock of land improvement from 1960–86 in Japan show an S-shaped curve, which is similar to that of the R&D knowledge stock. If this capital stock instead of the R&D stock were involved in the cost function, the estimation might have brought out almost the same results as in this study.

*[Other discussion of this paper and the author's reply appear on the following page.]*

# General Discussion—*Steve McCorriston, Rapporteur* (University of Exeter)

Questions addressed issues relating to both definitions used and points made by the presenters. Tanaka asked Shigeno to distinguish between the term "Noka" ("farm-attached household") commonly used in Japan and the term "farm household" that is frequently used in English translation. Shigeno agreed that one has to be careful in drawing a distinction between the definitions of farmer, farm household, and family farming, his definition being those "engaged in farming." Schmitt asked Tsuboi to clarify "successors," since many "successors" may participate in part-time farming, which may contradict the author's view concerning the future decline of family farming in Japan. Tsuboi's response was that his evidence only dealt with full-time farmers. Schmitt also addressed the first paper by questioning whether the economic determinants of aged farmers to supply labour differed from those affecting the labour force in nonfarm activity.

In dealing with the comments made by the discussion opener, Shigeno justified his observation of low opportunity cost of aged family labour in Japan by arguing that labour market frictions and other factors (such as the Japanese family system) prevented exit from farming. He nevertheless agreed that further research incorporating both economic and cultural factors is necessary, his study only dealing with the former. Tsuboi was not fully convinced by the discussant's view that smaller farms may be more damaging to the environment relative to large. Ito's response to his discussant focused on the effect of government land policies on R&D efficiency. In principle, he argued, the effects of government land policies could be empirically established by disaggregating his results, which could be done in future research, though he doubted whether it would affect his overall conclusion.

Participants in the discussion included Y. Tanaka (University of Tsukuba) and G. Schmitt (Universität Göttingen).

# Economic Impacts of the PROCISUR Programme: An International Study

*Robert Evenson and Elmar R. da Cruz*[1]

**Abstract:** The PROCISUR cooperative research programme is an agreement among Argentina, Brazil, Chile, Bolivia, Paraguay, and Uruguay. This paper analyses its impact in terms of rates of return, based on an analysis of the programme as it affects productivity changes in wheat, maize, and soyabeans in the member countries since the inception of the programme in 1978. Computed rates are high according to international standards. The paper stresses the importance of international cooperation.

## Introduction

In this paper we develop an analysis of the production data and an evaluation of the PROCISUR[2] programme as it affects productivity changes in wheat, maize, and soyabean production in the member countries since the inception of the programme in 1978.

## Methods and Data

### Methods

The methods used in this study require an extension of standard productivity decomposition methods in two dimensions. First, the PROCISUR investments must be modelled as being responsive to conditions in both sending and receiving regions and thus simultaneously determined with productivity growth. Second, the PROCISUR activities must be modelled as enhancing national research programmes.

Consider the basic productivity decomposition model:

(1) $\quad P_{it} = F(R_{it}^N, R_{it}^S, H_{it}, W_{it}, I_{it}, e_{it})$

where:

$P_{it}$ is an index of productivity. This may be an index of output per unit of total input (i.e., a "total factor productivity" index) or an index of output per hectare (a "partial factor productivity" index). It is measured for region $i$ and for different time periods.

$R_{it}^N$ is a research "stock" variable constructed from past expenditures on research directed towards improving $P_{it}$ for the region for which $P_{it}$ is measured (i.e., region $i$). Timing weights are used in the construction of $R_{it}^N$.

$R_{it}^S$ is a similar research "stock" variable constructed from past expenditures on research directed towards improving $P_{it}$ in other regions but where those improvements may potentially "spill-in" to region $i$.

$H_{it}$ is a measure of the human capital skills of farmers in region $i$. This may also include measures of extension services.

$W_{it}$ is a weather index measuring weather effects in region $i$, time $t$.

$I_{it}$ is a measure of public sector infrastructure investments in region $i$, time $t$.

$e_{it}$ is an error term.

Equation (1) is often estimated in logarithmic form with cross-section and time series data.

The most critical specification issue for the PROCISUR analysis is the specification of the spill-in variable $R_{it}^S$. The spill-in of technology is relevant to regions even where a local research programme exists. It is also relevant when the receiving region is in a different country from the originating region. Indeed, it is this spill-in that the PROCISUR programme seeks to facilitate.

The procedure entails defining three research variables: $R_{it}^N$ and $R_{it}^S$ as discussed above and an additional PROCISUR enhancement variable, $R_{it}^{SR}$.

The first variable, $R_{it}^N$, is the research stock variable, where the research activities are directed towards improving productivity in region $i$, is defined as:

(2) $\quad R_{it}^N = \Sigma_\alpha W_{t-\alpha} r_i, \ t-\alpha$

where the $W_{t-\alpha}$ are time weights reflecting the time relationship between research expenditure, $r_i$, $t-\alpha$, and productivity. Research conducted in time $t$ typically will not have an immediate impact on productivity. Many research projects do not have impacts for several years (some never do).

The second variable, $R_{it}^S$, is the basic spill-in variable. It is defined as:

(3) $\quad R_{it}^S = \Sigma_j \ G_{ij}^\beta R_{jt}^N$

where the $R_{jt}^N$ are research stocks defined in Equation (2) directed towards region $i$, but which can potentially spill-in to region $i$. The $G_{ij}^\beta$ are geo-climate spill-in weights measuring the proportionate value of research in region $j$ to productivity enhancement in region $i$ via direct, semi-direct, and indirect spill-in. These weights are designed to adjust for geo-climate impediments to technology spill-in.

The third variable is the PROCISUR enhancement variable. It is defined as:

(4) $\quad R_{it}^{SP} = \Sigma_j \ G_{ij}^\beta R_{jt}^N PR_{ijt}$

where the $G_{ij}^\beta$ and $R_{jt}^N$ are defined above. The $PR_{ijt}$ are the cumulated (to time $t$) expenditures on PROCISUR activities, where $i$ is the receiving region and $j$ is the sending region. Thus, $R_{it}^{SP}$ is an interaction variable designed to test whether PROCISUR activities increase or enhance the value of spill-in research. It is defined with respect to sending and receiving regions.

It can be reasonably argued that the time lag inherent in the $W_{t-\alpha}$ weights effectively creates a "recursive" structure between the research spending variables and productivity change. Since it takes time before research affects productivity, the current research stock is unlikely to be influenced by current productivity change. It cannot be argued, however, that the PROCISUR activities do not respond to the perceived opportunities for research enhancement. We would expect that PROCISUR activities, $PR_{ij}$, would respond positively to the past productivity performance in region $j$ and negatively to the current research capacity in region $i$. Accordingly, the $R_{it}^{SP}$ variable should be treated as an endogenous variable in a simultaneous system with Equation (1). We thus have the following two-equation system that we will estimate using Zellner's seemingly unrelated regression procedure:

(5) $\quad P_{it} = F(R_{it}^N, R_{it}^S, R_{it}^{SP} W_{it}, I_{it})$

(6) $\quad R_{it}^{SP} = F(R_{it}^{N*}, P_{jt}^*, I_{it})$

where $P_{it}^*$ is defined as $\Sigma_j \ G_{ij} P_{jt}$ and $^*$ indicates lagged values.

## Data and Variable Definitions

Data were assembled from a number of sources for 14 regions for the 1966–87 period. The regions include six states in Brazil (Mato Grosso, Minas Gerais, Paraná, São Paulo, Santa Catarina, and Rio Grande do Sul), four states in Argentina (Buenos Aires, La Pampa, Córdoba, and Santa Fé), Bolivia, Chile, Paraguay, and Uruguay. Table 1 reports variable definitions.

Table 1—Variable Definitions: PROCISUR Analysis

| Endogenous Variables | |
|---|---|
| LIYIELD | Natural logarithm of the commodity yield index. For each region, state, and commodity, this index was constructed as the ratio of yield in year $t$ to the 1966–70 average yield. Thus, regional differences in the 1966–70 average yields are not incorporated in this index. |
| LPRNGHI | Natural logarithm of the PROCISUR spill-in research stock, [Equation (4)]. This is the PROCISUR enhancement variable (see below for estimation of the $G_{ij}$ weights). PROCISUR data include the cumulated commodity data plus the general data. |
| Exogenous Variables | |
| LCRESEXP[1] | Natural logarithm of the state's research stock, $R_{it}^N$ [Equation (2)]. This variable is constructed from research expenditures in the state. |
| LRNGHI[1] | Natural logarithm of the spill-in research stock, $R_i BS_t$ [Equation (3)]. This is the basic spill-in research stock. |
| LSRNR[1] | $LCRESEXP \times R_{it}^S$, the spill-in research stock. |
| LEXTA[1] | Natural logarithm of field extension staff (for all crops) per hectare of cultivated land. The time weights are 0.25 for $\alpha = 0$, 0.5 for $\alpha = 1$, 0.25 for $\alpha = 2$, and 0 for $\alpha$ greater than 2. |
| LRESEX[1] | $LCRESEXP \times$ extension stock. |
| LSTRESA[2] | The average of $LCRESEXP$ for periods $t-1$, $t-2$, $t-3$, and $t-4$. |
| LNYIELDA[2] | Natural logarithm of the spill-in in weighted yield index averaged for periods $t-1$, $t-2$, $t-3$, and $t-4$. |
| YEAR[2] | A time variable, 1964, 1966, etc. |
| GOOD, POOR, BAD[1] | Dummy variables for weather effects: GOOD = 1 if yields are more than 1.5 standard deviations above trend; POOR = 1 if yields are from 1.5 to 2 standard deviations below trend; and BAD = 1 if yields are more than 2 standard deviations below trend. |
| BRMT, BRMG[1,2] | Dummy variables for Bolivian states. |

[1]Included in LIYIELD equation.
[2]Included in LPRNGHI equation.
Note: The $G_{ij}^\beta$ weights were estimated using geo-climate data from Papadakis (1975).

## Model Estimates

Table 2 summarizes estimates of the key parameters of the model for the third-stage simultaneous equations estimates for pooled data for all 14 states and for the 6 Brazilian states.

In all cases, the expected relationship between PROCISUR inputs and the key predicting variables is borne out. The sign on the lagged state research variable, LSTRESA, is always negative. The sign of the lagged productivity variable, LNYIELDA is always positive. All

Table 2—Third-Stage Estimates of Key Parameters: PROCISUR Analysis

| | Six Brazilian States | | | All PROCISUR States | | |
|---|---|---|---|---|---|---|
| | Maize | Wheat | Soyabeans | Maize | Wheat | Soyabeans |
| Parameter estimates: | | | | | | |
| LN (state research) × LCRESEXP | −0.0111** | −0.0049 | −0.0021 | 0.0135** | 0.0058* | −0.0003 |
| LN (state research) × spill-in research LSRNR | −7.613(12) | 6.831(10) | −2.375(10)** | −3.455(10) | 1.103(10) | −2.741(10)*** |
| LN (state research) × extension | 6.064** | 9.006** | 4.028 | 0.0002 | −0.0007 | −0.0065** |
| LN (spill-in research) LRNGHI | 0.0254** | 0.0061 | 0.0773*** | 0.0321** | 0.0502*** | 0.0669*** |
| PROCISUR enhancement LPRNGHI | 0.0061** | 0.0065*** | 0.0104*** | 0.0165*** | 0.0067*** | 0.0145** |
| LN (extension) LEXTA | 0.0131 | −0.054* | −0.045 | −0.061** | −0.083 | −0.044 |
| Weighted $R^2$ for system | 0.825 | 0.835 | 0.815 | 0.750 | 0.720 | 0.784 |
| Computed marginal elasticities: | | | | | | |
| State research | 0.0188 | 0.0258 | 0.0343 | 0.0096 | 0.0886 | 0.0238 |
| PROCISUR | 0.0061 | 0.0065 | 0.0104 | 0.0165 | 0.0067 | 0.0145 |
| Computed marginal products: | | | | | | |
| State research | 1.3 | 1.5 | 2.3 | 0.8 | 5.9 | 1.6 |
| PROCISUR | 12 | 11 | 20 | 33 | 11 | 29 |
| Computed marginal internal rates of return: | | | | | | |
| State research | 36 | 39 | 50 | 26 | 78 | 41 |
| PROCISUR | 115 | 110 | 148 | 191 | 110 | 179 |

Notes: Numbers in parentheses are $n$ in $10^{-n}$; i.e., the decimal point is moved $n$ places to the left. * $t$-ratio between 1.5 and 2.0. ** $t$-ratio between 2.0 and 3.0. *** $t$-ratio greater than 3.0. Elasticities are evaluated at mean levels of interacted variables. State research includes spill-in.

25

coefficients are statistically significant. This indicates that, as expected, PROCISUR activities respond positively to spill-in potential as measured by the productivity performance of spill-in geo-climate neighbours. These activities also respond positively to low research capacity in the recipient state. These results support the general validity of the model and lend credence to the PROCISUR enhancement estimates reported in Table 2.

The estimates reported in Table 2 are reported for Brazilian states and for the aggregate of all states. Aggregate results can be expected to be the most reliable generally because they capture the international effect of PROCISUR through cross-section variation. It would be much more difficult to measure a PROCISUR effect for a country with only a single time series (e.g., Paraguay) because of the limited number of observations. Nonetheless, it is of interest to disaggregate the data to some extent to investigate whether there are significant differences between groups of states.

Computed marginal productivity elasticities and marginal products are provided to enable the reader to interpret the net impacts of the research variable. The marginal elasticity for state research is computed as:

(7)  $d \ln(Y)/d \ln(RN) + d \ln(Y)/d \ln(RS)$

where the interacting variables entering into these derivatives are evaluated at mean levels in the relevant data set. Thus the fact that for maize and soyabeans the interaction terms ($LSRNR$) between state and spill-in research are negative (indicating that spill-in research is a substitute for state research) does not mean that the marginal product of research is negative. The negative term is more than offset by other positive terms.

The results are generally as expected for the agricultural research variable in all three commodities. Spill-in research is highly significant in all commodities for Brazil and for all states combined. Spill-in research is a substitute for state research in maize and soyabeans. State research is also significantly positive in maize and wheat. The combined effects of state research plus spill-in are significantly positive for all commodities in all regional groupings.

The results for extension are much weaker. Few significant extension coefficients are estimated.

The chief interest is in the PROCISUR enhancement variable, $LPRNGHI$. If PROCISUR has had an impact, we would expect first that spill-in research is a significant determinant of productivity and second that it has a higher impact when enhanced by PROCISUR activity. The estimates show significant PROCISUR enhancement effects for all three commodities for both data sets. This can be regarded as a strong result given the data and given the consistency of the second equation results. The finding of PROCISUR impacts of roughly similar magnitude in each commodity and data set lends further credence to the results.

## Economic Implications

Table 2 reports the calculated estimated marginal productivity elasticities for the state research programmes and for PROCISUR. These are computed as the logarithmic derivatives of the estimated equations. Where a variable is involved in the calculation, it is set to its mean value in the relevant data set. These elasticities are approximately comparable to those obtained in other studies of this nature. (See Evenson and da Cruz, 1989, for a review.)

It is possible to compute the marginal products from the elasticities by making use of the relationship:

(8)  $MP = Elasticity \times Average\ Product$

The average product must thus be computed as the ratio of the cumulated stock to the value of agricultural product. The average stock is approximately five times the average investment level in the PROCISUR data since research spending is rising. Data for Brazil and other PROCISUR countries indicate that research expenditures relative to commodity value were approximately 0.003 for maize and soyabeans and 0.0035 for wheat. PROCISUR spending as noted earlier is actually only 1 percent of national research expenditures for the recent years.

These factors are then used to convert the elasticity estimates into marginal product estimates in Table 2. These marginal products are to be interpreted as the annual benefit stream (adjusted for time weights) from a single one-dollar investment in time $t$. Thus a one-dollar investment in maize research in time $t$ will produce an income stream of 0.8 dollars that will be realized in future periods according to the time weights. They indicate that nothing will be realized in year $t+1$, 0.16 in year $t+2$ (0.2×0.8), 0.32 in year $t+3$ (0.4×0.8), 0.48 in year $t+4$ (0.6×0.8), 0.64 in year $t+5$ (0.8×0.8), and 0.8 thereafter (0.8×1). This can then be treated in an investment context and an internal rate of return to investment calculated (Table 2).

In the case of maize research, a one-dollar investment in time $t$ will yield an internal rate of return to investment of 26 percent. The comparable internal rate of return for wheat in all PROCISUR regions is a very high 78 percent. The internal rate of return for soyabeans is 41 percent.

For Brazilian research, the comparable internal rates of return are 36 percent for maize, 39 percent for wheat, and 50 percent for soyabeans. These returns (except for wheat) are somewhat lower than estimated in other studies but nonetheless represent high returns to investment. (See Evenson and da Cruz, 1989, for a review.)

The returns to PROCISUR research can also be computed. Note that the marginal products are extraordinarily high for PROCISUR impacts. Since PROCISUR enhances national research programmes and since there is a lag between PROCISUR spending and enhancement, the time lags are somewhat longer than for national research spending. Taking these time lags to be double those of national research spending, we find internal rates of return to PROCISUR are 191 percent for maize, 110 percent for wheat, and 179 percent for soyabeans. (The comparable figures for the six Brazilian states are 115 percent for maize, 110 percent for wheat, and 148 percent for soyabeans.) These are extraordinarily high rates of return. Even if they are overestimated by a factor of four, they are still extraordinarily high. They are higher than the rates of return of the International Agricultural Research Centres. For the case of IARC investment in maize, millet, and sorghum in Latin America, Evenson (1988) found rates of return above 80 percent.

It would seem reasonable to conclude that the marginal returns to PROCISUR appear to be extremely high. They indicate that the PROCISUR programme, which is actually a relatively small programme (only 1 percent of national research spending), has had an extraordinarily high "leverage" factor giving it very high returns. The programme has clearly been effective and has yielded large benefits. The signals presented by this study indicate that it can fruitfully be continued and expanded.

The relevance of PROCISUR-type programmes to other regions and countries will depend on the willingness of the research units to cooperate in the programme. Cooperation in the PROCISUR programme appears to have been very good and the programme appears to have been effectively administered.

# Notes

[1]Yale University and Empresa Brasileira de Pesquisa Agropecuária, respectively.

[2]PROCISUR is a cooperative programme for agricultural research, to which Argentina, Brazil, Chile, Paraguay, Uruguay, and Bolivia are signatories. It facilitates the exchange of agricultural research results, genetic materials, etc., among the countries involved.

# References

Evenson, R.E., "IARC Investment, National Research and Extension Investment and Field Crop Productivity," Yale University, New Haven, Conn., USA, 1988, mimeo.

Evenson, R.E. and da Cruz, E.R., "The Impacts of the PROCISUR Programme: An International Study," Yale University, New Haven, Conn. USA, 1989, mimeo.

Papadakis, J., "Agricultural Climates of the World," Salvat Publishing Company, Barcelona, Spain, 1975.

---

# Discussion Opening—*Wallace Huffman* (Iowa State University)

This paper presents an interesting economic analysis of the effects of interregional spillover effects of agricultural research on agricultural productivity. Spillovers can be of several types: direct transfers (e.g., varieties), modified transfers (e.g., locally modified varieties), pre-technology science transfers (i.e., the methods of science), or human capacity for research (i.e., scientists trained in one area can move to another area to work). This paper is interesting because it considers the effects of interstate transfers of technology within regions of countries and international transfers through PROCISUR. The PROCISUR programme was founded in 1978, which means that it is young in terms of the life of scientific institutions. The data for the Evenson-Cruz study are taken from six countries and a total of 14 regions or states during 1966–87.

A few characteristics of their analysis merit particular attention. First, the model is unusual for productivity analysis in the sense that it is a two-equation model where research stocks affect productivity and where past productivity affect research spill-ins. Most studies have used a single-equation methodology. Second, considerable evidence exists that there is a long lag, often as much as 25–35 years, in the effects on productivity of public agricultural research. Thus, the time period of their analysis is short, and the long-term success of the PROCISUR programme is still to be determined. Third, the paper does not describe how the PROCISUR programme fits into a multi-layered and interconnected organization or mix of science that seems most productive over the long term. For example, which institutions should focus on advances in general sciences, which on pretechnology sciences (the sciences that are linked upstream to general science and downstream to applied agricultural sciences), and which on borrowing and minor adaptation, and how should these institutions be most productively interrelated? These are the types of issues that are facing most agricultural science institutions today.

Science and technology policy has had significant impacts on agricultural productivity in many developed and developing countries during at least the past several decades. In some countries, there is a long history of developing crop varieties that are generally higher yielding than previous ones, and many have greater pest and drought resistance than old or traditional varieties. We might also expect a lot more from science and technology policy. For example, we might try to use science and technology policy to solve major income distribution inequalities, eliminate poverty, or stop environmental degradation, as was suggested in major addresses at the beginning of this conference. All of these seem to be complex economic, social, political, and biological issues. However, it seems more likely that good agricultural science and technology policy can at best be expected to lead to good science or, in practical terms, to steadily increasing agricultural productivity. Are we not misguided in expecting it to also solve major social and environmental problems in developing or developed countries? This is not meant to suggest that these are not important research issues for science, only that each agricultural research project should not necessarily be evaluated with respect to its contribution to these broad social/environmental issues.

*[Other discussion of this paper and the authors' reply appear on page 42.]*

# Economic Efficiency of Agricultural Research

*Chrabrin Bashev and Rumen Denchev*[1]

**Abstract:** The economic dimension of science has been one of the most treated problems in economics during the past three and a half decades. Over 6,000 publications appeared on the assessment of research results; several hundred formulae have been proposed to measure the efficiency of research workers, research institutions, and of science as a whole. The different aspects of research activity and its role in the realization of social goals have also been clarified. However, the overall economic theory of research efficiency is still in its formative stage. The number of studies of agricultural efficiency increased in parallel with agriculture's share in the economy, the growth of society, and with the amount of resources allocated to development of agricultural production. During the past decade, considerable progress has been made in clarifying the approach to estimation of research efficiency in agriculture, using agricultural economic analytical techniques, and identifying factors affecting efficiency growth in research and innovation. This paper presents some ideas related to scientific efficiency in agricultural research. It also estimates the economic efficiency of wheat selection in Bulgaria.

## Introduction

The problem of the economic dimensions of science has been given considerable attention in the economic literature during the past 35 years (Braunstein, 1980; Dobrov, 1978; Mansfield, 1968; and Twiss, 1986). During this period, over 6,000 publications appeared on the assessment of research results, several hundred formulae were proposed to measure the efficiency of research workers, research institutions, and of science as a whole (Zavlin, 1980). Various aspects of research activity and its role in the realization of social goals have been clarified. However, the economic theory of research efficiency is still at a formative stage.

The number of studies of agricultural efficiency increased, as did agriculture's share in economic growth and the resources allocated to development of agricultural production (Griliches, 1963; Latimer, 1965; and Matthews, 1964).

During recent decades, there has been considerable progress in clarifying the approach to estimation of research project efficiency in agriculture, using agricultural analytical techniques (Arnon, 1968; Griliches, 1958; Capablo, 1988; and Metcalf, 1970) and of the factors affecting studies of the growth of efficiency and the innovation process (Jones, 1963; and Metcalf, 1970).

The economic effects of the application of research achievements can be seen in the following three forms: in national income growth (additive value), in the size of averted losses, and in the increase of nonproductive (free) time. Often it occurs on a scale that exceeds the limits of economically differentiated groups of the economy. This requires that economic analysis of research take account not of its local effect in these separate groups within the scientific cycle but of its importance for the national economy.

This paper discusses techniques of evaluating and estimating agricultural research efficiency, with particular reference to determining the economic efficiency of wheat selection in Bulgaria.

## An Estimation of Research Efficiency of Wheat Selection in Bulgaria

Until the mid-1980s, wheat production in Bulgaria was based mainly on varieties selected abroad. During 1976–89, 35 varieties of wheat were created, established, and distributed in Bulgaria. They had to meet various economic criteria, and their application determined the scientific and technical level of production during the period.

The economic effects of introducing new varieties are estimated using (1) the results of annual, agro-ecological testing under the state system of variety testing (in 30 experiment stations) and (2) statistical information on the extent of introduction by main agro-ecological regions.

The optimum results in the corresponding agro-ecological regions are used to test the new varieties. Compared to the control (base), the test regions do not feel the economic effects of changes in production costs per unit, since new varieties are grown under similar technological conditions. Representative samples of the respective types and groups of wheat are used for comparison of grain quality. The complete comparability of all factors allows the annual economic potential created by selection to be differentiated. Under these conditions, the economic potential finds expression in the rate (or change) of incomes per unit of area as a result of increased productivity and quality of the new varieties compared to the basic ones.

The introduction of Bulgarian selected varieties has resulted in changes in the variety structure of wheat production. Compared with 16 percent in 1975, Bulgarian varieties of wheat have covered almost the entire country since 1983.

The speed of realization of economic potential varies between varieties from 1–2 years to 7–9 years for simple wheat, 3–4 years for strong wheat, and 2–4 years for hard wheat. The scale of introduction for the different varieties has different dynamics. A variety's share in the total volume of production also varies when its economic potential is exhausted and during the whole period of effective use. The percentage of wheat introduction in more than 10 percent of the area under wheat for the different years is 19 percent; strong and hard wheat is 100 percent. The importance of the different varieties also varies in terms of their contribution to improving the variety structure of wheat production. The share of the new varieties of soft wheat, introduced on over 1 million dka, is 50 percent. They account for 90.7 percent of the total volume of area under new varieties during the period.

Economic potential is realized on 22.5 percent of the area under new wheat varieties. This percentage gives an idea of the total quantity of land necessary to realize efficient scientific and technical progress in the field of wheat selection during the period of analysis.

The new varieties contribute to the formation of a variety structure to varying extents in the respective agro-ecological regions and subregions. Their largest share in total wheat production is found in the Danube-Dobruja and Central Northern Bulgarian regions, where they exceed national innovation levels by about 10–12 percent and 6–8 percent, respectively. The percentage and absolute scale of variety innovation is greatest in the Danube-Dobruja region, which contains over 50 percent (64 percent during 1976–80) of the total area in which new varieties were introduced. The distribution of the new varieties in the different agro-ecological regions and subregions gives an idea of the regional distribution of newly created varieties.

The total economic potential effect of introduction of new wheat varieties during 1976–89 can be valued at 52,169,913 leva (Table 1). The extent of actual realization of economic potential of the new varieties when widely introduced is 60–70 percent, according to experimental data (Panayotov, 1987). The economic effect of the introduction of new wheat varieties during the period ranges from 31,302,000 leva to 36,519,000 leva.

The dynamics of the total economic effect of introduction of new varieties vary for different years of the period. The rate of economic effect is characterized periodically by sharp drops as a result of the economic losses from introduction of new varieties during various years.

The economic effect of the introduction of new varieties shows different dynamics, share of total size, and absolute level during the different years of potential realization and for the period as a whole. Only as an exception do the changes and the levels of these indices coincide with the indices for production volume by variety.

The total effect of 37.5 percent of the new varieties is negative for the period of introduction as a whole. The economic effect of the introduction of four varieties (13 percent of the number of varieties) at the same time amounts to 88 percent of its total quantity for the period (i.e., real economic profit from improving the variety structure during the period has been realized for only 13.2 percent of the total area under new varieties).

The current level of efficiency of research costs of total variety selection during the period is 3.33 leva/lv, 3.83 leva/lv for soft wheat and 0.76 leva/lv for hard wheat. This indicates the return on total expenditure on selection when introducing new varieties in the year of introduction (for less than four months). During the first five years of the period, the level of efficiency of expenditure is higher, 0.76 leva for total expenditure and 0.77 leva for soft wheat selection. Increased biological and economic production potential is therefore achieved by increased investment of public research funds in variety selection. Production efficiency dynamics during the period arose through increased investment in science per unit of additional economic effect.

Table 1—Economic Effect of New Wheat Varieties Introduction in Bulgaria

| New varieties | 1976–80 | 1981–85 | 1986–89 | 1976–89 |
|---|---|---|---|---|
| | | Thousand leva | | |
| Wheat: | 12,491 | −9,034 | 43,035 | 46,492 |
| Positive effect | 21,679 | 15,359 | 45,125 | 82,163 |
| Negative effect | 9,188 | 24,393 | 2,090 | 35,671 |
| | | | | |
| Strong wheat: | 5,191 | 173,000 | −2,616 | 3,749 |
| Positive effect | 5,191 | 289,000 | 6 | 6,487 |
| Negative effect | 0 | 116,000 | 2,622 | 2,738 |
| | | | | |
| Hard wheat: | 0 | 654,000 | 1,275 | 1,929 |
| Positive effect | 0 | 654,000 | 1,275 | 1,929 |
| | | | | |
| New varieties total: | 17,682 | −7,207 | 41,695 | 52,169 |
| Positive effect | 26,870 | 17,302 | 46,407 | 90,578 |
| Negative effect | 9,188 | 24,509 | 4,712 | 38,409 |

The additional economic effect per unit of cultivated land in terms of realization of the economic potential of new varieties is 1.19 leva/dka. This indicates an increase in quantity of 330 percent, compared to the first five years of the period. It also means that the introduction of variety selection has brought about increased efficiency in the use of land resources during the period.

The economic effect of every research worker employed in the field of research in variety selection has increased to 1,169,000 leva and 223,000 leva, respectively, during the period. It has increased compared to the first five-year period by 145 percent and 140 percent, respectively. Thus the increasing efficiency of labour of those employed in this field of research can be expressed in terms of periods required for improving their level of qualifications.

The level of efficiency of fixed assets used in research during the period is 10.94 leva/lv, including, on the basis of their active part, −26.5 leva/lv, an increase of 227 percent and 203 percent, respectively, compared to the first five years. This expresses the increased return on fixed assets in the research field.

During 1976–89, the extent of the economic effect of the introduction of new varieties of wheat amounts to 1.34 percent of the size of the net product of the sub-branch during the period.

Figure 1 expresses the dynamics of the main indices of wheat production efficiency: where $A$ is average yield, $B$ is labour productivity, $C$ is production profitability, and $D$ is increased efficiency in the development of the variety structure during the period. $A$, $B$, and $C$ vary for the different years of the period and do not correspond to the dynamics of the economic effect of introduction of new varieties. $D$ is due mainly to other wheat production efficiency factors. The direct influence of the economic effects of selection development on the total efficiency level of production is comparatively small.

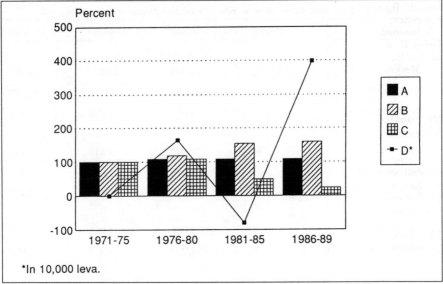

Figure 1—Efficiency of Production and Wheat Selection in Bulgaria

## Note

[1]Institute of Economics and Organization of Agriculture, Bulgaria.

## References

Arnon, I., *Organization and Administration of Agricultural Research*, Elsevier Publishing Company Ltd., New York, N.Y., USA, 1968.

Braunstein, Y., *et al.*, "The Economics of R&D," *Management Science*, 1980.

Capablo, S., *et al.*, *Agricultural Productivity Measurement and Explanation*, Resources for the Future, Washington, D.C., USA, 1988.

Dobrov, G., "Upravlenie na efektivnosta na nauchnata deinost," *Naukowa Dumka*, 1978.

Griliches, Z., "Research Costs and Social Returns: Hybrid Corn and Related Innovations," *Journal of Political Economy*, Vol. 66, No. 3, 1958.

Griliches, Z., "The Sources of Measured Productivity Growth: US Agriculture 1940–60," *Journal of Political Economy*, Vol. 21, No. 4, 1963.

Jones, G., "The Diffusion of Agricultural Innovations," *Journal of Agricultural Economics*, Vol. 15, No. 2, 1963, pp. 387–405.

Latimer, R., *et al.*, "Geographic Distribution of Research Costs and Benefits," *Journal of Farm Economics*, Vol. 47, No. 2, 1965, pp. 234–241.

Mansfield, E., *Industrial Research and Technological Innovation*, W.W. Norton, New York, N.Y., USA, 1968.

Matthews, R., "Some Aspects of Post-War Growth in the British Economy in Relation to Historical Experience," *Transactions of Manchester Statistical Society One Hundred and Thirty Second Session*, 1964.

Metcalf, D., *The Economics of Agriculture*, Penguin Education, London, UK, 1970.

Panayotov, I., "Postijenia na Selektiata na Pshenitsata Triticale v IPS 'Dobruja' prez Izminalite 25 Godini," *Selskostopanska Nauka*, Vol. 6, 1987.

Twiss, B., *Managing Technological Innovation*, Longman, London, UK, 1986.

Zavlin, P., *Nauchnia Trud v Usloviata na Nauchno-technicheskata Revolutia*, Ikonomika, Moskva, 1980.

---

**Discussion Opening**—*Katalin Daubner* (Budapest University of Economic Sciences)

I congratulate the authors on their challenging and far-reaching ideas that may provide relevant starting points for discussion. I suggest that the discussion focus on: (1) inter-linkages between research, on the one hand, and mass-scale market-production, on the other; (2) effective market demand as a final decisive criterion of research relevancy and its economic efficiency; and (3) a broader approach to and procedures for quantifying research efficiency.

*[This paper was shortened considerably by the editors, and much of the discussion opener's remarks dealt with portions of the paper that were deleted.]*

# The Structure of Research and Transfer Policies in International Agriculture: Evidence and Implications

David R. Lee and Gordon C. Rausser[1]

**Abstract:** This paper addresses the well-known paradoxes of high rates of protection, underinvestment in agricultural research, and relatively high productivity that characterize developed country agriculture, while developing country agriculture is typically characterized by taxation of the sector, research underinvestment, and low sectoral productivity. The paper tests the proposition emerging from political economy theory that productive policies (e.g., research) and redistributive policies (e.g., subsidies) can be viewed as complementary in that the latter compensate producers who lose from the price-reducing effects of the former. The economic relationships between agricultural research expenditure, total policy transfers, sector productivity, and other variables are examined for a sample of developed and developing countries. The results confirm the complementarity hypothesis and show that increased relative rates of research expenditure are associated with higher agricultural productivity, higher country incomes, and higher rates of agricultural protection found in developed countries. The reverse is shown to occur in low-income countries. The results suggest that both policy and trade reforms in developed countries and increased agricultural research allocation and sector productivity in developing countries may be harder to accomplish than previously thought due to the complementarity phenomenon.

## Introduction

Agricultural economics research has consistently demonstrated that governments significantly underinvest in publicly supported agricultural research (Ruttan, 1982; and Peterson and Hayami, 1977). Various reasons for this underinvestment have been advanced, ranging from a host of institutional and political factors (Ruttan, 1987) to theoretical arguments pertaining to the joint social provision of agricultural research and subsidies (Alston, Edwards, and Freebairn, 1988; and de Gorter, Neilson, and Rausser, 1990). With few exceptions, however (e.g., Pardey, Kang, and Elliott, 1989), empirical evidence on the underlying relationship between agricultural research expenditure and total government interventions in agriculture across countries, while widely speculated upon, has been left unexamined.

This paper reports the results of an examination into the empirical relationship between agricultural research support and total economic transfers in a sample of developed and developing countries. The evidence is shown to provide empirical support for recent theoretical contributions to the agricultural economics literature (Rausser and Foster, 1990; and de Gorter, Neilson, and Rausser, 1990) that have used political economic arguments to advance the notion that agricultural research and subsidy-type transfers can be viewed as jointly provided complementary policies. This argument suggests that agricultural subsidies serve, in part, to compensate producers for the potential losses induced by productivity-enhancing but price-reducing research policies. The international evidence reported here provides empirical foundation for the view that trade and policy reforms aimed at reducing agricultural subsidies may be harder to effect than is sometimes thought.

## Agricultural Research—Theory and Practice

Since Griliches' seminal work on hybrid maize in the late 1950s, a long tradition of agricultural economics research has established that the rates of return to agricultural research are typically very high. Arndt and Ruttan, for example, cite the results of 20 major studies of agricultural research in a variety of developing and developed countries covering periods ranging from 1880 to 1973. The annual internal rate of return estimates calculated in these studies range up to 90 percent and average in the 40–60 percent range. Many other studies, too numerous to mention, have demonstrated similar results.

Given rates of return of these magnitudes, the obvious policy prescription is for governments, as well as the private sector, to devote significantly more resources to agricultural research. Since "underinvestment" in agricultural research is chronic and widespread, the corollary question becomes, *why* do governments universally underinvest in agricultural research? Common answers to this question have been many (Ruttan, 1987): spillover effects to other countries, regions, and consumers; inefficient resource allocation in research; anticipated adverse socioeconomic effects from research-induced productivity enhancement; or difficulties in generating political support.

Despite the longstanding acceptance of the "underinvestment" hypothesis, recent research has begun seriously to question, largely from a theoretical perspective, both the extent of and reasons for research underinvestment. Lindner and Jarrett (1988) and Norton and Davis (1981), for example, show how analytical assumptions regarding the shape and shifts in underlying supply (and demand) functions will lead to widely varying estimates of research effects. Alston, Edwards, and Freebairn (1988) and Oehmke (1988) both show how interactions between agricultural research and subsidy programmes tend to increase the social costs of the latter, meaning that the benefits of research may be significantly overstated when measured in isolation from other policy effects. The implications of this and related research are that, while underinvestment may indeed exist and may be remedied by public policy changes, the gains from research suggested by many earlier studies may be both substantially overstated and lead to improper policy solutions.

A second very distinct line of economic research has, over the past decade, addressed what is typically considered to be an unrelated paradox in international agriculture; i.e., the simultaneous protection of agriculture in developed countries and taxation of agriculture in developing countries. A number of empirical studies, most notably the World Bank's 1986 *World Development Report*, have shown this pattern to occur widely, although these outcomes have been attributed to a variety of different contributing factors: the relative returns from protection gained by producers vs. consumers (Balisacan and Roumasset, 1987), existing comparative advantages (and disadvantages) in agricultural production (Honma and Hayami, 1986), and the potential employment effects of removing price distortions (Bale and Lutz, 1981). Research on agricultural protection and its removal has become important in recent years with market-oriented structural reforms in many developing countries and the current debate over multilateral trade reform in the GATT.[2]

A key to the resolution of these two paradoxes—i.e., underinvestment in agricultural research and protection vs. taxation of agriculture in developed vs. developing countries—is suggested by recent theoretical developments in the political economy of agriculture. Wicksell, Mueller, and Rausser have all recognized the usefulness of distinguishing between public policies designed to improve allocative efficiency ("productive" policies) on the one hand and policies designed to generate economic transfers to various groups ("redistributive" policies) on the other. Public agricultural research expenditure can be argued to belong in the former group (even though it may ultimately affect the distribution of welfare among producers and between producers and consumers) because they promote greater sector productivity and efficiency. Agricultural subsidies and related transfers clearly belong in the category of redistributive policies.

Based on this distinction, Rausser and Foster (1990) have shown that redistributive policies can be welfare *increasing* if they are treated not in an isolated fashion but combined with productive policies that, by themselves, would be impossible to implement due to insufficient political support. Their proposition is based on the notion of government maximizing a political preference function, $PPF = w (C) + (1-w) (F)$, where C and F are consumer and producer surplus measures, respectively, and $w$ and $(1-w)$ are the associated preference weights. They derive the further theoretical result that "the expansion of total social welfare biased towards one group ... leads to a change in the degree of wealth transfer in favour of the other group" (Rausser and Foster, 1990, p. 650). These results suggest that the coexistence or complementarity of both productive (e.g., research) and redistributive (e.g., subsidy) policies is not a perverse but a rational response to conflicting demands by support-maximizing governments.

Most recently, de Gorter, Neilson, and Rausser (1990) have developed a similar, though more comprehensive, theoretical argument specifically applied to the joint determination of

35

agricultural research and subsidies. Their comparative static results show that the observed policy mix can be shown to depend on the relative welfare weights of producers and consumers, relative elasticities of supply and demand for the affected commodities, and the marginal producer response to research inputs. They conclude that "by providing a vehicle through which to compensate producers for losses incurred as a result of research expenditures, production subsidies may be necessary components of potentially Pareto-improving portfolios of policy instruments" (de Gorter, Neilson, and Rausser, 1990, pp. 28–29). Their thesis of research and subsidy complementarity (for methodological details, see de Gorter, Neilson, and Rausser) is briefly applied to and finds partial support in US agriculture. They end by calling for an examination of the robustness of the complementarity hypothesis through application to the widely varying conditions characterizing developing countries.

This paper provides such an examination. Sample data for 23 developed and developing countries (see Table 1) are drawn together to examine cross-country relationships among agricultural sector performance, protection, research expenditure, and other key variables. These key variables and associated data sources include: agricultural research expenditure from the recently published ISNAR data base on national agricultural research systems; agricultural value-added data from the World Bank's *World Development Report* (recent issues); agricultural labour force data from FAO's *Production Yearbook* (recent issues); and agricultural protection data derived from Webb, Lopez, and Penn (1990). Further methodological details, including underlying regression estimates, are contained in Lee and Rausser (1991).

Table 1—Countries Used in Analysis

| | | | |
|---|---|---|---|
| Argentina | Colombia | Mexico | South Korea |
| Australia | Egypt | New Zealand | Thailand |
| Bangladesh | India | Nigeria | Turkey |
| Brazil | Indonesia | Pakistan | USA |
| Canada | Japan | Poland | Yugoslavia |
| Chile | Kenya | Senegal | |

# Results

Looking first at overall agricultural protection among the sample countries, a measure of protection that has been widely used in recent research by the US Department of Agriculture, OECD, and others is the "producer subsidy equivalent" (PSE). This measure estimates the value of direct and indirect government policy transfers to producers of specified commodities and has been calculated for a wide range of developed and developing countries (Webb, A.J., Lopez, M., and Penn, R., 1990). Figure 1 shows average aggregate PSEs for the sample countries in 1982–86 plotted against a measure of national wealth, the natural logarithm of each country's average per capita GNP. The general result of developed countries protecting and developing countries taxing their agricultural sectors (positive and negative PSEs, respectively) is confirmed for the sample countries as well.

In terms of agricultural research specifically, Figures 2 and 3 show two measures of research expenditure across the sample countries. Gross agricultural research expenditure (in log terms) is plotted against agricultural GDP in Figure 2, with the expected result that countries with larger agricultural sectors (in absolute terms) are shown to devote more resources to agricultural research. In Figure 3, a relative measure of research expenditure (i.e., research's proportion of agricultural GDP) is plotted against average per capita GDP for 1983–85. Boyce and Evenson (1975) and Pardey, Kang, and Elliott (1989) have termed this latter measure "agricultural research intensity." A clear positive correlation is evident, confirming, for this sample, those authors' earlier findings that wealthier countries are able to devote a greater share of the wealth generated by agriculture to reinvestment in its productive potential. Low-income sample countries, by contrast, are shown uniformly to exhibit an agricultural research intensity of less than two percent.

Before turning to the policy dimensions of agricultural research, it is necessary briefly to define and examine the behaviour of the specific measure of agricultural sector performance used in this analysis. Following recent research by Houck and Rossman (1990) and de Janvry and Sadoulet (1988), the productivity measure used here is "agricultural value-added per agricultural worker." The methodological limitations of not using a multiple factor productivity measure here are well known, but generating such measures requires extraordinarily extensive data, which in practical terms are virtually impossible to get for most low- and middle-income countries. Use of the value-added measure does avoid one of the major limitations of partial productivity measures by excluding the value of purchased inputs. Agricultural sector performance, using this measure, is shown in Figure 4 to demonstrate a very distinct linear relationship to per capita GNP.

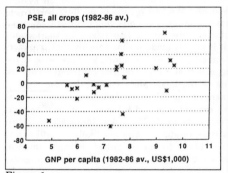

Figure 1

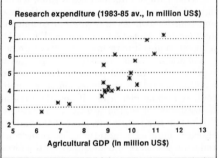

Figure 2

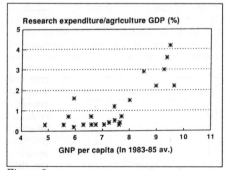

Figure 3

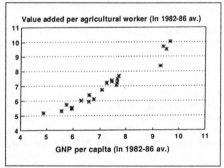

Figure 4

Using the same productivity measure, Figure 5 relates agricultural productivity to research expenditure across the sample countries. The demonstrated relationship is again predictable, given that agricultural research expenditure has long been argued to be a significant determinant of agricultural performance. While research impact on productivity has been argued to extend over as many as 30 years (Pardey and Craig, 1989), Figure 5 shows that even in the short run higher research allocations are strongly associated with higher agricultural productivity.

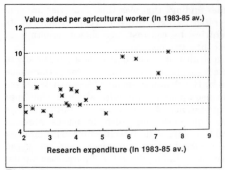

Figure 5

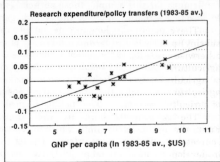

Figure 6

Given these intermediate results, let us consider the two results of primary interest in light of the earlier discussion. One of the key issues concerns the mix of agricultural research and transfer policies and whether complementarity between the two exists, for the reasons enumerated above. As seen above (Figure 1), it is clearly evident that subsidy-type transfers increase in absolute terms as country incomes increase. However, addressing the issue of the mix of productive and redistributive policies suggests that it is the *relative* contribution of each type of policy to total policy interventions that is the key. More specifically, the issue is whether the relative contribution of research increases simultaneously with subsidy levels as country income increases. If not, then research expenditure and subsidies can be viewed as "substitutes"; if so, the two can be viewed as "complements." The policy implications of this distinction are important, since if the latter is true, producers who are likely to suffer from research-induced long-term price reductions are likely to oppose policy reforms aimed at reducing transfers and subsidies. If the former is the case, then policy and trade reforms are likely to be far easier to accomplish.

The evidence in Figure 6 indeed shows that the ratio of research expenditure to subsidy-type transfers unambiguously increases with country income levels (the regression relationship is also given). The same result (not shown) can also be demonstrated to apply with respect to protection levels themselves. This lends strong support to the argument that agricultural subsidies can be viewed, at least in part, as mechanisms for compensating producers for their potential losses from productivity-enhancing but price-reducing agricultural research. The result is robust, extending over a wide range of low, middle and high-income countries.

The second key result relates to the relationship between agricultural performance (i.e., productivity) and the same relative measure of agricultural research intensity. Given the productivity effects of increased agricultural research, one would hypothesize that, as research expenditure increases in magnitude relative to welfare-reducing subsidy levels, agricultural sector performance itself should increase. Figure 7, which relates productivity to relative research expenditure, shows that this is in fact the case. This suggests that one reason for the record of strong productivity growth in developed country agriculture is that, despite a strong

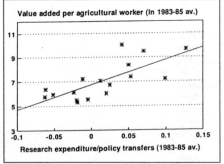

Figure 7

tendency towards increased protection, these countries show evidence of a complementary tendency towards support for agricultural research, with the resulting performance effects.

## Implications

The complementary provision of productive (e.g., research) and redistributive (e.g., subsidy) policies by governments provides perhaps the only consistent explanation for the otherwise paradoxical outcomes of developed countries (which generally *protect* their agricultural sectors while investing more in agricultural research and generating *higher* levels of agricultural productivity), compared to developing countries (which typically tax their agricultural sectors, invest *little* in agricultural research, and demonstrate generally *low* levels of agricultural productivity).

The implications of these outcomes are numerous, but two are particularly important. First, the obstacles to proposed policy reforms—particularly stemming from GATT-type sources—are likely to be even greater than is often thought, given that producers, primarily in developed countries, receive positive rates of protection in part as compensation for the adverse price and income effects induced by productivity-enhancing agricultural research. Given that the latter originates from both public and private sources and can be slowed but can never be "stopped," its adverse impacts are inevitable and can only be offset (if deemed necessary) by public policy interventions.

Second, for developing countries, the obstacles to increasing agricultural research and productivity are reinforced by a "vicious cycle," wherein low research allocations (along with other factors) lead to low agricultural productivity, creating no need for compensation for adverse effects. This, along with other factors, perpetuates the "taxation" of agriculture and provides little incentive (or political support) for increased research allocations.

## Notes

[1]Cornell University and University of California (Berkeley), respectively. The research reported herein was initiated while Lee was on leave with the US Agency for International Development and the US Department of Agriculture and while Rausser was at the US Agency for International Development. The senior author wishes to acknowledge the support of both institutions while holding neither responsible for the views or any errors or omissions contained within.

[2]It is worth noting that most GATT-related agricultural trade liberalization research has included research expenditure in aggregate measures of trade protection (e.g., Webb, Lopez, and Penn, 1990), although research and subsidy policies in fact have widely different functions and impacts.

## References

Alston, J.M., Edwards, G.W., and Freebairn, J.W., "Market Distortions and Benefits from Research," *American Journal of Agricultural Economics*, Vol. 70, No. 2, 1988, pp. 281–288.

Bale, M., and Lutz, E., "Price Distortions in Agriculture and their Effects: An International Comparison," *American Journal of Agricultural Economics*, Vol. 63, No. 1, 1981, pp. 6–22.

Balisacan, A.M., and Roumasset, J., "Public Choice of Economic Policy: The Growth of Agricultural Protection," *Weltwirtshaftliches Archiv*, Vol. 123, 1987, pp. 232–248.

Boyce, J., and Evenson, R., *National and International Agricultural Research and Extension Programs*, Agricultural Development Council, New York, N.Y., USA, 1975.

de Gorter, H., Neilson, D., and Rausser, G.C., "Productive and Predatory Public Policies: Research Expenditures and Producer Subsidies in Agriculture," Working Paper in Agricultural Economics, No. 90–17, Cornell University, Ithaca, N.Y., USA, 1990.

de Janvry, A., and Sadoulet, E., "The Conditions for Complementarity between Aid and Trade in Agriculture," *Economic Development and Cultural Change*, Vol. 37, No. 1, 1988, pp. 1–30.

Honma, M., and Hayami, Y., "Structure of Agricultural Protection in Industrial Countries," *Journal of International Economics*, Vol. 20, No. 2, 1986, pp. 115–129.

Houck, J.P., and Rossman, C.E., "Agricultural Productivity and International Food Trade: A Cross-Section Approach," in Vocke, G. (Ed.), *Trade and Development: Impact of Foreign Aid on U.S. Agriculture*, Staff Report No. AGES–9044, Economic Research Service, US Department of Agriculture, Washington, D.C., USA, 1990, pp. 9–25.

Lee, D.R., and Rausser, G.R., "Complementarity in International Agricultural Research and Subsidy Programs: A Cross-Country Analysis," Working Paper in Agricultural Economics, Cornell University, Ithaca, N.Y., USA, forthcoming.

Lindner, R., and Jarrett, F., "Supply Shifts and the Size of Research Results," *American Journal of Agricultural Economics*, Vol. 60, No. 1, 1978, pp. 48–58.

Mueller, D., *Public Choice II*, Cambridge University Press, Cambridge, UK, 1989.

Norton, G., and Davis, J., "Evaluating Returns to Agricultural Research: A Review," *American Journal of Agricultural Economics*, Vol. 63, 1981, pp. 685–99.

Oehmke, J., "The Calculation of Returns to Research in Distorted Markets," *Agricultural Economics*, Vol. 2, No. 4, 1988, pp. 291–302.

Pardey, P., and Craig, G., "Causal Relationships between Public Sector Agricultural Research Expenditures and Output," *American Journal of Agricultural Economics*, Vol. 71, No. 1, 1989, pp. 9–19.

Pardey, P.G., Kang, M.S., and Elliott, H., "Structure of Public Support for National Agricultural Research: A Political Economy Perspective," *Agricultural Economics*, Vol. 3, No. 4, 1989, pp. 261–278.

Peterson, W., and Hayami, Y., "Technical Change in Agriculture," in Martin, L. (Ed.), *A Survey of Agricultural Economics Literature*, University of Minnesota Press, Minneapolis, Minn., USA, Vol. 3, 1977, pp. 497–540.

Rausser, G.C., "Political Economic Markets: PERTs and PESTs in Food and Agriculture," *American Journal of Agricultural Economics*, Vol. 64, No. 3, 1982, pp. 821–833.

Rausser, G.C., and Foster, W.E., "Political Preference Functions and Public Policy Reform," *American Journal of Agricultural Economics*, Vol. 72, No. 3, 1990, pp. 641–652.

Ruttan, V., *Agricultural Research Policy*, University of Minnesota Press, Minneapolis, Minn., USA, 1982.

Ruttan, V., *Agricultural Research Policy and Development*, Food and Agriculture Organization of the United Nations, Rome, Italy, 1987.

Webb, A.M., Lopez, M., and Penn, R. (Eds.) "Estimates of Producer and Consumer Subsidy Equivalents: Government Intervention in Agriculture, 1982–87," Statistical Bulletin No. 803, Economic Research Service, US Department of Agriculture, Washington, D.C., USA, 1990.

---

## Discussion Opening—*Shankar Narayanan* (Agriculture Canada)

The large amount of positive protection for agriculture primarily in the developed countries in the mid-1980s (subsidy/transfer policy) emanated essentially from the market effects (i.e., price and income reducing effects) of a combination of policies: own productive policy, competing country's farm policy (trade war), and weather. Affordability of protection remained a non-issue for these countries with very high national income and a low agriculture GDP share because the absolute level of subsidy (protection) to agriculture, in spite of being very large, formed only a very minute fraction of the total GDP in relative terms, generally less than one percent. The inference that the positive protection is induced *in part* by productivity-enhancing agricultural research is valid, but is it robust? Further empirical validation is needed to determine the exact weights of the productive policy effects on protection. The data from the 1980s appear deficient in this respect as protection (subsidy and non-tariff barriers) during this period was driven mainly by the trade war. One should perhaps examine the mid-1970s data for North America, when the returns from investment in agriculture were higher than from investment in stocks.

Developing (low income) countries with a predominant agriculture sector providing a 30- to 40-percent share of GDP and a greater then 50-percent share of employment cannot *afford not to tax agriculture* as a source of revenue. The perpetuation of "taxation," however, is contingent upon the long-term continuation of a structure dominated by low income, and the result was a structure dominated by low income and by the agricultural sector, which may change as the economies develop. Historical growth models of industrialized countries also show that the agricultural sector provided the capital for their early economic development.

The limit to research allocation (agricultural research intensity) is set by its relative share to agricultural GDP. In absolute terms, this may lead to low allocations, especially in the developing countries, even where there is a predominant agriculture GDP contribution. It is therefore hard to generalize about what is and what is not low.

The suggestion of low productivity being a response to "low" research allocation in the developing countries may be dubious. Low productivity may be due rather to inappropriate application of research results (e.g., high capital and knowledge requirements generally constrain the effective implementation of new techniques in the developing countries).

The benefits to consumers, who are the majority nationally and globally, in this spectrum of complementarity paradoxes should be also taken into account.

What are the limits to protection under these paradoxical situations and how are they set? Should productive research be slowed down or redirected in order to eliminate its counter-productive market effects? Are there farm subsidy policies that do not distort the national or international markets? Does the root cause lie in the saturation of food demand, leaving farm commodity supply much in excess? If so, what are the diversification implications?

*[Other discussion of this paper and the authors' reply appear on the following page.]*

# General Discussion—*Ian M. Sheldon,* Rapporteur (Ohio State University)

Several comments and questions were raised in relation to the paper by Evenson and Cruz. Anderson questioned whether there was systematic bias in the results because of the exclusion of international spill-in contributions. Evenson agreed that this was the case, but indicated that it is difficult to obtain the relevant data. Sanint commented on the discrepancy in rates of return to agricultural research among the Brazilian states and compared to those of PROCISUR, as it raises questions about the effectiveness of EMBRAPA's role in Brazil and the possible duplication of research effort within Brazil. Belshaw wondered whether the PROCISUR experience provided an interesting example of what Eicher (ISNAR, 1989) has described as "technology-borrowing" activity as opposed to "technology producing" activity, which implies the possible centrality of "science and technology" policy within a multi-faceted agricultural research policy package. . In response, Evenson agreed that PROCISUR did facilitate a broad diffusion of technological gains, e.g., weaker programmes in Paraguay and Uruguay were able to take advantage of stronger programmes in Brazil. In addition, a country such as Paraguay recognizes that it is a "technology borrower," and sees open trade as an opportunity rather than a threat in this respect. However, to be a good borrower does require a layering of internal science and technology production. Thomson asked what types of technology transfer were effective; had any research of this type been conducted for the EC and Eastern Europe; do PROCISUR-type activities work directly or indirectly? Evenson responded that PROCISUR programmes were very sharply focused, and he did not know of any research on such transfers within Eastern Europe.

The paper by Lee and Rausser also elicited several points from the audience. Sanint questioned whether the result that producers may be harmed by successful agricultural research was valid in a dynamic, general equilibrium context where demand is also shifting and the affected commodity develops strong links with the rest of the economy as output increases. Lee responded by saying that the paper did not promote the view that agricultural research should be slowed down because of harmful effects on producer prices; the focus of the paper is whether the effects of under-investment in agricultural research can be militated against through the use of agricultural subsidies. Colman thought it an attractive notion that it might be efficient for the public sector to compensate farmers for their losses from research expenditure through agricultural subsidies, if the benefits from such a policy exceeded the costs of compensation. But he also wondered whether detailed examination of the political process would reveal any such tradeoffs; e.g., in the UK both research and subsidies are being mooted as areas for reductions in public expenditure. He also asked whether the data used in the paper referred only to public research expenditure. If not, why should the public sector compensate losers from private research expenditure? Lee agreed that explicit modelling of the policy process was required and that the comment about private research was valid, but the data were often unavailable. Parikh wondered whether poor countries "disprotect" their agriculture because rich countries protect theirs, i.e., food imports are cheaper. Lee suggested that developing countries do not use subsidies because of the depressant effect on price. Thomson questioned whether agricultural subsidies may be triggered more often by trade crises than by the effects of research expenditure. In response, Lee agreed that there were many possible reasons for the use of agricultural subsidies. Kislev questioned whether the problem of under-investment in agricultural research ought to be embedded in a more general discussion of public under-investment. Lee accepted that the agricultural research expenditure literature has not focused on the general equilibrium analysis of investment. Lee also responded to comments by the discussion opener, suggesting that agricultural protection was not just affordable in the 1980s—such trade distortions are and have been a chronic issue. In addition, many developing countries cannot afford not to tax agriculture, although many LDCs have implemented structural adjustment programmes as a means of getting rid of export taxes.

Participants in the discussion included J. Anderson (World Bank), D. Belshaw (University of East Anglia), D. Colman (University of Manchester), Y. Kislev (Hebrew University), K. Parikh (Indira Gandhi Institute of Development Research), L.R. Sanint (CIAT), and K.J. Thomson (University of Aberdeen).

# Cutting the Agricultural Price Pie: Power or Justice?

## *Ewa Rabinowicz*[1]

**Abstract:** Agricultural protection varies considerably among different commodities. To understand inter-industry protection within agriculture, the price-setting procedure needs to be analysed. The price-setting procedure in Sweden is organized as a two-stage negotiation procedure involving producers and consumers. This type of price setting gives farmers an excellent opportunity to affect relative prices. The question is whether the stronger groups have used this situation to affect relative prices to their advantage. Four different price-setting principles are discussed in this paper: inertia, fairness, monopolistic price discrimination, and power. The model is tested by pooling cross-section (13 commodities) and time-series data (1973–83). Fairness and inertia together are the major explanation of the trend in relative prices at the negotiation level. In spite of their strong influence on the price-setting procedure, powerful groups are not misusing the situation to their advantage. It is not possible for a farm organization to build its image on the notion of solidarity and at the same time exploit weaker groups. Neither is it possible for farmers to exploit consumers by behaving as price discriminating monopolists and keep the credibility of their organization intact.

## Introduction

Producers of agricultural products in Sweden have been beneficiaries of support since the 1930s. However, levels of support measured as PSEs vary considerably between products. Subsidies account for 16 percent of the income of sheepmeat producers and for more than 75 percent in the case of milk producers. Variation within the EC is even larger. It is therefore legitimate to ask for an explanation of such a big difference. High levels of protection in agriculture as a whole as compared with other sectors of the economy have been analysed recently in terms of political market-place theory (Anderson and Hayami, 1986; Gardner, 1989; and Paarlberg, 1989). Models of the political market have also been applied in the field of trade theory to analyse level of tariffs in different industries, changes of level of tariffs, etc. (Lavergn, 1983; Baldwin, 1985; Finger *et al.*, 1982; and Goldstein, 1986). Inter-industrial protection within agriculture has not attracted so much analysis; Gardner (1987) is a notable exception.

Variables used in models attempting to explain differences in inter-industry protection usually include the relative international competitiveness of each industry, pressure-group influence, efforts to minimize displacement costs, international negotiations, historical factors, and miscellaneous aspects of the public interest (Lavergne, 1983). Agriculture has one unique feature, however, as compared with other industries, particularily in the European context, namely that prices are set annually (or sometimes even semiannually) by an administrative procedure (this is the case for the EC and for all Nordic countries). In order to understand inter-industry protection within agriculture, the price-setting procedure needs to be analysed. The issue is not very well researched. Fearne (1989) analysed the price-setting procedure of the EC, but only for the agricultural sector. A related line of research is the issue of endogenization of policy in agricultural sector models, in particular trade models. This modelling is, however, often done in a very simplistic way, either by price wedges or by price transmission functions, without any explicit behavioural analysis (Whaley and Wigle, 1990). This paper attempts to analyse price setting in Swedish agriculture in order to explain inter-industry protection in the sector. The purpose of the paper is further elaborated in the next section, following a description of the Swedish price-setting procedure.

## Setting of Agricultural Prices in Sweden

During the period analysed (1973–85), price setting was organized as a two-stage semi-negotiation procedure[2] involving producers and consumers and supervised by the National Agricultural Marketing Board. Producers are represented by the leaders of the farmers' union. The Consumer Delegation, appointed by the government and consisting of, *inter alia*, representatives of labour unions and the agricultural processing industries, acts on behalf of the consumers. Consumer interests are not easily organized. The Consumer Delegation is, accordingly, an artificial, bureacratic structure, having at least partial vested interests in the preservation of the system rather than being a true "fighter" for consumer interests.

In the negotiations, prices (so-called "middle prices") of semi-processed products at the wholesale level are set. Border protection (variable levies) is adjusted accordingly.[3] In the first stage of the negotiations, a total compensation amount (TCA) to the whole farm sector is decided. The TCA consists of three parts: compensation for increased cost at the farm and processing levels and an income parity component. This stage of the negotiations was strongly governed by formal rules that gave more or less automatic compensation according to official indexes and statistics. There is thus not much to be analysed here as far as the level of compensation is concerned. One can, of course, ask the fundamental question of why this type of approach was chosen. This, however, takes us back to where we started, namely, to the question of why agriculture is supported, as distinct from other sectors.

The second stage of the negotiations, where the TCA is distributed among different commodities in the form of price increases is, on the other hand, not based on formal rules or official statistics. In other words, discretion rather than rules is used. According to several members of the Consumer Delegation (whom the author interviewed), farmers exert strong influence on the distribution. This is hardly surprising, taking into account the strength of the incentives to both negotiating parties involved, as pointed out above.

This type of price-setting procedure gives farmers an excellent opportunity to affect relative prices. The question to be asked here is whether the stronger farm groups have used this situation to affect the distribution and hence the relative prices to their advantage or whether other criteria were used. The purpose of this paper is to establish what factors determined the development of relative middle prices during 1973–85; i.e., to analyse the second stage of the negotiation process.

## Analytical Models of the Distribution Process

From a mathematical point of view, the second stage of the negotiations can be viewed as a problem with one equation (all the money has to be distributed) and several unknowns (changes in individual prices have to be determined). The process can be described by the following equation:

(1) $\quad TCA = \sum \Delta P_i D_i$

where the $D_i$s are the past levels of consumption, taken as fixed. Basing the distribution on the fixed levels of demand (i.e., disregarding the impact of price increases on demand) may seem strange at first sight but is not always unreasonable. If the level of agricultural prices, by and large, follows inflation, and if relative prices of different food commodities are not changing drastically, the impact on consumption can be expected to be limited, particularly if household incomes are rising as well. At the same time, the process of negotiation is facilitated by avoiding discussions of unknown and uncertain demand elasticities.

Many different sets of $\Delta P_i$ satisfy the above equation.[4] In order to make the system soluble, a "closure" of some kind has to be introduced. Different closures or distribution principles are discussed below.

A very simple way of arriving at a unique price solution would be to postulate that all prices should be increased by the same proportion. This distribution would leave the relative prices of different products unchanged, preserving the *status quo*. Bias towards the *status quo* is a very strong tendency in many social processes. On the other hand, this simple proportionality rule would make the second stage of negotiation redundant. It is conceivable that both farmer and consumer representatives might have an interest in making the negotiation process more complicated and thereby more "important."

Compensating farmers for increased cost (including an income parity component) on the sector level is based on the notion of justice or fairness. A natural way of prolonging this procedure would be to distribute the TCA in proportion to the development of costs of individual commodities. One could argue that a distribution based on relative cost development is the only one that preserves the logical consistency of the whole negotiation process from the point of view of society. If farmers are to be compensated for cost increases, the logical thing to do is to increase the prices of the products whose costs have increased. Otherwise, the outcome can be that, for instance, the price of milk is increased because production costs of pigmeat have increased (thereby contributing to the cost increase at the sector level). On the other hand, this type of distribution would totally disconnect the domestic price from market conditions on domestic and foreign markets.

Distribution of the TCA in relation to cost might be claimed to be fair from the point of view of different groups of producers. If different groups of producers are not distinguished and all farmers are treated and act as one group, the collective interest of the group would be to reach maximum revenue the following year. The distribution problem can then be formulated as the budget restriction:

(2) $\displaystyle\sum_{i=1}^{n} \Delta P_i D_i = TCA$

and the objective function:

(3) $Max: \sum (\Delta P_i + P_i)(\Delta D_i + D_i)$

Assuming that own-price elasticities are known and constant and cross-price elasticities are zero, the optimal solution of the problem is given by:

(4) $\Delta P_i = \dfrac{\lambda - e_i - 1}{2} \dfrac{P_i}{e_i}$ , $i = 1 \dots n$ $\qquad \lambda = \dfrac{TB + \sum \dfrac{D_i P_i}{2} + \sum \dfrac{D_i P_i}{2e_i}}{\sum D_i \dfrac{P_i}{2e_i}}$

The intuition behind the above solution is very simple. The price of least-price sensitive products should be increased and all other prices decreased. There are, however, obvious limitations to this type of behaviour insofar as the external and internal legitimacy of the process is concerned. Huge price increases on inelastic necessities could induce consumers (and politicians) to reconsider the whole idea of supporting farmers. Over-exploiting the system is dangerous for farmers since they are dependent on public sympathy. In other words, external legitimacy puts upper limits on price changes. The lower limits are set by internal considerations. If all farmers produced the same commodities in the same proportions, it would be rational for them to act as revenue- maximizing monopolists and to discriminate among commodities in relation to demand elasticities. If farmers are highly specialized, as is increasingly the case, there are obvious limitations on such behaviour. Producers of price-sensitive commodities benefit relatively less by price increases since demand is discouraged more, but they are still better off if prices are raised (since demand elasticities are, in absolute values, less than one). The level of revenue at the sector level can be of no interest to them. The winners could theoretically compensate the losers (since revenue is maximized), but it is difficult to find a compensation scheme that is workable in practice.

However, it is possible that farmers, at least to some extent, act on the illusion of a collective interest, in spite of the conflicting objectives in reality. There is some evidence of this being the case. Improving the prices of grains is often portrayed in farmer newspapers as a success story, in spite of the fact that this by no means benefits all farmers.

Since farmers strongly influence the distribution of the TCA, it is possible that more powerful groups use this situation to their advantage rather than in order to maximize the collective revenue. In such a case, price changes would be biased against products of weaker groups. However, distributing the TCA cannot evolve to a pure game of power. The weaker groups could revolt and the unity of the farm movement could be lost. If stronger groups can achieve more by preserving unity, as is probably the case, they will limit their claims or short-sighted interests so as not to endanger the future of the farm movement.

## Statistical Models to be Tested

Four different distribution principles, *status quo*, fairness, revenue maximization, and power, have been presented as mutually exclusive hypotheses. In reality, several criteria may be used at the same time, together with some additional variables. It has been argued that in the case of both revenue maximization and power-oriented distribution, behaviour is limited by social acceptability. It is reasonable to argue that limits on price changes are related to trends in relative costs of production. World market prices and domestic market conditions probably affect distribution as well, since it may be easier for the producers of relatively undersupplied commodities facing favourable world market prices to argue for price increases for their products. Furthermore, since a price change in product $i$ is dependent on all the other price changes, as products are competing with each other, a price index is introduced as an explanatory variable ($Pind_i$). As may be seen from Equation (1), the outcome of the price-setting procedure is a vector of price change in *absolute* and not percentage terms (as, for instance, is the case in the EC). However, this creates some problems as far as the pooling of data is concerned. $\Delta Pi$ are of very different orders of magnitude, so that the intercept and some of the slope coefficients cannot be expected to be the same for all variables. This is particularly the case for power indicators and elasticities. Pooling of data is necessary since there are only 13 observations (commodities) each year. The simplest way to avoid this problem is to reformulate the equations in terms of percentage changes. This reformulation is, however, made at the expense of the explanatory power of the model. Simple analysis of the correlation between price and cost changes shows a stronger relation between absolute than between percentage changes. This is due to the nature of the cost variable (see the next section for discussion).

(5) $\quad PMP_{it} = \alpha + b\, PATC_{it} + c\, PPW_{it} + d\, SSR_{it} + e\, Pind_{it} + f\, El_i + g\, Pow\, ind_i + e_{it}$

PMP, PATC, and PPW are percentage changes in prices, average total cost, and world market prices, respectively. SSR is the self-sufficiency ratio.

## Definition and Measurements of Variables

The prices analysed in the paper are the outcomes of price-setting decisions and not actual market prices. Implementation of price decisions has occasionally been postponed by the government's anti-inflationary measures. In the mid-1970s, price increases were partly replaced by food subsidies, which were later removed. Due to these factors, there is a discrepancy between the two price series. However, the discrepancies are due to factors outside the negotiation system and can be disregarded, as we are studying the process of price setting and not the development of market prices. Prices were changed twice a year, on 1 January and 1 July. Due to the lack of semi-annual data on cost variables, we analyse annual changes. The lag structure is accounted for by aggregating a price change in July of year $t$ with a price change in January of year $t + 1$. These two changes are related to changes in the explanatory variables in year $t$.

Production costs are composed of two parts: average total costs at farm level $(ATC_F)$ and average processing costs $(ATC_P)$. $ATC_F$ was calculated using the standard method of gross margin calculations commonly used in Swedish agriculture, in both the research and extension services. Existing estimates were revised to ensure consistency over time and across commodities. Calculations are based on the "Central Plains" region, which is considered to be representative of average conditions in Sweden. Gross margin calculations are to a great extent based on assumptions about input/yield levels, thereby becoming normative cost indicators rather than positive figures. The fact that the cost figures are partly normative does not necessarily pose a problem. The hypothesis of fairness does not say that the TCA is distributed in relation to the trend in real cost but only in relation to that trend as it is perceived by the negotiators. If farmers base the distribution proposal on cost figures, they use the same type of gross margin calculations, since information on the trend in true cost at the commodity level is not available. Gross margin figures are, however, probably treated as indicators and not as precise figures, which makes calculation of percentage changes questionable.

$ATC_P$ was estimated using the same background data as used in the negotiations. The cost of processing crop products was disregarded due to the lack of complete time series.

Strength or power of a group cannot be measured directly. Indicators of power used in the literature are strongly influenced by Olson (1965), who claims that it is difficult to organize a group for the purpose of achieving a common benefit. Any effort by a potential organizer has to be weighed against potental gains that will accrue to all, even to nonparticipants. Accordingly, many indicators relate to, on the one hand, the cost of a collective action (i.e., number of producers, geographical dispersion, rate of growth of production, etc.) and, on the other hand, to the benefits from it (i.e., average size of production, etc.). It follows from the theory that it is easy to organize small homogeneous groups facing a large potential benefit. A possible criticism is that Olson's theory describes a process of forming an organization or rather explains why certain groups (like consumers or taxpayers) never managed to organize themselves, while the issue is often one of measuring the strength of an existing organization with established channels of communication. Furthermore, it can be easily shown that the relationship between size and strength is not that simple. Small milk producers in northern Sweden (Norrland) appear to be a very powerful group, while the few large producers of eggs (80 percent of eggs are produced by less than 200 producers) are not. Regional support to milk producers has been considerably increased recently, while newly introduced environmental restrictions on egg production ("hens' liberation") resulted in income loss for producers without any compensation being paid. The explanation is purely ideological. While small farmers in Norrland are extremely popular with politicians of all persuasions, large-scale egg producers are not considered to be "true farmers."

Several different indicators of power were used: number of producers, average size of production, share of the commodity in total farm income at the price-negotiation level, number of full-time producers, and share of the commodity produced by large farmers. As a measure of domestic market conditions, the self-sufficiency ratio was used:

$$(6) \quad \frac{S_{it}-D_{it}}{S_{it}} - 1$$

World market conditions were described by border prices. To test the hypothesis that farmers behave as one group in putting price increases on inelastic products, a set of demand elasticities is needed. Unfortunately, existing estimates do not cover all commodities included. For the missing elasticities, some assumptions were made. Power indicators and elasticities vary between commodities but are constant over time, thereby resembling dummy variables. Price changes in remaining products were represented by a price index:

$$(7) \quad \frac{\sum \Delta P_j D_j}{\sum P_j D_j}, \ i \neq j$$

## Estimation Results

The model was estimated by pooling time-series and cross-section data. The intercept and slope coefficients for production costs, prices of other commodities, domestic market conditions, power index, and elasticities were assumed to be the same for all products. For world market prices, this assumption is not valid due to policy differences. A 1977 agricultural policy decision decreed that all arable land should be preserved and emerging surpluses should be exported in the form of grains. For remaining products, self-sufficiency was the stated objective. A dummy variable was used to account for this difference. The results are given below, with $t$ values in parentheses.

$$PMP_t = -0.023 + 0.081\ PATC_t + 0.157\ PMP_{t-1} + 0.096\ PPW_t - 0.018\ SSR_t +$$
$$\phantom{PMP_t = }(-2.21)\quad (2.19)\qquad\quad (2.99)\qquad\qquad (4.90)\qquad\quad (-1.75)$$

(8)

$$0.90\ Pind_t + Pow\ ind\ (3)_t$$
$$(8.66)\qquad\quad (2.76)$$

$$R^2 = 0.59 \qquad F(7,\ 122) = 25.3 \qquad DW = 2.12$$

The above equation covers grains (wheat, rye, barley, and oats). All coefficients have the expected signs. All are significant at the 5-percent level, with the exception of the coefficient for $SSR$, which is significant at the 10-percent level. The vector of demand elasticities was omitted as the coefficient proved insignificant (the sign of the coefficient was, however, correct). The power indicator used above is average size of production. Similar results were produced by using the logarithm of the share of the commodity in total farm income. Other indicators performed less well; in particular, the share of the commodity produced by the largest farmers. In the latter case, the sign is negative but not significant, indicating that a high degree of concentration in production has, if any, a negative impact on price changes. In the case of non-grains (oilseeds, potatoes, milk, beef, veal, sheepmeat, pigmeat, poultry, and eggs), the coefficient for $PPW$ is significantly negative. Since this can hardly be the causal explanation for the phenomenon in question, one may draw the conclusion that world market prices are not taken into consideration for these commodities. Lagged price change was introduced to avoid the problem of serial correlation of residuals.

$Pind$ is highly significant, explaining most of the variance. This is partly due to the fact that the cost variable, because of its normative character, may not really reflect cost changes at the sector level. It can be argued, however, that the high degree of correlation also reflects inertia of the system; i.e., a tendency for different prices to develop in a similar fashion.

## Summary and Conclusions

Returning to our previously formulated hypotheses, it seems that inertia and fairness are together the major explanation of trends in relative prices at the negotiation level. In spite of their strong influence on the price-setting procedure, powerful groups are not misusing the situation. Some of the power variables are significant but contribute only marginally to the explanation of total variance. It is obviously not possible for a farm organization to build its image on the notion of solidarity and to exploit weaker groups at the same time. Neither is it possible for it to exploit consumers by behaving as a price-discriminating monopolist and still keep the credibility of the organization intact.

Domestic market conditions have some influence on negotiated prices. World market prices affect domestic grain prices. A great deal of variance in the sample still remains unexplained. This may be due to the fact that some important variables may still be missing or to an error of measurement in variables used. The most likely reason is that we are dealing with a unique social process. By pooling the data, we are searching for consistent rules of behaviour over time and across commodities. Lack of continuity and consistency in behaviour may, however, be present as different persons have been participating at different times.[5] During the 1980s, prices of beef, veal, and sheepmeat increased by exactly the same amount on several occasions, suggesting the use of some rules of thumb. It is also conceivable that different rules were used for different commodities. It is more important to find the right price for beef than for veal and sheepmeat, which are minor commodities.

## Notes

[1] Swedish University of Agricultural Sciences.

[2] Since the final decision is taken by Parliament and not by the negotiating parties themselves, the negotiations should rather be seen as deliberations.

[3] Any surpluses of livestock products are exported with the aid of producer-financed subsidies, thereby lowering producer prices. Surpluses of crop products are financed by government.

[4] The number of solutions is not endless, since very small price changes (less than Skr0.01) are not practicable.

[5] Three different leaders of farm organizations, with a very different style of leadership, were participating on behalf of farmers during the period analysed in this paper.

## References

Anderson, K., and Hayami, Y., *The Political Economy of Agricultural Protection*, Allen and Unwin, Sydney, Australia, 1986, 185 pp.

Baldwin, R.E., *The Political Economy of US Import Policy*, MIT Press, Cambridge, Mass., USA, 1985.

Fearne, A.P., "A 'Satisficing' Model of CAP Decision-Making," *Journal of Agricultural Economics*, Vol. 40, No. 1, 1989, pp. 71–81.

Finger, J.M., Hall, K., and Nelson, D.R., "The Political Economy of Administered Protection," *American Economic Review*, Vol. 72, 1982, pp. 452–466.

Gardner, B., "Causes of US Commodity Programmes," *Journal of Political Economy*, Vol. 95, No. 2, 1987, pp. 290–310.

Gardner, B.L., "Economic Theory and Farm Politics," *American Journal of Agricultural Economics*, Vol. 71, No. 5, 1989, pp. 1165–1171.

Goldstein, J., "The Political Economy of Trade: Institutions of Protection," *American Political Science Review*, Vol. 80, No. 1, 1986, pp. 161–184.

Lavergne, R.P., *The Political Economy of US Tariffs: An Empirical Analysis*, Academic Press, London, UK, 1983.

Olson, M., *Logic of Collective Action*, Harvard University Press, Cambridge, Mass., USA, 1965, 176 pp.

Paarlberg, R., "A Political Economy of American Agricultural Policy: Three Approaches," *American Journal of Agricultural Economics*, Vol. 71, No. 5, 1989, pp. 1157–1164.

Whaley, J., and Wigle, R., "Terms of Trade Effects, Agricultural Trade Liberalization and Developing Countries," in Goldin, I., and Knutsen, O. (Eds.), *Agricultural Trade Liberalization*, Organization for Economic Cooperation and Development, Paris, France, 1990.

## Discussion Opening—*David Kelch* (US Department of Agriculture)

Farm groups in most developed temperate zone countries have become more politically astute over the past 10–12 years in dealing with the political consequences of surplus production. Consequently, it is not surprising that the author dismisses two of the hypotheses to explain the distribution of revenues to farmers across commodities in the annual price-setting negotiations in Sweden. These two hypotheses are based on the social unacceptability, or should we say political unacceptability, of (1) monopolistic price discrimination, which damages the farmer's image in the public's eye, or (2) the use of power through numbers of growers or percentage of production against a minority of farmers, which destroys farmer unity. Having already answered the question politically, given the political constraints in Sweden on farmer actions within the price-setting process, the author goes on to develop a model that results in the same conclusion; i.e., the price-setting process among farmers is based on fairness and inertia and not on monopolistic price discrimination or power.

The model developed to examine the relationships seems adequately specified with the appropriate variables and caveats about data limitations, although the author ignores the possible presence of heteroscedasticity in the pooled data variance and does not incorporate a technology variable. Notwithstanding the theoretical and statistical considerations incorporated into the model by the author, one word stands out in the paper that is more likely than any econometric model to explain total farm revenue, the origin of that revenue, and its distribution among farmers. The word is perception. The author points out that the distribution to farmers of the total compensation amount through prices is not determined by real cost trends in the various commodities, but rather by the cost trends as perceived by the negotiators. The author also points out that small dairy farmers receive greater support compared to factory egg producers who are not "considered" (i.e., perceived) to be "true farmers."

The words perception and justice seem crucial to understanding the price-setting process, yet the conclusions do not address them. Are we to conclude that the restraint of power in slicing the agricultural pie among farmers is "just" when it is based on perceptions among farmers? Are we to conclude that democratic societies determine farm support by the perception that one is a "true farmer"?

This reviewer is led to believe that Swedish farm groups could use some agricultural economists to provide measures of costs of production to farmers and the budgetary consequences to consumers and taxpayers of supporting inefficient producers while penalizing efficient producers. Both the size of the pie and how it is sliced might then change.

*[Other discussion of this paper and the author's reply appear on page 69.]*

# Impact of the EC's Rebalancing Strategy on Developing Countries: The Case of Feed

*Monika Hartmann and Peter Michael Schmitz*[1]

**Abstract:** This paper analyses the effects of a rebalancing policy for the EC feed sector on the less developed countries. The theoretical part of the study reveals that evaluation of changing world market prices from the developing countries' perspective depends on the trade position of the LDCs in the relevant markets, on home-made distortions in the poor countries, as well as on the degree of insulation of domestic markets from the world market. These findings are supported by the empirical results presented in the paper. Using the sequential approach based on the Hicksian compensated curves of the new welfare economics, the efficiency effects due to a rebalancing policy in the EC feed sector are calculated for 16 developing countries/regions. These countries/regions cover about 98 percent of the developing world. The results show that, depending on the assumed world price transmission elasticity as well as on the application or absence of national agricultural policies in the countries considered, the developing world as a group would have to bear new welfare losses of between US$519 million and US$789 million. These welfare effects would not be spread evenly over the countries under consideration. While most developing countries and regions would experience a substantial decrease in their net economic welfare, some countries would enjoy considerable welfare gains.

## Introduction

Sustainable agricultural development is heavily influenced by national agricultural policies. Both domestic and foreign policies directly or indirectly affect the performance of a national food and agricultural sector. The efforts at international coordination of agricultural policies within GATT negotiations deserve and obviously receive serious attention, especially those policies on the agenda of the negotiations that are supposed to have the most adverse effects on other countries. The Common Agricultural Policy (CAP) of the European Communities (EC) appears to belong to that category. The EC and some other industrialized countries have been blamed for pressing, distorting, and destabilizing world market prices through their highly protective policy measures. The USA and the Cairns group are therefore asked to significantly reduce their protection levels for temperate zone food products. The EC's response to this requirement from more free-trade-orientated countries was a rebalancing proposal, which implies slightly reduced protection levels for traditional surplus products (grain, sugar, milk and milk products, and beef) provided mainly by supply control measures and the creation of new tariffs and import quotas for nongrain feed and oilseeds. The rebalancing proposal and its likely impact on developing countries is the main topic of the paper.

A vast literature exists on the effects of unilateral or multilateral trade liberalization (Goldin and Knudsen, 1990; Koester, 1982; Matthews, 1985; Parikh *et al.* 1988; and Tyers and Anderson, 1988). Only a few studies, however, concentrate on the rebalancing strategy and its external effects (Commission of the European Communities, 1988; ABARE, 1990; and Mahé and Tavéra, 1988). Since the focus of these contributions is not on the Third World, it seems worthwhile to analyse in more detail the price, trade, and welfare effects of this expected EC policy reform on developing countries. The paper therefore aims to develop a theoretical framework (which allows the above-mentioned effects to be calculated and assessed) and to estimate quantitatively the impact on developing countries under different policy scenarios for seven commodities.

In contrast to previous studies on rebalancing effects, this paper disaggregates the developing world into 16 countries/regions that cover about 98 percent of the developing world. In addition, cross price effects among the seven commodities are considered; the equivalent variation concept is taken as a welfare measure using a modified approximation technique; aggregation of welfare effects follows the sequential approach avoiding path dependence of the calculations; consumer price and producer price effects are calculated separately; and endogenous policy responses in the developing countries are included in the paper as well as different price policies for consumers and producers.

## Theoretical Analysis

At first glance, rebalancing of the EC's agricultural protection seems to be in line with Corden's (1974, p. 367) request for harmonized protection rates and therefore a desirable strategy. To prevent intra-sectoral distortions, he advised the implementation of similar protection rates for different commodities so as to realize at least a second-best optimum. However, there are severe doubts that the EC's rebalancing proposal does meet the criteria required for this optimum. Without going into the details of rebalancing (Hartmann, 1991a), one can expect an increase of the average EC protection level across agricultural commodities to lead to a rise in distortion compared to both nonagriculture and international agriculture. Harmonized protection levels do not require eliminating imports as many politicians suggest. Rather, the nominal protection level necessary to provide complete import substitution very often exceeds the second-best level. The introduction of new protectionism on the oilseed and feed markets raises uncertainty about the future behaviour of decision makers and creates political property rights that are not that easy to remove.

Hence, from an international welfare point of view, the risk of more welfare losses exceeds the opportunity to move towards a more liberalized world market situation. Despite this general conclusion, individual developing countries may gain or lose welfare due to the rebalancing strategy. Thus, an impact analysis will reveal the country-specific welfare effects of a change in the world market price structure due to the EC's rebalancing strategy. In contrast to the empirical part of this paper, where an advanced welfare approach is applied, the following considerations are based on a simple graphical illustration of the problem. Figure 1 shows the most important determinants of the welfare effects. Given a single market without any relation to other markets, one can derive the surplus effects (changes in producer and consumer surpluses) and the budget effects of exogenous and endogenous price adjustments. The analysis starts with declining world market prices. Three cases can be examined.

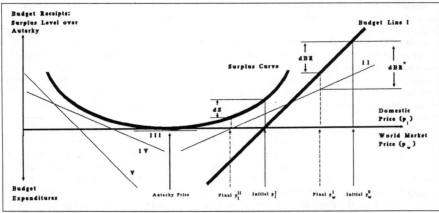

Figure 1—Welfare Effects of Exogenous and Endogenous Price Adjustments

**Case 1.** The country concerned does not pursue an own agricultural policy. World market prices equal domestic producer and consumer prices. If a price decrease from $p_i^I$ to $p_i^{II}$ is assumed, then the sum of producer and consumer surplus changes over the autarky welfare level declines by $dS$ in the case of an exporting country. These surplus changes equal the total welfare effects since neither budget receipts nor expenditures occur. The surplus curve indicates that in the case of an importing country the private welfare effect would increase due to falling world market prices. The surplus curve is a monotonic convex function with its lowest point at the level of the autarky price where no gains from trade emerge. Its degree of convexity depends on the supply and demand elasticities. Elastic responses of market participants induce a high degree of convexity of the surplus curve.

**Case 2.** The country has implemented an agricultural policy. Domestic prices deviate from world market prices. The former are fixed below the world market level at $p_i^l$, and the country exports the commodity. Falling world market prices from $p_w^o$ to $p_w^l$ then only affect the budget receipts (export taxes), whereas consumer and producer surpluses are not influenced. Export taxes decline by $dBR$ as the budget line I indicates. The budget line is a linear function with respect to changing world market prices and depends on a certain domestic price level that determines the point of intersection with the price axis. The slope of the budget line becomes flatter the closer the fixed domestic price approaches the autarky price. Depending on the desired level of fixed national prices, one gets different budget lines *I*, *II*, *III*, *IV*, and *V*. Their absolute slopes are also determined by demand and supply elasticities. The more responsive the market participants, the steeper the corresponding budget line. An importing country has a downwards sloping budget line, which implies decreasing budget expenditures when world market prices decline.

**Case 3.** Finally, it is assumed that there is an endogenous shift of domestic prices due to world market price adjustments. In concrete terms, this implies equivalent absolute changes in the corresponding prices. Thus the absolute price gap remains constant. With falling prices, the budget line *II* conditioned on $p_i^{II}$ indicates the relevant level of budget receipts at the world market price level $p_w^l$. Hence, receipts decrease by $dBR^*$. Simultaneously, the surpluses are reduced by $dS$, inducing an overall welfare loss of $dW = dS + dBR^*$, compared to the situation prior to the price change. If only domestic price were adjusted downwards at given world market prices, the surplus loss would be the same, whereas the loss of receipts would be lower, as the vertical difference between the budget lines *I* and *II* indicates at the world market price level $p_w^o$.

Extending these results to the multi-market case, one can draw the following conclusions. The final welfare effect for a single country depends on: the structure of production and consumption of the seven commodities for which world market prices adjust; the initial trade status of the country concerned in each commodity market; the application or absence of national agricultural policy measures driving a wedge between world market prices and domestic consumer/producer prices; the responsiveness of domestic prices to changes in world market prices (price transmission elasticity); own- and cross-price elasticities of demand and supply across agricultural commodities; and absolute level of exogenous and endogenous price changes that can reverse the trade position from one of exporter to importer or vice versa. Since the structure of the impact of these driving forces is of such a complex nature, one cannot arrive at simple theoretical conclusions. Hence, a first assessment of the EC's rebalancing strategy must and can only be given on the basis of an empirical analysis, which is the subject of the next section.

# Empirical Analysis

## Method and Data Used

To measure the international effects of policy changes in several different food commodity markets, a global multi-commodity model is required. In this paper, the world price effects due to a policy of rebalancing EC agricultural protection are taken from a simulation run of the SWOPSIM (static world policy simulation) model (ABARE, 1990, p. 14).[2] In this policy run, the average level of discrimination faced by consumers in the EC remains similar to the base case, while producer protection declines slightly. Table 1 summarizes the assumptions as well as the European and world market price effects of this policy simulation.

Given these multiple world market price changes (Table 1), the welfare effects in LDCs can be calculated. An adequate framework for policy evaluation in the case of multiple-price changes and market interdependencies is the sequential approach based on Hicksian compensated curves. This new welfare economics approach is also an exact welfare measure in the case of existing and/or changing market distortions. This latter aspect is very important, since in many developing countries the agricultural sector is directly and indirectly affected by a complex set of policies (Krueger, Schiff and Valdés, 1988; and World Bank, 1986, p. 61).

Table 1—Assumptions and Results of an ABARE Study on an EC Rebalancing Strategy using the USDA SWOPSIM Model of Percentage Change

| Commodity or Commodity Group | Assumptions | | Results | | Change in World Market Price |
|---|---|---|---|---|---|
| | Change in the EC's | | | | |
| | CSE | PSE | Consumer Price | Producer Price | |
| Wheat | −41.7 | −43.5 | −23.6 | −26.4 | 9.0 |
| Maize | −57.4 | −58.2 | −24.8 | −28.6 | 3.0 |
| Other coarse grains | −20.0 | −20.0 | −12.7 | −11.3 | −0.5 |
| Soyabeans | 46.1 | — | 44.7 | — | −2.5 |
| Soyabean meal | 46.1 | 46.1 | 41.5 | 41.5 | −7.5 |
| Other oilseeds | 46.1 | — | 44.1 | — | −5.0 |
| Other meals | 46.1 | 46.1 | 39.3 | 39.3 | −12.0 |

Source: ABARE (1990).

Using this extended applied welfare economics framework, the efficiency effects on LDCs due to a multiple world market price change consist of three components (Just, Hueth, and Schmitz, 1982, pp. 338–341). First, the benefits to consumers can be measured by a special approximation of the Hicksian equivalent variation.[3] Second, the welfare effects on producers are equal to the changes in producer rent. Both welfare measures are calculated sequentially. This implies that the resulting supply and compensated demand curves are successively conditioned on previously considered price changes. For the single-market case, the sum of producer and consumer surplus changes can be found in Figure 1 as $dS$. In addition to these total private welfare effects of an externally induced multiple price change, one has to consider the change in the government budget in the developing country (the budget line in Figure 1). This third component must be calculated on all markets where internal distortions exist in the initial situation or are introduced or altered as a consequence of the EC policy reform. The calculation of the budget effects does not follow the sequential procedure; rather, it is derived as the difference between the initial and the final situation after all adjustments have taken place.

The distributional and efficiency effects of externally induced world market price changes on developing countries depend very much on the existing home-made distortions and the degree of insulation of the poor countries. To incorporate the former determinant, the analysis considers internal agricultural policies where data on protection/distortion levels were available (Sullivan et al., 1989).[4] To capture the latter argument, the calculations are done using two different sets of world price transmission elasticities (WPTE). One extreme case assumes that world market price changes are fully transmitted into the developing country (WPTE=1). This implies that the absolute change in world market prices equals the absolute change in domestic prices (Figure 1). In an alternative scenario, no internal price adjustments due to externally induced world market price changes (WPTE=0) are considered. The analysis thus covers the whole range of possible adjustment policies in the LDCs due to an EC rebalancing policy.[5]

The effects of policy intervention are calculated for 16 developing countries and regions[6] for 1986. All data, price elasticities, and protection rates are taken from the USDA's SWOPSIM database (Sullivan *et al.*, 1989) and therefore coincide with the database used to identify the world market price effects due to an EC rebalancing reform. In accordance with the Slutsky-Schultz relationship for food products, income elasticities are set equal to the negative value of the sum of the own- and cross-price elasticities. Unfortunately, the data concerning the internal sector-specific distortions in the developing countries are very incomplete. Furthermore, all calculations are based on official time-series data and are therefore limited to the formal sector of the economy. The restriction of this analysis to only seven commodities is due to a lack of information concerning the world market price effects for other products as a consequence of an EC rebalancing reform. Given these limitations, the picture presented should be seen only as indicative of the effects a rebalancing policy in the EC feed sector would have on the agricultural sector and the overall economic performance of the developing countries.

## Results

The distributional and efficiency effects of an EC rebalancing reform on the developing countries are summarized in Table 2. Examination of these results shows that the net welfare effects are quantitatively and qualitatively very unevenly spread over the developing countries and regions. While, for example, Brazil and China are the major losers from an EC rebalancing reform, India might enjoy substantial welfare gains.

The welfare effects of an EC policy reform on a developing country depend heavily on the trade position of the LDC in grain on the one hand and oilseeds and oilmeals on the other. While exporters of wheat and maize and importers of oilseeds and oilmeals tend to gain in net economic welfare, the opposite holds for those countries that have an inverse trade structure. Thus, the substantial welfare losses to Brazil, China, and other sub-Saharan Africa, documented in Table 2, are not surprising, since these countries import considerable quantities of grain while they are at the same time exporters of oilseeds and oilmeals. For most of the countries/regions considered, the sign of the overall welfare effects is not that easy to identify, since these countries are more or less general exporters (importers) of the commodities considered, or have a very complex trade structure. In this case, the welfare effects depend on the size of the deficits, surpluses, and EC-induced world market price effects on the relevant markets (Figure 1).

In addition to the trade position the degree of internal agricultural protection/discrimination in the LDCs plays a considerable role in the determination of the welfare effects, if world market price changes are transmitted to the domestic market. Table 2 reveals that an omission of own agricultural policies in the countries considered not only changes the results relating to the distribution of welfare effects over consumers, producers, and taxpayers but might even be misleading with respect to the sign of the net social welfare effects due to the policy of rebalancing EC agricultural protection (India, Nigeria, and other South Asia).

A third major finding and confirmation of the theoretical analysis is that the impact of an EC-rebalancing policy on LDCs depends very much on the assumed world price transmission elasticity (Figure 1 and Table 2). If developing countries' governments fully transmit the EC-induced world market price changes to their domestic markets, three countries/regions (India, other South Asia, and Nigeria) will enjoy substantial welfare gains, while the majority of developing countries/regions will lose in net economic welfare terms. Besides Brazil, the main losers in this policy scenario are other Latin America, China, other low and middle-income Asia, Middle East, and North Africa oil producers, and other sub-Saharan Africa. The net social welfare loss per year to the developing countries as a group amounts to $519 million under the given domestic agricultural policy.

Turning to the alternative case in which LDCs are assumed to isolate their domestic market from world market price changes, all countries/regions except Mexico and South Africa have to bear welfare losses. The net loss of $789 million per year to the Third World is much

Table 2—Welfare Effects of a Rebalancing Policy in the EC Feed Sector for Selected Developing Countries and Regions ($ million)[1]

| Country/Region[2] | World Price Transmission Elasticity (WPTE) = 1 | | | | WPTE = 0 |
|---|---|---|---|---|---|
| | Equivalent Variation | Producer Surplus Change | Net-Budget Income Change | Welfare Change | Welfare Change |
| Mexico | -28.97 (-28.67) | 36.10 (35.63) | (-8.50) | 7.13 (-1.54) | 5.32 |
| Brazil | -7.74 (-7.54) | -142.18 (-142.30) | (-1.80) | -149.92 (-151.64) | -155.06 |
| Argentina | 34.20 (34.05) | -60.75 (-61.70) | (-14.31) | -26.55 (-41.96) | -34.35 |
| Venezuela | 0.75 (0.71) | -0.72 (-0.73) | (-3.02) | 0.03 (-3.04) | -0.49 |
| Other Latin America | -57.49 (-57.49) | -0.12 (-0.12) | – | -57.61 (-57.61) | -61.05 |
| Nigeria | -0.84 (0.01) | -8.95 (-8.95) | (36.30) | -9.79 (27.36) | -10.27 |
| Other sub-Saharan Africa | 14.48 (14.48) | -66.37 (-66.37) | – | -51.89 (-51.89) | -54.94 |
| Egypt | -71.20 (-70.62) | 8.32 (8.30) | (16.34) | -62.88 (-45.98) | -64.18 |
| Middle East and North Africa non-oil producers | -163.88 (-163.88) | 130.53 (130.53) | – | -33.35 (-33.35) | -37.79 |
| Middle East and North Africa oil producers | -225.70 (-225.70) | 129.94 (129.94) | – | -95.76 (-95.76) | -99.39 |
| South Africa | -20.08 (-20.08) | 28.23 (28.14) | (-3.17) | 8.15 (4.89) | 7.30 |
| India | -175.88 (-174.88) | 174.80 (176.99) | (98.77) | -1.08 (100.88) | -14.82 |
| Other South Asia | 141.60 (-140.22) | 120.96 (122.12) | (61.12) | -20.64 (43.02) | -28.23 |
| Indonesia | 17.88 (17.72) | -34.64 (-34.65) | (8.38) | -16.76 (-8.55) | -19.05 |
| Other low- and middle-income Asia | 4.73 (4.73) | -88.14 (-88.14) | – | -83.41 (-83.41) | -86.03 |
| China | -697.30 (-697.30) | 576.83 (576.83) | – | -120.47 (-120.47) | -135.96 |
| All developing countries | -1,518.64 (-1,514.68) | 803.84 (805.52) | (190.11) | -714.80 (-519.05) | -788.99 |

[1]Values in parentheses include own agricultural policies for those countries where data were available. [2]For the country composition of the regions, see Note 6. Source: Own calculations for 1986. Data are from Sullivan, Wainio, and Roningen (1989); and ABARE (1990).

greater than in the above scenario. Since, in the past, developing countries very much isolated their domestic markets from international price changes (Anderson and Tyers, 1990, p. 50), this scenario might be closer to the real welfare effects in the developing countries.

The effects on foreign exchange earnings in LDCs due to an EC rebalancing policy are shown in Table 3.

Table 3—Change in Foreign Exchange Earnings ($ Million)[1]

|  | WPTE=1 | WPTE=0 |
|---|---|---|
| Mexico | 150.50 | 5.32 |
| Brazil | −143.14 | −155.06 |
| Argentina | −32.63 | −34.35 |
| Venezuela | −3.61 | −0.49 |
| Other Latin America | −55.99 | −61.05 |
| Nigeria | 27.34 | −10.27 |
| Other sub-Saharan Africa | −34.29 | −54.94 |
| Egypt | −23.48 | −64.18 |
| Middle East and North Africa non-oil producers | 24.69 | −37.79 |
| Middle East and North Africa oil producers | −33.07 | −99.39 |
| South Africa | 15.09 | 7.30 |
| India | 345.86 | −14.82 |
| China | 20.17 | −135.96 |
| Other South Asia | 148.43 | −28.23 |
| Indonesia | −38.46 | −19.05 |
| Other low- and middle-income Asia | −109.55 | −86.03 |
| All developing countries | 157.86 | −788.99 |

[1]Values include own agricultural policies in the country considered.

Source: Own calculations for 1986. Data are from Sullivan et al. (1989); and ABARE (1990).

In the case of total domestic isolation (WPTE=0), the foreign exchange effects equal the net welfare changes in the last column of Table 2. These effects are similar to the welfare changes spread unevenly over developing countries. While in some countries/regions, such as Brazil, the foreign exchange earnings will decrease by a substantial amount, others, such as India and other South Asia, might experience considerable increases in net export earnings. Here again the sign and the size of the effects depend very much on the assumed world price transmission elasticity. Assuming a full price transmission, developing countries as a group would enjoy an increase in foreign exchange earnings of $158 million, thus easing their foreign exchange deficits by about 3 percent for the corresponding commodities. In contrast, under the assumption of zero WPTEs, net export earnings would decrease by $789 million. In this latter scenario, developing countries as a group would experience an aggravation of their foreign exchange deficits with respect to the considered commodity markets by about 15 percent.

Many studies analysing the impact of a liberalization of EC agricultural policy on the Third World conclude that these countries as a group would have to bear a welfare loss as a consequence of such a reform (Koester, 1982; Matthews, 1985; Parikh et al., 1988; Tyers, 1989;

and Tyers and Anderson, 1988). The results presented in this paper suggest that developing countries as a group would also not be beneficiaries in the case of a rebalancing policy in the EC feed sector. Thus, one may jump to the conclusion that a CAP liberalization and a rebalancing strategy point in the same direction and hence, from a developing countries perspective, it does not matter which strategy the EC pursues. This conclusion, however, neglects some important aspects.

First, some doubts that a rebalancing strategy will really reduce support on highly protected markets seem justified. It is likely that this policy leads primarily to an increase in protection for deficit products, thereby further increasing worldwide distortion and reducing the export opportunities for LDCs.

But even if one assumes that the EC really reduces protection for the CAP core commodities, there remains at least one major difference between an EC rebalancing strategy and a liberalization policy: the impact these reforms have on stability and predictability in world agricultural markets. While the latter leads to a general reduction of world market price risk (Tyers and Anderson, 1988) and future policy uncertainty, the former increases policy uncertainty and has at best an ambiguous impact on price risk (ABARE, 1990, p. 21). This holds even more if a rebalancing strategy were not the outcome of the present GATT round but the reason for its failure. In that case, world agricultural markets might be put in total disarray. Since developing countries in particular benefit from predictability, certainty, and stability in world agricultural trade (Balassa, 1988, p. 46; and Matthews, 1988, p. 32), it is in all their interests to ensure a liberalization of world agricultural markets.

# Notes

[1]Universität Frankfurt.

[2]Some additional empirical studies analyse the world market price effects of alternative EC rebalancing strategies (Commission of the European Communities, 1988; Mahé, 1984; and Mahé and Tavéra, 1988). Because of the different assumptions concerning the policy simulations and differences in product coverage and product aggregates, it is unfortunately not possible to compare the results.

[3]For the method to approximate the equivalent variation, see Hartmann (1991b).

[4]Because of a lack of data, the incorporation of sector-specific policies in the analysis was only possible for Mexico, Brazil, Argentina, Nigeria, Egypt, India, other South Asia, and Indonesia.

[5]Since the welfare effects in a multi-commodity model with existing distortions are not a monotonic function of the world price transmission elasticities, the two scenarios do not necessarily represent the upper- and lower-bound welfare effects in a developing country due to an EC rebalancing strategy.

[6]These 16 countries/regions encompass about 98 percent of the developing world. Virtually all the countries classified by the World Bank (1986, p. 180) as low-income and lower middle-income economies are included in the analysis. The country composition of the regional aggregates is: other Latin America—Belize, Costa Rica, El Salvador, Honduras, Guatemala, Nicaragua, Panama, Bahamas, Bermuda, Cuba, Dominican Republic, Haiti, Jamaica, Trinidad and Tobago, Barbados, Bonaire, Curaçao, French West Indies, Guadeloupe, Martinique, Turks and Caicos Islands, Cayman Islands, Aruba, British West Indies, Leeward-Windward Islands, St. Kitts, Netherlands Antilles, Antigua, Nevis, Montserrat, British Virgin Islands, Grenada, St. Vincent, St. Lucia, Dominica, Guyana, French Guiana, Surinam, Bolivia, Chile, Colombia, Ecuador, and Paraguay, Peru, and Uruguay; other sub-Saharan Africa— Angola, Benin, Botswana, Burkina Faso, Burundi, Cameroon, Cape Verde, Central African Republic, Chad, Comoros, Congo, Cote d'Ivoire, Djibouti, Equatorial Guinea, Ethiopia, Gabon, Gambia, Ghana, Guinea, Guinea-Bissau, Kenya, Lesotho, Liberia, Madagascar, Malawi, Mali, Mauritania, Mauritius, Mozambique, Namibia, Niger, Réunion, Rwanda, São Tomé/Principe, Seychelles, Sierra Leone, Somalia, Senegal, Sudan, Swaziland, Tanzania, Togo, Uganda, Zaire, Zambia,

and Zimbabwe; Middle East and North Africa (non-oil-producers)—Turkey, Cyprus, Lebanon, Israel, Gaza, West Bank, Jordan, North Yemen, South Yemen, and Morocco; Middle East and North Africa (oil producers)—Syria, Iraq, Iran, Kuwait, Qatar, Saudi Arabia, United Arab Emirates, Oman, Bahrain, Algeria, Tunisia, and Libya; other South Asia—Afghanistan, Bangladesh, Bhutan, Nepal, Pakistan, and Sri Lanka; and other low and middle-income Asia— Thailand, Malaysia, Philippines, South Korea, North Korea, Brunei, Burma, Fiji, Cambodia, Laos, Mongolia, and Vietnam.

# References

ABARE (Australian Bureau of Agricultural and Resource Economics), "Some Implications of 'Rebalancing' EC Agricultural Protection," Discussion Paper No. 90–5, Canberra, Australia, 1990.

Anderson, K., and Tyers, R., "How Developing Countries Could Gain from Agricultural Trade Liberalization," in Goldin, I., and Knudsen, O. (Eds.), *Agricultural Trade Liberalization for Developing Countries*, Organization for Economic Cooperation and Development, Paris, France, 1990, pp. 41–76.

Balassa, B., "Interest of Developing Countries in the Uruguay Round," *World Economy*, Vol. 11, No. 1, 1988, pp. 39–55.

Commission of the European Communities, *Disharmonies in EC and US Agricultural Policy Measures*, Brussels, Belgium, and Luxembourg, 1988.

Corden, W.M., *Trade Policy and Economic Welfare*, Oxford University Press, Oxford, UK, 1974.

Goldin I., and Knudsen, O. (Eds.), *Agricultural Trade Liberalization for Developing Countries*, Organization for Economic Cooperation and Development, Paris, France, 1990.

Hartmann, M., "Old Wine in New Bottles—Agricultural Protectionism in the EC," *Intereconomics*, Vol. 26, No. 2, 1991a, pp. 58–63.

Hartmann, M., "Wohlfahrtsmessung auf Interdependenten und Verzerrten Märkten," Die Europäische Agrarpolitik aus Sicht der Entwicklungsländer, Ph.D. thesis, 1991b.

Just, R.E., Hueth, D.L., and Schmitz, A., *Applied Welfare Economics and Public Policy*, Prentice-Hall Inc., Englewood Cliffs, N.J., USA, 1982.

Koester, U., *Policy Options for the Grain Economy of the European Community: Implications for Developing Countries*, International Food Policy Research Institute, Research Report No. 35, Washington, D.C., USA, 1982.

Krueger, A.O., Schiff, M., and Valdés, A., "Agricultural Incentives in Developing Countries: Measuring the Effect of Sectoral and Economywide Policies," *World Bank Economic Review*, Vol. 2, No. 3, 1988, pp. 255–271.

Mahé, L.P., "A Lower but more Balanced Protection for European Agriculture," *European Review of Agricultural Economics*, Vol. 11, No. 2, 1984, pp. 217–234.

Mahé, L.P., and Tavéra, C., "Harmonization of EC and U.S. Agricultural Policies," *European Review of Agricultural Economics*, Vol. 15 , No. 4, 1988, pp. 327–348.

Matthews, A., *The Common Agricultural Policy and the Less Developed Countries*, Gill and MacMillan Ltd. and Trócaire, Dublin, Ireland, 1985.

Matthews, A., "The Agricultural Negotiations in the Uruguay Round and the Developing Countries," *Trócaire Development Review*, 1988, pp. 21–36.

Parikh, K.S., Fischer, G., Frohberg, K., and Gulbrandsen, O., *Towards Free Trade in Agriculture*, Klumes, Dordrecht, Netherlands, 1988.

Sullivan, J., Wainio, J., and Roningen, V., *A Data Base for Trade Liberalization Studies*, Staff Report No. AGES 89–12, Economic Research Service, US Department of Agriculture, Washington, D.C., USA, 1989.

Tyers, R., "Developing Country Interests in Agricultural Trade Reform," *Agricultural Economics*, Vol. 3, 1989, pp. 169–186.

Tyers, R., and Anderson, K., "Liberalising OECD Agricultural Policies in the Uruguay Round: Effect on Trade and Welfare," *Journal of Agricultural Economics*, Vol. 39, No. 2, 1988, pp. 197–216.

World Bank, *World Development Report*, Washington, D.C., USA, 1986.

---

**Discussion Opening**—*Daniel V. Gordon* (Norwegian School of Economics and Business Administration)

In this paper, Hartmann and Schmitz (HS) attempt to measure a welfare change for some developing countries caused by a change in the tariff and quota structure for agricultural commodities in the EC. This is an interesting problem and a non-trivial empirical exercise.

It is common in applied welfare analysis to add up consumer surpluses associated with a price change and define this value as a money measure of a welfare change. The most favoured measures are Marshallian consumer surplus and Hicksian consumer surplus; i.e., compensating variation (CV) and equivalent variation (EV). Under the restriction of zero-income effects (i.e., indifference curves are parallel) all three measures coincide, but in general they provide different measures of welfare change. HS employ the EV index. Although no justification for the procedure is provided, this index is superior to a Marshallian approach. In the single-consumer case, EV also performs better than CV. However, EV measurement is path dependent; the way prices change matters in measuring welfare change. The problem is more serious in the HS application because many prices change simultaneously. HS assume the order of price change to be known and proceed with measurement. But, nevertheless, the welfare change calculated is not unique.

In addition to this, consumer surplus measures are seriously criticized for three reasons. First, they cannot account for income redistribution (i.e., they are silent on changes in economic inequality). Second, they suffer from the Boadway Paradox and cannot be used as efficiency indexes. Finally, they do not always represent a social binary ordering (i.e., the preference ordering may fail to be asymmetric). This criticism is so severe as to negate the use of consumer surplus measures in applied welfare analysis.

However, alternative measures are available to the applied economist. Money metric utility (i.e., the minimum income needed at fixed reference prices to acquire a commodity bundle that is at least as good as the one actually consumed) is one. Money metrics is an index of utility level, not change in utility, upon which consumer surplus measures are based. It is always a social ordering and allows inequality aversion. However, it does require homothetic preferences. Extended money metrics (i.e., the minimum expenditure a reference individual must have at reference prices to achieve the level of utility of each member of the household) is another measure. Extended money metrics improves on money metrics by accounting for the well-being of each person in the economy. However, it does require interpersonal comparisons of utility. Welfare ratios (i.e., the ratio of household income to the minimum expenditure needed for a reference level of utility) are another measure. The advantages are that family size is taken account of and the ratio economizes on interpersonal comparisons. The index is also easy to calculate with minimum data requirements. However, without homothetic preferences, it is not an exact index of well-being. Finally, equivalence scales may provide a useful methodology for making interpersonal comparisons and for applied welfare analyses. This measure requires no reference prices, has no problems of concavity, and is easily calculated from demand data.

These new welfare measures are an improvement over the old surplus measures. In applied work, the choices of index can be based on available data. Of course, it would be preferable to calculate several measures of well-being. Hartmann and Schmitz have an interesting problem and by including the new welfare measurements they would allow a comparison to the surplus measurements already calculated and add to the paper itself.

*[Other discussion of this paper and the authors' reply appear on page 69.]*

# EC Enlargement and Trade Liberalization in the Vegetable Oils Market

*Elisabetta Croci-Angelini and Secondo Tarditi*[1]

**Abstract:** The third EC enlargement to include Spain and Portugal raised substantial problems in the international olive oil market, where the EC–10 accounted for 48 percent of production and 52 percent of consumption and the new EC members for 30 percent and 24 percent, respectively. Olive oil was highly supported in the EC–10, where producer prices were more than twice the international prices. Extending (from 1986 to 1991) the EC–10 price support to Spanish and Portuguese producers would have meant huge EC–12 budgetary costs and a complete price collapse of the tiny world market.

This paper presents three simulations of the world market by means of a multi-product and multi-regional price equilibrium model of the oils and fats sector: (1) the impact of the EC enlargement to include Spain and Portugal without CAP changes; (2) the impact of the EC–12 offer in the GATT negotiations implying a 30-percent decrease in domestic support for olive oil, butter, animal fats, and a 6-percent import tariff for oilseeds and other vegetable oils; and (3) the decrease in the producer price support of olive oil that would be needed to offset the impact of enlargement on the EC–12 trade balance.

## Introduction

The relatively high degree of substitutability among oils and fats at the consumer level compels the policy maker to develop coordinated policy measures for this sector of agricultural production. The differences in price policies concerning vegetable oils between the EC–10 and Spain and Portugal created considerable problems in the third EC enlargement. Though Spain and Portugal joined the EC in 1986, a transitional period was established to delay the consequences of a unified market for oils and fats until 1991.

In order to assess in quantitative terms the impacts of the existing policy measures, a non-linear price equilibrium model was used for the analysis of international trade in oils and fats.[2] The world market was divided into five regional aggregates: the EC before its third enlargement (EC–10), Spain and Portugal grouped as the Iberian countries (ICs), the USA, Centrally Planned Economies (CPEs), and the Rest of the World (RoW). The oils and fats sector was divided into four sub-sectors: olive oil, other vegetable oils, butter, and other animal fats, reflecting the existing major differences in production, substitutability at the consumer level, and different policy measures implemented in the EC.

We chose 1986, when GATT negotiations started in Punta del Este, as a reference point, and the statistical information used was based largely on FAO estimates of production, utilization, and prices (FAO, 1986a, 1986b, and various years). Since olive oil supply is characterized by very high variability from year to year, production and consumption figures for this commodity were estimated on their long-term trends.

The quantitative analysis was carried out with reference to agricultural markets at the producer level. The policy analysis examined in the following pages refers to the world situation in 1986. As in most comparative static analyses, the model does not take into consideration time trends or productivity developments, but focuses on the likely impacts of the envisaged changes due to price policies.

## Economic Policy Measures in the Olive Oil Market

The common policy for the vegetable oils sector was laid down in 1966 when the EC–6 self-sufficiency ratio for these products was very low. Under the GATT agreement, no tariffs could be levied on oilseeds or seed oils at the EC border. Given the remarkable degree of substitutability between olive oil and other vegetable oils, the domestic market price for olive oil could thus not be raised as high as the target producer price without reducing consumption too much. Consequently, the support system had to be organized on the basis of a scheme of deficiency payments.

61

For each marketing year, the EC Council of Ministers sets a representative market price (RMP) at a level "permitting normal marketing of olive oil, account being taken of likely price trends of competing products" (European Communities, 1988). The domestic price level is prevented from downward movements by a threshold price (TP) for olive oil imported from third countries, which is "fixed in such a way that at the Community frontier crossing point the selling price will be the same as the representative market price" (*op. cit.*). The threshold price follows any adjustment in the representative market price during the marketing year.

A c.i.f. price is fixed with reference to the most favourable purchasing possibilities on the world market. The import levy makes up the difference between the threshold price and the c.i.f. price when the world market price is below the representative market price. The European Commission sets its amount periodically, roughly equivalent in principle to export refunds to allow EC traders to export on the world market.

The Council of Ministers establishes a target producer price (TPP), which "is fixed at a level which is fair to producers, account being taken of the need to keep Community production at the required level" (*op. cit.*) to provide olive oil producers with what it considers a fair income and retain the number of trees and the existing level of production. The target producer price is attained by providing the olive oil producer with a direct producer subsidy (PS), limited to the production of oil originating from trees planted before October 31, 1978, in the EC–9, before 1981 in Greece, and before 1984 in Spain and Portugal.

The production subsidy was the difference between the target producer price and the representative market price *(PS=TPP–RMP)* until April 1979, when the representative market price was lowered while the producer subsidy was not increased. An equivalent drop in the producer price was prevented by the institution of the so-called consumer subsidy (CS), covering the difference between the target price minus the producer subsidy and the representative market price *(CS=TPP–PS–RMP)*. Imported oil cannot benefit from the consumer subsidy, and, in order to avoid frauds, it must either be sold directly to consumers or be subject to an extra duty equivalent to the consumer subsidy granted to domestic production. Therefore, benefits of the so-called consumer subsidy accrue only to domestic producers. In practice, the difference between the producer target price and the representative market price (i.e., the sum of the producer subsidy and of the so-called consumer subsidy) is equivalent to a traditional deficiency payment.

Oilseed production is supported by deficiency payments measures as well, in order to guarantee producer prices and increase the EC–10 self-sufficiency level.

## Impact of EC Enlargement

Although the enlargement of the EC–10 to include Spain and Portugal for oils and fats is not yet fully effective, it is important to foresee the impacts that a unified market will have on supply, demand, and trade balance, as well as on EC budgetary expenditures, in order to implement appropriate countermeasures aimed at offsetting some undesired effects of economic integration between the EC–10 and the Iberian countries.

The unification of EC–10 and Iberian markets for oils and fats will not, under present CAP regulations, cause relevant changes in the EC–10 market where prices at both the producer and market levels will not change. The impact of EC enlargement will be mainly felt in the Iberian and third countries, while EC–10 countries will be concerned with the budgetary expenditure they will have to share with the newcomers. Table 1 summarizes the main results of the model.

In the Iberian countries, the enforcement of present CAP rules will mean much higher producer (109 percent) and market (31 percent) prices for olive oil together with higher producer prices (47 percent) and lower market prices (–56 percent) for other vegetable oils. The price of butter will be reduced, and the price of other animal fats will show minor changes.

The Iberian supply response to olive oil price changes will depend on how CAP policies are used to implement producer price increases. If the producer subsidy is limited to existing olive trees and strictly controlled, supply response will be much lower than in the case where increases in producer price are granted to all producers by means of a generalized consumer subsidy. The likely range of the Iberian supply response was examined by assuming a supply elasticity of 0.5.

According to this hypothesis, Iberian olive oil supply will expand by 210,000 t, while the converging effect of higher market prices for olive oil and lower market prices for other vegetable oils will reduce olive oil demand by 104,000 t. The result will be a huge increase in the exportable surplus (314,000 t), which, if dumped on the thin world market, would depress the world price for olive oil by as much as 50 percent. However, the effects are likely to be catastrophic for olive oil producers in third countries under all reasonable assumptions of Iberian supply response.

Taking into account that Tunisia, Turkey, and other relatively poor Mediterranean countries will face the most unpleasant consequences of this policy, severe political problems are likely to emerge. The EC Council of Ministers could develop some new and ingenious uses for olive oil extra surpluses, as it has done for butter. However, it will not be an easy task under present budgetary constraints.

The impact of EC prices on the Iberian market for other vegetable oils will have the opposite effect on the Iberian trade balance. The increase in demand due to lowered consumer prices will offset the increase in supply due to higher producer prices and expand by one third the present Iberian deficit in other vegetable oils.

Iberian olive oil producers will benefit most from a unified market for oils and fats under CAP regulations. Their economic surplus will increase by 1,000 million ECU, much more than the gain enjoyed by oilseed producers (136 million ECU).

The EC budget, on the other hand, will bear the largest burden due to the converging effects of new producer subsidies paid to the Iberian producers and higher export restitutions for the increased EC–12 exports to a world market where prices have dropped dramatically. The increase in EC–12 budget expenditures would be 1,338 million ECU in the olive oil sector.

In the Iberian countries, the impact on consumer surplus of lower prices for other vegetable oils (374 million ECU) and for butter will more than offset the effect of higher market prices for olive oil (–157 million ECU). The net consumer benefit for the whole oils and fats sector will be 224 million ECU.

The overall economic welfare impact of the unification of the EC–10 and Iberian oils and fats markets will be substantially negative (–479 million ECU), mainly as a consequence of increased distortions in the Iberian olive oil sector. In the other vegetable oils sector, the negative impact on overall economic welfare of higher producer prices is completely offset by the positive effect of lower market prices.

## The EC–12 Proposal in the GATT Negotiations

GATT negotiations focused on agriculture for the first time in 1986, which turned out to be a major obstacle in the effort to reach an agreement. In 1988, the mid-term review in Montreal failed to envisage a compromise on the most thorny issues, among which the agricultural sector was still outstanding. On the eve of the conclusion of the negotiations, the EC proposed a 30-percent cut in protection for most agricultural products (including oilseeds, olive oil, as well as livestock products), with reference to support levels existing in 1986. Scenario B in Table 1 shows how this proposal would affect the oils and fats sector. The results show that major changes are likely to occur vis-à-vis the situation existing in 1986, before Spain and Portugal joined the EC (reference scenario).

In the EC–10, prices for all oils and fats will drop, with the exception of the demand price for other vegetable oils, where the EC proposal suggests the introduction of a 6-percent tariff on imports in order to rebalance the price support system. While domestic supply shrinks for all goods, domestic demand shows an increase for both olive oil and butter to the detriment

Table 1—Impact of Alternative Policies

| | Reference Scenario | | | | New Scenario | | | | Change vs. Reference | | | | Percent Change vs. Ref. | | | |
|---|---|---|---|---|---|---|---|---|---|---|---|---|---|---|---|---|
| | EC-10 | Iberia | USA | RoW | EC-10 | Iberia | USA | RoW | EC-10 | Iberia | USA | RoW | EC-10 | Iberia | USA | RoW |
| | | | | | Scenario A: EC Enlargement without Change in the CAP | | | | | | | | | | | |
| **Olive Oil:** | | | | | | | | | | | | | | | | |
| Supply price (ECU/t) | 3226 | 1544 | 1387 | 1387 | 3226 | 3226 | 688 | 688 | 0 | 1682 | -699 | -699 | 0 | 109 | -50 | -50 |
| Demand price (ECU/t) | 1986 | 1520 | 1387 | 1387 | 1986 | 1986 | 688 | 688 | 0 | 466 | -699 | -699 | 0 | 31 | -50 | -50 |
| Supply protection rate (%) | 133 | 11 | 0 | base | 369 | 369 | 0 | base | 236 | 358 | 0 | 0 | 178 | 3158 | | |
| Demand protection rate (%) | 43 | 105 | 0 | base | 189 | 189 | 0 | base | 145 | 179 | -0 | 0 | 337 | 1865 | | |
| Supply ('000 t) | 800 | 490 | 1 | 359 | 800 | 700 | 1 | 205 | 0 | 210 | -0 | -154 | 0 | 43 | -43 | -43 |
| Demand ('000 t) | 860 | 390 | 40 | 340 | 860 | 286 | 57 | 483 | -0 | -104 | 17 | 143 | -0 | -27 | 42 | 42 |
| Trade balance ('000 t) | -60 | 100 | -39 | 19 | -60 | 414 | -56 | -278 | 0 | 314 | -17 | -297 | -0 | 3145 | 44 | |
| Budget impact (million ECU) | 956 | 25 | 0 | 0 | 914 | 1405 | 0 | 0 | -42 | 1380 | 0 | 0 | -4 | 5510 | | |
| Producer surplus (million ECU) | | | | | | | | | 0 | 1000 | -1 | -197 | | | | |
| Consumer surplus (million ECU) | | | | | | | | | 0 | -157 | 34 | 298 | | | | |
| Economic welfare (million ECU) | | | | | | | | | 42 | -537 | 33 | 91 | | | | |
| **Other Vegetable Oils:** | | | | | | | | | | | | | | | | |
| Supply price (ECU/t) | 1280 | 872 | 366 | 366 | 1280 | 1280 | 366 | 366 | 0 | 408 | 0 | 0 | 0 | 47 | 0 | 0 |
| Demand price (ECU/t) | 366 | 831 | 366 | 366 | 366 | 366 | 366 | 366 | 0 | -465 | 0 | 0 | 0 | -56 | 0 | 0 |
| Supply protection rate (%) | 250 | 138 | 0 | base | 249 | 249 | 0 | base | -0 | 111 | 0 | 0 | -0 | 80 | | |
| Demand protection rate (%) | 0 | 127 | 0 | base | 0 | 0 | 0 | base | 0 | -127 | 0 | 0 | 0 | -100 | | |
| Supply ('000 t) | 2416 | 318 | 11982 | 26427 | 2416 | 349 | 11991 | 26472 | 0 | 31 | 9 | 45 | 0 | 10 | 0 | 0 |
| Demand ('000 t) | 5449 | 724 | 6532 | 28283 | 5446 | 887 | 6523 | 28216 | -3 | 163 | -9 | -67 | -0 | 23 | -0 | -0 |
| Trade balance ('000 t) | -3033 | -406 | 5450 | -1856 | -3030 | -539 | 5468 | -1744 | 3 | -133 | 18 | 112 | -0 | 33 | 0 | -6 |
| Budget impact (million ECU) | 2208 | -176 | 0 | 0 | 2207 | 318 | 0 | 0 | -1 | 494 | 0 | 0 | -0 | -281 | | |
| Producer surplus (million ECU) | | | | | | | | | 0 | 136 | 4 | 10 | | | | |
| Consumer surplus (million ECU) | | | | | | | | | -2 | 374 | -2 | -10 | | | | |
| Economic welfare (million ECU) | | | | | | | | | -1 | 16 | 2 | -1 | | | | |

**Scenario B: EC-12 Reduction in Support and Rebalancing**

| Olive Oil: | Reference Scenario | | | | New Scenario | | | | Change vs. Reference | | | | Percent Change vs. Ref. | | | |
|---|---|---|---|---|---|---|---|---|---|---|---|---|---|---|---|---|
| | EC-10 | Iberia | USA | RoW | EC-10 | Iberia | USA | RoW | EC-10 | Iberia | USA | RoW | EC-10 | Iberia | USA | RoW |
| Supply price (ECU/t) | 3226 | 1544 | 1387 | 1387 | 2678 | 2678 | 1126 | 1126 | -548 | 1134 | -261 | -261 | -17 | 73 | -19 | -19 |
| Demand price (ECU/t) | 1986 | 1520 | 1387 | 1387 | 1805 | 1805 | 1126 | 1126 | -181 | 285 | -261 | -261 | -9 | 19 | -19 | -19 |
| Supply protection rate (%) | 133 | 11 | 0 | base | 138 | 138 | 0 | base | 0 | 1 | 0 | 0 | 4 | 1118 | | |
| Demand protection rate (%) | 43 | 105 | 0 | base | 60 | 60 | 0 | base | 0 | 1 | 0 | 0 | 40 | 529 | | |
| Supply ('000 t) | 800 | 490 | 1 | 359 | 708 | 643 | 1 | 303 | -92 | 153 | -0 | -56 | -12 | 31 | -15 | -16 |
| Demand ('000 t) | 860 | 390 | 40 | 340 | 906 | 304 | 44 | 379 | 46 | -86 | 5 | 39 | 5 | -22 | 11 | 12 |
| Trade balance ('000 t) | -60 | 100 | -39 | 19 | -199 | 339 | -44 | -76 | -139 | 239 | -5 | -95 | 231 | 239 | 12 | |
| Budget impact (million ECU) | 956 | 25 | 0 | 0 | 483 | 791 | 0 | 0 | -473 | 766 | 0 | 0 | -49 | 3060 | | |
| Producer surplus (million ECU) | | | | | | | | | -413 | 642 | -0 | -86 | | | | |
| Consumer surplus (million ECU) | | | | | | | | | 160 | -99 | 11 | 94 | | | | |
| Economic welfare (million ECU) | | | | | | | | | 220 | -223 | 11 | 7 | | | | |

**Scenario C: Offset of Enlargement on External Trade**

| Olive Oil: | Reference Scenario | | | | New Scenario | | | | Change vs. Reference | | | | Percent Change vs. Ref. | | | |
|---|---|---|---|---|---|---|---|---|---|---|---|---|---|---|---|---|
| | EC-10 | Iberia | USA | RoW | EC-10 | Iberia | USA | RoW | EC-10 | Iberia | USA | RoW | EC-10 | Iberia | USA | RoW |
| Supply price (ECU/t) | 3226 | 1544 | 1387 | 1387 | 2360 | 2360 | 1392 | 1392 | -866 | 816 | 5 | 5 | -27 | 53 | 0 | 0 |
| Demand price (ECU/t) | 1986 | 1520 | 1387 | 1387 | 1805 | 1805 | 1392 | 1392 | -181 | 285 | 5 | 5 | -9 | 19 | 0 | 0 |
| Supply protection rate (%) | 133 | 11 | 0 | base | 70 | 70 | 0 | base | -1 | 1 | 0 | 0 | -48 | 514 | | |
| Demand protection rate (%) | 43 | 105 | 0 | base | 30 | 30 | 0 | base | -0 | 0 | 0 | 0 | -31 | 209 | | |
| Supply ('000 t) | 800 | 490 | 1 | 359 | 648 | 603 | 1 | 359 | -152 | 113 | 0 | 0 | -19 | 23 | 0 | 0 |
| Demand ('000 t) | 860 | 390 | 40 | 340 | 907 | 304 | 40 | 341 | 47 | -86 | 0 | 1 | 5 | -22 | 0 | 0 |
| Trade balance ('000 t) | -60 | 100 | -39 | 19 | -259 | 299 | -39 | 18 | -199 | 199 | -0 | -1 | 332 | 199 | 0 | |
| Budget impact (million ECU) | 956 | 25 | 0 | 0 | 253 | 458 | 0 | 0 | -703 | 433 | 0 | 0 | -74 | 1731 | | |
| Producer surplus (million ECU) | | | | | | | | | -627 | 445 | 0 | 2 | | | | |
| Consumer surplus (million ECU) | | | | | | | | | 160 | -99 | -0 | -2 | | | | |
| Economic welfare (million ECU) | | | | | | | | | 236 | -86 | -0 | 0 | | | | |

of other vegetable oils, while other animal fats record minor variations in prices as well as in quantities.

The enforcement of the EC proposals in Spain and Portugal, which will coincide with the end of their transitional period, is likely to mean less dramatic changes as compared to the former scenario (i.e., extension of CAP regulations to the Iberian countries without policy changes). Still, producer prices will show a considerable increase (73 percent) for olive oil, not so striking for other vegetable oils (20 percent), matched by a decrease in price for butter and for other animal fats.

Under this hypothesis the third enlargement will still mean an increase in the olive oil protection rate in the entire EC-12 and will depress the world market price (-19 percent) though to a lesser extent than in the previous scenario (-50 percent). This implies a net increase in budgetary expenditures needed to support producers, as the savings in the EC-10 are more than offset by the additional funds needed for Spain and Portugal.

While consumers in EC-10 will face price adjustments less than 10 percent in both directions (-9 percent for olive oil and 7 percent for other vegetable oils) that will not greatly affect their demand, consumers in Iberian countries will experience higher prices for olive oil (in the range of 20 percent) and lower prices for other vegetable oils (by more than 50 percent). *Ceteris paribus*, this is likely to considerably modify their consumption patterns.

In order to limit EC budget expenditures, a maximum guaranteed quantity of olive oil for which production aid is payable was set in 1987 at 1.35 Mt per year. If actual production exceeds it, the unit producer subsidy should be reduced in proportion to the excess. With reference to a four-year average, according to the model, the impact of such a policy on supply would be approximately equivalent to the 30-percent cut in support proposed by the EC-12 at the GATT negotiations; i.e., a reduction of EC-12 supply to 1.35 Mt.

## Impact on Third Countries

One of the most disturbing effects of EC-10 enlargement to Spain and Portugal, as far as the olive oil market is concerned, is the disruption of the international market, where a few Mediterranean countries offer one of their most typical products from which a notable share of their farmers earn a living. The EC always avoided damaging these less-developed countries, some of which have applied for accession and others of which have special trade arrangements with the EC-12.

Taking into account these political relationships, a world market situation was simulated in which the impact of accession of the Iberian countries would be absorbed by internal adjustment of the EC-12, without affecting the external trade with third countries.

In the scenario B just described, the drop in producer price was 17 percent of the 1986 target price. In order to offset the increased Iberian supply by a reduction in EC-10 supply (scenario C), the olive oil production target price should be further decreased by 10 percent of the 1986 level. The resulting rate of protection with respect to the thin international market would then be 70 percent for the supply price and 30 percent for the demand price. Overall EC-12 welfare would then increase by 150 million ECU, as the welfare loss due to increased protection in the Iberian countries would be more than compensated by the welfare gain of reduced protection in EC-10 countries.

## Concluding Remarks

The EC oils and fats sector will necessarily undergo a strong adjustment process due to the EC enlargement to include Spain and Portugal. In particular, the olive oil market will be upset by the merging of the Iberian countries, which account for one quarter of the world market and where producer prices were about 50 percent lower than in the EC.

The impact of such a merger will be disruptive both in terms of world market prices for olive oil and of EC budgetary outlays. The negative effects of present EC agricultural policy, due to wide inter-commodity price distortions and to inconsistencies among policy measures within the oils and fats sector, will become fully apparent.

Harmonization of price policy alone is not likely to be the answer in a situation that is very uneven in many respects. With reference to the olive oil sector, producer incomes would be less affected by inevitable price reductions if appropriate structural policies were implemented, reducing present production costs in traditional and inefficient olive groves.

A differential treatment could be given to disadvantaged areas where olive production cannot be substituted by other economic activities. The rate of producer subsidy could be related to the positive externalities developed by agriculture in these areas, especially if it is granted according to existing olive trees in order to ease administrative controls.

## Notes

[1] Università di Siena.
[2] For more details on this model, see Tarditi and Croci-Angelini, 1988.

## References

European Communities, *Disharmonies in EC and US Agricultural Policy Measures*, Luxembourg, 1988.

FAO (Food and Agriculture Organization), *FAO Agricultural Commodity Projections to 1990*, FAO Economic and Social Development Paper No. 62, Rome, Italy, 1986a.

FAO (Food and Agriculture Organization), *Income Elasticities of Demand for Agricultural Products*, Rome, Italy, 1986b.

FAO (Food and Agriculture Organization), *Committee on Commodity Problems, Intergovernmental Group on Oilseeds, Oils, and, Fats*, Rome, Italy, various years.

Tarditi, S., and Croci-Angelini, E., *The EC Olive Oil Subsector and the International Trade for Oils and Fats*, DAP Background Paper, EC Commission, Brussels, Belgium, 1988, mimeo.

---

## Discussion Opening—*Hyunok Lee* (US Department of Agriculture)

This paper assesses the impacts of the third EC enlargement on vegetable oil markets. Particular attention is given to the olive oil markets in Spain and Portugal. Trade liberalization of the vegetable oil sector is especially important to Spain and Portugal because these two countries account for 30 percent of world olive oil production. The study's innovation is to assess the impacts on the international as well as EC–10 and Spanish/Portuguese olive oil markets of extending EC price supports to Spanish/Portuguese vegetable oil and fat producers. A simulation model is used.

This study provides some insight for policy makers on the magnitude of market changes and resulting budgetary outlays. With this information, it may be possible to design some policy options to minimize budgetary outlays. While this study is relevant and potentially important to the world olive oil sector, I have some reservations.

While empirical results are presented in an unequivocal manner, the underlying assumptions of the model are not clear. It suffices to say that model assumptions can dictate simulation results. However, the paper does not provide information on any of its hypotheses. The authors refer to a mimeo report of theirs to which readers do not generally have easy access. For example, the authors only give one value for the supply elasticity of the Iberian countries (0.5). Aside from the question of the reasonableness of this value (I understand it takes years to grow an olive tree), the supply elasticities of other vegetable oils and demand elasticities determine the magnitude of the shift of the supply and demand curves as described for scenario A. For the reader to judge the reasonableness of these results, he or she needs to know the relevant elasticities.

Finally, since olive oil and other vegetable oils are close substitutes, what is going to happen to the olive oil markets after the third enlargement critically depends on the relative rate at which the EC supports olive oil producers compared to other vegetable oil producers. Nominal supports are much less important.

My recommendation to the authors is to incorporate into the paper more information on the model. Furthermore, to check the stability of the results, I suggest the authors reproduce the runs with alternative parameter values.

*[Other discussion of this paper and the authors' reply appear on the following page.]*

**General Discussion**—*Zhang Cheng-Liang, Rapporteur* (Beijing Agricultural Engineering University)

In reply to Deaton's comment that the concept of justice in his paper is not clear and that there would be different understandings of it, Rabinowicz argued that it is not necessary for everyone to agree equally on the content of justice. She was also asked if she had considered using the theory of cooperative games as a basis for her analysis. She replied that the percent problem would make the theory of cooperative games difficult to apply in the paper. Doubts were also expressed about whether the information provided in the paper would have any impact on policy decisions.

In reply to Gordon's comment on welfare changes resulting from the EC's rebalancing strategy, Schmitz replied that welfare changes would involve measurement. Another question from the floor related to differences between the authors' results and those of Mahé and Tavéra in analysing the rebalancing issue. In reply, Hartmann pointed out two main differences: (1) their paper emphasizes developing countries, while Mahé *et al.* focus on the USA, and (2) the paper stresses unilateral policy change.

Replying to a question about whether the elasticity of demand would rise as the price drops in response to a rise in production, Tarditi indicated that there are possibilities of changing to a lower price level.

Participants in the discussion included G. Jones (University of Oxford), U. Koester (Universität Kiel), and A. Oskam (Agricultural University of Wageningen).

# Measuring Market Power for Marketing Firms: The Case of Japanese Soyabean Markets

*Konomi Ohno and Paul Gallagher*[1]

**Abstract:** This paper extends Bresnahan's market power measure, which can be estimated econometrically, to marketing firms that have potential for price discrimination. An investigation of Japanese soyabean markets during 1973–78 using the model reveals an episode in which Japanese importers exercised some market power for several years after the US soyabean embargo of June 1973. An analysis of welfare loss and exchange rate transmission is also presented.

## Market Power Coefficients

The firm's profit-maximizing rule is to set perceived marginal revenue equal to marginal cost. In a competitive market, any attempt by a firm to raise prices by supplying fewer commodities would result in increased supply by other firms. A single firm thus has no market power to influence market prices. Hence, perceived marginal revenue equals price and also equals marginal cost. When market power exists, both perceived marginal revenue and marginal cost are less than price.

Bresnahan (1982) argues that market power in an industry can be measured as a coefficient, $\lambda$, in the following relation between price ($P$) and quantity ($Q$):

$$(1) \quad P = MC - \lambda Q \, \frac{\partial P}{\partial Q}$$

This function postulates equality between perceived marginal revenue and marginal cost. When $\lambda = 0$, the market is perfectly competitive. When $\lambda = 1$, the market is monopolistic. In an oligopolistic market structure, $\lambda$ lies between zero and unity. In a case where inverse demand and marginal cost are represented by $P = G(Q, Y, \alpha)$ and $MC = C(Q, W, \beta)$, where $\alpha$ and $\beta$ are parameters, and while $Y$ and $W$ are exogenous consumer income and wages, respectively, then the price relationship becomes:

$$(2) \quad P = C(Q, W, \beta) - \lambda Q \frac{\partial G}{\partial Q}(Q, Y, \alpha)$$

Treating $P$ and $Q$ as endogenous variables, the demand function and price relationship are simultaneously estimated to reveal the market power coefficient, $\lambda$.

However, Bresnahan explains that the degree of market power, $\lambda$, cannot be identified unless an additional interaction between price and income is included in a system of linear demand and marginal costs. If a change in the exogenous variable on the demand side, $Y$, only shifts the demand function in parallel, the hypotheses of competition and monopoly are not differentiated.

However, when characteristics of marketing firms and their sales environment are recognized, Bresnahan's additional variable can become unnecessary. As an illustration, consider a general formulation of the marketing problem. Suppose firms buy from producers and sell the product to human consumers and large-scale processors. Further, marginal revenues in product markets differ due to differences in demand elasticities and market power. Marketing firms' costs rise due to material and processing costs. Costs are also higher for the human consumption market, owing to local distribution costs.

The demand functions are:

$$(3) \quad D_1 = \alpha_0 N + \alpha_1 \frac{P_1 N}{CPI_1} + \alpha_2 \frac{YN}{CPI_1}$$

(4) $D_2 = \beta_0 + \beta_1 \dfrac{Y_s^o P_s^o + Y_s^m P_s^m}{CPI_2} - \beta_1 \dfrac{P_2}{CPI_2} + \beta_2 \dfrac{MR}{CPI_2} + \beta_3 C$

where $D_i$ and $P_i$ are quantity demanded and price in market $i$.[2] Equation (3) shows that the demand for direct human consumption depends on real price and real income. Equation (4) tells us that the demand for processing is determined by real margins to process soyabeans and rapeseed and the capacity of factories. Perceived marginal revenues, in turn, depend on market power and the parameters of the demand functions:

(5) $MR_1 = P_1 + \lambda_1 \left( \dfrac{D_i CPI_1}{\alpha_1} \right)$, and $MR_2 = P_2 - \lambda_2 \left( \dfrac{D_2 CPI_2}{\beta_1} \right)$

A general formulation of the cost function specifies material and processing components and adjusts processing costs to wages ($W$):

(6) $C(Q_1, Q_2) = P^*(Q_1 + Q_2) + W \left\{ \left[ \alpha_s (Q_1 + Q_2) + \dfrac{\beta_s}{2}(Q_1 + Q_2)^2 \right] + \left[ \alpha_{s1} Q_1 + \dfrac{\beta_{s1}}{2} Q_1^2 \right] \right\}$

where the $Q_i$ are marketing firms' outputs for market $i$. Notice that costs are higher in the local market when $\alpha_{s1} \neq 0$. Also, marginal costs are different and increasing when $\beta_s$ and $\beta_{s1}$ are positive.

Pricing relationships for both product markets can be developed from solutions to the maximum profit problem for marketing firms. The profit function is:

(7) $\pi = P_1 D_1 + P_2 D_2 - C(Q_1, Q_2)$

This function can be expressed in terms of the $D_i$s by noting that $Q_1 = D_1$ and $Q_2 = D_2 + S$, where $S$ is the change in ending stocks. Then the first order conditions are:

(8) $\dfrac{\partial \pi}{\partial Q_1} = MR_1 - \{P^* + W[\alpha_s + \beta_s(D_1 + D_2 + s)] + W[\alpha_{s1} + \beta_{s1} D_1]\} = 0$

(9) $\dfrac{\partial \pi}{\partial Q_2} = MR_2 - \{P^* + W[\alpha_s + \beta_s(D_1 + D_2 + s)]\} = 0$

The implied pricing functions are:

(10) $P_1 = -\dfrac{\lambda_1}{\alpha_1} \left( \dfrac{CPI_1 D_1}{N} \right) + (\alpha_s + \alpha_{s1})W + (\beta_s + \beta_{s1})(WD_1) + \beta_s(WD_2) + \beta_s(WS) + P^*$

(11) $P_2 = \dfrac{\lambda_2}{\beta_1}(CPI_2 D_2) + \alpha_s W + \beta_s(WD_1) + \beta_s(WD_2) + \beta_s(WS) + P^*$

An econometrically useful form of the marketing system is given by Equations (3), (4), (10), and (11). There are 10 endogenous variables and 8 exogenous variables:

Endogenous: $P_1$, $P_2$, $D_1$, $D_2$, $\dfrac{P_1 N}{CPI_1}$, $\dfrac{P_2}{CPI_2}$, $\dfrac{CPI_1 D_1}{N}$, $CPI_2 D_2$, $WD_1$, and $WD_2$

Exogenous: $N$, $\dfrac{YN}{CPI_1}$, $\dfrac{Y_s^o P_s^o + Y_s^m P_s^m}{CPI_2}$, $\dfrac{MR}{CPI_2}$, $C$, $W$, $WS$, and $P^*$

The criterion for identifying an equation is that the number of included endogenous variables less one must be equal or less than the number of excluded exogenous variables. For instance, two endogenous variables are included in Equation (3) ($D_1$ and $P_1/CPI_1$). Six exogenous variables are excluded. Thus, Equation (3) is identified because $1 < 6$. Following the same rule, Equations (4), (10), and (11) are also identified. Furthermore, $\lambda_1$ and $\lambda_2$ can both be determined from the first coefficient of the respective price equations and demand price response parameters ($\alpha_1$ and $\beta_1$). Thus, the oligopoly solution is identified for marketing sectors with two product markets.

For subsequent empirical investigations, the capacity adjustments by marketing firms should also be included. Now the profit function is:

$$(12) \quad \pi = P_1 D_1 + P_2 D_2 - \left[ P^*(Q_1 + Q_2) + W\alpha_s(Q_T - \bar{Q}_T) + \frac{\beta_s}{2}(Q_T - \bar{Q}_T)2 + \alpha_{s1}(Q_1 - \bar{Q}_1) + \frac{\beta_{s1}}{2}(Q_1 - \bar{Q}_1)^2 \right]$$

where $\bar{Q}_T$ and $\bar{Q}_1$ are capacities and $Q_T = Q_1 + Q_2$. Now the pricing functions are:

$$(13) \quad P_1 = -\frac{\lambda_1}{\alpha_1} \frac{CPI_1 D_1}{N} + (\alpha_s + \alpha_{s1})W + (\beta_s + \beta_{s1})WD_1 + \beta_s WD_2 + \beta_s WS$$
$$- \beta_s W\bar{Q}_T - \beta_{s1} W\bar{Q}_1 + P^*$$

$$(14) \quad P_2 = \frac{\lambda_2}{\beta_1} CPI_2 D_2 + \alpha_s W + \beta_s WD_1 + \beta_s WD_2 + \beta_s WS - \beta_s W\bar{Q}_T + P^*$$

where (14) is identical to (11) except one term, $-\beta_s W\bar{Q}_T$, and there are more additional terms in (13) compared to (10). The four equations, (3), (4), (13), and (14), are still identified, as are $\lambda_1$ and $\lambda_2$.

The cost structure of marketing firms is an empirical issue. Short-run marginal cost functions could be constant ($\beta_s = 0$) in both markets when capital stock (handling and storage equipment) is fixed and variable costs are proportional to labour and energy used for handling. Further, Thompson and Dahl (1979) hypothesize economies of scale in transport, information network, risk bearing, and storage space for US grain exporters. As scale of operations increases and firms accumulate capital, the marginal cost of marketing firms could decrease over longer run periods. The inverse relationship between marginal cost and capacity in the above cost function potentially accounts for these long-run cost adjustments.

## Japanese Soyabean Markets

The two-market assumption is an alternative method to Bresnahan's demand notation for identifying the market power coefficients. Soyabean markets in Japan seem well-suited for testing this model. There are two primary soyabean usages in Japan. One is for direct human consumption as food (*tofu, natto*, etc.) except oil, and the other is for livestock feeds and oil. The former market accounts for 30 percent of all soyabean consumption in the Nation. More than 88 percent of soyabeans are imported, with the primary sources being the USA, China, and Brazil. Crushing mills are located on the coast to minimize transport costs. Other imported soyabeans are unloaded there and sent to urban areas where human consumption points are concentrated.

Point-of-import prices and urban wholesale prices have behaved differently. The unit value import price closely follows the US export price adjusted by the exchange rate and transport costs. That close relationship suggests that a constant margin model may be suitable. Similarly, Tokyo wholesale prices from the early 1970s and post-1979 period closely reflect import prices. However, there appears to have been an episode of extremely high wholesale prices during 1973–78. Supplies worldwide were short in 1973, and all import and wholesale prices increased. However, domestic wholesale prices increased more than proportionately and remained high even after world prices declined. This period of high domestic prices may have been triggered by the US soyabean embargo, which was in effect for 5 days from 21 June 1973. Afterwards, export licences were set at 50 percent of unfilled export contracts until 1 September 1973 (Kost et al., 1986).

There was an inventory buildup in anticipation of the embargo. However, consumption behaviour was not unusual; i.e., there was a consumption decrease in the presence of high domestic prices during the high-price era of the early 1970s.

## Estimation and Data for the Soyabean Market

Specification of demand relationships in Japan's soyabean markets and preliminary hypothesis testing produced a more precise system of demand and pricing functions. These functions are shown below as Equations (15)–(18).

The demand function for the human consumption market (3) is a per-capita function. Population then becomes a scaling factor for independent variables in the market demand function, as shown in Equation (15) below. Also, separability for food consumption is assumed, so $Y/CPI_1$ and $P_1/CPI_1$ in Equation (3) are the ratios of nominal household expenditure on food and nominal soyabean wholesale price to a consumer food price index (Phlips, 1983, p. 73). Finally, seasonal trends in soyabean consumption are taken into account with dummy variables, one for the second and third quarters and the other for the fourth quarter.

In market 2, rapeseed margins are included as an exogenous variable in Equation (16), since it is expected that soyabeans would be replaced by this important substitute if rapeseed profitability increased. A capacity measure is also included as an explanation for the secular increase in demand.

Several preliminary specifications of pricing equations were also examined. In particular, the data supported the notion of constant marginal costs for both markets. That is, the coefficients $\beta_S$ and $\beta_{S1}$ were not statistically significant. With regard to market power, the coefficient $\lambda_2$ was not statistically significant. Similarly, the market power coefficient $\lambda_1$ was not statistically significant in some preliminary specifications. However, $\lambda_1$ was found to be statistically significant when "an episode" of monopoly pricing between 1973 and the first half of 1978 was specified. Hence, Equation (13) is slightly changed as follows:

$$(13)' \quad P_1 = -\frac{CPI_1 D_1}{N\alpha_1}[\lambda_{11}D+\lambda_{12}(1-D)] + (\alpha_s+\alpha_{s1})W + (\beta_s+\beta_{s1})WD_1 + \beta_s WD_2$$
$$+ \beta_s WS - \beta_s W\hat{Q}_T - \beta_{s1}W\hat{Q}_1 + P^*$$

where $D = 1$ during 1973–78; otherwise $D = 0$.

Then the equations are simultaneously estimated and the hypotheses $\lambda_{12} = 0$, $\lambda_2 = 0$, $\beta_s = 0$, and $\beta_{s1} = 0$ are tested. The $\chi^2$ is 7.72, which is less than $\chi^2$ (4, 0.05). The hypothesis cannot be rejected at the 0.05 level.

A typical system of estimation equations for Japan's soyabean market is shown below:

$$(15) \quad D_1 = \alpha_0 N + \frac{\alpha_1 P_1}{CPI_1} N + \alpha_2 \frac{YN}{CPI_1} + \alpha_3 D_{23}N + \alpha_4 D_4 N$$

$$(16) \quad D_2 = \beta_0 + \beta_1 \frac{MS}{CPI_2} + \beta_2 \frac{MR}{CPI_2} + \beta_3 C$$

$$(17) \quad P_1 = -\frac{\lambda_{11}}{\alpha_1} \frac{CPI_1 D_1}{N} D + (\alpha_s+\alpha_{s1}) W + P^*$$

$$(18) \quad P_2 = \alpha_s W + P^*$$

The variable definitions are given in Table 1. Quarterly data for 1971–88 are used for each variable. Most data come from Japanese domestic sources.

## Results

Table 2 summarizes the empirical results. Two sets of estimates are shown. One is a full system while the other is separated. The latter system is added because of concerns about the import unit value as an accurate measure of transaction prices in the processing ($P_2$) market. Both sets of equations in the tables are similar. Quantities of soyabeans consumed in market 1 demonstrate a statistically significant negative relationship with relative prices of soyabeans and a positive relationship with household expenditure on food. Statistically significant seasonal trends show that direct human consumption of soyabeans is affected by seasonal factors, high in the fourth quarter and low in the second and third quarters. Food made from soyabeans, such as *tofu* and *aburaage*, are largely consumed during the New Year celebrations, the most important Japanese holiday, and high expenditure on food during the fourth quarter may be supported by the large additional income provided by December bonuses.

Table 1—Definition of Variables

| Variable | Definition | Unit |
|---|---|---|
| $D_1$ | Soyabean use for direct consumption (approximated as domestic production plus imports plus difference in stocks between the previous period and the current period minus quantity processed) | 1000 t |
| $D_2$ | Soyabean use for processing into meal and oil | 1000 t |
| $S$ | Change in ending stocks | |
| $P_1$ | Weighted average of wholesale prices of Japanese, US, and Chinese soyabeans by market shares ($P_1=P_1^{US}MS^{US}+P_1^J MS^J +P_1^C MS^C$, where $MS$ represents market share and superscripts $US$, $J$, and $C$ respectively represent USA, Japan, and China) | ¥/kg |
| $P_2$ | Unit value of imported soyabeans | ¥/kg |
| $P_R$ | Rapeseed price to large processors | ¥/kg |
| $M_S$ | Soyabean margin ($=Y_s^o P_s^o+Y_s^m P_s^m-P_2$) | |
| $M_R$ | Rapeseed margin ($=Y_r^o P_r^o+Y_r^m P_r^m-P_R$) | |
| $Y_I^J$ | $J$ yield from one ton of $I$ (meal and oil yields for soyabeans and rapeseed are calculated by dividing soyabean oil or meal production by soyabean use by processors for every quarter between the first quarter of 1971 and the fourth quarter of 1988 and regressed on the time from 1 to 72) | |
| $P_I^J$ | Wholesale price of $I$ $J$ ($I$ = soyabeans or rapeseed; $J$ = oil or meal) | ¥/kg |
| $P^*$ | Export price of American soyabeans, adjusted for freight and exchange rate | ¥/kg |
| $CPI_1$ | Consumer price index for food | % |
| $CPI_2$ | Consumer price index | % |
| $N$ | Population | million |
| $Y$ | Nominal per-household consumption of food, beverages, and cigarettes | ¥1,000 million |
| $C$ | Capacity to process soyabeans for oil or meal calculated from $Q$ | 1000 t |
| $W$ | Nominal wages in food industry | ¥1,000 |

The estimation of Equation (16) shows that quantities of soyabeans processed in market 2 are positively related to soyabean margins and capacity. There is a negative relationship between quantities and rapeseed margins, but it is not significant.

In Equation (18), the hypothesis that an intercept term equals zero was not rejected from Equation (17), $\alpha_{S1} \neq 0$ with $t = 7.2$. These results specify the cost functions for the Japanese soyabean marketing firms. Since:

(19) $TC = P^*(Q_1+Q_2) + \alpha_S W(Q_1+Q_2-\bar{Q}_T) + \alpha_{S1} W(Q_1-\bar{Q}_1)$

the marginal cost for each market is constant with marginal cost in market 1 and significantly higher than that in market 2. This result indicates that significant handling and transport costs are incurred in marketing for human consumption after unloading at the ports.

Table 2—Estimation Results for Equations (15)–(18)

| Coefficient/ Variable | Combined System | | Separated System | | | |
|---|---|---|---|---|---|---|
| | | | Human Consumption | | Processing | |
| | Estimate | $t$-ratio | Estimate | $t$-ratio | Estimate | $t$-ratio |
| $\alpha_0$ | 0.02385 | 1.49 | 0.02172 | 1.35 | | |
| $\alpha_1$ | −0.0046541 | −2.58 | −0.0043112 | −2.38 | | |
| $\alpha_2$ | 0.41388 | 0.17 | 0.68773 | 0.29 | | |
| $\alpha_3$ | −0.0048923 | −1.97 | −0.0050297 | −2.03 | | |
| $\alpha_4$ | 0.01133 | 2.29 | 0.01093 | 2.20 | | |
| $\beta_0$ | −153.33 | −1.61 | | | −146.58 | −1.53 |
| $\beta_1$ | 358.72 | 1.81 | | | 283.51 | 1.42 |
| $\beta_2$ | 81.01448 | 0.36 | | | 96.05749 | 0.42 |
| $\beta_3$ | 0.98441 | 12.91 | | | 0.98462 | 12.84 |
| $\alpha_S$ | 0.00018821 | 0.18 | | | 0.000184 | 0.18 |
| $\alpha_{S1}$ | 0.02122 | 7.19 | | | | |
| $\alpha_S+\alpha_{S1}$ | | | 0.02127 | 7.60 | | |
| $\lambda_{11}$ | 0.08137 | 2.53 | 0.07632 | 2.35 | | |
| | Summary Statistics for Combined System* | | | | | |
| | $R^2$ | $DW$ | | | | |
| $P^1$ | 0.62 | 1.4 | | | | |
| $P^2$ | 0.87 | 1.0 | | | | |
| $Q^1$ | 0.58 | 2.4 | | | | |
| $Q^2$ | 0.77 | 2.4 | | | | |

*Values for the combined and separated system are the same up to three significant digits. The statistically significant $\lambda_1$ ($t=2.53$) suggests that market power existed in the wholesale market between 1973 and the first half of 1978 and that the market was not competitive. The Japanese soyabean markets were segmented, and marketing firms might have exercised market power and pursued policies of price discrimination in the two soyabean markets during 1973–78.[3]

## Welfare Analysis and Exchange Rate Transmission

The market power coefficient $\lambda_1$ is small relative to its pure monopoly value, but it is statistically significant. Profit margins and consumer welfare losses, which are based on estimates of demand functions and market power coefficients, are presented in this section. Judgments on the importance of this market power episode are enhanced with these conventional performance measures.

Pricing behaviour and loss of consumer surplus are shown in Figure 1. The $MR$ schedule depicts the firms' perceptions of how revenues will change when price changes; it depends on $\lambda_1$. The condition that $MR = MC$ defines the equilibrium price and quantity, $P^o$ and $D^o$. The competitive solution is also given; as $\lambda_1$ approaches zero, $MR$ rotates to $D$. Then the price reduces to marginal cost ($\delta$) and consumption expands to $D^c$. The area of $P^o \delta BA$ is the consumer welfare loss. This area is calculated from the values of $P^o$, $MC$, $D^o$, and $D^c$ for each period from 1973 up to the first half of 1978.

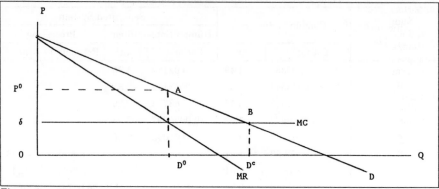

Figure 1

The estimated demand, marginal revenue, and marginal cost functions enable us to specify profit margins and to calculate algebraically the loss of consumer surplus. The inverse demand, marginal revenue, and marginal cost functions given from Equations (15) and (17) are:

$$(20) \quad P_{1t} = \frac{CPI_{1t}}{\alpha_1 N_t} D_{1t} + \alpha_{dt}, \quad MR_t = P_{1t} + \lambda_1 \frac{CPI_{1t}}{\alpha_1 N_t} D_{1t}, \quad MC = (\alpha_S + \alpha_{S1})W_t + P_t^* \equiv \delta_t$$

where: $\alpha_{dt} \equiv -\dfrac{CPI_{1t}}{\alpha_1}\left[\alpha_0 + \dfrac{\alpha_2 Y_t}{CPI_{1t}} + \alpha_3 D_{23} + \alpha_4 D_4\right]$

and $t$ shows that each variable depends on time. Each parameter follows the result in Table 2.

The values that define the welfare area can be calculated from the above marginal revenue, marginal cost, and price functions. The appropriate prices and quantities are given below:

$$(21) \quad P_t^o = \frac{\alpha_{dt}\lambda_1 + \delta_t}{1+\lambda_1}, \quad D_t^o = \frac{(\delta_t - \alpha_{dt})\alpha_1 N_t}{(1+\lambda_1)CPI_{1t}}, \quad \text{and } D_t^c = \frac{(\delta_t - \alpha_d t)\alpha_1 N_t}{(1+\lambda_1)CPI_{1t}}$$

Profit margins are measured as $(P_t^o - \delta_t)/P_t^o$ for each period. The average was 22 percent.

The loss of consumer surplus during the period was \$376 million, of which \$361 million were transferred to marketing firms and the rest was wasted as dead-weight loss.

## Exchange Rate Transmission

Another aspect of competitiveness in the market is the degree to which the exchange rate is transmitted to the wholesale price. A perfectly competitive market has an elasticity of wholesale price with respect to an exchange rate of unity, assuming that the pricing strategy of soyabean exporters is not affected by exchange rate changes (no pricing to the market) and that transaction costs from import points to wholesale markets are fixed. That is, the percentage change in the exchange rate is perfectly absorbed by the percentage change in the wholesale price. If the yen appreciates against the US dollar by 10 percent, then wholesale prices of US soyabeans decline by 10 percent under perfect competition. However, when the market is not competitive, a change in the exchange rate may be adjusted by the firm's profit margins as well as wholesale prices. So if the yen appreciates against the dollar by 10 percent, and wholesale prices of US soyabeans decline by 8 percent, then 2 percent is left in importers' pockets as their additional profit.

Suppose the marketing cost ($\delta$) includes the product of the export country price and the exchange rate, $P^* = Pe$, where $P^*$ and $P$ are import prices in yen and in dollars, and $e$ is the exchange rate (yen per dollar). Then, any changes in the exchange rate are perceived through changes in import prices in yen terms. When the exchange rate changes, the import price in yen terms will be affected as well, which will influence importers' marginal costs. Figure 1 suggests that the level of $P^o$ is determined by a combination of demand, marginal revenue, and marginal cost functions. The argument in Equation (21) clarifies that $P_t^o$ depends on demand conditions, marginal cost, and market power. An exchange rate transmission elasticity is obtained:

$$(22) \quad \frac{e}{p} \frac{\partial p}{\partial e} = \frac{e}{P} \frac{\partial P}{\partial \delta} \frac{\partial \delta}{\partial e} = \frac{1}{1 + \dfrac{\alpha_{dt}\lambda_1 + (\alpha_s + \alpha_{s1})W_t}{e_t P_t^e}}$$

where $0 < \lambda < 1$ and $pe$ is the export price of US soyabeans, assuming that US exporters are not price discriminating due to a change in the exchange rate (i.e., $\alpha pe/\alpha e = 0$).

The elasticities from 1973 up to the first half of 1978 were calculated for each period. The average is 60.4 percent. Elasticities for the same period with an assumed competitive structure ($\lambda = 0$) are 86.7 percent. The exchange rate transmission was incomplete in 1973–78 while the yen was in a long appreciating trend against the dollar and relatively stable. The small magnitude of the market power coefficient seems to have a relatively large influence on the exchange rate transmission elasticity.

## Conclusion

Bresnahan's method for measuring a market power coefficient was applied to marketing firms where an interactive exogenous variable with prices was not necessary. This two-market model was tested in the Japanese soyabean market. The data are consistent with an episode in which there was market power in the Japanese soyabean wholesale market after the US embargo. The estimates suggest that Japanese consumers lost $376 million during this episode, most of which was transferred to the importers. Also, the average exchange rate transmission was 68.9 percent, which indicates price stickiness in the wholesale market.

The market power episode ceased in late 1978, and the market has been competitive since then. This could be explained by increased domestic supplies and imports of soyabeans from China in the late 1970s. Further investigations might focus on the potential role of energy costs or marketing risks as factors contributing to the period of unusually high margins.

## Notes

[1] Iowa State University.

[2] Additional variable definitions are given in Table 1.

[3] The increase in price in the first quarter of 1973 might be due to unusual circumstances. The model was tested excluding this observation, but $\lambda_1$ was still statistically significant.

## References

Bresnahan, T.F., "The Oligopoly Solution Concept is Identified," *Economics Letters*, Vol. 10, 1982, pp. 87–92.

Kost, W.E., O'Brien, P.O., Sarko, R.N., and Webb, A., "History of Recent U.S. Embargoes and Trade Restrictions," in Kost, W.E., McCalla, A.T., and Webb, A. (Eds.), *Embargoes, Surplus Disposal and U.S. Agriculture*, Agricultural Economic Report No. 564, Economic Research Service, US Department of Agriculture, Washington, D.C., USA, 1986.

Phlips, L., *Applied Consumption Analysis*, North-Holland Publishing Co., Amsterdam, Netherlands, 1983.

Thompson, S.R., and Dahl, R.P., "The Economic Performance of the U.S. Grain Export Industry," Agricultural Experiment Station, University of Minnesota, Minneapolis, Minn., USA, 1979.

---

**Discussion Opening**—*Joyce A.S. Cacho* (Virginia Polytechnic Institute and State University)

During 1970–88, Japan imported soyabeans from the USA, Brazil, and China. Soyabean imports from the USA and Brazil are processed for vegetable oil and protein meal. The direct human consumption market is primarily supplied by Japanese-produced soyabeans and imports from China. China's soyabeans are closest in variety to Japanese soyabeans. The variety differences between soyabeans for feed and food use mean that soyabeans are not a homogeneous product in the Japanese market. The US embargo, which was in effect 21–26 June 1973, would be expected to affect the processing market and have a neutral effect in the human consumption market.

Between 1973 and the first half of 1978, an episode of high wholesale prices for soyabeans occurred in Japan. This sustained price level may represent the market power of the firms (approximately 10) involved in the for-feed market. In the estimation of the system of equations for Japan's soyabean market and testing for market power, the authors include a variable to represent this in the equation for soyabeans for food use. Since the demand patterns for feed-use and food-use markets are separate and different, the market power variable may be included more appropriately in the feed-use equation where market power is likely to have occurred after the US embargo.

The technique developed in this paper to measure market power is methodologically sound. In addition to providing an alternative technique to using time-series analysis of market power, price in Japan is comprised of US price and the US dollar/yen exchange rate. Thus the linkage between the competitiveness of the market and the exchange rate transmission may be analysed. However, the application of the technique to the Japanese soyabean market and the corresponding results need to be re-examined. The assumption that the feed-use and food-use markets are separate but linked is brought into question by the fact that a varietal difference exists between the soyabeans supplied to each market.

*[Other discussion of the paper and the authors' reply appear on page 93.]*

# A Comparative Study of Soyabean Import Demand in Taiwan and Japan

*Rhung-Jieh Woo*[1]

**Abstract:** Simultaneous equation models of the soyabean sectors in Japan and Taiwan are developed and estimated through seemingly unrelated regressions. The models integrate domestic supply and demand for soyabeans, soyameal, and soyaoil as well as the livestock market. Based on the established models, the impacts of economic growth and policy simulations on these markets are evaluated and compared by performing dynamic simulation analyses. Growth factors had greater impacts on soyabean import demand than did policy factors, and growth impacts themselves were more significant in Taiwan than in Japan. A 10-percent currency devaluation with respect to the US dollar would only decrease soyabean import demand in Japan and Taiwan by 0.17 and 0.35 percent, respectively. A 10-percent increase in the soyabean support price would stimulate growth in domestic soyabean import plantings of 16–54 percent in Taiwan and 9.93 percent in Japan. A 10-percent increase in Taiwan's net livestock exports or a 10-percent decrease in Japan's net livestock imports would increase soyabean import demand by 0.12 and 0.48 percent, respectively. Increased demand for soyameal and a decline in demand for soyaoil were also noticed. If the profitability of soyabean crushing were improved, soyabean import demand would increase and demand for meal and oil would decrease.

## Introduction

Japan and Taiwan are two of the most important and dynamic growth markets for US soyabeans in East Asia. US soyabean exports to these areas increased from 1.21 Mt in 1962 to more than 6 Mt (about 30 percent of total US soyabean exports) in 1988.

Many intriguing questions surround the past and future growth of soyabean import demand in these economies. For example, the main factors that influence import demand for soyabeans in these natural-resource-poor East Asian economies with different degrees of economic development are the extent to which these factors exert their influence, whether these soyabean markets will continue to expand as the economies continue to grow, and how policy interventions, such as domestic soyabean support prices and livestock import controls, would affect these soyabean markets.

The general objective of this study is to develop econometric models and give quantitative descriptions of these soyabean markets as well as to analyse and compare the impacts of economic growth and policy interventions upon these markets.

## The Conceptual Model

The conceptual model of the soyabean sector for these soyabean-importing economies is presented in Table 1. Partly for simplicity and partly because almost all the soyabeans imported by Taiwan and Japan are from the USA, a small-country assumption was adopted in building the model. The regional model specified is a recursive equation system comprising equations for domestic soyabean supply, soyabean crushing demand, soyabean food demand, soyameal demand, livestock demand, soyaoil demand, price linkages, and trade clearance identities. The salient characteristic of this model is that it integrates domestic supply and demand for soyabeans, soyameal, and soyaoil, as well as the livestock sector, for these important US soyabean markets in Asia.

## Model Estimation and Validation

Seemingly unrelated regressions, or joint generalized least squares, were adopted to estimate the empirical models (for details, see Woo, 1985). Four major conclusions can be drawn from the estimated results. First, the soyabean support price has a positive influence on domestic soyabean acreage in both economies. However, the acreage level in each economy is inelastic with respect to the support prices. The soyabean acreage of an ensuing crop year is significantly influenced by the acreage planted in the current year in each economy. These results indicate a slow soyabean production adjustment to changing economic incentives in these soyabean-importing economies. Therefore, although policy makers in Taiwan and Japan have been encouraging domestic soyabean production through price-supporting policies, it is unlikely that these policies will substantially expand domestic soyabean production or dramatically influence their soyabean import demand in the near future.

### Table 1—Conceptual Model of the Soyabean Markets in Japan and Taiwan

(1) Soyabean acreage = $f$ (deflated soyabean support price, lagged soyabean acreage, others)
(2) Soyabean crushing demand = $f$ (deflated soyabean crushing ratio, crushing capacity, others)
(3) Soyameal demand = $f$ (deflated soyameal retail price, deflated livestock price, livestock production, others)
(4) Livestock consumption, per capita = $f$ (real income per capita, deflated livestock price, others)
(5) Soyabean food demand, per capita = $f$ (real income per capita, deflated soyabean price, others)
(6) Soyaoil demand, per capita = $f$ (real income per capita, deflated soyaoil retail price, others)

Price Linkages:

(7) Soyabean import price = $f$ (US soyabean export price times the exchange rate, others)
(8) Soyameal retail price = $f$ (US soyameal export price times the exchange rate, others)
(9) Soyaoil retail price = $f$ (US soyaoil export price times the exchange rate, others)

Identities:

(10) Soyabean excess demand = soyabean crushing demand + soyabean food demand
+ soyabean seeds demand and waste
+ soyabean net stock change
− soyabean acreage times soyabean yield

(11) Soyameal excess demand = soyameal demand + soyameal net stock change
− soyabean crushing demand times soyameal yield

(12) Soyaoil excess demand = soyaoil demand + soyaoil net stock change
− soyabean crushing demand times soyaoil yield

(13) Livestock production = livestock consumption + livestock net stock change
+ livestock net exports

(14) Soyabean crushing ratio = (soyameal retail price times soyameal yield
+ soyaoil retail price times soyaoil yield)/
soyabean import price

Second, domestic livestock production is a decisive determinant of soyabean crushing demand in each economy. The results suggest that growth in soyabean crushing demand keeps pace with the growth of the livestock sector. Consequently, policies discouraging domestic livestock production, such as trade liberalization on livestock products, will decrease domestic soyabean crushing demand and import demand in both countries. The crushing ratio has a positive influence on soyabean crushing demand in each economy. Since the estimated elasticity in each economy is very small, policies increasing profitability in the soyabean crushing process have only limited effects on soyabean crushing demand and import demand.

Third, soyameal demand in each economy is inelastic with respect to its own price, but is influenced primarily by domestic livestock production. These findings imply both that soyameal is an essential ingredient of formula feed for livestock and that changes in soyameal price will not greatly influence demand.

Fourth, per capita real income and the price of meat are the major factors influencing per capita meat demand in each economy. If domestic livestock production is defined as the sum of total domestic meat consumption, changes in livestock ending stocks, net meat exports, and egg production, then it is the major factor influencing soyabean crushing demand and soyameal demand. Income and population growth therefore have important influences on demand for soyabeans and soyameal. These results indicate that per capita real income and population growth have been the major factors facilitating the rapid growth of these soyabean markets. As these economies continue to grow in the near future, these soyabean markets are likely to expand continuously because of the demand potential for meat among the large affluent population and the limited potential for soyabean production in these natural-resource-poor economies.

## Impacts of Economic Growth and Policy Intervention

To evaluate the impacts of economic growth and policy changes on these soyabean markets, exogenous changes in each economy were hypothesized, individually, to perform dynamic simulation analyses using the established models. The hypothetical changes, which were assumed to have been introduced since 1975, included a devaluation of the domestic currency, an increase in the soyabean support price, technological improvements in the domestic soyabean-crushing process, and an increase in net livestock exports (or a decrease in livestock imports).

After adjustment for each policy change, the dynamic simulation was repeated for the period of study. The difference between the new simulation results and the base simulation results represented the impacts of each policy change on the endogenous variables. The average percentage impacts of different policies on several interested endogenous variables are summarized in Table 2. In order to measure the impacts of economic growth on the soyabean market, population and per capita real income in these economies were, individually, assumed to be fixed at 1970 levels. The differences between the new simulation results and the base simulation results of the last period indicated the impacts of population growth and per capita real income growth since 1970. These results are also reported in Table 2.

These simulations of growth and policy impacts lead to six results. First, growth factors had greater impacts on soyabean import demand than did policy factors. Population and income growth stimulate growth in the livestock sector, which, in turn, is the major factor causing growth in soyameal demand and soyabean crushing demand. As a result, growth factors have an important influence on soyabean import demand in Japan and Taiwan. On the other hand, because domestic soyabean production is relatively small compared with total soyabean demand and because the supply elasticity with respect to the support price is low, agricultural production policies usually have only limited effects on domestic soyabean production and soyabean import demand. Moreover, the elasticities of demand for soyabeans and their products with respect to their own prices are generally low. Therefore, policies (such

Table 2—Relative Impacts of Economic Growth and of Policy Changes

| Variable | Country | Simulation Base | | 10-Percent Devaluation | 10-Percent Increase in Soyabean Support Price | 10-Percent Increase in Soyabean Crushing Ratio | 10-Percent Decrease in Livestock Net Imports | Population Growth Effect since 1970 | Real Income Growth Effect since 1970 |
|---|---|---|---|---|---|---|---|---|---|
| | | Average Value since 1975 | Last Period | | | | | | |
| Soyabean acreage | T | 21,680 ha | 7,967 ha | — | 16.54 | — | — | — | — |
| | J | 114,520 ha | 150,081 ha | — | 9.93 | — | — | — | — |
| Soyabean crushing demand | T | 809,340 t | 900,960 t | -0.19 | — | 1.25 | 0.14 | 14.87 | 22.73 |
| | J | 3,174,600 t | 3,594,460 t | -0.22 | — | 1.59 | 0.60 | 7.73 | 5.79 |
| Soyabean food demand (per capita) | T | 10.71 kg | 10.79 kg | -0.98 | — | — | — | — | 4.18 |
| | J | 6.44 kg | 6.06 kg | — | — | — | — | — | 23.39 |
| Soyabean import demand | T | 992,770 t | 1,089,210 t | -0.35 | -0.28 | 1.03 | 0.12 | 16.11 | 19.58 |
| | J | 3,929,030 t | 4,153,170 t | -0.17 | -0.43 | 1.28 | 0.48 | 8.79 | 9.07 |
| Soyameal demand | T | 641,440 t | 721,630 t | -0.79 | — | — | 0.14 | 14.87 | 22.73 |
| | J | 2,744,150 t | 3,114,620 t | — | — | — | 0.70 | 8.60 | 6.34 |
| Soyameal import demand* | T | 6,720 t | 18,900 t | -3,753 t | — | -72,005 t | 17 t | 2,810 t | 4,300 t |
| | J | 280,090 t | 316,730 t | 5,390 t | — | -288,720 t | 4,437 t | 54,560 t | 35,340 t |
| Soyaoil demand (per capita) | T | 8.27 kg | 8.98 kg | -0.78 | — | — | — | — | 32.13 |
| | J | 4.97 kg | 5.38 kg | -1.19 | — | — | — | — | 20.93 |
| Soyaoil import demand* | T | -901 t | 3,280 t | -867 t | — | -16,447 t | -191 t | 9,790 t | 16,290 t |
| | J | 5,805 t | 7,460 t | -5,354 t | — | -66,761 t | -3,437 t | 29,630 t | 97,760 t |
| Meat demand (per capita) | T | 46.51 kg | 50.87 kg | — | — | — | — | — | 31.32 |
| | J | 24.95 kg | 38.83 kg | — | — | — | — | — | 8.85 |
| Livestock production | T | 991,810 t | 1,118,680 t | — | — | — | 0.16 | 17.20 | 26.29 |
| | J | 4,642530 t | 5,215,880 t | — | — | — | 0.63 | 7.96 | 5.80 |

*Impacts upon soyameal imports and soyaoil imports are reported in quantity change; the other impacts are in percentage change.

as a devaluation in domestic currency) influencing domestic prices of soyabeans and their products cannot have much influence on soyabean import demand.

Second, the growth impacts on the soyabean markets in Taiwan were more significant than those in Japan. These simulation results indicated that, without per capita real income growth since 1970, soyabean import demand in the last period of study would have decreased by 20 and 9 percent, respectively, in Taiwan and Japan. The simulation results also revealed that population growth impacts on soyabean import demand in Taiwan and Japan were 16 and 9 percent, respectively, of their import demands. The results can be explained by noting that the elasticity of per capita livestock demand with respect to per capita real income in Taiwan is higher than in Japan. In addition, the elasticity of soyabean crushing demand with respect to domestic livestock production in Taiwan is also greater than that in Japan. Therefore, growth factors had greater effects on the livestock sector, soyabean crushing demand, and soyabean import demand in Taiwan than in Japan.

Third, a 10-percent devaluation of the Japanese yen or Taiwanese dollar with respect to the US dollar would decrease soyabean import demand in Japan and Taiwan by only 0.17 and 0.35 percent, respectively. Such a devaluation would raise domestic prices of soyabeans and products. The higher domestic prices would, in turn, decrease soyameal demand, per capita soyaoil demand, and per capita soyabean food demand. On the other hand, the soyabean crushing ratio in Japan and Taiwan would decrease, causing domestic soyabean crushing demand, soyameal production, and soyaoil production to decline. As a result, net imports of soyabeans, soyameal, and soyaoil in these economies (except for soyameal imports in Japan) would decrease. These results suggest that domestic soyaoil and soyameal supply elasticities with respect to their own prices in these economies (except for soyameal supply elasticity in Japan) were lower than the demand elasticities with respect to their own prices.

Fourth, a 10-percent increase in the soyabean support price would stimulate increases in domestic soyabean acreage planted in Taiwan and Japan of 17 and 10 percent, respectively. However, since the percentages of domestic soyabean production with respect to the total demand for soyabeans in these soyabean-importing economies were small, the increase in domestic soyabean production would have no significant impacts on the import demand for soyabeans. The estimated percentage impacts on soyabean imports were −0.43 and −0.28 percent in Japan and Taiwan, respectively.

Fifth, a 10-percent increase in Taiwan's net livestock exports or a 10-percent decrease in Japan's net livestock imports, assuming domestic meat demand is not affected, would increase soyabean import demand in Taiwan and Japan by 0.12 and 0.48 percent, respectively. Since domestic livestock production would increase, soyameal feed demand and soyabean crushing demand would be stimulated, so that soyabean import demand would rise in each economy. The simulation results also indicated an increase in soyameal import demand and a decline in soyaoil import demand in Taiwan and Japan. Since domestic soyaoil demand was unaffected, increased soyabean crushing demand would raise domestic soyaoil production and decrease soyaoil import demand in each economy. On the other hand, since domestic soyameal production and demand changed in the same direction in each country, the net impacts on soyameal import demand would depend upon the supply and demand elasticities. The simulation results implied that the soyameal demand elasticity with respect to livestock production was greater than the supply elasticity in each country. Furthermore, since net livestock imports (or exports) were relatively small compared with total production volume, the impacts of changes in net livestock imports (or exports) on soyabean markets would not be overwhelming, especially in Taiwan.

Sixth, an improvement in the profitability of the soyabean crushing process would increase soyabean import demand but decrease soyameal and soyaoil import demand in each economy. In Japan and Taiwan, the demand for soyabean meal as a livestock feed is growing more rapidly than the demand for soyabean oil as a food. The imbalance in growth of demand for meal and oil has resulted in excess stocks of soyabean oil and has reduced the profitability of crushing. Policies augmenting the profitability of the soyabean crushing industry, such as encouraging soyaoil exports, would thus raise soyabean crushing demand and import demand and make formula feed producers less dependent upon soyameal imports.

## Note

[1]National Taiwan University.

## Reference

Woo, R.J., "The Effects of Economic Growth and Policy Intervention in Japan, Taiwan, and South Korea upon the Import Demand for Soybeans," Ph.D. dissertation, Iowa State University, Ames, Iowa, USA, 1985.

## Discussion Opening—*Mary Bohman* (University of British Columbia)

The paper analyses soyabean import demand in Japan and Taiwan by estimating domestic supply and demand for related products. The absence of general information on agriculture in these countries makes it difficult to evaluate the model.

The estimated supply elasticity is an example of an important parameter for policy analysis that cannot be evaluated based on information provided in the paper. It would be helpful to know if land constraints or institutional factors cause the low supply response. Omission of relevant structural details would bias the estimate of the elasticity.

In the Japanese case, rice diversification policies play an important role in soyabean production. In order to reduce rice surpluses, farmers receive money for planting alternative crops. Commodities where Japan has a low level of self-sufficiency including soyabeans receive the most generous payments. The data used by USDA's Economic Research Service to calculate the PSE (producer subsidy equivalent) show that more money is transferred via this mechanism than by direct price supports for soyabeans. In the model, only past acreage and price are included to explain acreage response. The policy information suggests that changes in rice diversification payments affect soyabean acreage. Therefore, exclusion of this variable causes a biased estimate of the elasticity.

Misspecification of the model has resulted in biased parameter estimates. Although the model appears satisfactory for forecasting, I would be hesitant to use the results for policy purposes. A revised model could be interesting for domestic policy analysis of alternative policies. To facilitate this, the author should provide more complete results, including budgetary costs.

What does the model tell us about direction of future soyabean import demand? First, policy changes in the livestock sector make a large difference. Second, domestic production is not important, as shown by the fact that in Taiwan a 10-percent increase in the soyabean support price resulted in a 17-percent increase in acreage and only a 0.28-percent reduction in imports. This result is driven by the small share of domestic supply relative to consumption and not strongly affected by the supply elasticity. Finally, income growth will have an impact. Although I do not believe the parameters are precisely estimated, the numbers suggest that this is a major task.

*[Other discussion of this paper and the author's reply appear on page 93.]*

# Agricultural Policy Reform in the Pacific Rim: The Case of Japan

*Vernon O. Roningen and Praveen M. Dixit*[1]

**Abstract:** The world trade model used for this study suggests that Japanese agricultural policies have substantially depressed world rice prices. The study also indicates that such policies have imposed considerable distortions in Japan's domestic economy, costing consumers and taxpayers nearly $1.73 to provide a dollar of support to agricultural producers. The study concludes that while elimination of agricultural support could considerably lower Japanese agricultural incomes, such losses may not affect the living standards of most farmers because agriculture contributes less than 8 percent of total household incomes for nearly 70 percent of farmers.

## Introduction

The pressures on Japan to reform agricultural policies have been mounting over the last decade. With a current account surplus averaging over $80,000 million during the late 1980s, Japan is facing growing criticism for continuing to exploit an increasingly open manufacturing market while restricting agricultural imports. These pressures on Japanese agriculture can be expected to increase even more as the end of the Uruguay Round approaches.

What happens to agriculture in Japan is of great interest to the world agricultural community. Japan is one of the world's richest countries, with a per capita GNP 8–10 percent higher than that of the USA. With a population base about half that of the USA, Japan is also one of the largest consumers of agricultural products among the industrial countries. And, despite attempts to maintain self-sufficiency in basic foodstuffs, Japan is the largest national importer of agricultural products and accounts for nearly a tenth of world agricultural imports. Any change in Japanese policies can be expected substantially to affect world agricultural markets.

This paper addresses two issues that relate to Japan in an open economy: the extent of government intervention in Japanese agriculture and how it has distorted world agricultural prices and trade and the domestic and international costs of Japanese agricultural policies.

## Quantifying Agricultural Support in Japan

The policies and programmes used to support agriculture in Japan are complex and diverse across commodities (ABARE, 1988). A major practical step in understanding these measures has been the development and acceptance of a measurement methodology in the form of producer subsidy equivalents (PSEs) and consumer subsidy equivalents (CSEs).

A PSE is the level of subsidy that would be necessary to compensate producers for the removal of government programmes affecting a particular commodity (Ballenger, 1987). Similarly, a CSE is the level of subsidy that would be necessary to compensate consumers for the removal of government programmes. Figure 1 shows the extent of government protection to the Japanese agricultural sector in 1989/90 as

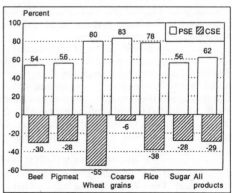

Figure 1—Rates of Support to Agriculture in Japan, 1989/90

represented by the producer subsidy equivalents (Webb *et al.*, 1990). The weighted average PSE for all commodities (62 percent) indicates that nearly two-thirds of total producer income is generated by transfers from government policies. The PSEs are highest for wheat, coarse grains (barley), and rice, followed by sugar and livestock products. Because imports of maize and soyabeans are relatively free, government support for these two commodities are virtually non-existent.

The costs of agricultural support have to be borne either directly by domestic consumers through higher food prices or by taxpayers through increased government expenditures. Policies that artificially raise (tax) prices to consumers account for about 90 percent of the support to the agricultural commodities. Consequently, the CSEs are relatively high (29 percent), indicating the costs of policies to Japanese consumers represented nearly a third of consumer expenditure on the commodities considered.

How do Japan's PSEs and CSEs compare with those of other countries? Figure 2 indicates that the extent of government intervention in Japan is the highest among the industrial market economies. It is

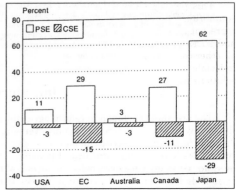

Figure 2—Rates of Support to Agriculture across Countries, 1989/90

more than twice that of the EC and Canada and at least five times that of the USA and Australia. Also, the source of support in Japan varies considerably from that in the USA and Australia. While policies that transfer incomes from consumers to producers form the primary basis of Japanese agricultural support, countries like the USA and Australia rely more on direct government budget support.

## The Modelling Framework

Summary measures of protection such as PSEs and CSEs are good indicators of the level of support to agriculture but do not incorporate the supply and demand responses to such policies. They therefore have little to say about the consequences of such protection on commonly used indicators of economic performance such as output and income of the sectors being supported or their effects on world agricultural markets. This requires a framework that would allow for economic responses.

The economic implications of agricultural policy reform in Japan are analysed using the static world policy simulation modelling (SWOPSIM) framework (Roningen, 1986). A SWOPSIM model is characterized by three basic features: it is a non-spatial price equilibrium model, it is an intermediate-run static model that represents world agriculture in a given year, and it is a multi-commodity, multi-region partial equilibrium model. In order to use this static, non-spatial partial equilibrium model to describe world agricultural trade, it is assumed that world markets are competitive, that domestic and traded goods are perfect substitutes in consumption, and that a geographical region, possibly containing many countries, is one market place.

The economic structure of SWOPSIM models includes constant domestic supply and demand equations. Trade is the difference between domestic supply and total demand (absorption). The policy structure is embedded in equations linking domestic and world prices. Policies (PSEs and CSEs) are inserted as subsidy equivalents at the producer, consumer, export, or import levels. (For details on the economic and policy structures and the use of summary support measures in the modelling framework, see Roningen and Dixit, 1989.)

The version of SWOPSIM used for this study (ST89) is based on 1989/90 marketing year data. The world is divided into 11 regions, 7 of which represent the industrial market economies, 3 developing countries, and 1 centrally planned economies. Included in the model are 22 agricultural commodities, representing mostly temperate-zone products. Tropical products, which account for a substantial proportion of agricultural trade of developing countries, are not included.

This paper presents the results of experiments using the ST89 model in which new equilibrium solutions were obtained by removing PSEs and CSEs. The new solutions represent an approximation of the resulting adjustments in production, consumption, trade, and prices of agricultural commodities expected after 5 years, with the important proviso that all other conditions remain the same as in the base year, 1989/90. This permits the analysis to isolate and identify the differences between the new solutions and the initial or reference solutions and to attribute them to the removal of distortionary agricultural policies.

## Eliminating Agricultural Intervention in Japan

The model and the aggregate measures of government intervention were used to simulate conditions that would exist if Japan unilaterally eliminated all agricultural policies as they existed in 1989/90. From this, the distortions in world prices and trade and the annual economic welfare costs of such policies were deduced.

### Effects on World Commodity Prices

Japanese agricultural policies have, on average, depressed world commodity prices by 2 percent by encouraging uneconomic production and inhibiting consumption (Figure 3). In other words, if Japan were to abandon its policy of supporting agricultural producers and taxing consumers, its production would contract and consumption would expand. As a result, world commodity prices would be higher. About 40 percent of this increase in world price would emanate from the liberalization of Japanese consumer demand.

In comparing the effects of Japan's policies on world agricultural markets with those of other countries, the results show that policies of industrial market economies, taken together, have, on average, depressed world commodity prices by 12 percent (Figure 4). This suggests that Japanese policies account for nearly a sixth of the depression in world prices. Only EC policies (8 percent) contribute more to aggregate world price changes than do those of Japan.

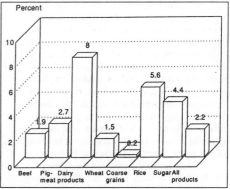

Figure 3—World Price Effects of Unilateral Japanese Liberalization

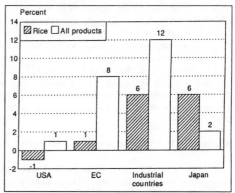

Figure 4—Comparing World Price Effects of Unilateral Policy Reform

The world price implications of Japan's policies are especially dramatic for rice. Japan's policies have depressed world rice prices (6 percent) more than the combined effects of all other industrial market economies' policies (1 percent). This reflects both the high levels of assistance to Japanese rice producers and the relatively small proportion (4 percent) of global rice production that is traded in the world market.

Despite high levels of protection, the price distortionary implications of Japanese policies do not have the same consequences for most other commodities. Japanese policies depress the world sugar price by 4 percent and livestock product prices by only 2–3 percent. The country's role in global production and trade, rather than the levels of support, appears to be the dominating factor in determining the extent of trade distortion.

## Implications for Japanese Imports

Perhaps the most contentious issue concerning Japanese agricultural policy is the import restriction on rice (Figure 5). The model results indicate that that policy has prevented nearly 4.7 Mt of rice from entering Japan annually. Hence, whereas there are virtually no imports of rice into Japan at present, imports would be nearly half (46 percent) of total consumption in a liberalized environment. Imports of other products, especially wheat and sugar, have also been affected by import restrictions, but their magnitudes are not very large. By contrast, imports of maize would actually decline by 4 Mt if Japan unilaterally liberalized its agricultural policies because of the contraction in the livestock sector.

The USA and Australia have been in

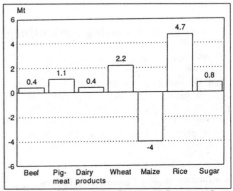

Figure 5—Changes in Japanese Imports from Unilateral Agricultural Reform

the forefront of international efforts to pressure Japan into liberalizing its beef import policies. In response, Japan recently agreed to eliminate all import quotas on beef by 1991 and replace them with higher *ad valorem* tariffs. The question is, How great an increase in imports can be expected from a liberalization of the Japanese beef market? Estimates from the model indicate that Japanese imports of beef would nearly double from base levels, an increase of 417,000 t annually, if all border restrictions were eliminated. Nearly half of the additional imports would originate from the USA, with the rest coming from Australia, the EC, and South America. Alston, Carter, and Jarvis (1989) contest this view and argue that countries such as Australia would gain greater market shares because of the elimination of "discriminatory import quotas" that have favoured US beef.

By denying access to their markets, Japan's agricultural policies cost other countries $6,800 million in export earnings. Nearly 25 percent of the loss in earnings occurs because of support to rice producers. Import restrictions on pigmeat and beef account for most of the rest.

## Economic Welfare Implications of Policy Reform

Agricultural support policies in Japan have reduced national income by encouraging inefficient use of resources. They have also transferred resources from consumers and taxpayers to agricultural producers.

The study shows that less than 60 percent of the costs to consumers and taxpayers in Japan are transferred to producers (Figure 6). The rest ($5,700 million) represents deadweight (income) losses to society arising out of misallocated resources.

In other words, Japan's policy of providing support to agricultural producers is inefficient because the costs to consumers and taxpayers of distortionary policies ($38,000 million) are considerably more than the benefits to producers ($22,000 million). For every $1.00 that producers in Japan gain because of protectionist agricultural policies, consumers and taxpayers lose $1.73. Consumers and taxpayers in the USA, on the other hand, forfeit only $1.05 in transfers for every dollar gained by producers.

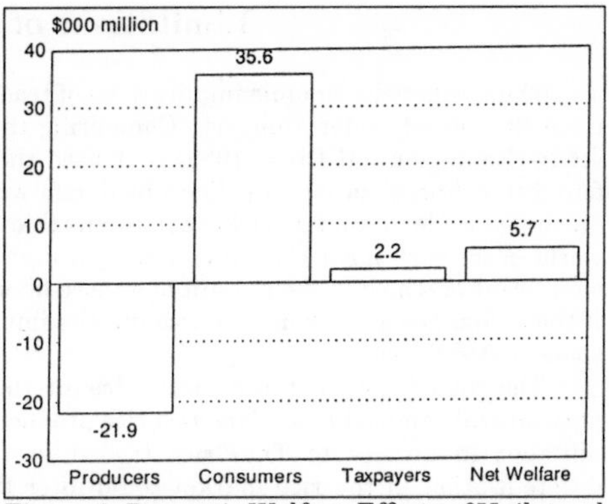

Japanese consumers have had to shoulder much of the burden of domestic agricultural support policies. Model results indicate that each consumer spends an additional $290 annually on food to maintain agricultural support. This is nearly twice the per capita costs that consumers in the EC have had to bear to support their agricultural sector.

Opponents of trade reform often point to the losses in producer incomes as an argument against liberalization. Indeed, these results indicate that producer surplus losses from unilateral policy reform would be about $22,000 million, or about 45 percent of the value of agricultural production in 1989/90. Rice producers would shoulder about two-thirds of this loss, while most of the other remaining losses would accrue to beef and pigmeat producers.

While these losses may appear large, consider the following. In 1986, agricultural income provided only about 2 percent of household income for the 40 percent of farm households with 0.5 hectare or less of cultivated land, and less than 8 percent of household income for the 28 percent of farm households with 0.5–1.0 hectare of cultivated land (ABARE, 1988). Therefore, for nearly 70 percent of farm households, even a considerable drop in farm income as implied by the present results would have little effect on their living standards. This is especially true given that average income of farm households in 1986 was nearly 30 percent higher than average income of other households. It should, however, be noted that because the size of an average farm household (4.53 persons) is much larger than that of urban households (3.00), the difference in income, when expressed on a per capita basis, is less than 15 percent.

The domestic costs of distortionary policies represent only a part of the welfare costs of such policies. Japan's policies not only affect producers, consumers, and taxpayers within the country but also those in other countries (Figure 7). While Japanese policies raise producer incomes by $22,000 million, they cost producers in other countries $10,000 million because of their price-depressing effects. Conversely, such policies save consumers in food-deficit countries nearly $6,000 million. The global real income gain from unilateral Japanese agricultural liberalization is $5,000 million.

Figure 6–Domestic Welfare Effects of Unilateral Japanese Liberalization

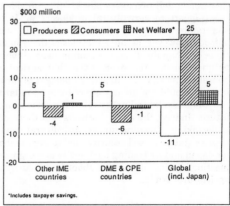

Figure 7—Global Welfare Effects of Unilateral Japanese Liberalization

## Limitations of the Analysis

The predicted economic implications of trade liberalization are likely to differ depending upon the period under analysis. Comparing the results of this study with those of another study (Roningen and Dixit, 1989) that used the 1986/87 marketing year as its base, it was found that liberalization of policies by Japan would have increased world agricultural prices much less under 1989/90 market conditions than under 1986/87 conditions. The Japanese and world economies have been undergoing a number of changes in the recent past, the most significant of which is the pursuit of a more liberal agricultural trading environment for beef. If these changes are taken into account, the implications of agricultural policy reform may be somewhat different.

The model does not take into account the substantial product differentiation among agricultural commodities. *Japonica* rice produced and consumed in Japan is, for instance, far different in quality to Thai rice traded in the world market. Recognition of product differentiation in the rice market could alter the economic implications of Japanese policy reform. Dixit and Roningen (1991) show that a proper representation of rice imports into Japan following unilateral liberalization may be somewhere between 3.6 Mt and 5.1 Mt, depending upon the degree of substitution assumed in Japanese consumption and foreign production of *Japonica* and *Indica* rice.

Finally, there is also the Lucas critique. Lucas (1976) argued that models estimated using data collected under a past policy régime may not be relevant to current or future conditions. This issue is of special concern when large shocks like trade liberalization occur. Should policy régimes change drastically, as would be the case with trade liberalization, a model based on historical parameters might not give the correct story.

## Conclusions

The objective of this study was to examine the distortionary costs of Japanese agricultural policies. It can be inferred from the analysis that Japanese policies have considerably distorted world rice prices but that the distortionary implications for other commodities are not as significant. The analysis also indicates that distortions introduced by agricultural policies have substantially benefited Japanese producers but hurt consumers and taxpayers. Consumers and taxpayers pay $1.73 to provide a dollar of support to agricultural producers.

Support to agricultural producers in Japan has often been justified as a means of maintaining parity between farm and non-farm incomes and of providing an assured quantity of food. However, because incomes of those in agriculture are nearly 30 percent higher than those of urban dwellers, the commitment to continue support to agriculture at past levels can be questioned. Moreover, given that agriculture contributes less than 8 percent of total household incomes for nearly 70 percent of farmers, even a substantial drop in agricultural incomes would not greatly affect living standards of most farm families.

Major unilateral policy reform in agriculture is one option that Japan could pursue in the future. Indeed, the analysis shows that there would be some real income gains from such an undertaking. Opening up the agricultural market, however, is not without risks. Problems of price stability, food security, and foreign exchange variability could abound. The challenge for Japan and its trading partners is to recognize the benefits of freer agricultural trade while meeting legitimate concerns about food security and rural prosperity without distorting world agricultural trade.

## Note

[1]Economic Research Service, US Department of Agriculture. Views expressed in this paper are those of the authors and not necessarily those of the US Department of Agriculture.

# References

ABARE (Australian Bureau of Agricultural and Resource Economics), *Japanese Agricultural Policies: A Time of Change*, Australian Government Publishing Service, Canberra, Australia, 1988.

Alston, J.M., Carter, C.A., and Jarvis, L.S., "Japanese Beef Trade Liberalization: It May Not Benefit Americans," *Choices*, Vol. 4, No. 4, 1989, pp. 26–30.

Ballenger, N. (Ed.), *Government Intervention in Agriculture: Measurement, Evaluation, and Implications for Trade Negotiations*, Foreign Agricultural Economic Report No. 229, Economic Research Service, US Department of Agriculture, Washington, D.C., USA, 1987.

Dixit, P.M., and Roningen, V.O., "Reforming Japanese Rice Policies: Importance of Product Differentiation," paper presented at an American Agricultural Economics Association meeting, Manhattan, Kans., USA, 1991.

Lucas, R.E., "Econometric Policy Evaluation: A Critique," in Bruener, K., and Meltzer, A.H. (Eds), *The Phillips Curve and the Labor Market*, North Holland Publishing Co., Amsterdam, Netherlands, 1976.

Roningen, V.O., *A Static World Policy Simulation (SWOPSIM) Modeling Framework*, Staff Report No. AGES–860625, Economic Research Service, US Department of Agriculture Washington, D.C., USA, 1986.

Roningen, V.O., and Dixit, P.M., *Economic Implications of Agricultural Policy Reforms in Industrial Market Economies*, Staff Report No. AGES 89–36, Economic Research Service, US Department of Agriculture, Washington, D.C., USA, 1989.

Webb, A.J., Lopez, M., and Penn, R. (Eds.), *Estimates of Producer and Consumer Subsidy Equivalents: Government Intervention in Agriculture, 1982–87*, Statistical Bulletin No. 803, Economic Research Service, US Department of Agriculture, Washington, D.C., USA, 1990.

---

**Discussion Opening**—*Nobuhiro Suzuki* (National Research Institute of Agricultural Economics, Japan)

The significance of support should be measured not only in terms of rates of PSEs but also in terms of total amount. The per-unit PSEs for wheat and barley were 80 and 83 percent respectively in 1989 (Figure 1). However, we cannot say that the subsidies to wheat and barley were the highest because only 16 percent of wheat and 17 percent of barley consumption were met by domestic supply in the same year. Japan's small effect on world price distortion, despite high per-unit PSEs, should be explained not only by the fact that Japan's share in global production and trade is small, but also by the fact that Japan's import share in total demand is already very large and the total amount of PSEs fairly small. This implies that per-unit PSEs are not good indicators of protection.

It is misleading to evaluate the effects of unilateral liberalization on world prices by only one time-period analysis because the results are too sensitive to the period under analysis. The same authors' previous study, based on 1986/87 data, indicated that the effect of US policy on the world rice price is an increase of 2.9 percent, whereas the present study, based on 1988/89 data, indicates a decrease of 1 percent. Similar differences can be noted for the cases of Japan's rice (from 19.6 down to 5.5 percent), Japan's dairy products (from 4.5 up to 8 percent), Japan's average (3.6 down to 2 percent) and the US average (5.9 down to 1 percent). The US effect is on average larger than that of Japan in the previous study, but smaller here. Moreover, it is doubtful that the US unilateral liberalization will have a negative effect on world rice prices (–1 percent) considering that the USA has already established the marketing loan system for its rice export expansion.

Trade liberalization may not be a very severe problem for farm households that earn a large proportion of their income from other sources. The concern is not about these farmers but about those who earn the majority of their income from farming and play a significant role in supplying food to consumers. We are also concerned that much of the demand for rice in Japan may be satisfied with imports. Rice is a staple food in Japan as are dairy products in the USA and EC. US imports of dairy products are kept at about 1 percent of total demand.

The simulation for unilateral liberalization is not effective in separating the country's effects on global economic welfare from those of other countries. For example, if Japan were to liberalize unilaterally under the situation where world dairy prices were distorted by US and EC policies, Japan's dairy imports might increase substantially. Japanese consumers' gains are generated not only as a result of Japan's policy changes but also as a consequence of the policies of some exporting countries that depress world prices. The magnitude of an increase in economic welfare by unilateral liberalization is not an appropriate indicator of how policies of the country concerned distort world markets.

Product differentiation is important, especially for beef and rice. The rice (beef) market should be divided into segments based on quality. It should then be examined carefully as to which segment of the domestic rice (beef) market can be substituted for the appropriate type of imported rice (beef).

Japanese consumers may prefer the present situation, considering the extra-market values of domestic farming and rural communities, even if food prices are rather high. It may be difficult to quantify such values and incorporate them into this kind of study. However, it would be better to consider external effects of agriculture in some way.

A reduction in PSEs as an aggregate indicator does not discern the various effects of different policy measures. For that purpose, a policy-specific version of the SWOPSIM model is required.

Japan will not import beef from South America because the beef is contaminated by foot and mouth disease. This study suggests that import restrictions on pigmeat are fairly effective in limiting the amount imported. This cannot be the case because Japan's only protection in this area is tariffs.

*[Other discussion of this paper and the authors' reply appear on the following page.]*

**General Discussion**—*R.D. Ghodake, Rapporteur* (Dept. of Agriculture and Livestock, Papua New Guinea)

In the general discussion, Ohno and Gallagher's use of data for 1971–88 but with conclusions on imperfect competition for 1973–78 was questioned, particularly whether this was a problem of degrees of freedom. Ohno replied that when the equations were estimated simultaneously, the hypothesis $H_1 = H_2 = 0$ was not rejected. However, $D_1$ was found to be statistically significant for the period 1973–78, and hence the episode was tested for that specific period.

In answer to a question about why US export prices are higher than import prices at some periods and lower at others, Gallagher replied that, as processing mills are generally located at points of import, import unit values (which are calculated on a c.i.f. basis) are a reasonable proxy for processors' transaction prices. Wholesale prices are reported for Tokyo locations, since this is regarded as a suitable price for evaluating users for human consumption like *tofu* shops. The higher import prices in Tokyo thus reflect local distribution costs and profits associated with wholesale distribution of imports.

In reply to the discussion opener, Woo explained that because of space limitation not much background information was provided, although such information is essential for developing economic models. The rice diversification payment was included in the model but was found to be statistically insignificant. From Theil's statistics, there is no serious problem about model misspecification. A misspecified model is not judged simply by income elasticities that are not consistent across individual soyabean products. However, there are no perfect models, and further modifications in the model according to the research objectives are really needed.

To comments that simultaneous equation bias should be checked, Nerlovian models bias supply elasticity downwards, and yields should be endogenous, Woo replied that, since the model specified is a recursive equation system, SUR is acceptable/valid for the estimation of such a simultaneous equation system. In addition, SUR estimates are quite close to 3SLS estimates in the primary study.

Asked how he dealt with the combined products soyaoil and soyameal, the author replied that soyabean meal and oil were treated as joint products. The model, as specified, integrated the market clearance identities of soyameal, soyabeans, soyaoil, soyabean crushing demand, and soyabean crushing ratio. The soyabeans, soyaoil, and soyameal markets are linked.

The inclusion of lagged dependent variables in the supply models, acreage equations in particular, was questioned, as it often causes a model to miss the turning point. Woo was asked whether this had been checked and the results with and without lagged dependent variables compared. Woo replied that, for the questions containing lagged dependent variables, Durban-Watson and $t$ statistics show no evidence of first-order serial correlations. For the behavioural equations that showed a problem of serial correlation, the Cochrane-Orcutt procedure was adopted to correct the first-order serial correlation and increase the efficiency of estimation. The conclusions might seem like common sense, but the model did exert evidence of common sense.

Woo was also asked why he chose to simulate the effects of devaluation, since the term "devaluation" is only used in cases where a managed currency is overvalued. Since the Japanese yen is freely tradeable and convertible and is not overvalued, it is inappropriate to consider a devaluation; it is simply a question of a price increase.

Roningen and Dixit were asked why, if Japanese rice imports under liberalization are 30 percent of world trade, world rice prices rise only by 6 percent? They replied that rice trade is small relative to the size of production and consumption. A small price change clears the world market, while creating a large trade change.

Asked to account for rice diversion policies in the PSE in their model, the authors replied that they did not do this in the PSE. They account for these policies for soyabeans and wheat on rice land by simultaneous shifts of the supply schedules for rice, wheat, and soyabeans as rice support is lowered.

Asked about comparative static analysis in relation to a dynamic real world, the authors acknowledged that it would be instructive and useful to carry out trade liberalization in a dynamic model, but the mechanics of such a model are much more complicated, and they made the decision to use as simple a model as possible. This is an area that should be developed in the future.

Asked to give further consideration to product differentiation, for example with different types of rice, such as *Japonica* and *Indica*, the authors indicated that they have considered differentiation for rice in another paper. Depending upon the assumptions about further substitution in production and consumption, the answers can be larger or smaller than theirs for Japanese rice imports, for example. They agree that differentiation is important for greater accuracy in some trade modelling situations for Japan, including rice and beef.

The model was considered very dangerous by one participant, based as it is on several assumptions for which there is very little knowledge and evidence, which could be misused by politicians and negotiators.

While this could happen, the authors believe that numbers openly calculated, checked, and debated help to formulate more rational policies that will ultimately foster rather than distort world trade. Economists should provide the best analysis and numbers possible; this makes progress possible but not inevitable.

Asked what exactly they meant by liberalization and its timing, the authors responded that they simply eliminated Japanese supply and calculated the new equilibrium. The elasticities in the model are medium term; i.e., 3–5 years, so it is assumed that the figures calculated will be fully realized on an annual basis after 3–5 years.

Participants in the discussion included J. Beghin (North Carolina State University), D. Colman (University of Manchester), S. Ito (Tokyo University), G.T. Jones (University of Oxford), J.B. Morison (University of New England), K. Oga (IFPRI), D. Pick (US Department of Agriculture), and H. Popp (Ministry of Agriculture, Switzerland).

# Problems of Transferring Crop-Water Production Function Knowledge to Developing Countries

*G.R. Soltani, S. Pandey, and W.F. Musgrave*[1]

**Abstract:** Knowledge of crop response to different moisture conditions is essential for analysing water allocation problems at the farm level. Inadequate experimental data and the time and cost involved in obtaining sufficient data for adequate estimations of yield-water relationships are probably the major bottlenecks in developing countries. One approach to this problem is to transfer water production function knowledge from advanced countries, where several specific forms of water production function models have been developed. It is not possible to transfer empirical production function models because they are site specific, but, among the synthetic partial crop-water response models, those that relate yield to evapotranspiration are transferable to developing countries. Simulation and physiological models are more realistic than partial models, that require a large amount of data.

## Introduction

A major agricultural problem facing developing arid regions throughout the world is the inadequate supply of water for irrigation. There are two alternative but not mutually exclusive approaches to the scarcity problem: increasing the capacity of the water supply systems where applicable and raising the efficiency of irrigation systems.

Opportunities for augmenting water supply are being sharply curtailed by increasing costs of water in many developing regions. This implies that the traditional ways of responding to general conditions of scarcity by building new storage and conveyance facilities and/or going deeper or farther for water may no longer suffice. Consequently, water planning will focus less on engineering problems and more on the problems associated with the more efficient use of existing water supplies.

In many developing countries, excessive water application at the farm level ranks high among the causes of water shortage (Soltani, 1977). The problem of low irrigation efficiency can be viewed as one of improper determination of how, when, and how much water to apply in irrigating cultivated lands. The "how" is a decision related to the choice of irrigation method. The "when" and "how much" are decisions that depend largely upon the yield-response to irrigation water.

The often pressing need for increasing food supply has meant that much irrigation research in developing countries has been motivated by a desire to determine the quantity of water that crops require to attain maximum yields. The design and operation of irrigation projects have also been guided by this criterion, resulting in low irrigation efficiency. Increasing scarcity of water for irrigation is, however, forcing many regions to raise irrigation efficiency as a logical option. In this respect, knowledge of crop response to different moisture conditions is desirable for analysing the costs and benefits of water projects and policies designed to promote increased efficiency of water use.

While there is a substantial body of theoretical and applied scientific knowledge about crop-water production functions, most of it has been generated in developed countries and assumes a level of data availability and research resources not usually found in developing countries. This lack of data and research resources limits the use of the available production function knowledge in developing countries. A possible way of overcoming this problem is to transfer (spatially generalize) crop-water production functions from developed to developing regions.

The objective of this paper is to investigate the possibility of transferring crop-water response models from developed to developing countries. A brief review of the theory of crop water production functions is followed by a discussion of the problems encountered in transferring such knowledge. Finally, ways of getting around such problems and feasible methodologies for estimating water production functions in developing countries are presented.

# Review of Crop-Water Production Functions

There are two basic types of water allocation problem at the farm level: determination of the optimal seasonal quantity of water to be applied to each crop and the optimal allocation of water over the growing season (irrigation scheduling).

To solve the first type of problem, a crop-water response function relating yield to various quantities of water and a cost function for the supply of irrigation water are necessary.[2] With continuous and differentiable functions, the optimal condition in a static world is attained when the marginal value product of water is equal to its marginal cost. However, the timing of water application is as important as the seasonal application rate. Therefore, for the determination of the optimal allocation of water, instead of a single response function, response functions for each growth stage are required (Yaron and Bresler, 1983). With such dated response functions, optimality is achieved by allocating the limited quantity of water in such a way as to equate expected marginal value products in all growth stages. A dated production function with all variables other than water held at a constant level can be written as:

(1) $Y = g \, [f_t \, (w_t)]$

where $Y$ is the crop yield, $w_t$ is the amount of water applied in stage $t$, and $f$ and $g$ are some arbitrary functions.

Modelling of crop-water production functions requires a theoretical framework concerning the relationships between the soil, irrigation water, and crop response. Detailed reviews of crop-water production function studies and the underlying theoretical framework are found in Hexem and Heady (1978), Vaux and Pruitt (1983), and Yaron and Bresler (1983).

For purposes of this paper, water production functions are grouped into various categories based on their transferability. Accordingly, water production function studies are grouped into two basic types: the empirical production function approach based on experimental or whole farm data and the synthetic (or deductive) approach in which plant response is modelled based on principles of plant growth.

The first type is further classified into studies based on data obtained from irrigation experiments conducted at one agro-climatic location and generalized (aggregate) production functions based on data generated at a number of locations. The second type is classified into soil moisture or evapotranspiration based models, mechanistic simulation models, and physiologically based models. Each of these categories of functions is discussed in turn.

## The Empirical Production Function Approach Based on Experimental Data

A conventional approach to estimating crop-water production functions is to correlate yield with applied water. In this approach, various functions are fitted to experimental data generated by varying the quantity of water applied (Hexem et al., 1976; Hoyt, 1982; and Ayer et al., 1983). The advantage of this approach is that, in relating yield to water applied, it focuses on the variable of most concern to planners and irrigators. An important difficulty associated with the approach, however, is that because the productivity of irrigation water depends on the timing as well as the quantity of water applied, any conclusion concerning the relationship between water quantity and yield is valid only so far as the distribution of water over the growing season conforms to the type of irrigation régimes applied in the experiments. This is a major limitation to the empirical application of this type of response function. While this problem is present in all locations, it is likely to be more critical in developing countries where actual irrigation practice probably deviates most substantially from experimental practice.

The so-called dated or micro-response function attempts to overcome this deficiency. Experimental data required for estimating this type of function are scarce (Minhas et al., 1974;

and Dinar *et al.*, 1986). In fact, the difficulty in empirical estimation of dated production functions is one of the major problems in determining optimal water use.

## Generalized Production Functions Using Experimental Data

One way to resolve the problem of site transferability is to develop a series of production functions corresponding to major classes of sites. This would entail a large amount of scientific work. An important contribution of this type is provided by the work of Hexem and Heady (1978), which attempted to aggregate and generalize site-specific yield-water-nitrogen relationships to develop production functions that could be used for predicting yield at different sites encompassing a great range of weather, soil, and spatial locations.

The estimation of such generalized production functions using some major variables to account for differences in soil, climate, and cultural practices is an attractive way of considering the possible transferability of production function knowledge from site to site. A problem with the approach, however, is that, again, it is usually very difficult to account for the effect of the timing of irrigation or rainfall on yield.

## Empirical Production Functions Approach Based on Whole Farm Data

All empirically derived production functions using experimental data share two common problems: they rely on knowledge derived from basic experiments that are generally expensive and time consuming and it is difficult to include consideration of all relevant factors in these experiments.

An aggregate production function using actual cross-sectional and/or time-series data on the quantity and timing of irrigation water as well as on other inputs has the potential to provide useful information for planners and irrigators in developing countries. The early work of Ruttan (1965) provides insight into the nature of such aggregated water production functions. The approach should be less costly and time consuming than the experimental approach. The kind of data needed is, however, not readily available in many developing countries and would have to be collected by means of extensive field surveys by experienced personnel or by a central statistical agency. There are also the normal and often substantial problems of production function estimation associated with this approach (Hexem and Heady, 1978).

## The Synthetic or Deductive Approach

An alternative to the direct estimation of yield-irrigation response is the synthetic (or deductive) approach where irrigation is linked to an intermediate variable representing some measure of crop water stress. Soil moisture and evapotranspiration are the commonly used intermediate variables. As these indices are more closely associated with crop water use than applied water, models using them are likely to be more transferable across locations. Within this approach, four classes of models are distinguished: soil moisture-based models, evaporation-based models, mechanistic simulation models, and physiological models.

**Soil-moisture-based models.** In this approach, an index of soil moisture stress, such as the number of stress days, percentage water depletion, or the level of available soil moisture is used as an intermediate variable (Moore, 1961; Palacios, 1981; and Yaron and Bresler, 1983). Accordingly, the production function is written as two separate relationships: one relating the soil moisture index to the irrigation decision variable and another relating crop yield to a soil moisture index. The relationship is expressed algebraically as follows:

(2)  $SMI = f(AW, \text{exogenous factors})$

(3)  $Y = g(SMI)$

where $SMI$ = index of soil moisture over growing season, $Y$ = yield, and $AW$ = a vector of irrigation decision variables.

**Evapotranspiration-based models.** Since evapotranspiration (ET) is more closely related to crop growth than are any of the aforementioned indices, it is used as a measure of the plant water stress in many yield-water response models. In such models, yield response is estimated with respect to ET, which is then related to irrigation by using a soil water balance model.

Two basic types of ET-based models are widely used: the multiplicative formulation developed by Jensen (1968) and the additive formulation developed by Stewart (1972). Algebraically these are the multiplicative model:

$$(4) \quad \frac{y}{y_p} = \Pi_i \left( \frac{ET_a}{ET_{p_i}} \right)_i^{b_i}$$

and the additive model:

$$(5) \quad y = \sum_{i=1}^{n} (1 - k_i ETD_i)$$

where $y$ = actual yield, $y_p$ = maximum potential yield, $ET_a$ = actual ET, $ET_p$ = potential ET, and:

$$(6) \quad ETD = 1 - \sum_j \left( \frac{ET_a}{ET_p} \right)$$

where $i$ = index of growth stage, $j$ = index for days in the $i$th growth stage, and $b_i$ and $k_i$ are weighting factors expressing sensitivity to water stress in the $i$th stage.

Despite the popularity of these models for irrigation scheduling (Rydzweski and Nairizi, 1979; and Bala et al., 1988), they have some limitations, particularly when used in developing countries. Two major limitations are that, in relating empirical yield to ET, all other factors are assumed to be non-limiting and all climatic and management variables are assumed to be reflected in the estimated value of potential yield achievable under the optimal management practice.

**Mechanistic simulation models.** The advantage of simulation models relative to previous models is that various plant processes such as photosynthesis, respiration, and water use are modelled and interlinked explicitly in a single formulation. These models can simulate daily growth in terms of its response to daily environmental conditions and can allow a systematic analysis of a complex production system (Pandey, 1986). As crop growth models are based on well-developed and robust theoretical principles and because their data requirements consist mainly of parameters that are well understood by biologists and that offer widespread applicability, simulation models are more transferable than most of the previous models.

Some examples of the use of simulation models for management decisions are given by Boggess et al. (1986), Hodges (1982), and Cuenca (1978). These models are relatively expensive to set up and are highly demanding of data (such as detailed meteorological and soil data). Furthermore, it is difficult to assess (validate) them due to a dearth of experimental data in developing countries. Despite these problems, simulation models are more explicit and complete representations of the theory of plant growth than partial crop-water production function models. This provides them with a greater flexibility and generality of application, which confer important advantages in trying to provide answers to water allocation problems in developing countries.

**Physiological models.** A major problem with soil moisture and evapotranspiration-based models is that determining the effect of water stress on photosynthesis (yield) and vegetative growth (dry matter) is difficult.

Crop-water responses are the result of the complex interaction of many physiological processes, each of which may be affected differently by plant water stress. Therefore, for yield-response models to be used to determine the relative effects of water stress on various physiological processes of plants, they should be based on the basic physiological principles that determine plant growth. A physiological approach focusing on the chemistry and physics of crop response to the environment may hold the most promise for producing transferable information. Despite the difficulty of constructing physiological models, they are potentially the answer to a number of problems indicated above, if only because they can be embedded in generalized simulation models that can incorporate many other complexities. It is not clear whether knowledge of physiological processes is sufficient to meet the demands of such models so long as the various difficulties relating to expense, data requirements, and validation already referred to with regard to simulation models cannot be diminished.

## Problems and Prospects

It is clear from the above review that the transferability of yield response models is enhanced if crop response is modelled by using knowledge of the plant growth processes, which are more or less universally applicable. In this context, mechanistic crop growth models hold out the most promise, as evidenced by their increasing popularity among researchers. However, validation of these models for developing countries can be expensive as this requires a large amount of location-specific data.

Due to their data-intensive nature, such models have not been widely used by farmers, even in developed countries. However, when appropriate data-gathering mechanisms are established for representative agroclimatic zones throughout the world, these models have the potential for being highly transferable. Substantial progress has been made in establishing such data-gathering mechanisms since the implementation of the International Benchmark Sites Network for Agrotechnology Transfer (IBSNAT) in 1982. It is planned that a minimum data set on weather, soil, and crops will be collected under the project to establish representative agroclimatic locations throughout the world (Uehara and Tsuji, 1991). These data, along with simulation models for 10 major crops, are expected to be made available in the form of a computerized decision-support system upon completion of the project. Thus, in the long run, it can be anticipated that such models will become vehicles for transfer of knowledge on crop response to irrigation from developed to developing countries.

For the more immediate run, the generalized production function approach of Heady and Hexem (1978) and the evapotranspiration model developed by FAO (Doorenbos and Kassam, 1979) seem promising. The generalized production function approach requires experimental data from a number of representative agro-climatic locations. Hence, the estimation of such functions could be quite expensive initially.

The procedure described in the FAO handbook (Doorenbos and Kassam, 1979) is probably the most transferable approach to yield-response modelling in the data-scarce developing countries. The report outlines simplified procedures for estimating actual and potential evapotranspiration and crop yield in the absence of moisture stress. Yield-reduction coefficients for various growth stages of a number of well-established crops grown in developing countries are also provided. The estimates are based on an extensive literature review and experimental work conducted by researchers throughout the world.

In conclusion, substantial potential exists for transferring knowledge on crop response functions from developed to developing countries. However, to make such transfers useful for irrigation decision making, initial investment in collecting and collating the kind of location-specific data required to adapt such knowledge to developing countries may be necessary.

If agricultural economists are to facilitate the more parsimonious water use that considerations of efficiency and the environment seem to require, then they need production function knowledge. Such knowledge is more abundant in the developed than the developing world. The brief review presented in this paper suggests that conventional production function knowledge based on experiments or on whole-farm data will not be as useful in the developing world as will deductive methods based on biophysical theories of growth. The issues here relate to model building, data collection, and validation. The FAO manual provides useful but rather crude knowledge. The planned provision of more biophysical data for representative locations and of simulation models for a number of major crops suggests a wider use of mechanistic crop growth models and perhaps physiological models. Whether the benefits of this will exceed the costs is an empirical question.

## Notes

[1]Shiraz University, University of New England, and University of New England, respectively.

[2]In fact, optimal cultural practice may vary as water application varies. Recognition of this possibility may call for a multi-dimensional response surface.

## References

Ayer, H.W., *et al.*, *Crop-Water Production Functions and Economic Implications for Washington*, Staff Report No. AGES–830314, Economic Research Service, US Department of Agriculture, Washington, D.C., USA, 1983.

Bala, B.K., *et al.*, "Simulation of Crop-Irrigation Systems," *Agricultural Systems*, Vol. 27, No. 1, 1988, pp. 51–65.

Boggess, W.G., *et al.*, "Evaluating Irrigation Strategies in Soybeans: A Simulation Approach—Irrigation Scheduling for Water and Energy Conservation in the 1980's," *Transactions of the ASAE*, Vol. 29, 1986, pp. 45–53.

Cuenca, R.H., "Transferable Simulation Model for Crop Soil Water Depletion," Ph.D. dissertation, University of California, Davis, Calif., USA, 1978, p. 310.

Dinar, A., Knapp, K.C., and Rhoades, J.D., "Production Function for Cotton with Dated Irrigation Quantities and Qualities," *Water Resources Research*, Vol. 22, No. 11, 1986, pp. 1519–1525.

Doorenbos, J., and Kassam, A.H., *Yield Response to Water*, Irrigation and Drainage Paper No. 33, Food and Agriculture Organization, Rome, Italy, 1979.

Hexem, R.W., *et al.*, "Application of a Two Variable Mitscherlich Function in the Analysis of Yield-Water Fertilizer Relationships in Corn," *Water Resources Research*, Vol. 12, No. 1, 1976, pp. 6–10.

Hexem, R.W., and Heady, E.O., *Water Production Functions in Irrigated Agriculture*, Iowa State University Press, Ames, Iowa, USA, 1978.

Hodges, T., *Yield Model Development: Second Generation Crop-Yield Model Review*, AGRISTARS, Columbia, Mo., USA, 1982.

Hoyt, G.H., *Crop-Water Production Functions and Economic Implications for the Texas High Plains Region*, NRE Staff Report, Economic Research Service, US Department of Agriculture, Washington, D.C., USA, 1982.

Jensen, M.E., "Water Consumption by Agricultural Plants," in Kozlowski, T.T. (Ed.), *Water Deficits and Plant Growth*, Vol. 2, Chap. 1, Academic Press, New York, N.Y., 1968, pp. 1–19.

Minhas, B.S., *et al.*, "Toward the Structure of a Production Function for Wheat Yields with Dated Inputs of Irrigation Water, *Water Resources Research*, Vol. 10, 1974, pp. 383–393.

Moore, C.V., "A General Analytical Framework for Estimating the Production Function for Crops Using Irrigation Water," *Journal of Farm Economics*, Vol. 43, 1961, pp. 876–888.

Palacios, E.V., "Response Functions of Crop Yields to Soil Moisture," *Water Resources Bulletin*, Vol. 17, 1981, pp. 699–703.

Pandey, S., "Economics of Water Harvesting and Supplementary Irrigation in the Semi-Arid Tropics of India: A Systems Approach," Ph.D. dissertation, University of New England, Armidale, NSW, Australia, 1986.

Ruttan, V.W., *The Economic Demand for Irrigated Acreage: New Methodology and Some Preliminary Projections*, Johns Hopkins University Press, Baltimore, Md., USA, 1965.

Rydzweski, J.R., and Nairizi, S., "Irrigation Planning Based on Water Deficits," *Water Resources Bulletin*, Vol. 15, 1979, pp. 316–325.

Soltani, G.R., "Economic Analysis of Water-Saving Techniques in Iran," *Water Resources Bulletin*, Vol. 13, 1977.

Stewart, J.I., "Prediction of Water Production Function and Associated Irrigation Programs to Minimize Crop Yield and Profit Loss Due to Limited Water," Ph.D. dissertation, University of California, Davis, Calif., USA, 1972.

Uehara, G., and Tsuji, G.Y., "Progress in Crop Modelling in the IBSNAT Project," in Muchow, R.C., and Bellamy, J.A. (Eds.), *Climatic Risk in Crop Production: Models and Management for the Semiarid Tropics and Subtropics*, CAB International, Wallingford, UK, 1991, pp. 143–156.

Vaux, H.J., Jr., and Pruitt, W.O., "Crop-Water Production Functions," in Hillel, D. (Ed.), *Advances in Irrigation*, Vol. 2, Academic Press, New York, N.Y., USA, 1983, pp. 61–97.

Yaron, D., and Bresler, E., "Economic Analysis of On-Farm Irrigation Using Response Functions of Crops," in Hillel, D. (Ed.), *Advances in Irrigation*, Vol. 2, Academic Press, New York, N.Y., USA, 1983, pp. 223–225.

---

## Discussion Opening—*Wen-yuan Huang (US Department of Agriculture)*

The Soltani and Musgrave paper addresses the problems associated with the selection of a model to transfer crop-water production knowledge from an advanced country to a developing country to improve irrigation efficiency (minimize excessive water application). The paper discusses various types of models and concludes with some general statements as to the relative applicability of the models.

To improve irrigation efficiency is a very site-specific task. Accordingly, a useful model would have to be site-specific. Many such models are reviewed in the paper. As such a model is transferred from one site to others, the transfer process can be time consuming and costly because it requires gathering a large amount of data and calibrating the model for the new site.

A selection of a specific model for the transfer would have to be based on the purpose of use and adoption cost. The purpose of use, for example, could be for pre-determining an irrigation schedule or for determining an irrigation schedule according to the daily field conditions. The adoption costs include costs associated with data collection, testing, and use of the model. Such costs can be reduced by matching the site-specific factors (soil, weather, and others) of the origin and target sites. This approach was used in a project at the University of Hawaii, which identified a soil family according to soil taxonomy. The project has shown that a direct transfer of technologies between sites of the same soil family can be achieved.

A successful transfer of irrigation technology has to be not only technically feasible but also economically viable and environmentally sound. The paper focuses mainly on the problems of transferring technical knowledge, which is an important first step towards the adoption of an irrigation technology.

*[Other discussion of this paper and the authors' reply appear on page 119.]*

# A Risk Analysis of Alternative Crop and Irrigation Strategies Using Biophysical Simulations

*Allen M. Featherstone, Adeyinka Osunsan, and Arlo W. Biere*[1]

**Abstract:** Combining risk programming with biophysical simulation offers potential benefits for helping farmers in developing countries choose cropping and irrigation strategies or for the study of farmer behaviour. Risk can have a significant impact on the way resources are allocated and should therefore be considered in empirical studies. This study uses risk programming and biophysical simulation models to find the expected utility-maximizing irrigation strategy and crop choice for southwestern Kansas farmers. Biophysical simulation models allow the researcher to obtain yield data for a longer time period than is typically available from agronomic studies, and to study risk on a very localized level. Direct expected utility maximization is used to determine the optimal strategies. Results from the study suggest that biophysical simulation models offer a promising avenue to further understanding of the impacts of risk on farm management decisions. Because biophysical simulation models are transferable to different regions of the world, biophysical simulation can be an attractive alternative to conducting risk research in developing countries.

## Introduction

Combining risk programming with biophysical simulation offers potential benefits for helping farmers in developing countries choose cropping and irrigation strategies. Risk can have a significant impact on the way resources are allocated and should therefore be considered in empirical studies. Risk arises due to uncertainty about output prices and yields because of biological lags in production, uncertain and uncontrollable weather conditions, and volatility in world grain markets. With the use of biophysical simulation models and risk programming, risk analysis can be done at substantially less cost. Research can be done on a site-specific basis without having to set up localized experiments. Cropping recommendations can then be tailored to a localized level. This paper uses risk programming and biophysical simulation models to find the expected utility-maximizing irrigation strategy and crop choice for southwestern Kansas farmers. Southwestern Kansas was used in this study so that comparisons could be made between actual experimental yields and yields generated by the biophysical simulation models.

While there has been extensive research on the allocation of irrigation water, research investigating water allocation under risk has been limited. Yaron and Dinar (1982) used a systems analysis approach to allocate water to cotton and fruit crops during peak irrigation seasons to maximize the farmer's income. Dudley, Howell, and Musgrave (1971) used an irrigation planning model and a simple crop growth model to choose acreage for irrigated crops. Yaron *et al.* (1973) examined wheat response to soil moisture and irrigation policy under conditions of unstable rainfall. Chanyalew, Featherstone, and Buller (1989) looked at the combination of irrigated maize, irrigated grain sorghum, and dryland sorghum under limited groundwater using a profit maximization model.

One of the few studies to incorporate risk into the analysis is that of Harris and Mapp (1986), who used stochastic dominance to compare water-conserving strategies for grain sorghum. Research examining irrigation strategies under uncertainty is limited due to the difficulty of finding adequate data on the risk variables. Few agronomic experiments are funded for more than five years. However, risk analysis using just five years of data would be considered highly suspect at the very least. Recently, biophysical simulation models have been refined to the point where they can produce fairly reliable yield estimates. Several studies have used crop growth models to evaluate production decisions (Mapp and Eidman, 1976; Boggess *et al.*, 1985; and Boggess and Ritchie, 1988).

# Theory

Typically, agricultural firms are assumed to follow the competitive economic model with determinant input prices and quantities and uncertain output prices and yields. Deterministic economic analysis assumes that a producer is indifferent to risk. Incorporation of risk into economic analysis considers the decision maker's perception and attitude towards risk. Economists generally assume that farmers make decisions consistent with the expected utility maximization hypothesis. That is, a farmer maximizes expected utility of profit ($EU(\pi)$), where $\pi$ is profit, $U$ is a nonlinear utility function, and $E$ is the expectation operator (Sandmo, 1971; and Iishii, 1977). The utility function is further assumed to be a concave, continuous and twice-differentiable function of profit, so that the first derivative with respect to profit is positive zero and the second negative. Because utility is ordinal, risk preferences are often modelled using the Pratt-Arrow risk aversion coefficient, which is the negative of the second derivative divided by the first derivative of the utility function with respect to profit. As the risk aversion coefficient increases, a farmer is more averse to risk.

In this study, the direct expected-utility maximization approach was used because crop yields are probably not normally distributed (Day, 1965; and Gallagher, 1987) and because possible diversification strategies were of particular interest and likely to be the strategy of choice by a risk-averse farmer. Other empirical approaches used to analyse risk have included mean-variance analysis, MOTAD, and stochastic dominance. These all have the expected-utility hypothesis as a base, and could also be used with biophysical simulation models, but the mean-variance and MOTAD analysis are based on normal distributions. The expected value of a negative exponential utility function of profit was assumed to be the decision maker's objective function. The expected utility maximization problem can be written as:

$$(1) \quad Max \ EU(\pi) = Max \ \sum_{i=1}^{N} p_i - e^{-\lambda \pi i}$$

subject to: $\displaystyle\sum_{j=1}^{T} x_j = 1$, (land constraint), and: $\displaystyle\sum_{j=1}^{T} r_{ij}x_j = \pi_i$, for all $i$

where $\pi_i$ is the net return for outcome $i$, $r_{ij}$ is total revenue minus variable cost for crop and irrigation strategy $j$ for year $i$, $\lambda$ is the Pratt-Arrow absolute risk aversion coefficient, the $x_j$s are the different crop and irrigation strategies, $N$ is the number of years, and $p_i$ is probability of outcome $i$ occurring. The land constraint was included so that all land was farmed.

# Biophysical Simulation Models

Biophysical simulation models are mathematical models based on the biological and physical processes of daily growth. Model inputs are soil type, date of planting, plant genotype, initial soil moisture, soil characteristics, daily temperature, daily rainfall, and daily solar radiation. Because of the detailed information needed on soil type, soil characteristics, plant genotype, and solar radiation, these models are very site specific. Changing the soil characteristics will change the distribution of crop yield. Thus, results can be tailored to a localized area.

Three crop growth models, the CERES maize growth model (Jones and Kiniry, 1986), the SORGF grain sorghum model (Arkin et al., 1976), and the PHOTO wheat growth model (Brakke and Kanemasu, 1979), were validated using field trials conducted at the Southwest Kansas Branch Experiment Station from 1974 to 1982 (Worman et al., 1988). The predictive accuracy of each of model was checked by comparing the experimental average, maximum, and minimum yields against simulated yields. The range of yields simulated with CERES maize and the range of yields harvested from the trials were quite close. PHOTO did not perform as well; the highest simulated yield was not as high as the highest actual wheat yield.

However, the mean simulated yield was almost the same as the mean actual yield. SORGF also had some difficulty in simulating extreme yield. Furthermore, SORGF produced yields that exceeded actual yields 66 percent of the time and the mean simulated yield was 10 percent higher than the mean actual yield. The standard deviation of the simulated yields and that of the actual experimental yields were quite close for each of the models.

Also, to check the accuracy of the models, experimental yields were regressed against the simulated yield. When experimental yields are regressed against simulated yields, the slope coefficients were 0.95 for maize, 1.009 for wheat, and 0.902 for sorghum when the intercept was constrained to zero. Only the slope coefficient for sorghum was statistically different from one. Overall, all three models reasonably simulated crop yields. Other factors such as disease, insects, wind, and other stress factors not accounted for in the biophysical models could account for unexplained variation.

The validated models were used with 28 years of meteorological data to simulate yields for 28 years for 29 alternative non-irrigated and irrigated strategies defined in Table 1. The strategies included continuous non-irrigated cropping, fallow, and various levels of irrigation, ranging from 4 to 16 inches of water applied at various points in the growing season for each crop. The same soil type was assumed for all cropping strategies simulated (Worman et al., 1988). The simulated yields were adjusted using the estimated slope coefficients discussed above.

Wheat is expected to be the most profitable crop, with returns over variable cost per acre ranging from $51.89 to $98.51 depending on the irrigation strategy (Table 2). WHT7 is the most profitable wheat production strategy. Sorghum is the second most profitable crop, with returns ranging form $10.57 to $57.48. SGM6 is the most profitable sorghum production strategy. Maize is the least profitable crop, with returns over variable costs ranging from –$31.10 to $44.73 per acre. CRN8 is the most profitable maize production strategy.

The last column of Table 2 contains a measure of skewness for the crop returns and yields. A normal distribution has a skewness measure equal to zero. Based on Table 2, several crop production strategies have skewed yield distributions.

# Results

The returns over variable costs for each of the 28 years were input into the risk programming model. Five different Pratt-Arrow risk coefficients were used, ranging from 0.01 to 1.0. The mean, standard deviation, and certainty equivalent value of the portfolio are listed in the top three rows of Table 3.

The optimal crop production strategy for a farmer with a risk aversion coefficient of 0.01 (nearly risk neutral) would be to plant all acreage in wheat. WHT7 is the production strategy that would be used. A slightly more risk averse farmer (Pratt-Arrow = 0.05) would plant 31.9 percent of cropland in wheat (WHT7) and 68.1 percent in sorghum (SGM6). The expected returns per acre would drop by almost $28 dollars per acre. The standard deviation of this portfolio is $19.20 per acre, a reduction from $45.59 for the most profitable. The certainty equivalent value of the portfolio is nearly $60. If the farmer were yet more risk averse (Pratt-Arrow = 0.1), 0.5 percent of land would be in WHT3, 12.7 percent in WHT7, and 86.8 percent in SGM6. The most risk averse farmer (Pratt-Arrow = 1.0) would plant 2.0 percent of land in WHT7 and 98.0 percent in SGM6. The expected return is $58.27 per acre with a standard deviation of $14.67 per acre. The standard deviation is slightly larger than for the farmer with a Pratt-Arrow risk aversion coefficient of 0.5. This is due to the positive skewness measure on SGM6. Using mean-variance analysis here with skewed distributions would have produced a result inconsistent with expected utility maximization.

The shadow prices of the crops not in the optimal solution are listed in Table 3. The shadow prices are in units of certainty equivalent of income per acre. They are derived using the method found in Preckel, Featherstone, and Baker (1987). The shadow prices provide useful information for farmers considering irrigation strategies for the individual crops. As

Table 1—Alternative Cropping Strategies Considered

| Variable | Maize Cropping Strategies |
|---|---|
| *CRN1* | 4" irrigation before planting |
| *CRN2* | 8" irrigation before planting |
| *CRN3* | 4" irrigation in mid-July |
| *CRN4* | 4" irrigation before planting, 4" irrigation before tasselling |
| *CRN5* | 8" irrigation before planting, 4" irrigation before tasselling |
| *CRN6* | 4" irrigation before planting, 4" irrigation before tasselling, 4" irrigation at beginning of ear growth |
| *CRN7* | 8" irrigation before planting, 4" irrigation before tasselling, 4" irrigation at beginning of ear growth |
| *CRN8* | 4" irrigation before planting, 4" irrigation before tasselling, 4" irrigation between tasselling ear growth, 4" at ear growth |
| *CRN9* | Maize fallow rotation |
| *CRN10* | Dryland production |
| | Grain Sorghum Cropping Strategies |
| *SGM1* | 4" irrigation before planting |
| *SGM2* | 8" irrigation before planting |
| *SGM3* | 4" irrigation in mid-July |
| *SGM4* | 4" irrigation before planting, 4" irrigation at 9-leaf stage |
| *SGM5* | 8" irrigation before planting, 4" irrigation at 9-leaf stage |
| *SGM6* | 4" irrigation before planting, 4" irrigation at 9-leaf stage, 4" irrigation at boot stage |
| *SGM7* | 8" irrigation before planting, 4" irrigation at 9-leaf stage, 4" at boot stage |
| *SGM8* | 4" irrigation before planting, 4" irrigation at 9-leaf stage, 4" irrigation at boot stage, 4" irrigation at flowering |
| *SGM9* | Sorghum fallow rotation |
| *SGM10* | Dryland production |
| | Wheat Cropping Strategies |
| *WHT1* | 4" irrigation before planting |
| *WHT2* | 8" irrigation before planting |
| *WHT3* | 4" irrigation before planting, 4" irrigation at boot stage |
| *WHT4* | 8" irrigation before planting, 4" irrigation at boot stage |
| *WHT5* | 4" irrigation before planting, 4" irrigation at boot stage, 4" irrigation at soft dough stage |
| *WHT6* | 8" irrigation before planting, 4" irrigation at boot stage, 4" irrigation at soft dough stage |
| *WHT7* | 4" irrigation before planting, 4" irrigation at jointing, 4" irrigation at boot stage, 4" irrigation at soft dough stage |
| *WHT8* | Wheat fallow rotation |
| *WHT9* | Dryland production |

Table 2—Distribution of Yields and Returns for the Alternative Crop
and Irrigation Strategies

| Crop | Mean Yield | Std. Dev. Yield | Mean Return[1] | Std. Dev. Return | Skewness |
|------|------------|-----------------|----------------|------------------|----------|
|      | bu/acre | | $/acre | | |
| CRN1 | 49.7 | 40.6 | −24.89 | 81.19 | 1.47 |
| CRN2 | 59.7 | 40.5 | −15.62 | 80.94 | 1.07 |
| CRN3 | 69.4 | 42.6 | 12.11 | 85.20 | 0.54 |
| CRN4 | 71.0 | 39.3 | −2.00 | 78.66 | 0.58 |
| CRN5 | 78.0 | 38.2 | −15.20 | 76.30 | 0.43 |
| CRN6 | 106.9 | 24.8 | 42.59 | 49.69 | −0.22 |
| CRN7 | 110.1 | 22.5 | 34.30 | 45.07 | 0.11 |
| CRN8 | 115.3 | 21.7 | 44.73 | 43.36 | 0.21 |
| CRN9 | 56.6 | 40.7 | 1.54 | 81.48 | 1.17 |
| CRN10 | 36.4 | 40.0 | −31.10 | 80.09 | 1.97 |
| | | | | | |
| SGM1 | 69.2 | 32.8 | 22.58 | 57.32 | −0.31 |
| SGM2 | 85.8 | 23.7 | 35.74 | 41.40 | −0.85 |
| SGM3 | 85.6 | 29.0 | 46.13 | 50.74 | −1.39 |
| SGM4 | 90.6 | 21.2 | 41.54 | 37.06 | −1.22 |
| SGM5 | 97.3 | 16.1 | 40.53 | 28.14 | −0.81 |
| SGM6 | 107.7 | 8.6 | 57.48 | 15.07 | 1.09 |
| SGM7 | 108.1 | 8.4 | 47.33 | 14.71 | 1.10 |
| SGM8 | 108.0 | 8.6 | 47.24 | 14.97 | 0.99 |
| SGM9 | 74.9 | 28.1 | 41.81 | 49.25 | −0.73 |
| SGM10 | 46.1 | 31.1 | 10.57 | 54.41 | 0.50 |
| | | | | | |
| WHT1 | 31.8 | 11.6 | 63.79 | 45.89 | 1.13 |
| WHT2 | 33.6 | 12.2 | 59.77 | 48.24 | 0.89 |
| WHT3 | 45.1 | 11.8 | 92.56 | 46.69 | −0.04 |
| WHT4 | 46.7 | 12.6 | 88.10 | 49.61 | −0.05 |
| WHT5 | 45.5 | 11.7 | 83.23 | 46.32 | −0.07 |
| WHT6 | 47.0 | 12.5 | 78.72 | 49.19 | −0.09 |
| WHT7 | 53.3 | 12.3 | 98.51 | 48.46 | −0.49 |
| WHT8 | 30.6 | 11.1 | 69.89 | 43.66 | 0.67 |
| WHT9 | 24.4 | 8.3 | 51.89 | 32.75 | 0.95 |

[1]Returns above variable cost.

Table 3—Risk Programming Results for Various Pratt-Arrow Risk Coefficients

| Variable | Pratt-Arrow Absolute Risk Aversion Coefficients | | | | |
|---|---|---|---|---|---|
| | 0.01[a] | 0.05[b] | 0.1[c] | 0.5[d] | 1.0[e] |
| Mean | 98.51 | 70.57 | 62.88 | 58.52 | 58.27 |
| Standard deviation | 47.59 | 19.20 | 14.95 | 14.63 | 14.67 |
| Certainty equivalent | 86.44 | 59.91 | 53.07 | 43.15 | 41.04 |
| | Shadow Prices for Activities Not in the Optimal Solutions[f] | | | | |
| CRN1 | −114.46 | −106.66 | −102.23 | −110.83 | −122.54 |
| CRN2 | −105.03 | −98.61 | −95.21 | −106.28 | −117.53 |
| CRN3 | −82.10 | −80.23 | −76.51 | −83.96 | −93.37 |
| CRN4 | −91.92 | −90.19 | −88.59 | −103.41 | −114.11 |
| CRN5 | −102.70 | −99.01 | −98.33 | −119.05 | −130.35 |
| CRN6 | −42.68 | −39.67 | −41.18 | −64.00 | −73.05 |
| CRN7 | −47.32 | −39.73 | −40.36 | −58.25 | −66.52 |
| CRN8 | −37.53 | −28.40 | −28.29 | −41.25 | −48.35 |
| CRN9 | −88.53 | −82.65 | −78.87 | −87.37 | −98.44 |
| CRN10 | −118.09 | −105.86 | −100.66 | −109.69 | −121.34 |
| SGM1 | −62.67 | −57.50 | −59.59 | −83.59 | −90.75 |
| SGM2 | −45.42 | −40.15 | −45.22 | −69.14 | −74.77 |
| SGM3 | −37.96 | −40.22 | −47.82 | −82.74 | −88.54 |
| SGM4 | −41.04 | −36.86 | −41.11 | −59.76 | −63.63 |
| SGM5 | −31.63 | −13.67 | −15.14 | −24.52 | −27.20 |
| SGM6 | −17.70 | − | − | − | − |
| SGM7 | −26.68 | −6.37 | −5.60 | −3.27 | −2.98 |
| SGM8 | −26.86 | −6.72 | −6.16 | −4.72 | −4.64 |
| SGM9 | −44.74 | −42.67 | −45.62 | −65.52 | −70.75 |
| SGM10 | −73.69 | −64.05 | −64.03 | −81.00 | −88.40 |
| WHT1 | −24.23 | −11.38 | −10.77 | −18.11 | −19.82 |
| WHT2 | −30.08 | −18.43 | −17.35 | −20.55 | −21.52 |
| WHT3 | −2.91 | −0.18 | − | −1.02 | −1.78 |
| WHT4 | −9.01 | −7.05 | −6.23 | −4.18 | −4.47 |
| WHT5 | −12.19 | −9.52 | −9.35 | −10.32 | −11.01 |
| WHT6 | −18.32 | −16.47 | −15.72 | −13.84 | −14.10 |
| WHT7 | − | − | − | − | − |
| WHT8 | −18.09 | −3.80 | −1.12 | −0.74 | −1.13 |
| WHT9 | −31.20 | −13.78 | −12.52 | −20.78 | −23.23 |

[a]Optimal portfolio: WHT7 = 100 percent.
[b]Optimal portfolio: WHT7 = 31.9 percent, SGM6 = 68.1 percent.
[c]Optimal portfolio: WHT7 = 12.7 percent, WHT3 = 0.5 percent, SGM6 = 86.8 percent.
[d]Optimal portfolio: WHT7 = 2.5 percent, SGM6 = 97.5 percent.
[e]Optimal portfolio: WHT7 = 2.0 percent, SGM6 = 98.0 percent.
[f]In units of certainty equivalent per acre.

a farmer gets more risk averse, the wheat fallow strategy (WHT8) becomes more attractive. Thus, under some price combinations, it is likely that wheat fallow would be in the optimal solution. SGM7 continues to be more attractive for more risk-averse strategies because the penalty cost continues to decrease as risk aversion increases. Full irrigation of sorghum is a risk-reducing strategy for sorghum, whereas no irrigation of wheat is a risk-reducing strategy for wheat. CRN8 in all cases has the smallest penalty cost for maize for inclusion in the optimal portfolio. Given different price expectations where maize is relatively more profitable, CRN8 would be the maize strategy most likely to be used.

The results illustrate that crop simulation models can be useful in generating distributions of yields over a longer time frame for risk analysis. Checking distributions generated by biophysical simulation models with actual experimental data suggests that yield distributions generated with biophysical simulation models do not differ greatly from experimental data. The results also show that the yields can then be input into risk programming models to take economic behaviour into account.

In addition, calculating certainty-equivalent values for those cropping activities that do not enter the optimal portfolio can further help producers understand the consequences of different cropping and irrigation strategies. These shadow prices can then be useful for altering recommendations as market conditions change.

## Conclusions

This study used biophysical simulation models and risk programming to investigate alternative crop production and irrigation strategies. The biophysical simulation models were used to generate crop yields based upon historical weather patterns. The simulation models were used because time-series data are often not available for a long enough period to use risk analysis. Comparing simulated yields to actual experiment station yields suggests that the simulation models perform reasonably well.

Risk programming was used to investigate the crops and irrigation strategies that maximize expected utility of income for a risk-averse farmer. Direct expected utility maximization was used because yield distributions and incomes were skewed. Direct expected utility maximization also allows for a detailed interpretation of shadow prices on those crops and irrigation strategies that do not enter the optimal portfolio.

The results from the paper suggest that biophysical simulation models may be useful for risk analysis. These models are able to use historical weather data to project crop yields in the past. Because of the nature of agronomic experiments, insufficient data are usually available for risk analysis. Biophysical simulation models offer a promising avenue to further understanding of the impacts of risk on farm management decisions.

## Note

[1]Kansas State University.

## References

Arkin, G.F., Vanderlip, R.L., and Ritchie, J.T., "A Dynamic Grain Sorghum Growth Model," *Transactions of the ASAE*, Vol. 19, 1976, pp. 622–626.

Boggess, W.G., Cardelli, D.J., and Barfield, C.S., "A Bioeconomic Simulation Approach to Multi-Species Insect Management," *Southern Journal of Agricultural Economics*, Vol. 17, No. 2, 1985, pp. 43–56.

Boggess, W.G., and Ritchie, J.T., "Economic and Risk Analysis of Irrigation Decisions in Humid Regions," *Journal of Production Agriculture*, Vol. 1, No. 2, 1988, pp. 116–122.

Brakke, T.W., and Kanemasu, E.T., "Estimated Winter Wheat Yields from LANDSAT MSS Using Spectral Techniques," Proceedings of the Thirteenth International Symposium on Remote Sensing of the Environment, Ann Arbor, Mich., USA, 1979, pp. 629–641.

Chanyalew, D., Featherstone, A.M., and Buller, O.H., "Groundwater Allocation in Irrigated Crop Production," *Journal of Production Agriculture*, Vol. 2, No. 1, 1989, pp. 37–42.

Day, R.H., "Probability Distribution of Field Crop Yields," *Journal of Farm Economics*, Vol. 47, 1965, pp. 713–741.

Dudley, N.J., Howell, D.T., and Musgrave, W.F., "Irrigation Planning 2: Choosing Optimal Acreages within a Season," *Water Resources Research*, Vol. 7, 1971, pp. 1051–1063.

Gallagher, P., "U.S. Soybean Yields: Estimating and Forecasting with Nonsymmetric Disturbances," *American Journal of Agricultural Economics*, Vol. 69, 1987, pp. 796–803.

Harris, T.R., and Mapp, H.P., "A Stochastic Dominance Comparison of Water-Conserving Irrigation Strategies," *American Journal of Agricultural Economics*, Vol. 68, No. 2, 1986, pp. 298–305.

Iishii, Y., "On the Theory of the Competitive Firm under Price Uncertainty: Note," *American Economic Review*, Vol. 67, 1977, pp. 768–769.

Jones, C.A., and Kiniry, J.R., *CERES–Maize: A Simulation Model of Maize Growth and Development*, Texas A&M University Press, College Station, Tex., USA, 1986.

Mapp, H.P., Jr., and Eidman, V.R., "A Bioeconomic Simulation Analysis of Regulating Groundwater Irrigation," *American Journal of Agricultural Economics*, Vol. 58, 1976, pp. 391–402.

Preckel, P.V., Featherstone, A.M., and Baker, T.G., "Interpreting Dual Variables for Optimization with Nonmonetary Objectives," *American Journal of Agricultural Economics*, Vol. 69, No. 4, 1987, pp. 849–851.

Sandmo, A., "On the Theory of the Competitive Firm under Price Uncertainty," *American Economic Review*, Vol. 61, 1971, pp. 65–73.

Worman, F.O., Biere, A.W., Hooker, M.L., Vanderlip, R.L., and Kanemasu, E.T., "Simulation Analysis using Physiological Crop-Response Models: Alternative Cropping Strategies for Southwest Kansas," Agricultural Experiment Station Bulletin No. 653, Kansas State University, Manhattan, Kans., USA, 1988.

Yaron, D., and Dinar, A., "Optimal Allocation of Irrigation Water during Peak Seasons," *American Journal of Agricultural Economics*, Vol. 64, 1982, pp. 681–689.

Yaron, D., Strateener, G., Shimshi, D., and Weisbrod, M., "Wheat Response to Soil Moisture and the Optimal Irrigation Policy under Conditions of Unstable Rainfall," *Water Resources Research*, Vol. 9, 1973, pp. 1145–1154.

---

## Discussion Opening—*Chung L. Huang* (University of Georgia)

Various risk models and mathematical programming techniques have been developed to address risk and uncertainty in decision making involving crop production, pest management strategies, farm programmes, and rural bank portfolio behaviour. A common feature of these economic analyses is the incorporation of the decision maker's perception and attitude towards assuming risk into the framework of expected utility maximization hypotheses for modelling expectations of possible outcomes and their probabilities. Methodologically, a wide range of approaches has been used to solve decision-making problems involving risk and uncertainty of future events, varying from using conservative estimates for the uncertain elements to methods that explicitly incorporate probability density functions for the uncertain parameters. More recently, plant growth simulation models that consider the interactions of stochastic weather conditions, soil type, plant growth, moisture stress and irrigation decisions within an integrated bioeconomic framework have become important research tools. The present paper contributes appropriately to this class of growing risk analysis literature.

While the paper can be considered as generally outstanding, it lacks explanation of the theoretical model and methodology used. The paper only indicates that the decision maker's objective function was maximized assuming a negative exponential utility function of profit. The forms of the utility function represent different attitudes towards risk. The specification of the model implies that the decision maker is risk averse and has an attitude towards risk unrelated to wealth. It is not clear why this is the case. Although the probabilities of possible outcomes for different production strategies were explicitly incorporated in the objective function, there is no explanation of how the probability function was derived. Most importantly, the model assumes a profit-maximization framework. We all know that prices influence farmers' decisions, especially production adjustments. In the short run, farmers adjust by changing current inputs; in the longer term, solutions come by changing technologies and scale dimensions. It is not clear how price variations among crops and over time were considered in the formulation.

The authors rightly point out that biophysical simulation models are plant genotype and location or site specific. Any changes in input data such as soil characteristics result in different distributions of crop yields. Thus, it is suggested that the application of biophysical simulation models offers the advantage of transferability to different regions of the world and the capability of generating distributions of yields over a longer time frame for risk analysis. Given its site-specific nature, it seems that some sensitivity analyses reflecting differences in soil and weather should be performed to measure the stability of model results. In fact, the requirement of very detailed and specific input information would be an obstacle that limits the transferability. In the absence of experimental data, I wonder how the researchers would validate and calibrate the performance of the simulation models. Furthermore, if the growth simulation models are sensitive to plant genotype, then changing cultivars would appear to render the generation of yields distribution over a long period of time unnecessary or invalid.

More specifically, the results show no differences between risk aversion coefficients of 0.5 and 1.0, and they should be reported as such. Apparently, many farm plans are similarly organized. Hence, changes in risk aversion coefficients between 0.5 and 1.0 do not reduce net return or entail changes in the optimal portfolio.

The authors are to be commended for the enormous efforts devoted to bridging the gap between economic analysis and biophysical simulation techniques in an application to determine optimal crop choice and irrigation strategies.

*[Other discussion of this paper and the authors' reply appear on page 119.]*

# A General Approach for Evaluating the Economic Viability and Sustainability of Tropical Cropping Systems

*Simeon K. Ehui and Dunstan S.C. Spencer*[1]

**Abstract:** This paper presents a methodology for measuring economic viability and agricultural sustainability for new technology evaluation. The approach is based on the concept of interspatial and intertemporal total factor productivity, paying particular attention to the valuation of natural resource stocks and flows. Using a set of data available at the International Institute of Tropical Agriculture, the model is demonstrated by computing the intertemporal and interspatial total factor productivity indices for four cropping systems in southwestern Nigeria. Results show that the sustainability and economic viability measures are sensitive to changes in the stock of nutrients as well as to changes in material input uses and outputs. When common property resource flows are important, the measures provide markedly different results from conventional total factor productivity approaches.

## Introduction

Sub-Saharan Africa currently faces serious food problems, manifested in declining per capita food production, growing food imports and accelerating ecological degradation (Ehui and Hertel, 1989; and CGIAR, 1989). New technologies must therefore be developed that not only enhance food production but also maintain ecological stability and preserve the natural resource base; i.e., technologies that are both economically viable and sustainable (BIFAD, 1988). However, there is little guidance in the literature as to what practical methods are to be used for measuring the sustainability and economic viability of tropical cropping systems (CGIAR, 1989; and Lynam and Herdt, 1989). This paper uses recent advances in productivity measurement and economic index numbers to develop a model for measuring economic viability and agricultural sustainability for new technology evaluation.

The next section presents the conceptual framework. First, a generalized model for the measurement of total factor productivity (TFP) is developed. It is followed by the specification of intertemporal and interspatial TFP indices, which are used to measure sustainability and economic viability. In the third section, an empirical example is considered, and, in the fourth section, the paper closes with a summary and some concluding qualifications and comments.

## Conceptual Framework

### A Generalized Model for Measurement of TFP

The conventional approach to growth accounting uses TFP indexes to measure the residual growth in outputs not accounted for by the growth in factor inputs. Agriculture, however, is a sector that uses common-property natural resources (e.g., air, water, soil nutrients, etc.). The stock of these resources affects the production environment but is in many cases beyond the control of the farmer. For example, soil nutrients are removed by crops, erosion or leaching beyond the crop root zone, or other processes such as volatilization of nitrogen. Agricultural production can also contribute to the stock of some of the nutrients, particularly of nitrogen by leguminous plants. When the stock of resources is reduced, the farmer faces an implicit cost in terms of productivity foregone. Conversely, when the stock of resources is increased during the production process (e.g., via nitrogen fixation), the farmer derives an implicit benefit from the system. If these implicit costs and benefits are not accounted for when TFP is measured, results will be biased (Squires and Herrick, 1990). One way to account for this bias is to treat the resource stock as a technological constraint. Accounting for natural resource stocks and flows in sustainability measurement is particularly important since a desirable component of sustainable soil and crop management systems in

111

the tropics is a mechanism that can replenish soil nutrients removed by crops, erosion, and leaching.

Assuming that current prices are known, the maximization problem when changes in common property resources stocks are positive is stated as:

(1) $\quad Max \; \pi_p = P_{yt}Y_t + P_{zt}Z_t - G(Y_t, Z_t, W_t, B_t, t)$

where $\pi_t$ is a measure of aggregate profit in period $t$, including all benefits and costs of resource exploitation, and $B_t$ is a technology shift variable representing the level of resource abundance in period $t$. Equation (1) represents the case of open access in which $B_t$ is not a choice variable. The resource stock is beyond the control of farmers and they thus ignore its opportunity cost. $Z_t$ is an externality denoting the net resource flow (i.e., $B_{t+1}-B_t$) in period $t$. When changes in resource abundance levels are positive, there is a positive externality, and the resulting net resource flow, $Z_t$, is treated as an output, thus contributing positively to the aggregate profit. $Y_t$ is an index of crop outputs; $P_{yt}$ and $P_{zt}$ are the product and resource flow prices; $G()$ is the variable cost function for the optimal combination of variable of variable inputs, where $\delta G()/\delta B < 0$ and $\delta G()/\delta Z > 0$. $W_t$ is a vector of variable input prices; $t$ is the time trend representing the state of technical knowledge.

When the production process is depleting the resource at a rate faster than that required for sustainability, net changes in resource abundance levels are negative (i.e., $B_{t+1}-B_t=-Z_t$). We have thus a negative externality, and $Z_t$ is treated as an input, hence contributing negatively to the aggregate profit. This requires modification of the objective function (1) by replacing the sign before $P_zZ_t$ with a $(-)$ sign and, in this case, $\delta G()/\delta Z < 0$.

Using the first order conditions of (1), development of the continuous time Divisia index using the growth accounting approach gives:

(2) $\quad - \dfrac{\delta Ln\, C}{\delta t} = \dfrac{P_yY}{C}\dot{Y} + \left( \dfrac{P_zZ}{C}\dot{Z} - \Sigma_j \dfrac{W_jX_j}{C} \right)\dot{X}_j - \dot{B}$

where $C = \Sigma_j W_jX_j = P_yY + P_zZ =$ total revenue, assuming constant returns to scale. Dots above variables imply the logarithm derivative of the associated variable with time.

When changes in the resource stock are negative, the productivity index becomes:

(3) $\quad - \dfrac{\delta Ln\, C}{\delta t} = \dfrac{P_yY}{C}\dot{Y} - \left( \dfrac{P_zZ}{C}\dot{Z} - \Sigma_j \dfrac{W_jX_j}{C} \right)\dot{X}_j - \dot{B}$

where $C = \Sigma_j W_jX_j + P_zZ = P_yY$, assuming constant returns to scale.

Equations (2) and (3) indicate that TFP is measured as the residual after the growth rate of output has been allocated among changes in inputs and resource abundance and flows. It is clear from (2) and (3) that productivity measures are biased unless variations in the resource stocks and flows are accounted for.

## Intertemporal and Interspatial TFP Measures

Let us assume that the agricultural production process of cropping system $i$ in period $t$ can be represented by the dual variable cost function:

(4) $\quad G_{it} = G(Y_{it}, Z_{it}, W_{it}, B_{it}, T_t, D_i)$

where $G_{it}$ is the cost of production, $Y_{it}$ is crop output, $Z_{it}$ is the change in resource stocks, $W_{it}$ is a vector of input prices, $B_{it}$ is the resource stock, and $T_t$ and $D_i$ denote the intertemporal and interspatial efficiency difference indicators. Derivation of the intertemporal and interspatial TFP indices depends critically on the proper specification of the total cost function

$C_{it}$, which in turn depends on the nature of $Z_{it}$ (i.e., whether the change in the resource stock is positive or negative). We therefore consider two cases:

**Case 1: Net positive change in the resource stock.** Assuming constant returns to scale and competitive factor markets, application of Diewert's (1976) quadratic lemma to a logarithmic approximation of (4) gives:

$$
(5) \quad
\begin{aligned}
Ln\ C = \ & \frac{(R_{yis}+R_{yot})(LnY_{is}-LnY_{ot})}{2} + \frac{(R_{zis}+R_{zot})(LnZ_{is}-LnZ_{ot})}{2} \\
& + \frac{\Sigma_k(S_{kis}+S_{kot})(LnW_{kis}-LnW_{kot})}{2} - (LnB_{is}-LnB_{ot}) + \theta_{io} + \eta_{st}
\end{aligned}
$$

where $i$ and $o$ represent two distinct cropping systems and $s$ and $t$ represent two distinct time periods. $S_{kis}$ and $S_{kot}$ are the $k$th input factor cost shares, $R_{yis}$, $R_{yis}$, and $R_{yot}$ are the revenue shares for product $Y$, and $R_{zis}$ and $R_{zot}$ are (implicit) revenue shares for resource flow $Z$. $\theta_{io}$ and $\eta_{st}$ denote the interspatial and intertemporal effects and are defined as:

$$
(6) \quad \theta_{io} = \frac{\left. \dfrac{\delta Ln\ G}{\delta D} \right|_{D=D_i} + \left. \dfrac{\delta Ln\ G}{\delta D} \right|_{D=D_o}}{2} \ (D_i - D_o)
$$

$$
(7) \quad \eta_{st} = \frac{\left. \dfrac{\delta Ln\ G}{\delta T} \right|_{T=T_s} + \left. \dfrac{\delta Ln\ G}{\delta T} \right|_{T=T_t}}{2} \ (T_s - T_t)
$$

Equation (5) states that the cost difference across cropping systems and time periods can be broken into six terms: an output effect, a resource flow effect, an input price effect, a resource stock effect, an interspatial effect, and an intertemporal effect.

Following Denny and Fuss (1983), if we want to measure the intertemporal TFP (thus sustainability) of a particular technology, we set $D_i = D_o = 0$. Solving for $\eta_{st}$ in (5) yields the dual measure of intertemporal productivity for periods $s$ and $t$:

$$
(8) \quad
\begin{aligned}
\eta_{st} = \ & (Ln\ G_s - Ln\ G_t) - \frac{(R_{ys}+R_{yt})(Ln\ Y_s - Ln\ Y_t)}{2} - \frac{(R_{zs}+R_{zt})(Ln\ Z_s - Ln\ Z_t)}{2} \\
& - \frac{R_k(S_{ks}+S_{kt})(Ln\ W_{ks} - Ln\ W_{kt})}{2} + (Ln\ B_s - Ln\ B_t)
\end{aligned}
$$

Similarly, the dual measure of interspatial productivity between system $i$ and reference system $o$ at a particular point in time ($T_s = T_t = 0$) is:

$$
(9) \quad
\begin{aligned}
\theta_{io} = \ & (Ln\ G_i - Ln\ G_o) - \frac{(R_{yi}+R_{yo})(Ln\ Y_i - Ln\ Y_o)}{2} - \frac{(R_{zi}+R_{zo})(Ln\ Z_i - Ln\ Z_o)}{2} \\
& - \frac{R_k(S_{ks}+S_{ko})(Ln\ W_{ki} - Ln\ W_{ko})}{2} + (Ln\ B_i - Ln\ B_o)
\end{aligned}
$$

If we now turn to the primal space, the Tornqvist approximation to the logarithm change of the cost equation, $G = \Sigma_i W_i X_i$, with respect to time, yields (for periods $s$ and $t$ and systems $i$ and $o$):

$$(10) \quad \begin{aligned} Ln\ G &= (Ln\ G_{is} - Ln\ G_{ot}) \\ &= \frac{R_k(S_{is} + S_{kot})(Ln\ X_{kis} - Ln\ X_{kot})}{2} + \frac{\Sigma_k(S_{kis} + S_{kot})(Ln\ W_{kis} - Ln\ W_{kot})}{2} \end{aligned}$$

Equating (5) and (10) and solving for $(-\eta_{st})$ and $(\theta_{io})$ gives measures of intertemporal and interspatial productivity in the primal space (Ohta, 1974):

$$(11) \quad \tau_{st} = -\eta_{st} = \frac{(R_{sy} + R_{yt})(Ln\ Y_s - Ln\ Y_t)}{2} + \frac{(R_{zs} + R_{zt})(Ln\ Z_s - Ln\ Z_t)}{2} - \frac{\Sigma_k(S_{ks} + S_{kt})(Ln\ X_{ks} - Ln\ X_{kt})}{2} - (Ln\ B_s - Ln\ B_t)$$

$$(12) \quad \rho_{io} = -\theta_{io} = \frac{(R_{yi} + R_{yo})(Ln\ Y_i - Ln\ Y_o)}{2} + \frac{(R_{zi} + R_{zo})(Ln\ Z_i - Ln\ Z_o)}{2} - \frac{\Sigma_k(S_{ki} + S_{ko})(Ln\ X_{ki} - Ln\ X_{ko})}{2} - (Ln\ B_i - Ln\ B_o)$$

**Case 2: Net negative change in the resource stock.** Following the same procedure as in case 1, intertemporal and interspatial productivity measures in the primal space are, respectively, given by:

$$(13) \quad \tau_{st'} = (Ln\ Y_s - Ln\ Y_t) - \frac{(S_{zs} + S_{zt})(Ln\ Z_s - Ln\ Z_t)}{2} - \frac{\Sigma_k(S_{ks} + S_{kt})(Ln\ X_{ks} - Ln\ X_{kt})}{2} - (Ln\ B_s - Ln\ B_t)$$

$$(14) \quad \rho_{io'} = (Ln\ Y_i - Ln\ Y_o) - \frac{(S_{zi} + S_{zo})(Ln\ Z_i - Ln\ Z_o)}{2} - \frac{\Sigma_k(S_{ki} + S_{ko})(Ln\ X_{ki} - Ln\ X_{ko})}{2} - (Ln\ B_i - Ln\ B_o)$$

where $S_{zs}$ and $S_{zt}$ in Equation (13) and $S_{zi}$ and $S_{zo}$ in Equation (14) denote the (implicit) cost shares for depleted resource $Z$.

## An Empirical Example

This section demonstrates how the intertemporal and interspatial total factor productivity measures developed in Equations (11)–(14) can be used to measure the sustainability and economic viability of tropical cropping systems. A set of data is used that was generated during a four-year study by the United Nations University and the International Institute of Tropical Agriculture on the effects of deforestation and land use on soil, hydrology, microclimate, and productivity in the humid coastal belt of Nigeria (Lal and Ghuman, 1989). Four cropping systems, denoted A, B, C, and D, are evaluated for 1986 and 1988, two years for which a complete and balanced data set is available. In System A, land was cleared manually and cropped by a local farmer. Yams, melons, and plantains were grown in 1986, and plantains, melons, and cassava in 1988.

In all other systems, the land was cleared by a tractor equipped with a shear blade, and cropped by the researchers. In System B, cassava, maize, and cowpeas were planted in 1986, and cassava only in 1988. In System C, maize and cassava were planted in 1986, and rice in

1988. All crops in System C were grown in alleys formed by hedgerows of nitrogen-fixing trees or shrubs. In this system, known as alley cropping, the hedgerows were pruned periodically during the cropping season to prevent shading and reduce competition with food crops (Kang et al, 1989). In System D, plantains were grown in both years. No fertilizer was used in any of the cropping systems.

Since the cropping systems have multiple crop outputs, an implicit output index is calculated by dividing the total value of all output by a price index obtained by weighting the individual output prices by the revenue share of each crop. A corresponding implicit input quantity index is computed as the ratio of total expenditures on inputs to the weighted material input price. The latter is measured by an index of all material input prices weighted by the cost share of each input. A quantity index for implements used is computed as the ratio of total annual expenditure on capital input and the implicit capital service price. To create an aggregate capital service price, the price of each category of implement is share weighted in the same manner as the aggregate material price index.

To construct the division index for the soil nutrient stock, the total quantities of main soil nutrients—nitrogen, phosphorus, and potassium (in metric tons per hectare)—available in the top soil (0–10 cm) are share weighted. In determining the cost share for the resource stock, the opportunity cost of each soil nutrient was approximated with its replacement cost; i.e., market price of chemical fertilizer. Resource flows are derived as the difference between nutrient abundance levels for a given cropping system between 1986 and 1988 (intertemporal productivity) or between two competing crops.

Intertemporal and interspatial productivity indices for the four cropping systems were calculated and are reported in Tables 1 and 2. In column I, there is no adjustment for changes in resource stocks and flows. Column II provides productivity measures allowing for variations in the resource stock only. In column III, full correction is made by accounting for both changes in resource stocks and flows.

Column III in Table 1 shows that total factor productivity increased for Systems B and C and declined for Systems A and D. Systems B and C produced 6.25 and 11.58 times as much output in 1988 as in 1986 using the 1986 input bundle. Systems B and C can therefore be said to be sustainable over the two-year interval since, after properly accounting for temporal differences in input quality and quantity and resource stocks and flows, they produce more than in the reference year. Systems A and D produced only 0.22 and 0.88 as much output in 1988 as in 1986 using the 1986 input bundle. Thus, A and D can be said to be non-sustainable.[2]

Table 1—Intertemporal Total Factor Productivity (Sustainability) Indices for Four Cropping Systems under Experimental Conditions, in Southwestern Nigeria, 1986–88

| System | No Correction I | Resource Stock Only II | Resource Stock and Flow III |
|--------|------|------|------|
| A | 0.20 | 0.19* | 0.22* |
| B | 6.38 | 6.14* | 6.25* |
| C | 0.02 | 0.01* | 11.58* |
| D | 3.27 | 4.23** | 0.88** |

Note: Numbers with one star (*) indicate the case of a net positive change in resource abundance, while those with two stars (**) indicate the case of a net negative change in resource abundance levels.

The economic viability of Systems B, C, and D relative to A is compared in Table 2. In 1986, after accounting for changes in resource abundance and flows, Systems B and C are shown to be relatively less productive than the reference base system. The interspatial TFP indices are estimated to be 0.73 and 0.76 for Systems B and C, respectively, indicating that these systems use relatively more resources and produce a comparatively lower output than

System A. Only System D (in which only plantains were grown) is more productive. In 1988, productivity indices for all the systems show a different pattern. With interspatial TFP indices of 9.26 and 1.12, Systems B and C are now found to be economically more viable than System A. Similarly, with a TFP index of 0.14, System D is found to be economically less viable than the reference base system. The changes in productivity measures in 1988 compared to 1986 are attributable to the changes in soil nutrient status over the two-year period. For example, in System C (where crops are grown in association with leguminous trees), soil nutrients increased by 2.3 percent in 1988 compared to System A, with a revenue share of about 12 percent. In System D, where only plantains are grown, chemical fertility is depleted over time and this is reflected in the lower 1988 productivity measure. Soil nutrients decreased by 21 percent in this system compared to System A, representing about 14 percent of the total cost faced by the farmer in 1988.

Table 2—Interspatial Total Factor Productivity (Economic Viability) Indices for Four Cropping Systems under Experimental Conditions in Southwestern Nigeria, 1986 and 1988

| System | 1986 | | | 1988 | | |
|---|---|---|---|---|---|---|
| | No Correction I | Resource Stock Only II | Resource Stock and Flow III | No Correction I | Resource Stock Only II | Resource Stock and Flow III |
| A | 1 | 1 | 1 | 1 | 1 | 1 |
| B | 1.73 | 2.02** | 0.73** | 68.50 | 81.34** | 9.26** |
| C | 5.37 | 6.68** | 0.76** | 0.37 | 0.36* | 1.12* |
| D | 0.06 | 0.18* | 2.40* | 1.04 | 1.31** | 0.14** |

## Concluding Comments

A model for measuring economic viability and agricultural sustainability for new technology evaluation was presented. The approach was based on the concept of total factor productivity and the growth-accounting procedure modified to accommodate changes in natural resource stocks and flows.

First, using standard optimization techniques, a generalized model of productivity measurement was developed. It was shown that, when common property resource stocks are used, productivity measures using conventional approaches are biased unless changes in resource stocks and flows are fully accounted for. To measure economic viability and sustainability, Denny and Fuss's (1983) interspatial and intertemporal productivity measures were used, which are defined in terms of the productive capacity of a system over space and time. A system is said to be sustainable if, after fully accounting for natural resource stocks and flows, it produces at least the same amount of output as in the reference year. Similarly, System A is said to be economically more viable then System B if, after completely accounting for natural resource stocks and flows and conventional inputs, A produces relatively more output than B.

A set of data available at the International Institute of Tropical Agriculture were used to compute the intertemporal and interspatial total factor productivity indices for four cropping systems in southwestern Nigeria. Results show that the sustainability and economic viability measures are sensitive to changes in the stock of soil nutrients as well as to changes in material input uses and outputs. Where common property resource flows are important, the measures provide markedly different results from conventional TFP approaches. The alley cropping system in which crops are grown between rows of leguminous fixing trees is shown to be sustainable and economically more viable than other systems after completely accounting for (positive) changes in natural resource stocks and flows.

## Notes

[1]International Livestock Centre for Africa and International Institute of Tropical Agriculture, respectively.

[2]Note from Table 1 that completely accounting for changes in resource levels and flows substantially alters the productivity measures. This is particularly true for System C in which the hedgerow trees fix atmospheric nitrogen and recycle nutrients, and System D, where the plantains heavily deplete the soil of its nutrients. Note that in System C, if the nitrogen contribution of the trees is not accounted for, the intertemporal productivity index is lower than unity (column I), leading to the erroneous conclusion that the system is not sustainable.

## References

BIFAD (Board for International Food and Agricultural Development), *Environment and Natural Resources: Strategies for Sustainable Agriculture*, task force report, US Agency for International Development, Washington, D.C., USA, 1989.

CGIAR (Consultative Group on International Agricultural Research), "Sustainability Research in the CGIAR: Its Status and Future," Document No. MT/85 114, Agenda Item 5, World Bank, Washington, D.C., USA, 1989.

Denny, M., and Fuss, M., "A General Approach for Intertemporal and Interspatial Productivity Comparison," *Journal of Econometrics*, Vol. 23, 1983, pp. 315–330.

Diewert, W.E., "Exact and Superlative Index Numbers," *Journal of Econometrics*, Vol. 4, 1976, pp. 115–145.

Ehui, S.K., and Hertel, T.W., "Deforestation and Agricultural Productivity in the Côte d'Ivoire," *American Journal of Agricultural Economics*, Vol. 71, No. 3, 1989, pp. 703–711.

Kang, B.T., van der Kruijs, A.C.B.M., and Couper, D.C., "Alley Cropping for Food Production in the Humid and Subhumid Tropics," in Kang, B.T., and Reynolds, L. (Eds.), *Alley Farming in the Humid and Subhumid Tropics*, proceedings of an international workshop held at Ibadan, Nigeria, 10–14 March 1986, International Development Research Centre, Ottawa, Canada, 1989.

Lal, R., and Ghuman, B.S., "Effects of Deforestation and Land use on Soil, Hydrology, Microclimate, and Productivity in the Humid Tropics," International Institute of Tropical Agriculture, Ibadan, Nigeria, 1989.

Lynam, J.K., and Herdt, R.W., "Sense and Sustainability as an Objective in International Agricultural Research," *Agricultural Economics*, Vol. 3, No. 4, 1989, pp. 381–398.

Ohta, N., "A Note on the Duality between Production and Cost Functions: Rate of Return to Scale and Rate of Technical Progress," *Economic Studies Quarterly*, Vol. 25, 1974, pp. 63–65.

Squires, D.E., and Herrick, S.F., Jr., "Productivity Measurement in Common Property Resource Industries: An Application to the Pacific Coast Trawl Fisheries," *Rand Journal of Economics*, 1990.

---

**Discussion Opening**—*Miguel A. López-Pereira* (Centro Internacional de Mejoramiento de Maiz y Trigo)

The economics of sustainability of farming systems in developing countries has become a popular research topic as the concept becomes relevant to more countries where natural resources are dwindling due to, *inter alia*, natural causes, population growth, and inefficient farming methods, especially by small farmers on low fertility land (hillsides, forests).

In this paper, a formal treatment of the economics of sustainability is attempted. In other studies on sustainability, a more empirical approach to measuring the long-term feasibility of

117

farming systems has been followed. The authors attempt to develop a model in which flows of nutrients and other factors determining the long-term viability of farming systems are accounted for when assessing their economic feasibility. However, the complicated set of equations seems to indicate that the model has some basic shortcomings that should be addressed.

With regard to the intertemporal component, the model does not appear to account for time when measuring the costs and benefits of inputs and products (discount rate). If the time dimension is to be included in the analysis, there has to be a discount rate to make estimations over time consistent.

The empirical example used to demonstrate the application of the conceptual model is also inadequate. Is two years of farming enough to provide a measure of the sustainability (i.e., long-term viability) of a given farming system? There are many factors in different farming systems that need much more time to have any effect on sustainability, such as the residual effect of nutrients added to the soil and the effects of certain rotations on soil fertility and disease and pest resistance. Many of these factors have an effect on the sustainability of the system that is not linear over time, so that long-run projections based on only a few seasons are not useful. A full cropping cycle including a fallow period, usually requiring seven to ten years, would be the minimum required to measure sustainability. If the resource and product flows have not decreased over the cycle, then the system could be considered sustainable.

How is the model applied in practice; i.e., how are the $Z$s measured and translated to monetary values or to what the authors call "implicit revenue shares"? Soil nutrient flows such as nitrogen fertilizer (as used by authors in the example) are probably easy to measure. But one may run into measurement problems when including changes in other important factors such as soil structure, or the effect of a farming system on the genetic diversity of species in a forest (if we were analysing a system that requires the clearing of forests). Valuing these more intangible factors may be very difficult. Even given limitations regarding length, it would have been useful for researchers interested in applying the model elsewhere if the paper had a section or an appendix showing how one moves from the model equations to the figures shown in the tables of their example.

How is interspatial productivity compared among more than two cropping systems? Are these comparisons transitive; i.e., if A > B and B > C, is then A > C? Also, can it be concluded that System C is more sustainable and economically viable than System B? Note that interspatial TFP is greater in B than in C in Table 2, and from the intertemporal TFP of Table 1, System C is better than System B.

The model, which is an aggregate model, may be useful for selecting the best of a group of systems and may provide useful information to policy makers. However, many of the policy measures necessary to implement the socially optimal system at the farm level may pose real challenges. Individual farmers may act differently from what is considered a social optimum. Cases of rented land, for example, come to mind. If there are no long- term rental agreements, which is the rule in developing countries, farmers will not care for positive nutrient flows if they will not be able to use them. On the contrary, they will try to extract as much as possible from the soil in the form of crop products, with as few inputs as possible.

*[Other discussion of this paper and the authors' reply appear on the following page.]*

# General Discussion—*Xiao Hui, Rapporteur* (Beijing Agricultural University)

On the Musgrave paper, the danger of transferring site-specific production functions from developed countries to developing countries was raised. Even though the generation of site-specific production functions is very expensive, this should be an ideal objective. Agricultural economists can and should use existing agronomic data for economic specification of production response. It was also suggested that the use of crop models of the type developed by the authors has not led to significant successes. The authors were asked if they see scope for future breakthroughs based on the use of such a model.

In reply, Musgrave indicated that while they are aware that site specificity is very important, constructing so many site-specific production functions is very time consuming. Farmers have been told how to irrigate; this cannot wait for the generation of site-specific production functions. Therefore, transferring models from developed countries is the only way. While the production function is less site specific for decision making, the transferred production function has proved very useful for decision making in Australia.

On the Featherstone, Osunsan, and Biere paper, since the research work has potential for better decision making, the authors were asked where the pay-offs lie, and what the implications are for decision making at the farm or policy level. The crop simulation models are very useful. The authors were asked if the model can be applied to developing countries, and whether the crop growth component can be applied to the mixed cropping systems common in LDCs. In Africa, for example, crops have many disease and insect problems, so that to make results more realistic, the authors need to consider disease and insects, which are endogenous variables in production. Since evapotranspiration is a more meaningful parameter than soil moisture in the model, the authors were asked why they use soil moisture rather than evapotranspiration.

Featherstone replied that the models can be applied in developing countries by using different input variables. The generation of the probability function is based on 29 years of daily data. The problem of diseases and insect pests can be solved by introducing more constraints to the models. Water moisture is a more basic input than evapotranspiration and is therefore estimated in the model.

Ehui was asked about the implications for research priorities and for designing research strategies. The authors were strongly advised to spend less time on issues of definition and much more on clarifying the rather complicated methodology and interpretation of the results.

In reply, Ehui emphasized that the model was derived for actual uses, based on a short comparison year by year. Two years' data are not enough to show the results; the minimum time is three years. Moreover, the model is very practical and can be used very easily.

Participants in the discussion included R. Dumsday (La Trobe University), D.A.G. Green (University College of Wales), W. Grisley (Centro Internacional de Agricultura Tropical), and H. Jansen (AVRDC).

# Measuring External Effects of Agricultural Production: An Application for the Netherlands

*A.J. Oskam*[1]

**Abstract:** Although the external effects of agricultural production receive considerable attention, the measurement of these effects is not well developed. In this paper, after introducing a theoretical approach, the methodology is applied to the Netherlands. The method used requires information about the quantities and the shadow prices of external effects. Different approaches were used to derive these shadow prices. Although many uncertainties still exist, incorporating the environmental effects of agricultural production reduces the gross value added by about 5–15 percent. Internalizing these external effects gives producers the opportunity to adjust inputs, output, and technology. It can thus be considered as a maximum effect.

## Introduction

Agricultural production technology has changed considerably during recent decades. Some technologies have been displaced completely by new ones; the substitution of labour by capital equipment and delivered inputs and services is the most impressive example of this. Although the literature shows that several studies have been devoted to these changes in technology, the measurement and explanation of technological change, and the effects of changing price relations between inputs and/or outputs, there has been very limited attention to the external effects.

Increased attention to the effects of agricultural production on the environment, landscape, and nature makes it clear that these external effects can play a very important role in the future development of prices, production technology, and level of production. There is a strong tendency to internalize the effects of undesirable outputs of present and new production technology and also of the scale and intensity of production. These changes will affect future gross and net production.

The external effects of agricultural production could be incorporated in a revised production measure. Such a procedure would throw some light on the importance of these effects of agricultural production in the long term. Two studies illustrate such external effects (RIVM, 1988; and Ministerie van Landbouw en Visserij, 1989):

- air pollution, either by well-known ingredients like ammonia, nitrogen oxide, etc., or due to smell;
- surface water pollution, due to minerals (such as chemicals) or organic material in the surface water;
- ground water pollution, due to leaking of minerals or other materials that have their origins directly or indirectly in agricultural production;
- soil pollution, which can mean that the soil attracts a certain amount of pollutants that could potentially enter the ground water, surface water, or air;[2]
- the visual attractiveness of the landscape, due to its composition and variety; and
- the conservation of nature, so that a large variety of different species are retained.[3]

The empirical analysis here will be limited to environmental effects due to manure, fertilizer, and pesticides resulting in pollution of the air and ground and surface water. These external effects might be considered as having a quantity component and a related price component. A difficulty with external effects is that the implicit price or shadow price is mostly unknown. Shadow prices of external effects (measured in quantity units) can depend heavily on the particular levels. But as this is comparable with economic markets, where prices depend on quantities, we should not be too concerned that the shadow prices are only given in a marginal sense.

Before starting on the empirical analysis, the measurement of external effects from a theoretical perspective and the difference between private and social output values are considered.

## Theoretical Aspects of Measuring External Effects

Here we start with a transformation function with two groups of output variables, normal outputs and external effects.[4] Input variables are divided into variable inputs and quasi-fixed inputs. The technology variable explicitly is retained in the transformation function:

(1) $F(Y, Z, X, U, T) = 0$

where $Y$ = the vector of normal outputs with elements $Y_i$ and prices $p_i$ ($i=1, ..., I$), $Z$ = the vector of positive or negative external effects with elements $Z_k$ and shadow prices $v_k$ ($k=1, ..., K$), $X$ = the vector of variable inputs with elements $X_j$ and prices $w_j$ ($j=1, ..., J$), $U$ = the vector of quasi fixed inputs with elements $U_h$ and prices $q_h$ ($h=1, ..., H$), and $T$ = a technology variable.

Short-term profit maximizing implies that the difference between the normal output value and the cost of variable inputs is maximized under the restriction of the transformation function. This gives the following Lagrange multiplier:

(2)  $\max_{Y, X} L = p'Y - w'X + \mu F(Y, Z, X, U, T)$

where row vectors are given with a prime.

Optimal variable input ($X^*$) and output ($Y^*$) levels are functions of prices, quasi-fixed inputs, and technology (Chambers, 1988, chap. 3, p. 7):

(3)  $X_j^* = f_j(p, w, U, T)$  ($j=1, ..., n$)

(4)  $Y_i^* = g_i(p, w, U, T)$  ($i=1, ..., m$)

Under the assumption of profit-maximizing producers in a competitive market for outputs and variable inputs where the external effects are not of concern, the optimal input and output levels do not depend on the shadow prices of external effects. These external effects only influence production decisions under the condition of technical or legal restrictions on the level of the external effects $Z$ (Pittman, 1983).

The resulting private rent for quasi-fixed inputs ($D_p$) is equal to:

(5)  $D_p = p'Y^* - w'X^* \equiv q^{*'}U$

where $q^*$ is the vector of shadow prices of quasi-fixed inputs that results from an optimal choice of variable inputs and outputs.

For society, however, the external effects are important, and the resulting difference between output value and input value ($D_s$) is:

(6)  $D_s = (p'Y^* + v'Z^*) - w'X^* = q^{*'}U + v'Z^*$

where $Z^*$ is the level of external effects that belongs to input level $X^*$, the output level $Y^*$, quasi-fixed inputs $U$, and technology T.

If, however, producers were confronted with prices of external effects, $v$, they would adjust their input level $X$ and output level $Y$, resulting in a new short-term equilibrium, derived from:

(7)  $\max_{X, Y, Z} L^+ = p'Y + v'Z - w'X + \mu F(Y, Z, X, U, T)$

with optimal levels of outputs and variable inputs:

(8)  $X_j^+ = f_j^+ (p, v, w, U, T) (j=1, ..., J)$

(9)  $Y_i^+ = g_i^+ (p, v, w, U, T) (i=1, ..., I)$

(10)  $Z_k^+ = h_k^+ (p, v, w, U, T) (k=1, ..., K)$

The resulting levels of private and social rent for quasi-fixed inputs ($D_p^+$ and $D_s^+$, respectively) are equal to:

(11)  $D_p^+ = D_s^+ = p'Y^+ + v'Z^+ - w'X^+ = q^{+'}U + v'Z^+$

where $q^+$ is the vector of resulting shadow prices for quasi-fixed inputs.

Because producers can adjust their input and output levels to the new situation, $D_p^+ \geq D_s^+$. This approach gives a maximum difference between the social and private levels of remuneration for the quasi-fixed production factors.

## Data Description and Preliminary Analysis

### Data Description

The methodology is applied to the Netherlands. A long-term set of inputs and outputs for the Netherlands agricultural sector is used (Oskam, 1991). The data set is derived mainly from statistics compiled by the Central Bureau of Statistics (CBS) and the Agricultural Economics Institute. In addition, a number of aggregations, prices, and procedures are specifically developed for this data set.

### Quantity Component of External Effects

A second set of information is related to the external effects of agricultural production. Here a number of technical indicators are used, together with statistical information from the CBS, to calculate the quantity levels of each of these external effects. Because the analysis focuses on incorporating the external effects in productivity measurement, a short description is given of how these data were generated. A selection of these data is given in Table 1.

The levels of air, surface-water, ground-water, and soil pollution are derived entirely from the level of organic fertilization (manure), the level of use of chemicals (fertilizer and pesticides), minus the net use of these materials by plants. As this does not give the distribution of the net effects over the particular type of pollution, net losses of nitrate, phosphate, calcium, and ammonia were calculated.

Table 1—Estimated Quantities of Emission of Different Chemicals
in Netherlands Agriculture (1000 t/year)

| Year | Nitrogen (N) | Phosphate ($P_2O_5$) | Calcium ($K_2O$) | Ammonia ($NH_3$) | Pesticides (Active Ingredient) |
|------|------|------|------|------|------|
| 1950 | 39 | 53 | 108 | 98 | 3 |
| 1960 | 116 | 49 | 130 | 123 | 5 |
| 1970 | 349 | 78 | 218 | 162 | 9 |
| 1980 | 510 | 89 | 324 | 217 | 18 |
| 1985 | 543 | 92 | 375 | 238 | 20 |
| 1988 | 462 | 74 | 320 | 230 | 20 |

Levels for particular years were used to establish relations between the number of different types of animals and the particular nutrients in manure (CBS, 1989; RIVM, 1988; and Wijnands *et al.*, 1988). Efficiency factors and information on real livestock capital during 1949–88 were used. Ingredients from fertilizers were added, while net use of plants was related to a base year, size of arable land and grassland, and a factor for crop production per hectare. Some checks were used on available data for other years.

The use of pesticides is based on CBS (1989) and data for the use of pesticides on arable farms. No correction was introduced for the use of pesticides by plants.

## Price Component of External Effects

The price component of external effects (shadow prices) is the most difficult element in the whole analysis. Here, some heroic assumptions were made to prepare the statistics.

First, 1986 prices were sought for each type of external effect. In fact, these are meant to be marginal prices, representing the implicit damage of the last unit. Several sources were used. Three different methods are available to derive indications for shadow prices of external effects.

**Estimated costs per unit.** Because reductions of environmental pollution have been planned, a number of indications are given on marginal costs. A specific example is ammonia, where marginal costs per unit of reduction are available (Stolwijk, 1989, p. 18; and Oudendag and Wijnands, 1989, p. 9).

**Marginal costs of environmental measures in other parts of the economy.** This follows from the opportunity cost principle: the value of one unit of reduction follows from the costs of a similar unit of reduction in another part of the economy (Bressers, 1988).

**Direct valuation of environmental effects such as contingent valuation.** This method is often very specific and hardly any practical indications can be derived from empirical research in this area (Hanley, 1990; and Johansson, 1987).

In Table 2, the implicit marginal costs of 1 ton of average manure are given for three different levels of shadow prices.

Table 2—Calculated External Effects (guilders/kg) per Average Ton of Manure in 1986

| Ingredient | Quantity in kg/ton | Shadow Price/Value | | | | | |
|---|---|---|---|---|---|---|---|
| | | Low | | Medium | | High | |
| | | Price | Value | Price | Value | Price | Value |
| N | 4.98 | 0.5 | 2.49 | 1.2 | 5.98 | 2 | 9.96 |
| $P_2O_5$ | 2.63 | 1.0 | 2.63 | 1.8 | 4.73 | 3 | 7.89 |
| $K_2O$ | 6.10 | 0 | 0 | 0.1 | 0.61 | 0.3 | 1.83 |
| $NH_3$ | 2.50 | 1.5 | 3.75 | 2.5 | 6.25 | 4 | 10.00 |
| Total | | | 8.87 | | 17.57 | | 29.68 |

Sources: CBS (1989), Stolwijk (1989), Table 3, and own calculations.

In calculating prices of external effects for different quantities and in other years, the following general approach was used:

$$(12) \quad v = \alpha \, z^\beta y^\sigma$$

where $v$ = the real shadow price of a particular external effect (e.g., N in the form of nitrate, phosphate, ammonium, and pesticides), $z$ = the quantity of the external effect, $y$ = the real net product of the Netherlands, $\beta$ = (shadow) price flexibility of the external effect, $\sigma$ = (shadow) price flexibility with respect to real net product, and $\alpha$ = constant equalizing Equation (12) in the year 1986.

The implicit income elasticity of an external effect $(e_y)$ can be derived from Equation (12):

(13) $\quad e_y = -\dfrac{\sigma}{\beta}$

This income elasticity reflects the percentage of reduction of a negative external effect required to compensate for a 1-percent increase in real net product. Because external effects are considered to be luxury goods, $|e_y| > 1$ (Baumol and Oates, 1988). Here different values were used, but 1.25 was the central assumption.

The price flexibility of the external effect is more difficult to establish. There are critical values where an increase in the quantity of an external effect will have a large influence on the shadow price of this effect, while in other ranges, (e.g., under a particular level), changes in quantities have no or nearly no influence on shadow prices.

Because the analysis relates to the agricultural sector, which generates important negative external effects, we are in a situation where quantities influence shadow prices. The quantity effect therefore has to be incorporated; values of $\beta$ of 0.4, 0.2, and 1 are used as three guesses for the particular quantity effect of environmental goods.

## Empirical Analysis

Because of the uncertainty about the shadow prices of environmental effects, three series of measures were generated using the price levels and additional parameters, as given in Table 3. Gross value added was used as measure for the quasi-fixed inputs.

Table 3—Different Guesses of Shadow Prices (guilders/unit) and Parameters of Environmental Effects of Agricultural Production in 1986

| Variable/Parameter | Unit | Medium | Low | High |
|---|---|---|---|---|
| Nitrogen | kg N | 1.2 | 0.5 | 2 |
| Phosphate | kg $P_2O_5$ | 1.8 | 1 | 3 |
| Calcium | kg $K_2O$ | 0.1 | 0 | 0.3 |
| Ammonia | kg $NH_3$ | 2.5 | 1.5 | 4 |
| Pesticides | kg active ingredient | 10 | 5 | 25 |
| $\beta^1$ | | 0.4 | 0.2 | 1 |
| $\sigma^1$ | | 0.5 | 0.2 | 1.5 |

[1]These parameters influence the development of the shadow price (see Equation 12).

Focusing first on the medium guess of shadow prices, the resulting difference between private and social rent for quasi-fixed inputs $(D_p^+ - D_s)$ cannot be neglected (Table 4). This is due to two different elements. One is the increasing levels of environmental damage by the agricultural sector. This element consists of a quantity component and a price component: more pollution increases the shadow costs per unit of pollution—the quantity effect. The other element is an increasing level of real net product, generating higher shadow costs per unit of pollution—the income effect.

Both effects increased considerably during 1949–88. But their cross effect is also important: a higher level of pollution will be perceived as more damaging at a higher income level. Negative environmental effects of agricultural production increased from about 1 percent in 1949 to more than 11 percent at the beginning of the 1980s. Later on they levelled off somewhat. This is due to the approach used, where emissions directly affect the environment.

Table 4—External Effects of Agricultural Production in the Netherlands, Starting from Three Sets of Shadow Prices and Parameters

| Year | Low | | Medium | | High | |
|---|---|---|---|---|---|---|
| | External Effect (million guilders) | Gross Value Added (percent) | External Effect (million guilders) | Gross Value Added (percent) | External Effect (million guilders) | Gross Value Added (percent) |
| 1949 | 19 | 1.0 | 18 | 1.0 | 4 | 0.2 |
| 1955 | 36 | 1.2 | 40 | 1.4 | 13 | 0.4 |
| 1960 | 55 | 1.5 | 66 | 1.8 | 28 | 0.8 |
| 1965 | 91 | 1.9 | 123 | 2.6 | 72 | 1.5 |
| 1970 | 182 | 2.8 | 296 | 4.6 | 274 | 4.2 |
| 1975 | 354 | 3.7 | 630 | 6.6 | 763 | 8.0 |
| 1980 | 631 | 5.6 | 1,214 | 10.7 | 1,866 | 16.5 |
| 1985 | 821 | 5.0 | 1,635 | 9.9 | 2,782 | 16.8 |
| 1988 | 749 | 4.5 | 1,452 | 8.7 | 2,552 | 15.2 |

Incorporating the uncertainty about shadow prices of environmental effects leads to a rather wide range: 5–15 percent of gross value added in 1988.

## Concluding Remarks

The inclusion of environmental effects in measuring agricultural production gives some idea of possible changes in future rents of quasi-fixed factors, due to the incorporation of external effects. These estimates generate an upper bound because producers will use more efficient technologies after the introduction of quantity restrictions or taxes/levies (Pearce and Turner, 1990). Moreover, an increase in the prices of agricultural products due to the introduction of quantity restrictions or taxes/levies can partly or completely offset these effects. Here, either the Netherlands should be an important producer of a particular product or other countries should introduce similar regulations. As a last factor, the government might bear a part of the costs of incorporating external effects in the production decisions of the agricultural sector. This is, however, only a shift of the burden from the agricultural sector to the taxpayer.

Due to an increasing level of real net production in the economy, shadow prices of most external effects will increase. This will boost the (negative) value of the environmental effects unless quantities are sufficiently reduced.

The methodology is applied to the most important environmental effects of agricultural production over the past 40 years. The methodology is not very different from the earlier work of Hueting (1974) developed for a national economy. Empirical estimates of external effects of agricultural production are very scarce, which makes it difficult to compare these results with other research. This analysis focused on a limited number of external effects. More effects could be included, but the environmental effects are currently receiving the most attention.

## Notes

[1]Wageningen Agricultural University.
[2]There are two stages in this process, where the soil is still "attracting" these materials and where there is already leakage. Here, the other types of pollution are already mentioned.
[3]This can embrace direct impact on nature itself, as well as indirect effects, due to potential living circumstances for wildlife and birds.
[4]One could also add a set of input variables in the category of external effects, but these effects seem to be less important and are discarded in this analysis.

## References

Baumol, W.J., and Oates, W.E., *The Theory of Environmental Policy*, Cambridge University Press, Cambridge, UK, 1988.

Bressers, H., "Effluent Charges Can Work: The Case of Dutch Water Quality Policy," in Dietz, F., and Heijman, W. (Eds.), *Environmental Policy in a Market Economy*, Pudoc, Wageningen, Netherlands, 1988, pp. 5–39.

CBS (Central Bureau van Statistiek), *Algemene Milieustatistiek 1989*, Staatsuitgeverij, The Hague, Netherlands, 1989.

Chambers, R.G., *Applied Production Analysis: A Dual Approach*, Cambridge University Press, Cambridge, UK, 1988.

Hanley, N., "The Economics of Nitrate Pollution," *European Review of Agricultural Economics*, Vol. 17, 1990, pp. 129–151.

Hueting, R., *Nieuwe schaarste en economische groei*, Agon, Amsterdam, Netherlands, 1974.

Johansson, P.O., *The Economic Theory and Measurement of Environmental Benefits*, Cambridge University Press, Cambridge, UK, 1987.

Ministerie van Landbouw en Visserij, *Structuurnota Landbouw*, The Hague, Netherlands, 1989.

Oskam, A.J., *Macro-Economic Data of Dutch Agriculture*, Agricultural University, Wageningen, Netherlands, 1991.

Oudendag, D., and Wijnands, J., "Beperking van de ammoniak-emmissie uit dierlijke mest," *Onderzoekverslag*, No. 56, Landbouw-Economisch Institut, The Hague, Netherlands, 1989.

Pearce, D.W., and Turner, R.K., *Economics of Natural Resources and the Environment*, Harvester Wheatsheaf, New York, N.Y., USA, 1990.

Pittman, R.W., "Multilateral Productivity Comparisons with Undesirable Outputs," *Economic Journal*, Vol. 93, 1983, pp. 883–891.

RIVM, *Zorgen voor morgen*, Alphen aan den Rijn, Samsom, Netherlands, 1988.

Stolwijk, H., "Economische gevolgen voor de veehouderij van een drietal milieu-scenario's," *Onderzoekmemorandum*, No. 57, Central Bureau van Statistiek, The Hague, Netherlands, 1989.

Wijnands, J., Luesink, H., and van der Veen, M., "Impact of the Manure Laws in the Netherlands," *Tijdschrift voor Sociaal Wetenschappelijk Onderzoek van de Landbouw*, Vol. 3, No. 3, 1988, pp. 242–262.

---

## Discussion Opening—*Zhigang Chen* (Université Laval)

Oskam takes on a very ambitious task, that of measuring the environmental effect of agricultural production at the industry level. He makes the provisional assumption that monetary values can generally be placed on given physical changes in the environment. This assumption is extremely convenient, since it implies that environmental losses or gains and the cost of averting or creating them are measurable in the same dollar unit, and thus one can determine how much can justifiably be spent on environmental control.

A very interesting result of the paper is that incorporating the environmental effects of agricultural production reduces the gross value added in the Netherlands by 5–15 percent. However, the significance of this result depends on the reliability of the calculation and, unfortunately, there are reasons for questioning the accuracy and relevance of these calculations, given the method used to construct the quantity and shadow price of the environmental effects. Since estimating the shadow price of the environmental effect is an essential part of the paper, as is the whole measurement issue, more explanation is needed.

The difficulties inherent in quantifying the environmental impact of agricultural production may be conveniently summarized in a Coasian framework. Let us assume that the world consists of two groups—farmers and the rest of society. As farmers increase their production, their marginal net benefit falls to zero at some point beyond which they will not rationally wish to produce. Meanwhile, society must bear increasing marginal external costs related to the level of production. In practice, the shape and location of these marginal curves is unlikely to be known, and they will move under the impact of changing prices and available technology, so that it is extremely difficult to measure the amounts. Given the diversity of the resource base (i.e., soil type) and underlying technology in agriculture, it is likely that substantial survey and estimation work will be needed before marginal external benefits can be estimated with any accuracy. One may suggest the use of farm-gate prices, but this could result in an over-estimate if the social value of the extra output is less than the private value. This is especially important for agricultural products where price or production is subsidized.

The uncertainty in estimating marginal external benefits is probably insignificant compared to that related to the estimation of marginal external costs. There are several methods available, such as contingent valuation and hedonic prices, each with its own merits and shortcomings. Fortunately, agricultural and resource economists have been working on these problems for a number of years, so there is a considerable body of literature to consult.

Despite its simplification of the issues, Oskam's paper does provide some useful indications of the appropriate methodological direction to take and points towards future research needs. For example, the environmental impact of agricultural production is simply overlooked in the current analysis of agricultural productivity and comparisons across national boundaries, probably due to the difficulty of incorporating environmental effects in productivity measurement. Oskam's paper provides some ideas on how to deal with this problem as well as with other issues relating to agriculture and the environment.

*[Other discussion of this paper and the author's reply appear on page 143.]*

# Political Economy of International Pollution

*Marie L. Livingston and Harald von Witzke*[1]

**Abstract:** International pollution poses special problems for economic analysis. National governments can, in principle, be successful in improving environmental quality when the sources of pollution are located within the government's jurisdiction. This is not possible, however, when pollution originates abroad. In essence, the rules that govern transboundary pollution represent an international public good. Any solution to the problem requires international coordination and explicit recognition of both political and economic aspects involved. This paper presents a public choice model that captures the incentive structures faced by resource users and the marginal political and economic benefits and costs of regulation. Hypotheses derived and discussed concern the relative political strength of producers and consumers, transactions costs faced by each group, structure of the input market, relative size of the polluting industries involved, level of economic development, and amount of the externality "exported" to or "imported" from other countries.

## Introduction

Around the globe, the demand for environmental quality is growing. Air and water pollution that crosses international boundaries constitutes an increasing fraction of the problem. Transboundary pollution poses special problems for analysis. National governments can, in principle, be successful in improving environmental quality when the sources of pollution are located within the government's jurisdiction. This is not possible, however, when the domestic pollution originates abroad. The rules that govern the regulation of transboundary pollution represent an international institution and thus an international public good. No government can supply itself with such a good except in cooperation with other countries.

Transboundary pollution has gained some attention in the economic literature (D'Arge and Kneese, 1980; Baumol and Oates, 1986; and Kneese, Rolfe, and Harned, 1971). And, in recent years, models have been developed to accommodate problems that have both economic and political aspects (Bromley, 1982; Buchanan and Stubblebine, 1962; Mueller, 1981; and Olson, 1965). However, these models are rarely applied to international pollution. The objective of this paper is to model the political economy of international pollution using a public choice approach. First, a theoretical framework that captures the incentives of governments for regulation of pollution by domestic industries is developed. Then, specific hypotheses derived from the model and their policy implications are discussed.

## A Public Choice Approach to International Pollution

The theoretical framework focuses on domestic pollution control policy decisions in the presence of trans-frontier movements of pollutants. The nomenclature of economics is adopted, and pollution is referred to as an externality. The model represents supply-side approaches to policy modelling in that it is based on the political economic calculus of the regulator as the supplier of environment policy.[2] A non-cooperative model based on Nash behaviour is presented below, using the following symbols:

$W$ = policy maker's utility
$V$ = political support
$U_c$ = utility of consumers
$\pi_b$ = profit of producers
$\pi_c$ = income of consumers
$b$ = externality (bad)
$b_d$ = domestic externality consumed domestically
$b_e$ = domestic externality exported to and consumed in third countries
$b_m$ = externality from abroad; imported externality
$b_t$ = total externality consumed domestically

In this paper, the amount of the externality from abroad considered is given by the regulator $(b_m = \bar{b}_m)$. As will become evident, this assumption results in a Nash equilibrium from which the regulator cannot deviate unilaterally without being worse off. The domestically produced externality consumed abroad will be referred to as "exported" and the domestically consumed externality from abroad as "imported."

Assume a single regulator's strictly concave utility function that contains as arguments the political support from consumers and from a group of producers who also produce an externality in the form of pollution. The political support of the regulator from producers and consumers can be thought of as votes. Campaign contributions and other lobbying activities can be seen as generating votes from these two groups.

(1) $\quad W = W\,(V_b,\, V_c)$

The regulator maximizes utility subject to the following two constraints which are assumed to be concave:

(2) $\quad V_b = V_b\,[(\pi^*_b + \pi_b\,(b)]$

(3) $\quad V_c = V_c\,\{U_c\,[\pi_c(\pi_b),\, b_d,\, \bar{b}_m]\}$

where:

(4) $\quad b_d = \beta b$

(5) $\quad 0 \le \beta \le 1$

Equations (2) and (3) represent the political economic constraints that the regulator faces. According to Equation (2) the political support of producers is a function of their total profits, where $\pi^*_b$ denotes the actual profits when the externality is internalized, $\pi_b$ denotes the additional profits that result from the production of the externality, and $\pi_b$ is a positive function of $b$. If the externality is denoted at the private optimum as $\hat{b}$, then total profits at the private optimum are the profits at the social optimum (where $b=s$) plus the additional profits that result if the output of the externality is not regulated and the industry produces at the private optimum.[3]

(6) $\quad \hat{\pi}_b = \pi^*_b + \pi_b\,(\bar{b})$

In Equation (3), the regulator's political support from consumers is a positive function of consumer utility, where the utility is a positive function of consumer income and a negative function of the amount of the externality that is consumed domestically. Consumer income is related to producer income. How close this relationship is depends on the share of total inputs owned by consumers and used in the externality-producing industry. It also depends on the structure of the markets for production factors, as this determines how much the price and/or use of a production factor changes when profits change.

In Equation (4), the total externality produced is consumed in fixed proportions by domestic and foreign consumers. That is, $b = b_d + b_e$, where $b_d$ is defined as in Equation (4) and, therefore, $b_e = (1-\beta)b$. If $\beta = 1$, the externality is only consumed domestically; in this case, the maximization problem is reduced to one of optimal regulation of domestic pollution. If $\beta = 0$, the externality is entirely exported to third countries.

The solution to this maximization problem is:

$$(7) \quad \left(\frac{\partial W}{\partial V_b}\right)\left(\frac{\partial V_b}{\partial \pi_b}\right)\left(\frac{\partial \pi_b}{\partial b}\right) + \left(\frac{\partial W}{\partial V_c}\right)\left(\frac{\partial V_c}{\partial U_c}\right)\left(\frac{\partial U_c}{\partial \pi_c}\right)\left(\frac{\partial \pi_c}{\partial \pi_b}\right)\left(\frac{\partial \pi_b}{\partial b}\right) = -\left(\frac{\partial W}{\partial V_c}\right)\left(\frac{\partial V_c}{\partial U_c}\right)\left(\frac{\partial U_c}{\partial b_d}\right)\beta$$

Equation (7) has an obvious political economic interpretation. It can also serve as a basis for the formulation of hypotheses about the political economic optimum amount of the externality. The two terms of the sum on the left hand side of Equation (7) represent the marginal political economic benefits of deviating from the social optimum. These benefits arise via increased political support from producers and/or consumers as their incomes grow with increasing $b$. The right-hand side of Equation (7) represents the marginal political economic costs of increasing the output of $b$, as consumer utility is negatively affected by an increase in the consumption of the externality. Hence, the optimal amount of $b$ is chosen such that the marginal political economic benefit of an increase in $b$ equals its marginal political economic cost.

The political economic optimum condition for the regulator's control variable $b$ in Equation (7) can be illustrated graphically. Denote:

$$(8) \quad A = \left(\frac{\partial W}{\partial V_b}\right)\left(\frac{\partial V_b}{\partial \pi_b}\right)\left(\frac{\partial \pi_b}{\partial b}\right)$$

$$(9) \quad B = \left(\frac{\partial W}{\partial V_c}\right)\left(\frac{\partial V_c}{\partial U_c}\right)\left(\frac{\partial U_c}{\partial \pi_c}\right)\left(\frac{\partial \pi_c}{\partial \pi_b}\right)\left(\frac{\partial \pi_b}{\partial b}\right)$$

$$(10) \quad C = -\left(\frac{\partial W}{\partial V_c}\right)\left(\frac{\partial V_c}{\partial U_c}\right)\left(\frac{\partial U_c}{\partial b_d}\right)\beta$$

In Figure 1, the horizontal axis denotes the quantity of the externality and the vertical axis denotes the marginal political economic costs and benefits of deviating from the social welfare optimum(s).

In Equation (8), $A$ is positive, as are all partial derivatives of $A$. The regulator's utility is positively affected by an increase in political support from producers, their political support grows with increasing profits, and producer profits are a positive function of $b$. Therefore, $A$ is in the first quadrant.

Convexity of the constraint in Equation (2) implies that the private optimum in production is finite, that is, the marginal profit of an additional amount of the externality must be declining with increasing $b$, and $A$ (in Figure 1) has a negative slope.

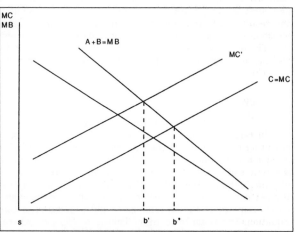

Figure 1—Marginal Political Economic Benefits and Costs of Government Regulation of a Negative Externality

In Equation (9), $B$ represents the marginal political economic benefits to the regulator that result from an increase in $b$ via the consumer income effect. The sum of $A$ and $B$ represents the total marginal political benefits ($MB$) of an increase in the externality. All partial derivatives of $B$ are non-negative. The only partial derivative that in reality may be zero is the change in consumer incomes as a consequence of a change in producer incomes. This would be the case when the price of consumer-owned inputs or their quantity is not affected by the change in producer profits, either because of a lack of consumer market power on input markets or because all inputs of producers are owned by non-consumers (e.g., foreigners). As long as none of the partial derivatives is negative, $B$ is in the first quadrant.

The slope of $B$ is negative for the same reason that $A$'s slope is negative. Hence, the total marginal political economic benefit of deviating from the social optimum $(A+B)$ declines with increasing $b$.

$C$ represents the marginal political economic costs $(MC)$ of a growing deviation from the social optimum via the loss in political support from domestic consumers that is the consequence of the increasing disutility of consuming $b_d$. The first two derivatives are positive, while $\partial U_c / \partial b_d$ is negative. As the expression on the right-hand side of Equation (10) is negative, $C$ must be in the first quadrant. The slope of $C$ is determined by $\beta$; it is positive as long as $\beta > 0$.

All other things being equal, the slopes of the curves are given by $\partial \pi_b / \partial b$ and $\beta$, respectively. The position of the curves in space is determined by the other components that determine the political economic costs and benefits of government regulation of the externality; i.e., these partial derivatives act as shifters of one or more of the curves in Figure 1. According to Equation (7) the political economic equilibrium is determined by the intersection of $MB$ $(=A+B)$ and $MC$ $(=C)$. In Figure 1, this is the case at $b^o$.

The model discussed here has several implications for the amounts of the externality produced domestically.

**Political weights** $(\partial W / \partial V_b; \partial W / \partial V_c)$. The marginal political weight of consumers does not determine, *a priori*, how much of the externality $b$ will be produced at the political economic optimum. This is the case because consumer utility declines with increasing $b$. However, consumers may also benefit from the production of $b$ via $\pi_c(\pi_b)$; i.e., $\partial W / \partial V_c$ affects both $MB$ and $MC$. For instance, a growing marginal political weight of consumers would not only shift $MC$ to the left but would also shift $B$ and thus $MB$ $(=A+B)$ to the right. Whether this results in an increase or decline of the optimal $b$ depends on the magnitude of these shifts, which are also affected by the other components of $B$ and $C$.

The effect of the marginal political weight of producers is unambiguous: the larger the weight, the larger will be the optimum $b$.

This can be illustrated by rewriting Equation (7), as follows:

$$(7a) \quad \left( \frac{\partial W}{\partial V_b} \right) \left( \frac{\partial V_b}{\partial \pi_b} \right) \left( \frac{\partial \pi_b}{\partial b} \right) = - \left( \frac{\partial W}{\partial V_c} \right) \left( \frac{\partial V_c}{\partial U_c} \right) \left[ \left( \frac{\partial U_c}{\partial \pi_c} \right) \left( \frac{\partial \pi_c}{\partial \pi_b} \right) \left( \frac{\partial \pi_b}{\partial b} \right) + \left( \frac{\partial U_c}{\partial b_d} \right) \beta \right]$$

In Equation (7a), the left-hand side depicts the marginal political economic benefits of deviating from the social optimum via growing support from producers, whereas the right-hand side contains the net cost of doing so via changing political support from consumers. As the first three partial derivatives in parentheses are larger than or equal to zero while $\partial U_c / db_d < 0$ and $0 \le \beta \le 1$, the sum in brackets can be positive or negative and thus can be the net political support from consumers; i.e., whether the net support from consumers is positive or negative is determined by the expression in brackets on the right-hand side of Equation (7a), where the marginal political weight attached to consumers acts as a multiplier (as does $\partial V_c / \partial U_c$). Of course, a regulator who is indifferent with regard to the origin of the votes will attach the same weights to producers and consumers.

**Political influence of producers** $(\partial V_b / \partial \pi_b)$. All other things being equal, the more sensitive the political support from producers to changes in profits $(\partial V_b / \partial \pi_b)$, the farther $A$ (and thus $MB$) will be to the right in Figure 1 and the more $b$ will exceed the social optimum. According to a central hypothesis of public choice theory, any group can be expected to react in a more pronounced way with political support and thus will be more influential the more efficiently it can organize its lobbying efforts. Typically, relatively small groups, groups with fairly homogenous interests, groups that can supply their members with selective benefits, or those which have low costs of organizing a lobby for other reasons (e.g., because they are regionally concentrated) are more successful on political economic markets (see Olson, 1965).

**Political influence of consumers** $(\partial V_c / \partial \pi_c)$. Arguments similar to those for political weights and political influence of consumers hold for the determinants of the marginal change in political support from consumers when their utility changes. Group characteristics determine the sensitivity of political support to changes in consumer utility. However, the direction of impact on the optimum $b$ cannot be determined *a priori*; $\partial V_c / \partial U_c$ acts as a multiplier, and the direction of its impact depends on whether the expression in brackets on the right-hand side of Equation (7a) is positive or negative.

**Income level** $(\partial U_c / \partial b_d; \partial U_c / \partial \pi_c)$. The direct effect of consumer incomes on the politically optimal output of the externality is unambiguous. The higher the income, the larger the marginal disutility of consuming the externality $(\partial U_c / \partial b)$ and the smaller the marginal utility of income $(\partial U_c / \partial \pi_c)$ generated by an additional unit of the externality. In Figure 1, the higher the income level, the further to the left will be both MC and MB and thus the lower will be the optimal $b$, all other things being equal. Hence, one can expect the regulation of a negative externality to become tighter when incomes rise.

**Structure of input market** $(\partial \pi_c / \partial \pi_b)$. The marginal change in consumer incomes as a consequence of a profit change in the industry that produces the externality and thus the position of $B$ is affected by the structure of the input market and the amount of production factors of the industry that is owned by consumers. The latter is, of course, also influenced by the size of the industry in terms of employment.

The structure of the input market directly affects the incidence of consumer incomes and the profit of producers and thus $\partial \pi_c / \partial \pi_b$. Curve B will be further to the left and the optimum b will be lower the less factor prices and/or total factor inputs increase with growing profits. For instance, if producer capital is predominantly owned by foreigners and/or its share in total employment is small, a change in profits will only marginally affect domestic consumers. Therefore, such industries will face relatively tight environmental regulation, all other things being equal.

**Sensitivity of producer incomes to environmental regulation** $(\partial \pi_b / \partial b)$. The more sensitive producer profits are to changes in $b$, the more inelastic will be both A and B in Figure 1. With increasing sensitivity of producer profits, environmental regulation will be less affected by a shift of MC to the left. Hence, one can expect that those industries that are crucially dependent on a process that results in the externality will face looser environmental regulation than those that can more easily substitute such a production process, *ceteris paribus*.

**Domestic consumption and export of the externality** $(\beta)$. In Equation (10), $\beta$ represents the share of the total output of the externality that is consumed domestically. If $\beta$ is zero (i.e., if the externality is consumed entirely by foreigners), the marginal political economic costs of environmental regulation are zero unless either altruism or some form of strategic behaviour with regard to mutually exported externalities is introduced. With increasing $\beta$, less of the externality is exported and more is consumed domestically, and MC in Figure 1 is further to the left, *ceteris paribus*. Consequently, the optimum will be at a lower $b$. For $\beta = 1$, the externality is entirely consumed domestically. If, in addition to this, there is no import of the externality from abroad, the problem is reduced to one of the political economic optimal environmental regulation in a closed country with no trans-frontier movements of the externality.

**Import of the externality** $(b_m)$. When domestic consumers are affected by an exogenously given externality from abroad that cannot be avoided, MC in Figure 1 shifts to MC′, where the difference between MC′ and MC results from the loss in political support by consumers who also consume the imported externality. As a consequence, the political economic optimum would shift to the left ($b'$); i.e., the optimal domestic output of the externality is lower in the presence of a given externality from abroad. From Figure 1, it is also clear that any reduction in the externality from abroad increases the domestic political economic optimum output of $b$.

## Summary and Conclusion

The model developed in this paper suggests several reasons for the existence of policies that allow the private sector to deviate from the social optimum if there are externalities in production. The foregoing analysis suggests that, *ceteris paribus*, one would expect greater regulation of transboundary pollution in jurisdictions that are net importers of pollution and in higher-income countries.

The incentives for producers are clear. Pressure for regulation from producers decreases when ownership of factor inputs is largely domestic, when profits are sensitive to regulation, and as the political influence of producers grows. The impact of transboundary pollution on consumers is uncertain, in terms of its impact on regulation. Even if consumer political influence is equal to that of producers, regulatory pressure from this group still hinges on which predominates: the direct disutility of pollution or the indirect utility of additional income via factor ownership.

Any agreement on international pollution policy consists of a set of rules that specifies the signatories' rights and obligations. Such an agreement represents a global public good. Public goods are frequently difficult to supply efficiently because of free riding. The free-rider problem can be solved in principle, however, through a system of conditional commitments to contribute to the production of a public good (Sugden, 1984). The key for international agreements on transboundary pollution is that they must be perceived as fair (Baumol, 1982) and provide the assurance that everybody plays by the rules (Sen, 1967). This assurance is crucial for the production of any public good (Runge, 1984). The model presented here provides a basis for predicting which countries are likely to pursue and abide by international pollution agreements in specific cases.

## Notes

[1]University of Northern Colorado and University of Minnesota, respectively.

[2]Notice that political economic models typically result in optima different from social welfare optima.

[3]As we have formulated the model such that the externality may partially or in total affect foreign countries, the term social optimum refers to a global social optimum.

## References

Baumol, W.J., "Applied Fairness Theory and Rationing Policy," *American Economic Review*, Vol. 72, 1982, pp. 639–651.

Baumol, W.J., and Oates, W.E., *The Theory of Environmental Policy*, Cambridge University Press, Cambridge, UK, 1986.

Bromley, D.W., "Land and Water Problems: An Institutional Perspective," *American Journal of Agricultural Economics*, Vol. 64, 1982, pp. 834–844.

Buchanan, J.M., and Stubblebine, W.C., "Externality," *Economica*, Vol. 29, 1962, pp. 371–384.

D'Arge, R.C., and Kneese, A.V., "State Liability for International Environmental Degradation: An Economic Perspective," *Natural Resources Journal*, Vol. 20, 1980, pp. 427–450.

Kneese, A.V., Rolfe, S.E., and Harned, J.W. (Eds.), *Managing the Environment: International Economic Cooperation for Pollution Control*, Praeger, New York, N.Y., USA, 1971.

Mueller, D.C., *Public Choice*, Cambridge University Press, Cambridge, UK, 1981.

Olson, M., *The Logic of Collective Action*, Harvard University Press, Cambridge, Mass., USA, 1965.

Runge, C.F., "Institutions and the Free Rider: The Assurance Problem in Collective Action," *Journal of Politics*, Vol. 46, 1984, pp. 154–181.

Sen, A.K., "Isolation, Assurance, and the Social Rate of Discount," *Quarterly Journal of Economics*, Vol. 81, 1967, pp. 112–124.

Sugden, R., "Reciprocity: The Supply of Public Goods through Voluntary Contributions," *Economic Journal*, Vol. 94, 1984, pp. 772–787.

---

## Discussion Opening—*László Kárpáti* (Agricultural University of Debrecen)

The paper discusses a very important problem: how to incorporate the effect of transboundary pollution in a neoclassical econometric model that determines the economic versus social optima for a given country. The importance of the topic cannot be underestimated. Transboundary pollution is an even more delicate issue than intracountry environmental control since the whole topic is embedded in the general set of relationships among independent countries. The paper concentrates mainly on questions relating to a hypothetical country. The other main problem of how to formulate an effective international agreement is not discussed, however, in any detail.

The topic suggested by the title of the paper is the political economic aspects of allocation of pollution among different countries. In this sense, the export/import ratio of pollution for a given country plays the most important role since this is the figure that represents one politically independent country in the material interchange with another politically independent country.

The ratio between the exported and imported "bad" externalities plays an important role in determining how the domestic regulations should be modified. This ratio is especially important in the case of European countries where the level of environmental contamination is high and, in addition, the size and geography of the countries make "pollution exports" unavoidable. Larger countries, like the USA or Canada, have two advantages over the smaller ones; they have a smaller "pollution transaction" ratio (because of their geography) and greater political economic bargaining power, which has an important role in the model in any case.

Three main topics can be suggested for discussion in connection with the paper. First, it is worthwhile considering the implications of different "pollution transaction" ratios on domestic pollution control policy in the case of countries of different sizes, based on the assumptions of the Livingston-Witzke model. The relative importance of total/traded ratio should also be taken into account. Secondly, the foundations of an effective bilateral pollution control agreement are worthy of review: balancing the interests of the two countries, reallocation of public goods, and simultaneous levelling of polluting sources. This question cannot be separated from the political economic (and sometimes, the military) power of the countries in question.

Applying the Livingston-Witzke model macro-regionally, a third question can be generated: whether the model can be applied as a simultaneous model system for a set of different countries on a multilateral base.

The theory of international negotiations, for example Raiffa's work at Harvard University, deals with the exploration of mutually advantageous elements for the participating countries. A theoretical model system that assists a group of independent countries find a mutually acceptable solution among themselves and partially suggests a domestically advantageous environment controlling policy would certainly be welcomed. In this, case domestic and international regulations may be connected, and the best tool for testing a new environmental treaty would be one in which the individual interests of the single countries and the common public goods ("total externalities") are better harmonized. Technically, it can be solved in other ways: large-scale multi-period linear programming, input-output analysis, and a simultaneous model system with a bargaining simulation framework.

The ideas in the paper support and demand a substantial rethinking in this area of science.

*[Other discussion of this paper and the authors' reply appear on page 143.]*

# Ecological versus Economic Objectives: A Public Decision Making Problem in Agricultural Water Management

*Slim Zekri and Carlos Romero*[1]

**Abstract:** The planning of a new irrigated area is a complex problem where a multiplicity of very different criteria (economic, ecological, social, etc.) have to be taken into account. A lexicographic goal programming model capable of handling this multiplicity is formulated. The methodology is applied to the planning of the irrigated lands of the village of Tauste in Aragón (Spain). An important result generated by the model is the conflict between economic criteria and environmental effects such as "salt-load," which affects the water quality in the basin. This matter is thoroughly analysed by determining the transformation curve between "salt-load" and the investment outlay in the irrigation systems discussed.

## Introduction

In many geographical areas, water is the main limiting factor for agricultural development. Consequently, the transformation of large rain-fed areas into irrigated lands has been and will continue to be a common practice to increase regional wealth in many parts of the world. The positive effects of irrigation in agricultural production, primarily in terms of increased profitability and employment, are well known.

However, the massive use of water for irrigation is not exempt from negative effects. Water is a multi-purpose natural resource essential for agriculture, industry, human consumption, recreational activities, etc.; hence, its intensive use in a single activity (e.g., agriculture) can generate an important opportunity cost in the other activities. Moreover, intensive irrigation practices are also one of the main causes of salinization of soils and river basins, as well as an important source of energy consumption (Aragüés *et al.*, 1985; and Golley *et al.*, 1990). For these reasons, the use of water for irrigation must be carefully planned, thus avoiding the very common practice of over-irrigation.

This paper has a twofold aim. First, a methodology is proposed that simultaneously determines the allocation of agricultural enterprises and irrigation systems in a newly irrigated area. Traditional criteria regarding private profitability, as well as social (e.g., employment levels) and ecological criteria (e.g., degradation of water quality due to "salt-loading" effects), are taken into account. The methodology demonstrates how a lexicographic goal programming (LGP) model is a suitable approach to these multi-criteria problems. Second, the proposed methodology is applied to a water management problem in the irrigated lands of the village of Tauste in Aragón (Spain). The irrigation efficiency level (65 percent) in this micro-region of more than 11,000 ha was found to be inadequate, generating important salinity problems for the waters of the Arba river.

## Irrigation Systems

To improve irrigation efficiency in the micro-region of Tauste, five feasible alternative systems were considered. The first system (hereafter System A) consisted of improvement of the existing infrastructure by covering the ditches with cement and levelling the land using laser techniques. This method leads to an irrigation efficiency of 72 percent (i.e., an increase of 7 percent over current efficiency). The second, third, and fourth systems (hereafter Systems B, C, and D) involved three different sprinkler systems with an irrigation efficiency of 81 percent. The differences between these three systems lie in the amount of energy and labour required. The fifth system (hereafter System E) consisted of a trickle irrigation system with an irrigation efficiency of 93 percent. Table 1 shows the characteristics of each of the five irrigation systems considered.

Table 1—Technical and Economic Characteristics of the Five Irrigation Systems Considered

| Characteristics | Improvement of the Current Infrastructure | Sprinkler Irrigation | | | Trickle Irrigation |
|---|---|---|---|---|---|
| | System A | System B | System C | System D | System E |
| Validity | All crops | All crops | All crops | Not for maize and sunflower | Only maize, vegetables, and fruit trees |
| Labour use (hours/ha/year) | Depends on crop | 6 | 3 | 60 | 8 |
| Energy (ptas/ha/year) | 0 | 4,628 | 7,518 | 4,630 | 4,420 |
| Cost (ptas/ha) | 138,800 | 371,510 | 260,160 | 136,850 | 228,010 |
| Planning horizon (years) | 15 | 25 | 15 | 15 | 20 |
| Annual maintenance and operational costs (ptas/ha) | 200 | 3,530 | 2,416 | 1,775 | 4,189 |
| Irrigation efficiency (percent) | 72 | 81 | 81 | 81 | 93 |

Table 2—Sensitivity Analysis Results

| Criteria | Scenario 1 | Scenario 2 | Scenario 3 |
|---|---|---|---|
| Water consumption (Hm³/year) | 59.5 | 59.5 | 62.8 |
| Energy use (million ptas/year) | 17.5 | 17.5 | 12.2 |
| NPV (million ptas) | 81 | 372 | 609 |
| Employment (work units/year) | 565 | 565 | 565 |
| Seasonal labour (work units/year) | 283 | 283 | 304 |
| Salt-load (t/year) | 20,619 | 20,619 | 30,584 |
| Loan (million ptas) | 1341 | 1099 | 2081 |
| Irrigation system (percent of whole area) | SA = 55, SF = 45 | SA = 55, SF = 45 | SA = 70, SF = 30 |

Currently, the whole area (approximately 11,000 ha) is irrigated by a furrow system. The five other systems have been successfully tested in the area. The irrigation scheduling system was not considered as it is a sophisticated system requiring an accurate data base and highly qualified manpower, not available in the region (Dudek *et al.*, 1981). Another method not considered consists of the establishment of a progressive water consumption levy, a policy that cannot be implemented because the current infrastructure of the area does not allow the measurement of individual water consumption. Consequently, farmers pay a fixed amount per hectare regardless of the water they use.

## Methodology

The following attributes are essential for policy-making purposes in the context of the decision-making problem under consideration and consequently endogenously introduced into the model (Zekri and Romero, 1990).

**Water consumption**. The reduction in water consumption for irrigation allows an increase in the use of water for other purposes, the irrigation of new areas, and the reduction of "salt-load" effects in return flows, thus improving water quality.

**Energy use for irrigation**. Public decision makers are interested in efficient irrigation systems that minimize the use of energy.

**Net present value** (NPV). The NPV represents a measurement of private profitability in the allocation of agricultural enterprises and irrigation systems.

**Employment**. This is important because farming is virtually the only economic activity in Tauste and its current unemployment rate is very high.

**Seasonal labour**. For the reasons mentioned, it is important to obtain high employment in terms of both daily wages and permanent workers.

"Salt-load" has not been explicitly considered. It is, however, complementary to water consumption; that is, when water consumption is minimized, "salt-load" is also minimized. Consequently, "salt-load" is considered an exogenous variable that is calculated once water consumption is obtained using the hydrosalinity model proposed by Aragüés *et al.* (1985) and adapted for Tauste.

Given the substantial number of attributes considered as well as the relative complexity of the constraint set, most of the potential benefits of some multi-criteria approaches such as multi-objective programming or compromise programming vanish. In fact, a problem of the size of the case presented in the paper is computationally almost intractable using multi-objective programming. However, such a problem can be easily treated through goal programming (Romero and Rehman, 1989, p. 102). Moreover, within the GP framework, the lexicographic variant is particularly relevant for decision making in the field of natural resource management where many attributes of a very different nature (economic, social, environmental, etc.) measured in different units have to be considered simultaneously (Romero, 1991, pp. 43–46).

In LGP, preemptive or absolute weights are attached to the achievement of various goals, which are grouped into a set of priorities. The problem corresponding to the highest priority level is solved first, and only then are the lower priorities considered. The structure of an LGP model can be summarized in the following way:

(1)   Lex min $a = [g_1(n, p), ..., g_i(n, p) ..., g_k(n, p)]$,

   subject to: $f_i(x) + n_i - p_i = b_i$, $\forall_i$, and $x \in F$,

where "Lex min" means a lexicographic optimization process; $g_i(n, p)$ is a function of the deviation variables; $f_i(x)$ is a mathematical expression for the $i$th attribute; $b_i$ is the target for the $i$th attribute; $n_i$ and $p_i$ are negative and positive deviational variables measuring the under-achievement and over-achievement of attribute $i$ with respect to its target; $x$ is a vector of decisional variables; and $F$ is the feasible set.[2]

In the problem under consideration, two priority levels were considered. In the first, the five attributes considered were included, attaching to them the following targets: 65 Hm$^3$/year for water consumption, 17.50 million ptas/year for use of energy, a zero value for the NPV (financial feasibility of the investment), 565 work units/year (1 work unit = 1920 hours) for the level of employment, and 565 work units/year for seasonal labour. For the attributes water consumption, energy use, and seasonal labour, the unwanted deviational variables are positive (i.e., over-achievements are not wanted), whereas for the attributes NPV and employment, the unwanted deviational variables are negative (i.e., under-achievements are not wanted).

The above vector of targets represents a situation that can be considered as an acceptable compromise for the policy maker and consequently should be satisfied in a Simonian sense, as far as possible. At the second priority level, the attributes water consumption, use of energy, and seasonal labour are again considered, although the targets are now set at more demanding levels: 54 Hm$^3$/year for water consumption, 5.5 million ptas/year for energy use, and 280 work units/year of seasonal labour. This second vector of targets represents what would satisfy the policy maker's needs as closely as possible once the achievement corresponding to the goals included in the first priority is maintained. The structure of the LGP model can be summarized as follows:

(2)   Lex min $a = [(\propto_1 p_1 + \propto_2 p_2 + \propto_3 n_3 + \propto_4 n_4 + \propto_5 p_5), (\propto_6 p_6 + \propto_7 p_7 + \propto_8 p_8)]$

subject to:

| | |
|---|---|
| Water ($w$) | $+ n_1 - p_1 = 65$ |
| Energy ($E$) | $+ n_2 - p_2 = 17.5$ |
| NPV | $+ n_3 - p_3 = 0$ |
| Employment ($EM$) | $+ n_4 - p_4 = 565$ |
| Seasonality ($S$) | $+ n_5 - p_5 = 565$ |
| Water ($w$) | $+ n_6 - p_6 = 54$ |
| Energy ($E$) | $+ n_7 - p_7 = 5.5$ |
| Seasonality ($S$) | $+ n_8 - p_8 = 280$ |

$x \in F$

The feasible set is formed for constraints related to land occupation, crop rotation, water requirements during the peak months, financial requirements, etc.[3] The weights $\propto_i$ play a double role: they represent the relative preferences of the decision maker and are also normalizing factors. Thus, for instance, $\propto_i = w_i / k_i$, measuring the relative importance attached by the decision maker to the $i$th goal with respect to the other goals included in that priority, and $k_i$ is equal to the difference between the ideal and the anti-ideal for this attribute. In this way, all traditional normalizing problems are avoided.

In order to compute the model, a planning horizon of 25 years was considered. It is assumed that the autonomous government will subsidize 30 percent of the outlay on the project and that the rest of the investment will be financed through a loan at a subsidized interest rate of 4 percent payable over 10 years. The discount rate chosen was 11 percent. All prices and costs are expressed in constant pesetas for 1988.

## Results

The LGP problem (2) was solved by resorting to the sequential linear method (Ignizio and Perlis, 1979) hybridized with the method proposed by Romero and Rehman (Romero, 1991, Chap. 2), thus avoiding the generation of inferior solutions. By attaching the same relative importance to the normalized goals placed in the same priority level, the following solution in the goal space was obtained:

$w$ = 65 Hm³/year; $E$ = 5.3 million ptas/year; $NPV$ = 346 million ptas

$EM$ = 565 work units/year; $S$ = 283 work units/year

The above solution permits a complete achievement of the goals included in the first priority. With respect to the goals included in the second priority, there is a slight discrepancy (with no real interest) for the goals energy and seasonal labour and an important deviation of 11 Hm³/year in water consumption. The corresponding irrigation structure is: 85 percent for System A (i.e., an improvement of the current infrastructure) and 15 percent for System E (trickle irrigation).

The "salt-load" corresponding to the above solution was calculated with the help of the hydrosalinity model. In this way, a "salt-load" of 36,209 t/year was obtained. This figure is important from an ecological point of view and is caused by the high water consumption of 65 Hm³/year.

As the project analysed is heavily financed with public funds, it is of great interest to investigate the possibility of reducing the high "salt-load" obtained by an increase in the outlay on the project. This task is accomplished by resorting to the Non-Inferior Set Estimation Method (NISE) (Cohon et al., 1979), which permits the determination of the transformation curve between outlay on the project and "salt-load." The transformation curve between the outlay on the project and water consumption is obtained. The figures for water consumption are transformed into "salt-load" figures using the hydrosalinity model (Figure 1).

Point A corresponds to the current situation; i.e., no improvement in the irrigation system. The corresponding "salt-load" is of 142,781 t/year. As usually occurs with this kind of analysis, the slopes of the straight lines connecting the corresponding extreme efficient points measure the opportunity cost of one objective ("salt-load" or water consumption) in terms of the other objective (outlay on the project or government subsidy).

In the southwest part of the transformation curve, the opportunity costs are not too high (e.g., an increase of 15,100 ptas in the outlay on the project for one ton of salt mass reduction along segment AB). However, in the northwest of the transformation curve, they are very high (e.g., an increase of 38,280 ptas in the outlay on the project for one ton of salt mass reduction along segment EF). It is possible to obtain sensible compromises in the neighbourhood of point C (i.e., acceptable figures of salt mass and water consumption) without greatly increasing outlay or government subsidy.

## Sensitivity Analysis

The uncertainty inherent in many of the parameters used makes it necessary to implement a sensitivity analysis to assess the impact of the uncertainty in the conclusions derived from the research. Several scenarios were built for this purpose, each of which attempts to reflect a realistic situation in terms of loan conditions, weights attached to the goals, costs of the irrigation systems, etc. The most significant results are shown in Table 2.[4]

To investigate the effects of a reduction in water consumption—and consequently a reduction in "salt-load"—in the three scenarios considered, the weight attached to the water consumption goal is 50 percent higher than the weights attached to the other goals.

The first scenario corresponds to the situation previously studied, with increased weight attached to the water consumption goal. The second scenario corresponds to an optimistic situation in which the cost of irrigation System A decreases 45 percent while the cash flow of the investment increases 2 percent per year. The third scenario corresponds to a pessimistic situation in which the cost of the different irrigation systems increases by 20 percent, no government subvention is considered, and the cash flow of the investment decreases by 2 percent per year.

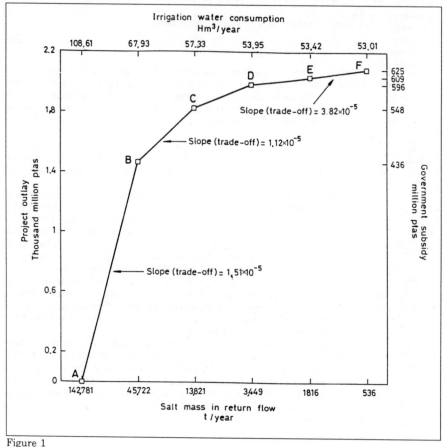

Figure 1

The results shown in Table 2 are instructive for the public decision maker as well as self-explanatory. However, the results in scenario 3 present some apparent oddities which should be clarified; e.g., without government subsidies, the NPV improves with respect to the other two scenarios. However, there is a simple explanation for this apparent anomaly. In scenario 3, the farmers must pay back the loans at an interest rate of 11 percent, which makes the financial constraints of the model more severe, so that cheaper combination irrigation systems are considered. These cheaper combinations generate higher NPV figures and at the same time considerably increase water consumption and hence "salt-load" figures.

## Concluding Remarks

Most of the goals relevant to the problem analysed are fully achieved (employment, NPV, etc.). However, there is a significant deviation between the target fixed for water consumption and the level actually achieved. This deviation is important because water is a scarce resource necessary for many purposes and also because the "salt-load" affecting the water quality of the river basin is directly proportional to the level of water consumed. In this sense, one of the conclusions derived from the model is particularly relevant: government subsidies can considerably improve environmental goals such as water consumption (e.g., "salt-load" instead of private goals such as NPV).

## Notes

[1] Universidad de Córdoba.
[2] For a more detailed description of the technical aspects of LGP, see Ignizio, 1976.
[3] For details, see Zekri 1991.
[4] For more details, see Zekri, 1991.

## References

Aragüés, R., Tanji, K.K., Quilez, D., Alberto, F., Faci, J., Machin, J., and Arrue, J.L., "Calibration and Verification of an Irrigation Return Flow Hydrosalinity Model," *Irrigation Science*, Vol. 6, 1985, pp. 85–94.

Cohon, J.L., Church, R.L., and Sheer, D.P., "Generating Multiobjective Trade-Offs: An Algorithm for Bicriterion Problems," *Water Resources Research*, Vol. 15, 1979, pp. 1001–1010.

Dudek, D.J., Horner, G.L., and English, M.J., "The Derived Demand for Irrigation Scheduling Services," *Western Journal of Agricultural Economics*, 1981, Vol. 6, No. 2, pp. 217–227.

Golley, F.B., Ruiz, A.C., and Bellot, J., "Analysis of Resource Allocation to Irrigated Maize and Wheat in Northern Spain," *Agriculture, Ecosystems and Environment*, Vol. 31, No. 4, 1990, pp. 313–323.

Ignizio, J.P., *Goal Programming and Extensions*, Lexington Books, Lexington, Mass., USA, 1976.

Ignizio, J.P., and Perlis, J.H., "Sequential Linear Goal Programming," *Computers and Operations Research*, Vol. 6, 1979, pp. 141–145.

Romero, C., *Handbook of Critical Issues in Goal Programming*, Pergamon Press, Oxford, UK, 1991.

Romero, C., and Rehman, T., *Multiple Criteria Analysis for Agricultural Decisions*, Elsevier, Amsterdam, Netherlands, 1989.

Zekri, S., "Modelos Decisionales Multicriterio en Planificación Agraria: Objetivos Económicos versus Objetivos Ambientales," Ph.D. dissertation, Universidad de Córdoba, Córdoba, Spain, 1991.

Zekri, S., and Romero, C., "A Multiobjective Methodology for the Assessment of the Status Quo Situation in Agriculture Planning: An Application to the Village of Tauste in Saragossa (Spain)," *VI European Congress of Agricultural Economists*, The Hague, Netherlands, Vol. 5, 1990, pp. 35–52.

# Discussion Opening—*A.K. Kashuliza* (University of London)

The role of irrigation in agricultural development and the Green Revolution, especially in geographical areas where water is the main limiting factor to agricultural production, is very clear and needs no over-emphasizing.

However, for sustainable agriculture, the increase in production has to be considered not only against economic and social costs, but also the costs to the environment (e.g., levels of waterlogging, salinity, etc.). Also, as rightly pointed out by Zekri and Romero in their paper, as water is a multipurpose natural resource essential for agriculture, industry, human consumption, etc., its intensive use in a single activity (e.g., agriculture) can generate an important opportunity cost in the other activities.

There are thus important trade-offs (economic, social, environmental, etc.) in the planning of water management projects or programmes. The decision maker is faced with a complex problem, where a multiplicity of criteria of very different nature have to be taken into account.

The Multi Criteria Decision Making (MCDM) techniques, of which LGP is a part, which have also evolved over the last two decades, are virtually the only way of satisfactorily investigating the type of problem described above.

The capital outlay attribute is not considered in the basic analysis (LGP) along with other attributes; i.e., water consumption, energy, NPV, employment, and seasonal labour. Capital outlay is an important attribute because the analysis is comparing technical and economic characteristics of five irrigation systems. Although the authors mention this attribute in a later part of the paper (i.e., that the government will subsidize 30 percent of the outlay on the project and that the rest of the investment can be made through a loan), this does not eliminate the need to have this attribute in the base model to enable both economic (social) and private decisions to be made about the projects. It is therefore surprising that this important attribute is only considered in passing in the sensitivity section of the paper.

In the LGP model, preemptive or absolute weights are attached to the achievement of various goals on an *ex ante* basis, which are then grouped into a set of priorities. Is there a standard methodology of deciding which relative weight to give to which attribute? Otherwise the process is very subjective and results could vary considerably depending on who the decision maker is on any particular occasion.

The logical sequence of the LGP model is that higher priority goals are satisfied before the lower ones (hence the lexicographic order). Similarly, the priority sets will have different attributes, with the higher priority sets having the most preferable attributes. However, in the formulation of Zekri and Romero's LGP model, both the priority sets have the same attributes (at different levels). I therefore feel that the authors are conducting a sensitivity analysis of the first priority set of attributes rather than solving the classical LGP problem (which requires several sets of priorities).

It is clear from Zekri and Romero's paper that results of LGP analysis can be greatly enhanced if water pricing (in the form of a progressive water consumption levy or other method) is also considered. This was not possible in this case because the infrastructure in the area does not allow measurement of individual consumption. It would be interesting to hear from those who have worked in areas where the pricing criteria could or has been incorporated in the LGP model on how it could have influenced the results presented in Zekri and Romero's paper.

**General Discussion**—*Geert Thijssen, Rapporteur* (Wageningen Agricultural University)

In relation to a central element of Oskam's paper, the measurement of external effects of agricultural production, many questions were raised about how the shadow prices were calculated, given that several methods are available, such as contingent valuation, travel costs, hedonic prices, etc. Oskam's method is one of estimated costs per unit of measure to be taken for the future; for example, for ammonia, a list of marginal costs per unit reduction was calculated for the Netherlands. The question was raised of whether these prevention costs are a good basis for estimating damage costs resulting from agricultural pollution. The human health effects of pesticide residues and, for example, salmonella in manure, food, or water are important damage costs, which should be taken into account. In reply, Oskam pointed out that in the Netherlands, the government has set up standards for adverse externalities, so that the health effects are implicitly taken into account in measuring the damage costs. There was some concern as to exactly how the shadow prices are measured, an aspect which Oskam will be taking up in further work. Another comment referred to calculating the positive externalities of agriculture, even though it is very difficult to value the landscape, for example.

Some fundamental questions were raised concerning the limitations of the type of model used by Livingston and Witzke; for example, whether it can be used to analyse the exploitation of tropical woods, in the Amazon region, and whether the neoclassical approach, which is continuous, can be used to analyse the political process, which can be characterized by jumps.

In reply, Livingston said the model can handle different types of institutional agreements between countries. Both countries can gain by the agreement: win-win. One country can compensate another country for reducing its pollution (for example, Sweden pays to Poland to reduce the pollution of the Baltic Sea): win-win. One country pays another country for reasons of justice, for example the developed countries finance pollution reduction by developing countries. The cost of organizing political support is incorporated in a new version of the model, which is also estimated. However, future generations are not considered in the model and only a democratic society was considered as a political structure.

Participants in the discussion included T. Haniotis (Commission of the EC), T. Hasebe (Tohokan University), E. Rabinowicz (Swedish University of Agricultural Sciences), T. Roberts (US Department of Agriculture), K. Thomson (University of Aberdeen), E. van Ravenswaay (Michigan State University), U. Vasavada (Université Laval), and T. Veeman (University of Alberta).

# Price and Subsidy Policy for Grain in China: Performance, Problems, and Prospects for Reform

## Ke Bingsheng[1]

**Abstract:** Based on a case study, this paper analyses China's current price and subsidy policy for grain and discusses some questions relating to the reorientation of this policy. First, the pricing system for the grain market, especially that for the state-run grain marketing agencies, is illustrated. Ways of subsidizing grain and the scale of the subsidy are then analysed. After discussion of the three major problems inherent in the system and a brief look at recent reform efforts, some points to be considered in terms of further reform are examined, including the need to modify the way of deciding priorities among conflicting policy objectives as well as the relationship between the state-set price and the open-market price. Finally, a proposal is made for gradual abolition of the subsidy and an operational cost-efficient approach is suggested.

## Introduction

Grain policy is the core of China's agricultural and food policy. In spite of the reform measures of the past 10 years, the grain sector, as with those for a few other farm products, still remains under strong direct government control, accompanied by a heavy burden on the state budget in subsidy payments on the one hand and declining supply and motivation among farmers engaged in grain production on the other hand. The Chinese government has been forced to tackle these and other problems and is seeking ways out of the dilemma. This paper, based on a case study, first illustrates China's current price and subsidy policy system for grain and then discusses some aspects of the reorientation of the policy.

## Double-Track System and Subsidy

At present, a so-called double-track system is applied in China's grain market, which means that there are two kinds of trade activities (one carried out according to the state-set plan and the other guided by the market mechanism) two kinds of prices (the state-set price for the first kind of trade and the open-market price for the second) and two kinds of marketing agencies (the state-run grain marketing agencies (SGM) and other marketing enterprises).

The state-set plan and price apply only to the SGMs. But not all the SGM trading activities are covered by central plan control. The local SGMs, after fulfilling the state plan, also operate according to the market situation, the so-called "negotiated" purchase, and resale at a "negotiated price." This part of the trading activities of SGMs is in fact of an open-market nature, for both the amount traded and the price applied are determined by the market situation, and the decisions involved are made by the local SGMs rather than by the government.

In practical terms, the negotiated price and the free market price are at the same level. Popular opinion regarding the negotiated price as a third type of price in addition to the state-set price and the open-market price is the result of misunderstanding.

The grain subsidy is paid only to the SGMs and only for the trade volume agreed under the state plan and at the state-set price. In practice, there are many kinds of subsidies for grain in China. Theoretically, however, they can be summarized into two categories: a subsidy to cover the "negative marketing margin"[2] or the differential between the state-set farm price and the consumer price, and a subsidy to cover marketing costs. This price and subsidy system is illustrated in Figure 1.

The trade of an SGM thus consists of three elements. The first (Part I) is farmer quota delivery, which is in turn rationed by the SGM to urban consumers. A total deficit of $(p_1{}'-p_2)q_1$ arises from this element. The same amount of state subsidy is needed to cover it. This would be the total subsidy if the quantity required for rationing were equal to $q_1$.

The fact is, however, that the quantity to be rationed is larger than $q_1$, because the farmer delivery quota has somehow been kept constant during recent years, while the rationed amount has expanded continuously due to the growth of the urban population. The increasing gap between $q_2$ and $q_1$ is bridged in practical terms by turning some of the SGM "negotiated purchasing" (purchasing on the open market) into "planned rationing." This is the second element of SGM trade (Part II), which also results in a deficit, and further subsidy has to be granted, which should be at least enough to cover the shaded area $(p_3'-p_2)(q_2-q_1)$. For the third element (Part III), which represents purely the open-market trade, no subsidy is granted.

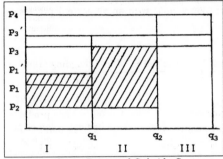

$q_1$: Quota purchasing at state-set farm price $p_1$
$q_2$: Amount needed for rationing at state-set consumer price $p_2$
$q_3$: Total amount traded
$q_2 - q_1$: Amount purchased at open-market purchasing price $p_3$ and rationed at $p_2$
$q_3 - q_2$: Resale at open-market price $p_4$ (retail or wholesale price)
$q_3 - q_1$: Open-market purchasing at $p_3$
$p_1 - p_2$: Negative marketing margin for Part I
$p_3 - p_2$: Negative marketing margin for Part II
$p_1' - p_1 = p_3' - p_3$: Average marketing cost

Shaded area: Amount of subsidy

Figure 1—Grain Price and Subsidy System

## A Case Study

A much clearer picture of the subsidy system can be gained by examining the results of a case study of the SGM of Zhengding County, Hebei Province. The major farm products grown there are wheat and maize, together taking up over 90 percent of the cultivated area. The case of wheat is used as an example.

**Quantity relationship.** As shown in Table 1, the traded volume of wheat in 1989 totalled 42,200 t, of which trade on the open market accounted for 40 percent at the purchase level and 25 percent at the resale level. The three elements of SGM trade account for 60 percent, 15 percent, and 25 percent, respectively.[3]

Table 1—Wheat Trade of the SGM in Zhengding (1989, 1000 t)

| Purchase | Ration | Open Market | Total |
|---|---|---|---|
| Quota | 25.4 (I) | — | 25.4 ($q_1$) |
| Open market | 6.3 (II) | 10.5 (III) | 16.8 |
| Total | 31.7 ($q_2$) | 10.5 | 42.2 ($q_3$) |

**Price relationship.** The corresponding prices are presented in Table 2. The negative marketing margin was large; 88 percent of the rationing price in the first element and 282 percent in the second. This means that the same high degree of government subsidy is required. There was also a great price disparity in the double-track system. At the farm level, the open-market price was twice as high as the state-set quota prices, while at the resale (retail or wholesale) level, the open-market price was more than four times the state-set rationing price.

**Subsidy.** It is relatively easy to calculate the subsidy needed to cover the negative market margin; i.e., 257 yuan/t for Part I and 827 yuan/t for Part II. The subsidies to cover marketing costs, however, are much more complicated. The major features of this kind of subsidy are listed in Table 3.

Table 2—Prices for Wheat in Zhengding (1989)

| | State-Set Price | | Open-Market Price | | Negative Marketing Margin | |
|---|---|---|---|---|---|---|
| | Purchase | Retail | Purchase | Resale | I | II |
| | $p_1$ | $p_2$ | $p_3$ | $p_4$ | | |
| Yuan/t | 550 | 293 | 1,120 | 1,200 | 257 | 827 |
| Percent | 188 | 100 | 382 | 410 | 88 | 282 |

Note: $p_2$ is derived from the price of flour (366 yuan/t) with an 80 percent processing ratio; $p_4$ is the wholesale price.

Table 3—Subsidy to Cover Marketing Costs of Wheat to the SGM in Zhengding (1989)

| | |
|---|---|
| Operating costs of retailing | 30.00 yuan/t |
| Operating costs of storage | 33.00 yuan/t/year |
| Capital (measured as the value of stored wheat) costs | |
| (derived from a subsidy rate of 0.0042 yuan/t month) | 27.72 yuan/t/year |
| Total (assuming an average storage time of 6 months) | 60.36 yuan/t |

To sum up, the average subsidy was 820 yuan/t for Part I and 890 yuan/t for Part II. The subsidy to cover the negative marketing margin was the major part of the total subsidy, making up 80–90 percent of the total sum.

To give an overview, the whole situation for wheat is displayed in Figure 2(a). Using the same method, the situation for maize is shown in Figure 2(b). The price relationships and the degree of subsidy for maize were similar to those for wheat. The major difference is that the share of Part III was much larger for maize than that for wheat, i.e., 70 percent for maize and 25 percent for wheat.

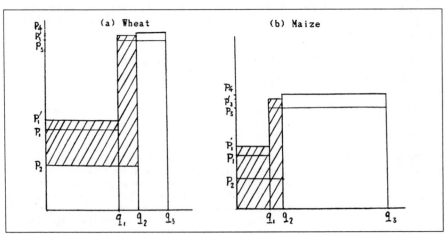

Figure 2—Grain Prices and Subsidies in Zhengding, 1989

## Problems

There are three major problems inherent in the above-mentioned price and subsidy system, which exert increasing pressure upon the Chinese government and call for long-term solutions.

The first is the disproportionate growth of the state financial burden, which has two dimensions. With a constant purchasing quota the increase in grain demand due to urban population expansion can only be met with Part II trade, which means, as in the example of wheat, that a more than 2-percent increase in subsidy is needed for every 1 percent of growth in the urban population. This alone suggests an 8-percent increase in subsidy annually, assuming a 4-percent annual growth rate of the urban population, as was actually the case during 1982–89. With a constant state-set consumer price and an increasing purchase price on both "tracks" (see Table 4), the average subsidy for each unit of grain rationed is increasing. If the price trend shown in Table 4 continues, this will mean an annual subsidy increase of 30 percent. Even if the actual price increase in the coming years is only half of this, the resulting subsidy increase will still be as high as 15 percent.

Table 4—Development of Purchasing Price for Grain in Zhengding (yuan/t)

| Year | 1986 | 1987 | 1988 | 1989 |
|---|---|---|---|---|
| Wheat: | | | | |
| State-set | 488 | 488 | 520 | 550 |
| Open market | 540 | 620 | 680 | 1,120 |
| Maize: | | | | |
| State-set | 322 | 342 | 342 | 362 |
| Open market | 420 | 420 | 470 | 680 |

The second problem is the increasing operational difficulty of quota purchasing. With the wide and widening gap between the open-market price and quota purchase price, increasingly greater operational and political costs are involved in persuading and forcing farmers to fulfil their delivery quota.

Besides its operational feasibility, the system is also questionable in social equity terms. The differential between the state-set quota price and the open-market price can be seen not only as a tax, but also as a subsidy paid by the farmer to the consumer. In the case of wheat, such a subsidy paid by the farmer for Part I trade amounted to 570 yuan/t, even larger than the government's share (320 yuan/t). This means that urban consumers are substantially subsidized by rural farmers, although the latter have only half of the income level of the former.

The third major problem is the high degree of waste in the marketing and consumption sectors. According to various sources, post-harvest waste of grain in China averages 12 percent of production (i.e., 50 Mt), while the waste in the consumption sector amounts to 20 Mt. Both figures greatly exceed China's annual grain imports. The high degree of waste is mainly due to inefficiency in the marketing system and the low price in the consumption sector, which is in turn closely related to the price and subsidy system discussed above.

## Recent Reform Efforts

Although there have been no centrally guided or fundamental changes in recent years, some individual experimental efforts have been made by various local governments. However, these efforts are mainly confined to changes in the subsidy-granting methods. They have tried to reduce or totally eliminate the negative marketing margins but not the subsidy. Neither the subsidy nor the price disparity between the two "tracks" has been abolished. The quota delivery obligation also remains. For example, in Guangdong Province, the state-set purchasing price for paddy rice has been increased from 191.60 yuan/t to 250 yuan/t, while the rationing price for rice has been raised from 148 yuan/t to 306 Yuan/t. The increase in the purchasing price is financed by cutting down the amount of subsidy previously granted to the farmer for fertilizer, while the increase in the rationing price is compensated with the same amount of direct subsidy to the consumer.

In all these experiments, the underlying problems remain basically unsolved, and new methods need to be sought.

## Aspects of Further Reform

**Policy objectives and instruments.** Ultimately, the problems are rooted in the conflict over current grain policy objectives, basically between the objectives of a low consumer price and sufficient supply. A low consumer price is in fact the central and starting point of the whole food policy system in China (Figure 3).

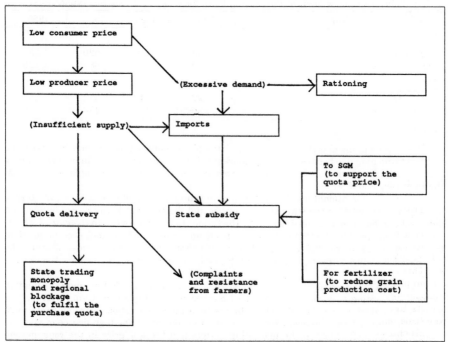

Figure 3—Grain Policy System

It is easy to see that compulsory quota purchasing, state subsidies, and imports are the major means for resolving the conflict between the two objectives. Since it is increasingly difficult to intensify the use of these means, consideration has to be given to whether one of the two objectives should be given up, and, if so, which.

Obviously, the price objective is less important than the quantity objective. With an improved income situation, people are much more concerned with the sufficiency of food supply than with the price level. Urban households that will be in a difficult situation if asked to pay for their food grain at the open-market price are nowadays exceptions in China. It is unconvincing to argue that people cannot afford food grain at the market price when they already own colour television sets, refrigerators, and washing machines. So, if the government is forced to cut down the subsidy, it should and has to give up the priority of retaining a low state-set consumer price. This will also induce a positive effect on improvement in marketing efficiency and reduce waste.

**Quantity guarantee.** Based on the above considerations, it can be proposed that the state-set price and subsidy should be abolished. In order to avoid or reduce the negative psychological effects and still give the urban consumer a feeling of security, the coupon system for rationing food grain can be further used. This, however, implies no more subsidy, only a quantity guarantee. This means that the urban consumer would receive the same amount of grain coupons as at present with which to buy grain from the SGM, but no longer at the subsidized price. Instead, they would have to pay the prevailing market price in the locality concerned.

**Open-market price.** Many people fear that there will be a strong increase in the open-market price if quota purchases and the subsidy are abolished. In effect, exactly the opposite will happen. Figure 4 helps to illustrate this. It suggests that the open-market price is neither directly influenced by the quota purchase price nor affected by marginal changes in the quota quantity. This clearly indicates that there is no direct correlation between the state subsidy and the open-market price.

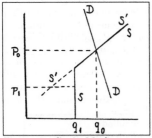

$q_1$: Quota purchasing at state-set price $p_1$.
$q_0 - q_1$: Open market purchasing at open market price $p_0$.
$S'S'$: Supply curve in the case of no quota purchasing.
$SS$: Supply curve with quota purchasing.
$DD$: Demand curve with subsidy.

Figure 4—Grain Market

There must be some indirect effects of the subsidy on the level of the open-market price due to the income effect of the subsidy on the demand curve. The abolition of the subsidy means a decline in the purchasing power of urban households. With a positive income elasticity of food grain demand (0.10–0.20), the demand curve will shift to the left. This implies a drop rather than an increase in the open-market price.

This is, of course, a conclusion based only on theoretical analysis. Its realization needs certain preconditions. The most fundamental of these are free competition and good market transparency. These are two decisive factors in achieving an efficient marketing system. Both aspects have been improving since the mid-1980s. The recent establishment of a grain wholesale market in Zhengzhou, Henan Province, is a significant effort in this direction.

**Mechanics of transition.** From both the political and the operational-technical viewpoints, the reform target cannot be accomplished in one move, neither can the subsidy be abolished overnight. It has to be carried out step by step.

At present, food grain coupons are issued monthly. The quantity of such subsidized grain per person per month varies from person to person according to age, sex, and occupation, covering a wide range from 5 kg for a one-year-old child to 13.5 kg for a dependent adult and from 15 kg for a white-collar employee to 30 kg for a heavy-industry worker such as a miner. So there is a technical problem of how gradually to reduce the amount of subsidized grain without involving too much operational cost. A possible solution is proposed that seems easy to implement: monthly reduction year by year. In the first year, the subsidy for one month (e.g., December) would be abolished, and for the remaining months remain unchanged. In the

149

second year, the subsidy for a further month (e.g., November) would be abolished, etc. Using this system, the whole subsidy can be eliminated in about 10 years.

For the low-income group that really has difficulties in buying food grain at the market price and needs government help, other solutions ought to be developed such as establishment and improvement of the social welfare system. The poor will be better and more efficiently subsidized by social policy, but not by market and price policy, which in fact benefits the upper-income group much more than the lower-income group.

## Notes

[1]Beijing Agricultural University.

[2]It is termed "negative" because the state-set consumer price is lower than the farm price.

[3]The corresponding figures for maize were 21 percent, 9 percent, and 70 percent, respectively.

---

## Discussion Opening—*Zbigniew Kowalski* (Academy of Technology and Agriculture, Poland)

Problems addressed in the paper bear a close resemblance to changes encountered very recently in Polish agriculture. Polish policy makers faced such problems at the beginning of the 1990s. Similar measures to those suggested by the paper were employed with impressive success. This places Polish farming in a position one step further on the way towards a market-oriented agriculture (a more general term that refers not only to a price setting mode but also to farmers seen as active agents able to benefit from the market). This discussion will therefore refer to some Polish experiences, which are common to all economies that change from central planning to the free market. An equitable food market, though naturally seen as the ultimate objective at the stage of reforms discussed in the paper, is in fact only a part of a broader package of problems that appear as prices are decontrolled.

Generally, any change in an agricultural price-setting mechanism seems to have two main aspects. First, it leads to a new situation on the food market for both producers and consumers (a new model of behaviour appears). Second, it affects the income distribution between rural and urban sectors. The first problem, as Polish experiences have shown, can be solved smoothly even if the market reform comes overnight (a situation which, according to the author, should be avoided in the case of China). The Polish food market has gone from central control to a free-price system within several months, turning from a 30-year-long period of shortages (with a food rationing system very similar to that in China) to the problem of overproduction. Having balanced the food market, the focus has moved on to the second question that emerged unexpectedly, namely the influence of the free market on farm income. It is known that under a central planning system that keeps food retail prices low, an essential part of farm income is being transferred out of agriculture as a hidden subsidy to consumers. But turning to a free market does not necessarily protect farmers from that. In centrally planned economies with family-type agriculture (as in China and formerly in Poland), the economy has assumed a curious dual structure—a highly centralized industrial sector versus a fragmented, scattered peasant agriculture. For central planners, the interest lies with agriculture. To control it, compulsory deliveries are often necessary. For market-eager reformists, such an agriculture is almost an ideal textbook model of free competition. But after freeing prices, agriculture is confronted by another textbook-model-like structure—a fully monopolized industry. As a result, agricultural incomes are again low (in Poland they are now, apparently, much lower then under central planning).

150

# An Intersectoral Perspective on the Relationship between the Agricultural and Industrial Sectors in Chinese Economic Development

## Won W. Koo and Lin Jinding[1]

**Abstract:** In a developing economy characterized by economic dualism, the interrelationship between the growth of the agricultural sector and that of the industrial sector is crucial for overall development. Theoretically, the agricultural and industrial sectors are closely linked. Agricultural progress would depend increasingly on the growth of industrial development and vice versa. However, no mutual dependency occurred in the Chinese economic development process. A causality test between the agricultural and industrial sectors of the Chinese economy indicates no cause-effect relationship. Growth models for the agricultural and industrial sectors were estimated using two-stage least squares. Labour productivity was low in China's agricultural sector before 1979, and the marginal productivity of labour was negative. Labour productivity and capital productivity in the industrial sector were also low. China's industrial development was mainly capital intensive and took place at the expense of the traditional agricultural sector. Labour productivity in the agricultural sector increased significantly after 1979, while productivity in the industrial sector decreased. This indicates that economic reform positively affected the agricultural sector in terms of labour productivity but negatively affected the industrial sector. Rural peasants have supported market-oriented economic reform more enthusiastically than urban people.

## Introduction

The interrelationship between the growth of the traditional indigenous agricultural and the modern industrial sectors of the economy are critical for overall development. Policy makers in most developing countries have realized its importance to industrialization and more recently the importance of the agricultural sector. However, economic analysis has largely neglected intersectoral links, concentrating instead on either macroeconomic or single-sector and subsector issues (Bacha, 1980). The primary objective of this paper is to evaluate the intersectoral perspective for China's agricultural and industrial economies.

Before 1949, the Chinese economy was very underdeveloped. The rural areas in China were destitute. From 1949, when the Communist party came to power, the Chinese leadership has promoted a nationwide industrialization programme. For a long time, priority in economic development was given to industry, especially heavy industry, and emphasis was placed on large-scale, state-owned industry, which was highly capital-intensive and concentrated mostly in urban areas. Although the Chinese leadership recognized agriculture's important contribution to economic growth and seemed to support a policy of concurrent growth, in practice they sought to achieve agricultural growth primarily through organizational changes and to accelerate industrial development through a high level of state investment financed largely through direct and indirect taxes on agricultural commodities. Consequently, great progress was made in China's industrial development, while agriculture was at a very low ebb. In terms of gross industrial and agricultural output value, the value of industrial output climbed from 30 percent of national income in the early 1950s to 74 percent in 1987, with agricultural output values falling from 70 percent to 25 percent. However, no corresponding changes took place in the employment structure. About 76 percent of the total labour force is still engaged in agriculture.

A series of reform programmes was launched on a large scale. The contract "responsibility system," with remuneration linked to output based on publicly owned land, was introduced and eventually gave way to individual household farming. Rural markets were free, and agricultural procurement prices rose significantly. The government switched from the take-grain-as-the-foundation policy to promoting a diversified development policy. Consequently, agricultural production grew rapidly. During 1978–88, agricultural output value increased at an average annual growth rate of 6.2 percent, and rural per capita incomes increased from 134 to 545 yuan, an average annual growth rate of 7.6 percent. Urban per-capita income rose from

316 yuan in 1978 to 1,119 yuan in 1987, an average annual increase rate of 5.9 percent (Zhong, 1989).

According to the World Bank (1985), Chinese agriculture will remain one of the largest and most important sectors of China's economy for the next two or three decades. By the year 2000, food will account for about 50 percent of the household budget and more than 50 percent of the total labour force still will work in agricultural activities. This implies that the Chinese agricultural sector will play an important role in the Chinese economy. It is therefore both interesting and challenging to study the agricultural and industrial economies in intersectoral terms, including patterns of sectoral development of the past four decades.

## Development of the Growth Model

Growth-promoting interactions between the agricultural and industrial sectors have been reviewed in the literature and accepted by many policy makers. The "theology" of development has emphasized that agricultural progress contributes to the support of great productivity throughout the economy. Agricultural progress will increasingly depend on growth of the industrial demand for agricultural commodities. In a dual economy, the ultimate question for future development of the economy is how the modern exchange sector can expand while the indigenous agricultural sector contracts. This requires an analysis of the interrelationship between the two sectors.

Using the methodology of Ranis and Fei (1964) to evaluate the interrelationship between the Chinese industrial and agricultural sectors, a growth model can be expressed as follows:

(1) $\quad AY_t = \alpha_0 AL_t^{\alpha 1} AB_t^{\alpha 2} IY_t^{\alpha 3}$

(2) $\quad IY_t = \beta_0 IK_t^{\beta 1} IB_t^{\beta 2} AY_t^{\beta 3}$

where $AY_t$ = gross national income in the agricultural sector, $AL_t$ = acres of arable land, $AB_t$ = the quantity of labour in the agricultural sector, $IY_t$ = gross national income in the industrial sector, $IK_t$ = the total amount of capital in the industrial sector, and $IB_t$ = the quantity of labour in the industrial sector.

In this model, $AY_t$ and $IY_t$ are treated as endogenous variables under an assumption that the two sectors of the economy help each other in the process of economic development and that the other variables ($AL_t$, $AB_t$, $IK_t$ and $IB_t$) are treated as exogenous. Equations (1) and (2) are a static model in which changes in the value of independent variables (i.e., $AL$ and $AB$ in Equation (1)) affect gross national income at the same time. There is, however, some evidence that indicates that changes in the value of independent variables in time $t$ affect gross income in $t$ and several periods in the future. Assuming that the dynamics take place under the partial adjustment hypothesis (Nerlove, 1958), Equation (1) can be rewritten as:

(3) $\quad AY_t^* = \alpha_0 AL_t^{\alpha 1} AB_t^{\alpha 2} IY_t^{\alpha 3}$

(4) $\quad \left( \dfrac{AY_t}{AY_{t-1}} \right) = \alpha \left( \dfrac{AY_t^*}{AY_{t-1}} \right)$

where $AY_t^*$ is desired or optimal gross income in the agricultural sector, and $\alpha$ is a dynamic adjustment coefficient. Combining Equations (3) and (4) yields:

(5) $\quad AY_t = \lambda \, \alpha_0 AL_t^{\lambda \alpha 1} AB_t^{\lambda \alpha 2} IY_t^{\lambda \alpha 3} AY_{t-1}$

Similarly, Equation (2) is rewritten, using the partial adjustment hypothesis, as follows:

(6) $\quad IY_t^* = \beta_0 IK_t^{\beta 1} IB_t^{\beta 2} AY_t^{\beta 3}$

(7) $\quad \left(\dfrac{IY_t}{IY_{t-1}}\right) = \lambda \left(\dfrac{IY_t^*}{IY_{t-1}}\right)$

Combining Equations (6) and (7) yields:

(8) $\quad IY_t = \lambda\beta_0 IK_t^{\lambda\beta 1} IB_t^{\lambda\beta 2} AY_t^{\lambda\beta 3} IY_{t-1}$

where $IY^*$ is desired or optimal growth income in the industrial sector and $\lambda$ is the dynamic adjustment coefficient. Equation (5) is a dynamic growth model for the agricultural sector and Equation (8) for the industrial sector.

Equations (5) and (8) are derived under an assumption that one sector of the Chinese economy influences the growth of the other sector. The causal direction between the agricultural and industrial sectors of the Chinese economy is tested using the procedure of Nelson and Schinert (Granger and Newbold, 1986). To test the null hypothesis that the growth of the industrial sector $(IY_t)$ does not cause the growth of the agricultural sector, the following equation is specified (Nelson and Schinert):

(9) $\quad AY_t = \displaystyle\sum_{j=1}^{k} d_{ij} AY_{t-j} + \sum_{i=1}^{n} d_{2i} IY_{t-i} + e_t$

(10) $\quad AY_t = \displaystyle\sum_{j=1}^{k} d_j AY_{t-j} + e_2 t$

Let us assume that $\hat{\sigma}_1^2$ and $\hat{\sigma}^2$ denote the residual estimates from Equations (9) and (10), respectively. The test statistic is:

(11) $\quad T = n(\hat{\sigma}^2 - \hat{\sigma}_1^2)/\hat{\sigma}_1^2$

which has an asymptotic $\chi^2$ distribution with $k$ degrees of freedom under the null hypothesis that the economic growth of $IY_t$ does not cause that of $AY_t$.

To test the null hypothesis that $AY_t$ does not cause $IY_t$, Equations (9) and (10) are respecified as:

(12) $\quad IY_t = \displaystyle\sum_{j=1}^{k} h_{ij} IY_{t-j} + \sum_{i=1}^{n} h_{2i} AY_{t-i} + e_{1t}$

(13) $\quad AY_t = \displaystyle\sum_{j=1}^{k} h_j IY_{t-j} + e_2 t$

The test statistics in Equation (11) are calculated from estimated residuals from Equations (12) and (13) and are used to test the null hypothesis.

## Empirical Results

Time-series data for 1952–88 were used to estimate the models. Most of the data used in this study were obtained from the 1988 *Almanac of China's Economy*. Chinese official economic statistics (except for 1958–60) are generally reliable. Other data such as those on the agricultural labour force and land came from Crook (1988). Land index data were adjusted based on Tang's index (Tang, 1981). National income is the value added to the country's material wealth from industry, agriculture, construction, transport, and trade. Industrial income in the model includes net material product from productive sectors other than agriculture. As an indicator of capital in the industrial sector, accumulated capital is the part of national income used to increase fixed capital assets, working capital, and material reserves.

### Relationship between the Agricultural and Industrial Sectors

Equations (12) and (13) are estimated as follows:

(14) $IY_t = -2.661 + 0.895\ IY_{t-1} - 0.127\ IY_{t-2} + 0.921\ AY_{t-1} - 0.344\ AY_{t-2} + e_{1t}$
$\qquad\ \ (2.480)\ (4.918)\qquad (0.882)\qquad\quad (3.046)\qquad\quad (0.915)$

$\quad R^2 = 0.9755,\ s.e. = 0.117$

(15) $IY_t = 0.182 + 1.257\ IY_{t-1} - 0.270\ IY_{t-2} + e_{2t}$ $\qquad R^2 = 0.9629,\ s.e. = 0.144$
$\qquad\ \ (0.572)\ (7.402)\qquad (1.603)$

Equations (9) and (10) are estimated as follows:

(16) $AY_t = 0.637 + 1.545\ AY_{t-1} - 0.677\ AY_{t-2} - 0.036\ IY_{t-1} + 0.089\ IY_{t-2} + e_{1t}$
$\qquad\ \ (1.213)\ (10.444)\qquad (3.679)\qquad\quad (0.404)\qquad\quad (1.263)$

$\quad R^2 = 0.9685,\ s.e. = 0.057$

(17) $AY_t = 0.123 - 1.531\ AY_{t-1} - 0.544\ AY_{t-2} + e_{2t}$ $\qquad R^2 = 0.9685,\ s.e. = 0.060$
$\qquad\ \ (0.405)\ (9.952)\qquad (3.354)$

where numbers in parentheses are the $t$-values for the corresponding parameters and $s.e.$ represents standard error.

The value of the $\chi^2$ statistic calculated from Equations (14) and (15) is larger than the critical value of the statistics at the 5-percent significance level, rejecting the null hypothesis that growth of the agricultural sector has not caused the growth of the industrial sector in the Chinese economy.

The $\chi^2$ test with Equations (16) and (17) accepts the null hypothesis that growth of the industrial sector has not caused growth in the agricultural sector in the Chinese economy.

The causality test indicates that growth of the agricultural sector has contributed to growth of the industrial sector, but that the industrial sector has not contributed to the growth of the agricultural sector. The following factors may explain this result.

**Industry has developed at the expense of an "agricultural squeeze."** In the 1950s, the Chinese leadership adopted many aspects of the Soviet model of economic development. The agricultural sector was a resource base to be "exploited" to serve development strategies. To accumulate capital to serve the development of the country's weak and underdeveloped industry, the government adopted the practice of monopolized state procurement and marketed farm and "sideline" products at low prices. The state purchased these commodities at extremely low prices in rural areas and marketed them at similar or slightly higher prices to urban residents and enterprises. This policy kept wage expenditure and cost of raw materials for its major industries low and created exceptional profits in the industrial sector and the necessary contribution of funds for its industrial development. Relevant statistics show that during 1949–78 the differentials between industrial and farm and "sideline" product prices have meant a "gratis contribution" of 600,000 million yuan from the peasants or 45 percent of their total income for this period (Jiang and Luo, 1989).

**An "urban bias" discriminated against agriculture.** The Chinese leadership, particularly Mao, recognized the distinct forms that agriculture's contribution could take.

**The rural areas became isolated from the urban areas**. A strict system of resident registration divided the country's urban and rural residents. The peasants had to work on limited arable land and perceived no possibility of improving their circumstances in this closed or semi-closed economy. Agricultural development lost vigour and vitality. Egalitarian distribution practices reduced the peasants' enthusiasm for production and productivity.

## Growth Model for Agricultural and Industrial Sectors

Growth models for the agricultural and industrial sectors (Equations (5) and (8)) were estimated using two-stage least squares. Following the causality test described in the previous section, the growth model for the agricultural sector does not include the growth measures in the industrial sector as an independent variable, but the industrial growth model contains growth measures for the agricultural sector. A dummy variable $(D_t)$ representing economic reform since 1978 and a variable interacting with the labour variables are included to investigate the impacts of the policy on labour productivities. The agricultural growth model also includes a trend variable to capture effects of improvements in farming technology. The estimated equations are as follows:

$$\text{Log } AY_t = - 1.854 + 1.292 \text{ Log } AL_t - 0.149 \text{ Log } AB_t + 0.709 \text{ Log } AY_{t-1} - 21.33 \ D_t$$
$$(0.419) \quad (1.613) \qquad\quad (0.339) \qquad\qquad (6.329) \qquad\qquad (1.950)$$

(18)

$$+ \ 1.694 \ (D_t \text{ Log } AB_t) + 0.0046 \ TR \qquad R^2 = 0.9622$$
$$(1.955) \qquad\qquad\quad (0.618)$$

$$\text{Log } IY_t = - 1.885 + 0.361 \text{ Log } IK_t + 0.381 \text{ Log } IB_t - 1.174 \text{ Log } AY_{t-1} + 0.381 \text{ Log } IY_{t-1}$$
$$(1.263) \quad (4.878) \qquad\quad (2.894) \qquad\qquad (2.897) \qquad\qquad (5.314)$$

(19)

$$+ \ 1.308 \text{ Log } AY_t + 4.379 \ D_t - 0.380 \ (D_t \text{ Log } IB_t) \qquad R^2 = 0.9880$$
$$(3.957) \qquad\qquad (2.488) \qquad (2.494)$$

where $D_t$ is a dummy variable representing the 1979–88 period in which the Chinese government used a semi-market-oriented economic policy. This dummy variable is used to evaluate the effects of economic policy on growth of gross national income in the agricultural and industrial sectors. The dummy variable interacting with the labour variable is used to evaluate changes in labour productivity in the agricultural and industrial sectors during 1979–88 compared to 1953–78.

The values of $R^2$ are 0.96 for the growth model of the agricultural sector and 0.99 for the growth model of the industrial sector, indicating that economic growth in both the agricultural and industrial sectors can be explained very well by the variables used in the models. In the growth model for the agricultural sector, the estimated coefficients are not highly significant except for the lagged dependent variables, although the model has a high $R^2$. This is due mainly to the high multicollinearity among the independent variables. The estimated coefficients in the growth model for the industrial sector all differ significantly from zero at the 5-percent significance level.

The dummy variable and the variable interacting with the labour variables can be adjusted to the intercept term and to the estimated coefficients for the labour variable for the models for 1979–88, while the coefficients are the same as those of Equations (18) and (19) for the models for 1952–77. The coefficients of labour for 1952–78 are –0.149 for the agricultural sector and 0.381 for the industrial sector and 1.545 and 0.001 for 1979–88. These coefficients are interpreted as marginal products of labour.

Three implications can be drawn from comparing these coefficients of the industrial and agricultural sectors models between these two time periods.

1. The marginal productivity of labour was negative and increased substantially in 1979–88. The economic institutions and strategy developed in China since the 1950s repeated the major features of the traditional Soviet model with only minor variations. Planners

attempted to extract the maximum level of surplus agricultural product to meet the demands of planned growth in the industrial sectors. During the period of collectivization of agricultural production, all the agricultural labour was kept on the farmland. Peasants could not work in non-agricultural lines of production, nor in forestry, animal husbandry, or fisheries.

The steady natural growth of the agricultural labour force and the sharp decline in the available arable land per capita produced an army of surplus agricultural workers. In 1978, the number of people of working age totalled 528 million, of which 298 million were employed, leaving labour resources of 230 million available (Yeh, 1984).

The situation regarding the rural labour surplus seems to have been more severe. A detailed study of 30 population teams in Nantong County, Jiangsu Province, concluded that this county had surplus labour with only 1.6 mu (about 0.107 ha) per head of the agricultural labour force. The study reports that about 4 mu (about 0.267 ha) per worker would be needed to avoid surplus labour (Song, 1982). This is a substantially higher estimate of labour requirements than many others have used. The Ministry of Agriculture, Animal Husbandry, and Fisheries uses an estimated average cropping intensity of 9 mu (about 0.6 ha) per worker in crop production to forecast labour requirements. An estimated one-third of the agricultural labour force is superfluous (World Bank, 1985). Although 1,000 million person-days of labour input were mobilized in China's agriculture, particularly in rural labour-intensive construction work campaigns since the 1950s, agricultural production per person-day fell. Consequently, China's success in absorbing rural surplus labour through collectivization brought with it a substantial decline in the average and marginal productivity of labour.

Since 1978, the new system of production responsibility in rural areas and the higher prices for state purchases of major farm products have encouraged peasants to engage in "sideline" production, revived free markets so that peasants can sell their privately produced products, and increased their incentive to work for the collective and for themselves. In 1979, the first year the new agricultural policies were put into effect, total output value from agriculture rose 8.6 percent over the 1978 level. Grain production increased by 6.1 percent, reaching 333.12 Mt, a record high. Cotton production rose by 1.8 percent, and the three oil-bearing crops (groundnuts, sesame, and rape) increased by 23.5 percent. Each peasant's average income rose from 117 yuan in 1977 to 170 yuan in 1980. Peasants' savings deposits in banks increased from 4,650 million yuan in 1977 to 12,660 million yuan in 1980 (Lin and Chao, 1982).

2. Both labour and capital productivity in the industrial sector are low, indicating that China's industrial development is based mainly on capital intensity with low efficiency of workers.

3. Unlike labour productivity in the agricultural sector, that in the industrial sector decreased in the 1979–80 period, indicating that economic reform since 1978 has affected the agricultural sector positively in terms of labour productivity but the industrial sector negatively.

## Summary and Conclusion

In a developing economy characterized by dualism, the interrelationship between growth of the agricultural and industrial sectors is crucial for overall development. Theoretically, the agricultural and industrial sectors are closely linked. Agricultural progress depends increasingly on the growth of industrial development and vice versa. However, this did not happen in the Chinese economic development process. Empirical testing of a dual growth model indicates that growth of the agricultural sector increased growth of the industrial sector, but growth in the industrial sector did not increase growth in the agricultural sector. Chinese planners followed Soviet economic development strategies of developing the industrial sector by an "agricultural squeeze." The government monopolized state procurement and marketed farm and "sideline" produce at low prices to accumulate enough capital to develop modern industry. Agriculture has been discriminated against by an "urban bias." A strict resident registration system, which divided the country's urban and rural residents into two parts and forced peasants to remain on limited arable land, also contributed to the interrelationship between agricultural and industrial development.

Growth models for the agricultural and industrial sectors were estimated using two-stage least squares. Labour productivity was low in the agricultural sector before 1979, and the marginal productivity of labour was negative. Since both labour productivity and capital productivity in the industrial sector were low, China's industrial development was based mainly on intensity of resource use. While labour productivity in the agricultural sector increased significantly after 1979, that in the industrial sector decreased, indicating that economic reform positively affected the agricultural sector in terms of labour productivity but affected the industrial sector negatively. Rural peasants have supported market-oriented economic reform more enthusiastically than urban dwellers.

## Note

[1]North Dakota State University and Xiamen University, respectively.

## References

Bacha, E.L., "Industrialization and Agricultural Development," in Cady, J., *et al.* (Eds.), *Policies for Industrial Progress in Developing Countries*, Oxford University Press, London, UK, 1980.

Crook, F.W., *Agricultural Statistics of the People's Republic of China, 1949–86*, Statistical Bulletin No. 764, Economic Research Service, US Department of Agriculture, Washington, D.C., USA, 1988.

Granger, C.W.J., and Newbold, P., *Forecasting Economic Time Series*, 2nd Ed., Academic Press, Orlando, Fla., USA, 1986.

Jiang, J.Y., and Luo, X.P., "Changes in the Income of Chinese Peasants since 1978," in Longworth, J.W. (Ed.), *China's Rural Development Miracle*, University of Queensland Press, St. Lucia, Australia, 1989.

Lin, W., and Chao, A., *China's Economic Reform*, University of Pennsylvania Press, Philadelphia, Pa., USA, 1982.

Nerlove, M., *The Dynamics of Supply: Estimation of Farmers' Response to Price*, Johns Hopkins University Press, Baltimore, Md., USA, 1958.

Ranis, G., and Fei, J.C.H., *Development of the Labor Surplus Economy*, Richard Irwin, Homewood, Ill., USA, 1964.

Song, L., "Village Labor Surplus and Its Outlet," *Social Science in China*, Vol. 5., 1982.

Tang, A., "Chinese Agriculture: Its Problems and Prospects," Working Paper 82–WO9, Department of Economics, Vanderbilt University, Nashville, Tenn., USA, 1981.

World Bank, *China: Long-Term Development Issues and Options*, Johns Hopkins University Press, Baltimore, Md., USA, 1985.

Yeh, K.C., "Macroeconomic Changes in the Chinese Economy during the Readjustment," *The China Quarterly*, Vol. 100, 1984.

Zhongguo Jingji Nianjian (1988 Almanac of China's Economy), Jingji Guanli Chubashe, Beijing, China, 1989.

Zhong J., "40 Years of Socialist Economic Construction," *China Reconstructs*, Vol. 38, 1989.

## Discussion Opening—*Petri Ollila* (University of Helsinki)

In the real world, the effects of any sector of economic activity cannot be isolated from other sectors. Koo and Lin have made a contribution to agricultural economic research in widening the view of analysis beyond agriculture. The complex process of interaction between the agricultural and industrial sectors has been captured in a relatively simple form of analysis.

Because my personal experience of Chinese circumstances is extremely limited, it is very difficult to evaluate how well the researchers have succeeded in their task. The evaluation is probably also hard for some other readers, because many things obviously well known to China experts are not defined in the paper. Knowledge of the exact contents of "industry" and "agriculture" would have helped in understanding exactly what has been analysed. The borders between these two and their relationship to other sectors are undefined. For instance, how is the income from self-sufficiency agriculture evaluated in the gross national income of agriculture? What is the unit of income, and has it remained comparable during the period discussed? How is the description of the state buying agricultural commodities at extremely low prices included into the model? What does it tell about the volume?

How stable have the categories "agriculture" and "industry" been during 1952–88, the period of analysis? In many countries, the following reasoning would be possible: The development of agriculture has had some technical and income effects. Adoption of the mattock, the steel plough, a better variety of rice, or an improved irrigation system actually shifts tasks from agriculture both downstream and upstream. This means that the development of agriculture has actually become visible in other economic sectors, which is also the finding of the present analysis. The limited population migration may be among the reasons for the finding that development of industry does not contribute to agriculture. Even if these categories had been stable and the data usable, some further clarification other than simply "Chinese official economic statistics (except for 1958–60) are generally reliable" should have been presented.

The authors make many choices about data and factors in the model, its shape, and the method of estimation, without much explanation of their choices. Would consumption have been a relevant factor? Although it is perhaps obvious, why was a Cobb-Douglas type model with two-stage least squares estimation chosen?

The implications of the results seem to me quite strong. If the null hypothesis that agricultural growth has not caused the growth of industrial sector is rejected, the opposite may not necessarily be true. The link between results and the explanation also remain to some extent unclear.

The description leads well into the problem area under analysis. The key finding that the development of agriculture has supported the development of the industrial sector but not vice versa is an interesting one. Some discussion about the data, estimation techniques, and the meaning of the results could have been expected.

*[Other discussion of this paper appears on page 166.]*

# Changes in China's Meat Consumption Patterns: Implications for International Grain Trade

*Praveen M. Dixit and Shwu-Eng H. Webb*[1]

**Abstract:** A world net trade model is used to study the consequences of changes in meat consumption patterns in China. Results suggest that such changes would considerably depress world grain prices, especially those for maize and rice. The study also shows that grain consumption requirements in China would fall by 8 percent and China would improve its agricultural balance of trade by $2,000 million. The study concludes that while self-sufficiency in grains may improve with consumption realignment, there would be real income losses because of the consumption distortions introduced.

## Introduction

To meet its commitment to staple urban food supply and to maintain self sufficiency, the Chinese government has consistently stressed grain production as a top policy priority, especially after the economic reforms of 1978, which allowed peasants to sell their surpluses after meeting their procurement quota. However, with the increase in freedom that followed reform, Chinese farmers shifted production into cash crops (e.g., fruit and vegetables) and livestock products, and grain production fell short of targeted production. Production stayed below the record 1984 levels for four years so that, in 1989, the government decided to make grain production again a top priority.

Procurement prices of food grains have been raised many times since 1989. The domestic wheat price is now well above the international price. Rice and maize prices have similarly also been raised. The government also adopted many measures to facilitate use of land for grain production. These measures, along with recentralization of production, raised grain production to a record level of 408 Mt in 1989, exceeding the 1984 level of 407 Mt. Available data indicate that 1990 grain production could even have reached 435 Mt.

Despite recent successes in grain production, China continues to have a lingering fear of its ability to feed a growing population and seeks to achieve the maximum possible level of food self-sufficiency. The Chinese Ministry of Finance estimates that grain requirements for the year 2000 could be 500 Mt. Several options are therefore being talked about to remedy the potential grain problem.

Greater market orientation is one such option. Several studies, including Dixit and Webb (1990) and Gunasekera *et al.* (1991), show that elimination of policies that tax grain producers and subsidize consumers could make China completely self-sufficient in grains. In addition, because more than 15 percent of the government's budget goes into subsidizing urban consumption of grains and edible oils, unilateral policy reform would also generate budgetary savings of almost $8,000 million.

Another option, which is being increasingly discussed as a means to solve China's grain problem, is to persuade the population to adjust its meat consumption behaviour. The objective of this option would be to steer consumption away from meats that have high feed-grain requirements to those with low feed-grain requirements. This latter scheme may be especially attractive if it can be achieved with minimal budgetary costs.

Adjusting Chinese meat consumption behaviour has another advantage. Given China's low per-capita meat consumption levels, any increase in per capita incomes would be likely greatly to increase meat consumption and generate additional demands for grains. Hence, any measure to guide consumption away from meats that use grains intensively could be extremely beneficial.

Either of these options can be expected to have a tremendous effect on Chinese agriculture. And what happens to agriculture in China is of great interest to the world agricultural community. China is the world's largest producer of grains, accounting for nearly 20 percent of global output in recent years. With 22 percent of the world's population, China is also the largest consumer of agricultural products. Any change in China's agricultural production and consumption can therefore be expected significantly to affect the world agricultural markets.

## Meat Consumption and Feed-Grain Requirement

China consumes a considerable amount of pigmeat. Nearly 82 percent of China's meat consumption (excluding fish products) is pigmeat, and its pigmeat total/meat consumption ratio is nearly twice the world average. Pigmeat, however, has a much higher feed-grain requirement than poultry. While each kg of pigmeat production requires 4.14 kg of grain, the ratio is only 3.2 for poultry (Table 1).

Table 1—Meat/Protein Ratio, Feed-Grain/Meat Ratio, and Per-Capita Meat Consumption in China, 1986

| Commodity | Meat/Protein Ratio[1] | Feed-Grain/ Meat Ratio[2] | Per-Capita Meat Consumption[3] | New Per-Capita Meat Consumption |
|---|---|---|---|---|
| | gm/kg | kg/kg | kg/year | kg/year |
| Pigmeat | 117.00 | 4.14 | 14.41 | 6.86 |
| Beef | 117.00 | 6.26 | 0.67 | 0.34 |
| Sheepmeat | 117.00 | 1.16 | 0.67 | 0.00 |
| Poultry | 156.00 | 3.20 | 1.72 | 6.86 |
| Fish | 103.00 | 0.87 | 5.40 | 7.33 |
| Average | 116.63 | 3.27 | 22.87 | 21.39 |

[1]Piazza (1986), p. 74.  [2]Tuan (1987).
[3]Unpublished data, Economic Research Service, US Department of Agriculture.

The first task in this study was to design meat consumption patterns that minimized feed-grain requirements. A linear programming model was built using the GAMS software to deduce the optimal meat consumption patterns. Because any unconstrained linear programming problem is more than likely to generate a corner solution that may seem rather unrealistic, a number of restrictions were imposed in determining the extent of changes in consumption habits.

First, it was assumed that, despite changes in meat consumption patterns, minimum protein intake from meats as represented by current levels would still be maintained. Second, it was assumed that pigmeat would account for at least 30 percent of Chinese meat consumption (including fish products, which are important in the Chinese diet). The figure of 30 percent was chosen because it reflects the pigmeat/meat consumption ratios in a number of industrial countries, including the USA. Third, because poultry products have low feed-grain requirements relative to other meats but about the same protein conversion ratio, it was assumed that at least 30 percent of meat consumption would be poultry. Fourth, because beef has the highest feed-grain conversion ratio, it was assumed that Chinese beef consumption would be no less than half the current levels. Mathematically, the procedures for calculating grain requirements under the proposed meat consumption patterns can be expressed as:

(1) minimize: $FG' = \Sigma (PC'_i \cdot FC_i) \cdot POP$

subject to: $\Sigma (PC'_i \cdot PT_i) \cdot POP \geq \Sigma (PC_i \cdot PT_i) \cdot POP$
$PC'_p \geq 0.30 \cdot \Sigma PC_i$
$PC'_c \geq 0.30 \cdot \Sigma PC_i$
$PC'_b \geq 0.50 \cdot PC_b$

where:  $FG'$ = feed-grain requirement with change in meat consumption
$PC'_i$ = per-capita consumption of meat $i$ after change
$PC_i$ = per-capita consumption of meat $i$ before change
$FC_i$ = feed-grain/meat conversion ratio
$PT_i$ = meat/protein conversion ratio
$POP$ = total population in China in 1986 (1,057,210,000)
$i$ = meat products, with $p$ for pigmeat, $b$ for beef, and $c$ for poultry

Given the assumptions on behaviour changes but maintaining current levels of protein nutrients in the meat diet, pigmeat consumption would decrease by 52 percent to 6.86 kg per capita, and beef consumption by 50 percent to 0.34 kg per capita, while poultry consumption would increase by 300 percent to 6.86 kg per capita, and fish product consumption by 36 percent to 7.3 kg per capita. With these changes in consumption patterns, the protein content per kg of meat consumed would increase by 7 percent. Consequently, less meat (a decline of 6 percent) is required to maintain the same protein level, and feed-grain requirements would decline by 16 percent per kg of meat production.

Chinese policy makers may view changes in meat consumption patterns as a means to reduce grain use, but changes in meat demand will not in themselves lower China's grain requirements. Associated changes in livestock production have to follow if feed-grain use is to decline; otherwise, China—a net exporter of meat products—would simply produce the same quantity of livestock products but export surpluses to the world market. Because China's objective is to minimize grain use, the targeted consumption patterns were achieved by instituting production controls so as to meet modified domestic demand levels and yet maintain existing levels of meat exports.

## The Modelling Framework

The economic implications of consumption reform in China are analysed using the Static World Policy Simulation (SWOPSIM) modelling framework (Roningen, 1986). A SWOPSIM model is a non-spatial price equilibrium model, an intermediate-run static model that represents world agriculture in a given year, and a multi-commodity, multi-region partial equilibrium model. In order to use this static, non-spatial price equilibrium model to describe world agricultural trade, it is assumed that world markets are competitive, that domestic and traded goods are perfect substitutes in consumption, and that a geographic "region," possibly containing many countries, is one market place.

The economic structure of SWOPSIM models includes constant domestic supply and demand equations. Trade is the difference between domestic supply and total demand (absorption). The policy structure is embedded in equations linking domestic and world prices. Policies (PSEs and CSEs) are inserted as subsidy equivalents at the producer, consumer, export, or import levels. Details on the economic and policy structures and the use of summary support measures in the modelling framework are presented in Roningen (1986) and Roningen and Dixit (1989).

The version of SWOPSIM used for this study (CH86) is based on 1986/87 marketing year data. The world is divided into 13 regions—7 of which represent the industrial market economies, 3 characterize developing countries, and 3 describe the centrally planned economies. Twenty-two agricultural commodities representing mainly temperate zone products are included in the model. Fish products, which could be a major source of relief for concern about Chinese grain self-sufficiency, are not included.

This paper presents the results of experiments using the CH86 model in which new equilibrium solutions are obtained by pegging consumption levels as specified earlier. The new solution represents an approximation of the resulting adjustments in production, consumption, trade, and prices of agricultural commodities expected after five years, with the important proviso that all other conditions remain the same as in the base year, 1986/87. This permits the analysis to isolate and identify the differences between the new solution and the initial or reference solution and to attribute them to the changes in consumption patterns.

## Realigning Chinese Consumption Patterns

Using the model and the modified grain requirements described earlier, we studied the economic implications of imposing changes in Chinese meat consumption behaviour. Two issues were of particular interest: how changes in consumption patterns would affect world commodity prices and how these changes would affect Chinese agricultural trade.

### Effects on World Prices

Model results indicate that if China were to realign its consumption patterns so as to minimize production of grain-intensive meats but fulfil current protein requirements, world agricultural prices, on average, would fall by 3 percent (Figure 1). The drop in world prices would be large for rice (16 percent) and maize (8 percent), which together account for the majority (73 percent) of total feed use in Chinese agriculture. There would also be some drop in the price of wheat (3 percent) and oilseed products (4 percent). By contrast, world prices for beef, pigmeat, and poultry would change very little because Chinese production and consumption patterns alter such that net trade remains the same.

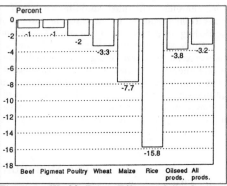

Figure 1—World Price Effects of Production/Consumption Realignment

How do these changes compare with those resulting from unilateral Chinese policy reform (Dixit and Webb, 1990)? Figure 2 indicates that even though, on average, the declines in world prices are very similar, there are at least two major differences. Whereas there is no change in the world pigmeat price under consumption realignment, the increase in price is more than 10 percent under unilateral policy reform. Secondly, the grain price changes, especially for rice (4 percent), are relatively moderate under unilateral reform compared with consumption realignment (16 percent).

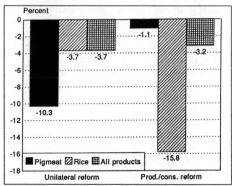

Figure 2—World Price Effects of Reforms

### Effects on Chinese Trade

Chinese demand for cereals would fall by 8 percent (35 Mt) if consumption and production patterns were realigned along the lines proposed in this study (Figure 3). As a result, Chinese cereal exports would increase by 19 Mt. Maize (11 Mt) and rice (8 Mt) would account for most of the increase. In other words, China would be self-sufficient in cereals and switch from being a net importer to a net exporter cereals. Exports of oilseed products would also increase slightly (2 Mt), while trade in most livestock products would remain unchanged.

These changes in import patterns are somewhat different than those that would be achieved under unilateral agricultural policy reform. Dixit and Webb (1990) show that, with unilateral reform, Chinese cereal exports would remain virtually unchanged while exports of pigmeat would expand by 2.6 Mt. Given that pigmeat is a high-value product, China's agricultural balance of trade would improve by $6,400 million under unilateral policy reform but by only $2,300 million under consumption reform. From a foreign exchange perspective, therefore, the gains in export revenues from trade liberalization are more appealing.

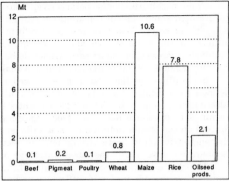

Figure 3—Changes in Chinese Net Trade from Consumption Realignment

Such differences in trade patterns between the two reform options occur largely because livestock trade is maintained at existing levels under consumption reform, forcing all adjustments vis-à-vis the world grain market. Under unilateral policy reform, on the other hand, both the grain and livestock sectors in China can respond to changes in world market conditions and so the adjustment for any one sector is much less.

## Limitations of Analysis

The economic implications of changes in Chinese meat consumption behaviour are likely to be different depending on the year under consideration. This study uses the 1986/87 marketing year as the basis of analysis. In many respects, it is probably more reasonable to examine the issue at a future date—the year 2000 or later—when China could be expected to import grains if current population and production trends continue.

This study assumes that modified Chinese consumption patterns are achieved through production and trade control measures. But whether this can be achieved in reality is doubtful; much of what is consumed in China is based on on-farm production and monitoring and maintaining proposed production régimes may be difficult. Rationing or taxing meat consumption could be another means of achieving the desired levels, but whether such rationing could be successful in practice is quite another question. Political unrest over the price of staple foods is a feature of many developing countries that the political hierarchy in China may not wish to import.

Finally, there is also the issue of how one establishes the minimum levels of consumption. In this analysis, it is assumed that at least 30 percent of meat consumption would be pigmeat, based on levels existing in some industrial countries. Realistically, 30 percent could be a rather difficult target to achieve, and minimum levels more reasonable than this may have implications very different than those presented in this paper.

## Conclusion

One of China's main concerns is to achieve the maximum possible level of grain self-sufficiency. The objective is to ensure that China does not have to rely too heavily on international markets for grains. The results indicate that the self-sufficiency ratio for grains increases from 0.98 to 1.05 with consumption realignment. Under free trade, the ratio rises to 1.

From a self-sufficiency perspective, therefore, there is little doubt that the Chinese government should favour consumption realignment. Consumption realignment, however, is not without costs. Whereas unilateral reform would generate real income gains of $2,500 million to the Chinese economy, consumption realignment would cost $1,600 million. Similarly, increases in export earnings would be nearly $6,000 million under unilateral reform but only $2,000 million under consumption realignment.

Self-sufficiency, however, can be viewed in a number of respects. Traditionally, it is defined as a country's ability to produce that quantity of food necessary for subsistence. Alternatively, it may be defined as a country's ability to acquire enough foreign exchange earnings to purchase necessary food requirements from the world market. The analysis focused on the former definition of self-sufficiency. If the latter approach were adopted, then the emphasis would be to strengthen the sector in which China would appear to have a comparative advantage—the labour-intensive pigmeat industry.

The choice for China is pretty clear. China can either pursue consumption realignment and achieve greater food self-sufficiency or undertake unilateral policy reform and achieve lower levels of self-sufficiency with increased real income (efficiency) gains. The challenge for China is to recognize the benefits of freer agricultural trade and yet not compromise on some of its domestic food concerns.

# Note

[1]US Department of Agriculture.

# References

Dixit, P.M., and Webb, S.-E.H., "Liberalizing Agricultural Policies in China: Effects on World Markets," paper presented at the American Agricultural Economics Association annual meeting, August 4–8, 1990, Vancouver, B.C., Canada.

Gunasekera, H.D.B.H., Andrews, N.P., Haszler, H.C., and Chapman, J.N., "Agricultural Policy Reform in China," Discussion Paper No. 91.4, Australian Bureau of Agricultural and Resource Economics, Canberra, Australia, March 1991.

Piazza, A., *Food Consumption and Nutritional Status in the People's Republic of China*, Westview Press, Boulder, Colo., USA, 1986.

Roningen, V.O., "A Static World Policy Simulation (SWOPSIM) Modeling Framework," Staff Report No. AGES–860625, Economic Research Service, US Department of Agriculture, Washington, D.C., USA, 1986.

Roningen, V.O., and Dixit, P.M., "Economic Implications of Agricultural Policy Reforms in Industrial Market Economies," Staff Report No. AGES–8936, Economic Research Service, US Department of Agriculture, Washington, D.C., USA, 1989.

Tuan, F., "China's Livestock Sector," Foreign Agricultural Economic Report No. 226, Economic Research Service, US Department of Agriculture, Washington, D.C., USA, 1987.

# Discussion Opening—*Thomas Wahl* (Washington State University)

This paper provides an interesting and useful analysis of the effects of changing China's meat consumption pattern upon international grain markets. The results are a beginning point for discussion. However, the analysis falls short of adequately examining the problem using the SWOPSIM model.

Alternative consumption patterns that are more realistic for China need to be considered. The consumption patterns imposed on the model reflect western biases and, further, they would force tremendous adjustments in China's livestock infrastructure to meet them.

Fish consumption and production are nearly ignored in the paper, despite the popularity of fish with consumers, the Chinese government's commitment to increase fish production, and the much greater efficiency of fish production. From Table 1, dividing the meat/protein ratios by the feed-grain/meat ratios should suggest the number of grams of protein produced from 1 kg of feed grain. The ratio is 28.3 gm/kg for pigmeat, 48.8 gm/kg for poultry, and 118.4 gm/kg for fish. These ratios clearly suggest that fish production uses less feed grain to produce protein than does poultry production.

The elasticities and PSEs used in the model are not presented; nor is an analysis of the sensitivity of the results to the assumed elasticities. The relative prices of poultry and pigmeat are not discussed. Currently, the poultry price is as high as that of pigmeat. The relative prices necessary to induce consumers drastically to alter their consumption pattern are not presented.

In summary, the paper is the beginning of a useful analysis of Chinese meat consumption. However, alternative consumption patterns, fish production levels, and elasticities need to be considered.

*[Other discussion of this paper appears on the following page.]*

# General Discussion—*Csaba Forgács, Rapporteur* (Budapest University of Economic Sciences)

Several issues connected with intersectoral analysis of the agricultural and industrial economies were discussed. It was argued that the conclusions drawn by Koo and Lin indicated no cause and effect relationship between the two sectors of the Chinese economy. Two methodological aspects were mentioned. First, if industrial goods such as fertilizers or chemicals had been chosen for analysis, a stronger correlation might have been observed. Second, if the industrial sector had been divided into two parts, rural industry and the rest of industry, the results would have been different, and the positive impact of the agricultural machinery industry could have been estimated.

The second key point of the discussion was Chinese meat consumption patterns. Although the usefulness of long-term analysis was emphasized, counter arguments, connected with model assumptions applied, related to their consistency and reality. It was also mentioned that the meat consumption level of one country cannot be applied automatically to another country. The questionable model assumptions may lead to results that cannot give a basis for the conclusions taken. Although models always simplify the real world, model assumptions have to be set up carefully, and variables should be defined exactly to obtain reliable results.

Participants in the discussion included He Xiping (Beijing Agricultural University), J.Y. Lin (Development Research Center of the State Council, China), Liu Xiaoqiu (Chinese Academy of Agricultural Mechanical Sciences), F.C. Tuan (US Department of Agriculture), Xiao Hui (Beijing Agricultural University), and Zhang Cungen (Chinese Academy of Agricultural Sciences).

# The Relationship between Food Aid and Food Trade: Theoretical Analysis and Quantitative Results

*Roland Herrmann, Carlo Prinz, and Patricia Schenck*[1]

**Abstract:** This paper clarifies linkages between food aid and food trade, both theoretically and empirically. A theoretical model first shows that domestic policy is crucial for the trade effects of food aid. Food trade may fall, remain constant, or even rise due to increased food aid. The important issue is whether the recipient country's government spends the counterpart funds to subsidize demand or supply. Estimated cereal import demand functions for Botswana, Egypt, Morocco, Peru, and Sudan indicate very different reactions of food imports to food aid across countries. The paper also investigates from the donor's point of view how a country's food import position affects the amount of food aid it receives. A cross-country analysis of the allocations of the EC's food aid reveals that per-capita food aid is higher the lower the per-capita income of a recipient country, the worse its balance-of-payments situation, and the more it depends on food imports.

## Introduction

Although there is an extensive economic literature on the pros and cons of food aid (Dearden and Ackroyd, 1989; Srinivasan, 1989; Lachmann, 1988; Clay and Singer, 1985; and Isenman and Singer, 1977), analytical and quantitative studies on many aspects of food aid are lacking. This holds true, for example, for the relationship between food aid and food trade. This paper aims to elaborate important linkages between food aid and food trade, both theoretically and quantitatively. Two questions are examined: how the food trade of recipient countries responds to food aid and whether the distribution of food aid by donors is responsive to the food-deficit situation of potential recipients, and, if so, to what extent.

## Impact of Food Aid on Food Trade

### Theoretical Analysis

If the answer to the question, "Does food aid lead to less food imports and, thus, contribute to a solution of balance-of-payments problems in developing countries?," is yes, the next question is, "To what extent does food aid reduce food imports?" In order to tackle these questions, a stylized model of the market for a food staple in a developing country is used. The analysis is comparative statics and hence concentrates on the direct and short-run rather than on the indirect and long-run effects of food aid on food trade.

Despite the stylized nature of the model, it contains a structure sufficient to draw some important theoretical conclusions on the impact of food aid on food trade. In particular, it incorporates one essential feature of food aid policy—the use of counterpart funds in the recipient countries. Counterpart funds arise when food aid is given in the form of bulk supply. In quantitative terms, bulk supply is the dominant type of food aid, whereas emergency food aid, project aid, and triangular transactions are less important (FAO (a)). Under bulk supply, food deliveries flow to the recipient country and are sold by the government through the normal marketing channels. Counterpart funds are equal to the quantity sold multiplied by the domestic market price. The government may use the counterpart funds in either the food or the non-food sectors of the economy. It can be shown that the trade impact of food aid depends crucially on the use of these counterpart funds. This implies that domestic policy in the recipient country is important for the trade effect of food aid.

The argument can be shown within a theoretical model. The benchmark situation without food aid is characterized by the following equations:

(1) $\quad S^{PR} = a + bp$

167

(2) $S^{GO} = 0$

(3) $D^{PR} = c + dp$

(4) $p = e + fp_w$

(5) $p_w = \bar{p}_w$

(6) $S = S^{PR} + S^{GO} + M$

(7) $S = D^{PR}$

Food supply $(S)$ in this country is composed of private supply $(S^{PR})$, supply by the government $(S^{GO})$, and imports $(M)$. Supply by the government is zero in the non-food-aid case. Food demand in the country consists only of private demand $(D^{PR})$. As Equations (1) and (3) show, private food supply and private food demand are a function of the price on the domestic market $(p)$. The domestic food price is a function of the world food price $(p_w)$. Equation (4) characterizes a price transmission equation often used in agricultural policy analysis. Differences between the domestic and the world price may be due to transport costs and national food price policies. Equation (5) describes the small-country case; changes in quantities traded by the respective country do not alter the world price. $a, b, c, d, e,$ and $f$ are coefficients of the model. The theoretical expectations on the signs are: $a > 0, b > 0, c > 0, d < 0, e > 0,$ and $f > 0$.

Food imports are included in Equation (6), which can be rewritten after including (7) as:

(8) $M = D^{PR} - S^{PR} - S^{GO}$

After introducing (1), (2), and (3) into (8), food imports in the non-food-aid situation can be written as:

(9) $M_0 = c + dp - a - bp$

Equations (4) to (8) remain valid in the model for the food-aid situation. Equations (1)–(3), however, are replaced by:

(1′) $S^{PR} = a + bp + h(\beta p \bar{FA})$

(2′) $S^{GO} = \bar{FA}$

(3′) $D^{PR} = c + dp + g(\alpha p \bar{FA})$

Equation (2′) indicates that an exogenously given amount of food aid $(\bar{FA})$ will be sold by the government on the domestic market. This yields counterpart funds in the magnitude of $(p\bar{FA})$. Counterpart funds will now be used partly to stimulate food demand and partly to stimulate food supply. We posit in Equation (3′) that a share of these counterpart funds is given to the population that demands food, whereby $0 \le \alpha \le 1$. Basically, this leads to an income effect. The coefficient $g$ $(>0)$ indicates how food demand increases when income rises by one monetary unit. In Equation (1′), it is assumed that a certain share of the counterpart funds is used for stimulating technical progress in domestic food production. $\beta$ ranges between 0 and 1. The additional assumption $(\alpha + \beta) \le 1$ guarantees that counterpart funds can also be partly spent in the non-food sector. $h$ indicates the responsiveness of domestic food production to an additional monetary unit spent on technical change in agriculture. The theoretical expectation is that $h > 0$.

In the system of Equations (1′)–(3′) and (4)–(8), food imports can be derived as:

(10) $M^* = c + dp + g(\alpha p\overline{FA}) - a - bp - h(\beta p\overline{FA}) - \overline{FA}$

A comparison of (9) and (10) yields the impact of food aid on food imports:

(11) $\Delta M = M^* - M = g(\alpha p\overline{FA}) - h(\beta p\overline{FA}) - \overline{FA} >$ or $\leq 0$

Equation (11) shows the very general result that food imports may be lowered or raised by food aid when the use of counterpart funds in the food sector is captured by the model. A one-to-one substitution of food imports by food aid ($\Delta M = -\overline{FA}$) is a special case when counterpart funds are not spent in the food market ($\alpha = 0$, $\beta = 0$). When counterpart funds are only used for stimulating food production ($\beta > 0$, $\alpha = 0$) (e.g., by fostering technical change in agriculture or by subsidizing inputs), an additional amount of food imports will be replaced by domestic production. In that case, increasing food aid by one ton leads to decreasing imports by more than one ton. The economic explanation is straightforward: food imports are directly substituted by food aid. The use of counterpart funds for shifting domestic production to the right will further reduce food imports. When counterpart funds are only used to subsidize demand ($\alpha > 0$, $\beta = 0$) (e.g., by transferring income to the poor), additional food imports will occur. In that case, an increase in food aid of one ton will lower food imports by less than one ton. Food imports may even rise when the shift in the food demand curve due to food subsidization overcompensates for the increased availability of food as a consequence of food aid. When counterpart funds are used to subsidize demand and supply ($\alpha > 0$, $\beta > 0$), the direction and size of the impact of food aid on food imports depend upon the relative shifts of the demand and supply curves. It can be derived that food aid is more likely to lead to a substitution of food imports:[2] the lower the share of counterpart funds used to subsidize demand, the lower will be $\alpha$; the weaker the response of the population to increasing incentives for food demand, the lower will be $g$; the higher the share of counterpart funds used to subsidize supply, the higher will be $\beta$; and the stronger the response of food producers to increasing incentives for food production, the higher will be $h$.

When the supply shift is larger than the demand shift, food aid will lead to a stronger substitution of food imports: the higher the amount of food aid, the larger will be $\overline{FA}$; and the higher will be the domestic price of food.

## Empirical Results on the Impact of Food Aid on Food Trade

Import demand functions are estimated for five LDCs that receive significant amounts of food aid: Botswana, Egypt, Peru, Sudan, and Morocco. The objective is to quantify the actual impact of food aid on food imports. The computations refer either to cereals or wheat, as cereals are the main food aid products with regard to delivered quantities (FAO (a)). The estimations are based on the same theoretical import demand model. The hypothesis is that net cereal imports ($M$) depend upon the national import price of cereals ($p_w$), a domestic income variable ($Y$) measured by GDP data, domestic cereal production ($PR$), and cereal food aid ($FA$):

(12) $M = f(P_w, Y, PR, FA)$

Table 1 shows the empirical results on the food aid-food trade linkage in the five countries. Coefficients of linear models, which are in line with the theoretical framework shown above, are presented, as well as those of log-linear models.

From Table 1, it can be derived that the impact of food aid on food imports follows no uniform pattern across the countries. For Peru and Botswana, the estimated coefficients indicate that commercial imports are substituted by food aid. This effect is commonly expected. The strongest degree of substitution exists in Botswana: increasing food aid deliveries by one ton causes a decline in commercial imports of nearly two tons. The empirical results show for Egypt, Sudan, and Morocco, however, a positive impact of food aid on food imports. In the case of the linear models, the positive coefficients are statistically significant at the 95-percent level for all three countries. A rise of 1 kg in food aid per capita leads to and increase in imports per capita of 0.47–1.03 kg. The estimated coefficients of $> -1$ lend support to the hypothesis that counterpart funds are primarily used to subsidize food demand in developing countries. Only in Botswana, where a more than one-to-one substitution between food aid and food trade was measured, are counterpart funds likely to be used for production subsidies. The regression coefficients of the log-linear models again suggest that a positive impact of food aid on food trade occurs in several countries.

Table 1—Estimated Reactions of New Cereal Imports to Variations in Cereal Food Aid, Selected Developing Countries, 1971–87[a]

| Country | Linear Models (1) | Log-Linear Models (2) |
|---|---|---|
| Botswana | $-1.9854^* \ FA_{t-1}$ <br> $(-2.40)$ | $0.2087 \ FA_{t-1}$ <br> $(1.37)$ |
| Egypt | $1.0343^* \ FA/C_t$ <br> $(2.39)$ | $0.3820^{**} \ FA/C_t$ <br> $(3.83)$ |
| Morocco[b] | $0.9651^* \ FA/C_t$ <br> $(2.35)$ | $0.0749 \ FA/C_t$ <br> $(1.24)$ |
| Peru | $-0.8631^* \ FA/C_t$ <br> $(-2.28)$ | $-0.0609^* \ FA/C_t$ <br> $(-2.24)$ |
| Sudan[b] | $0.04697^* \ FA/C_t$ <br> $(2.85)$ | $0.2662^{**} \ FA_t$ <br> $(4.46)$ |

[a]The levels of statistical significance of the estimated coefficients are indicated by * for 95 percent and ** for 99 percent. The values in parentheses are $t$-values. In most cases, the food aid variable was measured in per-capita values (FA/C). The indices $t$ and $t-1$ are the two periods considered.

[b]Referring to wheat food aid.

Sources: Authors' computations with data from FAO (1987), FAO (a), FAO (b), FAO (c), FAO (d), International Monetary Fund, and World Bank.

# The Influence of Food Import Dependence on the International Allocation of Food Aid: A Quantitative Analysis of EC Food Aid Policy

So far, food aid has been treated as an exogenous variable from the individual recipient country's point of view. We now analyse how the country's food imports are affected by changes in food aid. In the following analysis, food aid is treated as an endogenous variable on which the donor country decides according to certain criteria. Determinants of the international allocation of food aid are then elaborated for EC food aid policy. EC food aid consists of bilateral member and Community action in which the latter's share is roughly 60 percent (FAO (a), 1989). As well as officially declared criteria for the distribution of food aid, we test whether food import dependence affects the amount of EC food aid a country receives. The officially postulated criteria for EC food aid are: fundamental food requirements, per-capita income, the balance-of-payments situation (European Communities, 1982), and the economic and social effects, as well as the cost of the proposed measure (European Communities, 1986).

As this paper covers the 1983–85 period and the last criterion was added by the EC Commission in 1986, the cross-country analysis that follows concentrates on the first three criteria. Computations using the variable "fundamental food requirements"—measured by the daily per-capita calorie supply as a percentage of the daily calorie requirements—resulted in no statistically significant coefficients. This variable was therefore dropped from later regressions. Two additional variables are included that capture the recipient country's dependence on food imports.

The following alternative equations are estimated:

(13) $FA_i = f(GNP_i, CAB_i, CIMP_i)$

(14) $FA_i = f(GNP_i, CAB_i, SSR_i)$

$FA$ is the amount of EC cereal food aid per capita committed to the recipient country $i$. $GNP$ is the gross national product per capita and is used as a measure of per-capita income in the recipient country $i$. $CAB$ represents the current-account balance and captures the balance-of-payments situation in country $i$. $CIMP$ stands for cereal imports per capita in country $i$. $SSR$ is the self-sufficiency ratio for cereals in country $i$ and is computed as domestic cereal production divided by the sum of domestic cereal production and cereal imports. In general, the independent variables are lagged 2 years, because, it is argued, the allocation decision of the EC bureaucracy is likely to be based on the most recent data set available. Comparisons with cross-country data included in the World Bank's *World Development Report* show a time-lag for the relevant data of 2–3 years.

The empirical results are presented in Table 2. They confirm the importance of the postulated allocation criteria, $GNP$ and $CAB$. There is a negative influence of per-capita income and the balance-of-payments situation on the amount of food aid a country receives. This influence is significant at least at the 95-percent level in all model specifications. The estimated reaction to a $100 change in $GNP$ is 400–900 g of food aid per capita. A deterioration of the current-account balance by $10 per capita raises the deliveries of EC food aid 300–900 g per capita.

Besides the strong impacts of $GNP$ and $CAB$, the donations of food aid to a country are significantly negatively related to the country's self-sufficiency ratio in all equations and significantly positively influenced by cereal imports in all equations except in that for 1984. The regression coefficient indicates that an increase of one percentage point in the self-sufficiency ratio leads to a reduction in EC food aid of 75–100 g per capita. On the other hand, each additional kg of cereal imports per capita increases food aid by roughly 30 g per capita. The corrected coefficients of determination of the models are rather high: 50–82 percent of the variance in EC food aid shipments can be explained in all periods under consideration. These values of $\bar{R}^2$ are noteworthy, given a cross-country analysis.

## Summary and Conclusion

The linkages between food aid and food trade play a central role in the economics of food aid. It was the objective of this paper to study these linkages theoretically and in a quantitative analysis. The theoretical analysis revealed that domestic policy is crucial for the trade effects of food aid. Food trade may fall, remain constant, or even rise due to increased food aid. The important issue is whether the recipient country's government spends the counterpart funds on subsidizing demand or supply. Estimated cereal import demand functions for Botswana, Egypt, Morocco, Peru, and Sudan indicated very different reactions of food imports to changes in food aid. Positive as well as negative linkages were found. From a donor's point of view, the allocation of food aid is not exogenous, as it is for the recipient country, but depends on certain decision criteria. EC food aid in cereals was shown to depend heavily on two officially postulated allocation criteria, the income and the balance-of-payments situation. Moreover, the larger the amount of food aid a country receives from the EC, the more the country depends on food imports.

Table 2—Determinants of the International Allocation of the EC's Food Aid, 1983–85[a]

| Dependent Variables | Independent Variables | | | | | Test Statistics | |
|---|---|---|---|---|---|---|---|
| | CONST | $GNP_{t-2}$ | $CAB_{t-2}$ | $CIMP_{t-2}$ | $SSR_{t-2}$ | F | $\bar{R}^2$ |
| FA 1983 | 1.6000 (2.86)* | -0.0077 (-4.63)** | -0.0601 (-4.14)** | 0.0258 (3.62)** | | 10.98 | 0.70 |
| | 9.4650 (6.69)** | -0.0072 (-5.80)** | -0.0451 (-3.92)** | | -0.0833 (-5.47)** | 21.36 | 0.82 |
| FA 1984 | 3.0214 (2.94)** | -0.0090 (-3.59)** | -0.0803 (-4.08)** | 0.0270 (1.59) | | 9.74 | 0.51 |
| | 12.1805 (4.07)** | -0.0080 (-3.69)** | -0.0645 (-3.52)** | | -0.1016 (-3.10)** | 14.69 | 0.62 |
| FA 1985 | 2.1020 (2.89)** | -0.0055 (3.86)** | -0.0350 (-2.33)* | 0.0358 (3.47)** | | 9.19 | 0.54 |
| | 8.9939 (4.05)** | -0.0047 (-3.23)** | -0.0340 (-2.13)* | | -0.0751 (-3.09)** | 7.94 | 0.50 |
| FA 1983-85 | 2.3334 (4.51)** | -0.0071 (-6.10)** | -0.0575 (-5.46)** | 0.0300 (3.94)** | | 22.07 | 0.51 |
| | 10.0752 (6.50)** | -0.0065 (-6.01)** | -0.0487 (-4.69)** | | -0.0842 (-5.00)** | 27.43 | 0.57 |

[a]$FA$ = food aid; $CONST$ = constant term; $GNP$ = gross national product; $CAB$ = current account balance; $CIMP$ = cereal imports; $SSR$ = self-sufficiency ratio; $\bar{R}^2$ = corrected coefficient of determination; and $F$ = $F$-value. The levels of statistical significance of the estimated coefficients are indicated by * for 95 percent and ** for 99 percent. The values in parentheses are $t$-values. 14 countries are considered for 1983 and 26 and 22 for 1984 and 1985, respectively. The pooled-sample regression considered 62 observations.

Sources: Authors' computations using FAO (b), FAO (c), FAO (d), World Bank, and Commission of the European Communities (1986), pp. 10–11.

## Notes

[1]Universität Giessen.

[2]The following results can be derived by differentiating Equation (11) with regard to its determinants: $\delta(\Delta M)/\delta\alpha = gp\overline{FA} > 0$; $\delta(\Delta M)/\delta g = \alpha p\overline{FA} > 0$; $\delta(\Delta M)/\delta\beta = -(hp\overline{FA}) < 0$; $\delta(\Delta M)/\delta h = -(\beta p\overline{FA}) < 0$; $\delta(\Delta M)/\delta\overline{FA} = (g\alpha-h\beta)p - 1 < or \geq 0$; $\delta(\Delta M)/\delta p = g\alpha-h\beta > or \leq 0$.

## References

Clay, E.J., and Singer, H.W., "Food Aid and Development: Issues and Evidence," World Food Programme, Occasional Paper No. 3, Rome, Italy, 1985.

Commission of the European Communities, "Nahrungsmittelhilfe der Europäischen Gemeinschaft," Grünes Europa, No. 216, 1986.

Dearden, P.J., and Ackroyd, P.J., "Reassessing the Role of Food Aid," Food Policy, Vol. 14, No. 3, 1989, pp. 218–231.

European Communities, "Verordnung (EWG) Nr. 3331/82 des Rates vom 3 December 1982, über die Nahrungsmittelhilfepolitik und -verwaltung und zur Anderung der Verordnung (EWG) Nr. 2750/75," Amtsblatt der Europäischen Gemeinschaften, No. L352, 1982, pp. 1–4.

European Communities, "Verordnung (EWG) Nr. 3973/86 des Rates vom 22 December 1986, über die Nahrungsmittelhilfepolitik und -verwaltung," Amtsblatt der Europäischen Gemeinschaften, No. L370, 1986, pp. 1–4.

FAO (Food and Agriculture Organization), World Crop and Livestock Statistics, 1948–85, Rome, Italy, 1987.

FAO (a) (Food and Agriculture Organization), Food Aid in Figures, Rome, Italy, various years.

FAO (b) (Food and Agriculture Organization), Production Yearbook, Rome, Italy, various years.

FAO (c) (Food and Agriculture Organization), The State of Food and Agriculture, Rome, Italy, various years.

FAO (d) (Food and Agriculture Organization), Trade Yearbook, Rome, Italy, various years.

International Monetary Fund, International Financial Statistics: Yearbook, Washington, D.C., USA, various years.

Isenman, P.J., and Singer, H.W., "Food Aid: Disincentive Effects and Their Policy Implications," Economic Development and Cultural Change, Vol. 25, No. 2, 1977, pp. 205–237.

Lachmann, W., "Wirtschaftstheoeretische überlegungen zu möglichen Wirkungen der Nahrungsmittelhilfe in Entwicklungsländern," in Körner, H. (Ed.), Probleme der ländlichen Entwicklung in der Dritten Welt (Schriften des Vereins für Socialpolitik, Neue Folge, Vol. 173), Berlin, Germany, 1988, pp. 101–122.

Srinivasan, T.N., "Food Aid: A Cause of Development Failure or an Instrument for Success?," World Bank Economic Review, Vol. 3, No. 1, 1989, pp. 39–65.

World Bank, World Development Report, Washington, D.C., USA, various years.

---

**Discussion Opening**—*K.N. Ninan* (Institute for Social and Economic Change, India)

The paper raises interesting issues that merit a detailed discussion. While technically competent, the economic reasoning behind the relationship hypothesized between food aid and food trade needs to be placed on a sounder footing.

Food aid and food imports can be positively related only superficially when a scarcity situation stimulates both; i.e., when food aid is inadequate, it is supplemented by imports. Also, if aid increases incomes and there is a high income elasticity of demand for food in LDCs, some increase in imports may take place. But a positive association in either case does not amount to an absence of substitution of imports or at least an avoidance of scarcity. In some economies that do not normally import, deficits in domestic supply may be made up solely through food aid.

In the EC's allocation of food aid, the authors consider food imports as an important variable. The economic logic behind how imports lead to increased food aid needs to be spelt out clearly.

The analysis considers only the linkage between food aid and food trade when, in fact, it would have been more meaningful and useful if trade in general had been considered. If food aid replaces commercial food imports, the foreign exchange so saved could be used for importing non-food or investment goods. If food aid affects the recipient country by inducing a shift from production of food staples to production of exportable cash crops, this, too, could stimulate trade.

On the question of how food trade of recipient countries responds to food aid, the authors' findings revealed that, out of five countries studied, in two (Peru and Botswana), commercial imports were substituted by food aid, while in three (Egypt, Sudan, and Morocco), increased food aid led to increased food imports. However, in Botswana, the log-linear function showed a positive association as against a negative one in the linear function case. Not much light is shed on the factors that explain these diverse results. The authors argument that, wherever positive associations arose, counterpart funds may have been used to subsidize demand, and, where negative, to subsidize production, is too simplistic and not based on empirical support.

The authors argue that domestic policies are crucial for realizing the trade effects of food aid. This presupposes that only domestic policies matter and not other factors such as food aid distribution arrangements, institutional and structural constraints in the global market, etc. A recent study pertaining to Somalia revealed that food aid resulted in greater import dependency owing to ill-formulated food aid programmes, apart from unsound domestic policies. Moreover, surely donor countries themselves have some control over the use of these counterpart funds, particularly when they advocate policies against subsidies and in favour of private enterprise and a market-friendly approach, etc.—these will surely have some impact on the use of counterpart funds by recipient countries. Hence, external factors, too, are relevant in understanding the linkages between food aid and trade.

The paper considers only the short-run effects of food aid and trade linkages. It would be useful also to know the long-run effects, especially in the light of popular notions about food aid having strong negative effects on the economy of aid receivers, resulting in a "dependency syndrome."

Other issues of interest are: whether food aid displaces commercial imports or domestic agriculture; food aid's effects on consumption, nutrition, and the balance of payments position; food aid's role in inducing structural adjustments and reforms in recipient countries; the costs and benefits (to donors and recipients) of bilateral versus multilateral food aid programmes; the impact of counterpart funds on macroeconomic parameters such as inflation, domestic revenue mobilization, economic development; and above all, the political economy of food aid.

*[Other discussion of this paper and the authors' reply appear on page 190.]*

# Searching under the Light: The Neglect of Dynamics and Risk in the Analysis of Food Trade Reforms

## Rod Tyers[1]

**Abstract:** The substantial investment in models of global food markets immediately prior to and during the Uruguay Round of international trade negotiations has been a mixed blessing so far as the prospects for reform are concerned. At worst, results from these models have misled the negotiations, first because they have served the losers from reform better than the gainers and second because they have tended not to address a primary concern lending domestic political support to food market interventions, namely the avoidance of risks borne of dependence on international markets. The paper reviews some errors that have stemmed from the application of "standard" but inappropriate models and examines the implications of extending the standard methodology to include the combination of explicit food price risk with dynamic behaviour and market-insulating policies.

## Introduction

My title refers metaphorically to the story about the man who, having dropped his keys on a dark street, returns and chooses to search first beneath the street lights. In many fields of endeavour, this approach is clearly rational. It is efficient to eliminate the easy options before approaching the difficult ones. But the possibility remains that only the carriageway is illuminated and not the pavement down which the man had travelled. In this case, the "search under the light" strategy is misleading and, ultimately, inefficient.

It is my fear that some analyses of the domestic and international effects of agricultural policies and of alternatives for their reform have been thus misguided. The models used to characterize market and government behaviour have employed standard, rather than frontier, methodology, the scope of which seriously limits the power of the models to address the policy issues at hand. This can be particularly problematic in the area of economic policy since early results from "standard" models can mislead the processes of policy formation and institution building. The resulting mistakes can result in new policy régimes and institutions, the lives of which are not simply terminated when new research suggests a change of direction.

Global models covering multiple interacting commodity markets and incorporating endogenous policy formation have recently become standard practice in the analysis of agricultural trade policy (Roningen, 1986; and OECD, 1990). But the improvements they offer still leave important deficiencies that, in my view, must be addressed before we have a truly useful characterization of global food markets. In particular, they ignore the roles of intertemporal changes and uncertainty, which are critical to both policy formation and the behaviour of private agents. The consequences of policy, as measured in comparative static terms, are thereby confused with the motivation for policy formation.

The emphasis in early work was on such questions as: "Who benefits from the existing market distortions, by how much, and at whose expense?" Although rent seeking by immediate beneficiaries might be influenced by this, wider support for some distortionary policies need not depend solely on static measures of economic surplus gained and lost. In my view, the broad political agenda behind most distortionary agricultural policies is insulation against changes abroad rather than an activist redistribution of preexisting domestic wealth, yet insulation as an aspect of agricultural policy has thus far been only weakly addressed by research on agricultural trade.

## The Evolution of Food Trade Modelling Methodology

Since the commodity boom years of the 1970s, there have been substantial investments in research on food trade policy. Interest in the subject has since been further enhanced by the onset of the Uruguay Round of international trade negotiations and the important role assigned to agricultural reform therein. The bulk of the policy analysis thus stimulated has employed readily available and easily interpreted partial equilibrium analysis in a comparative static mode. The early work of this type addressed the effects of distortionary policies in single countries and single markets for homogeneous commodities, assuming that either the quantity traded or the border price is exogenous (Thompson, 1981).

Later, still in a comparative static mode, new approaches emphasized international interactions, most popularly in non-spatial partial equilibrium models of individual world commodity markets (in the manner of Valdés and Zeitz, 1980). Interactions between separate commodity markets were incorporated in some single models that retained the comparative static approach and the partial equilibrium assumption; i.e., that the totality of the markets represented is small compared with the economy as a whole (as in Rojko *et al.*, 1978).

Even while all these partial equilibrium comparative static models were informing the policy process, important deficiencies were recognized by some, including the assumption that policy is exogenous (Rausser and de Gorter, 1989). New models were built to experiment with endogenous policy formation, and empirically based price transmission equations have become standard (Roningen and Dixit, 1989). The failure of the partial equilibrium assumption in the case of developing countries has also been addressed with the adoption of general equilibrium techniques (Hertel, 1990) and their application to global models (Parikh *et al.*, 1988; and Burniaux and Waelbroeck, 1985).

But despite a very large academic literature, most current global models still ignore the role of intertemporal changes and uncertainty, both of which are critical to food market behaviour and policy formation. Our own efforts have crudely addressed food market dynamics (Tyers, 1985; and Tyers and Anderson, 1992), while work proceeds on more sophisticated representations (that simulate dynamic games), none of which has been available in time to inform the Uruguay Round.

## Food Trade Distortions Primarily Serve Market Insulation

As foreshadowed in the introduction, it is my contention that the broad political agenda behind most distortionary agricultural trade policies is insulation against changes abroad rather than an activistic redistribution of pre-existing domestic wealth. If this is true, the comparative static results fail to address the fundamental motivation for the policy. Since it is unlikely that reforms would be embarked upon simply in the interests of such dispersed groups as consumers and taxpayers, the results have served to better inform the likely losers from reform and hence to galvanize the forces against it (see the critique of Roningen and Dixit, 1989, by the Center for Rural Affairs, 1990). The remainder of this section addresses the veracity of my premise. Why should domestic market insulation be a primary motivation for food market distortions?

Following Kindleberger (1986) and Runge *et al.* (1989), one can readily characterize the risk-spreading role of world food markets as an international public good. Its use is not restricted to those countries that share risk by exposing their domestic agents to price fluctuations; neither is it characterized by direct rivalry. Countries that insulate their domestic markets, using trade to eliminate residual excess demands or supplies and thereby stabilizing domestic prices, might then be portrayed as free riders even if their levels of protection, averaged through time, are comparatively small. By exposing domestic agents to international price instability, countries help to spread risk and thereby contribute to the supply of the international public good. Typically, as with all public goods, the inability to exclude non-contributors leads to undersupply and to excessively risky world food markets.

But insulation need not be directed solely at short-run price fluctuations. Much of the agricultural protection we observe and that has been the subject of extensive comparative static analysis in recent years can be seen as a consequence of market insulation. Real food commodity prices have maintained a declining trend throughout this century, one which has steepened since the early 1970s (Grilli and Yang, 1988; and Tyers and Anderson, 1992). Simply by retarding the transmission of this decline to domestic markets, many governments have caused rates of protection to rise through time to substantial levels. Others have fully transmitted declines but with a lag, leading to continuous, though lower, levels of protection.

It is my argument, then, that the insulation of domestic markets is perceived by most governments to be in the national interest and that they thereby exploit the risk-spreading capacity of world markets, collectively enhancing international price risk. The literature supporting the Uruguay Round has tended not to focus on the effects of policy on international price risk or on the reasons why governments choose to avoid it. It has therefore failed fully to inform the negotiations as to the collective risk benefits from truly multilateral reform. More specifically, it has not provided enough quantitative evidence that, if enough countries choose to insulate less, better spread price risk would reduce the need for insulation by others. Progress in reducing agricultural trade barriers has therefore been limited in this round.

Resting as the above argument does on the premise that the insulating components of domestic food policies are indeed valued by governments for their own sake, it is appropriate that the economics behind the premise be explored further. It is sufficient to establish either that, for given international price risk, insulating policies yield a net improvement in the aggregate welfare of all domestic agents when border prices are risky or that insulating policies benefit those groups with the greatest political influence and that governments therefore perceive political benefits from their implementation. In one sense, it is surprising that such a premise should be accurate. Insulation is distortionary, creating efficiency losses in every year in which border prices depart from desired domestic levels. To make insulation worthwhile, domestic agents must be sufficiently averse to price risk to offset the efficiency losses.

In all countries, some agents can be expected to have stronger preferences for price stability than others. Since market insulation occurs in both developing and industrialized countries, we might hypothesize that this preference would be strongest among the groups with most apparent influence over agricultural policy in each case; i.e., broadly, consumers and industrial capital owners in developing countries and farmers in industrial countries (Anderson, Hayami, et al., 1986). What, then, are the directions of the welfare impacts of price stabilization on these groups?

Reviews of the theoretical literature on the subject highlight the ambiguous nature of the consumer and producer welfare effects of price stabilization (see, for example, Tyers, 1990). The preferences for price stability of these predominant groups thus remain matters for empirical analysis. To examine these effects in some illustrative cases, I briefly report the results from an elementary model of a single open commodity market (detailed mathematically in Tyers and Anderson, 1992, chap. 3).

The model might apply to the market for a key food commodity such as rice. It assumes that the focus country is a small trader in the commodity and cannot therefore influence the level of the international price. That price is, however, subject to random disturbances due to fluctuations in demand and supply in the wider international market. Domestic production of the commodity is also subject to random disturbances such as might be caused by weather and pest infestation. Together, these two sources of randomness generate the price and income risk from which the government seeks to insulate domestic agents.

To illustrate the magnitudes of the welfare effects of insulation, consider two small archetypal economies. One is a poor country that imports rice, and the other is an industrial country that exports rice. In the poor country, farmers consume half the rice but earn only a quarter of national income. Workers consume the rest and receive wages that are compensated for rice price fluctuations. They and industrial capital owners earn three quarters of the national income. Household incomes differ between farmers and workers and so, therefore, do their rice consumption parameters. In the industrial country, on the other hand, farmers consume only a small fraction of their total output and earn a small share of the national income. Their household incomes are similar to those of workers, however, and the parameters governing their consumption behaviour are therefore identical.

All agents in both the developing and the industrial country are assumed to be averse to risk to degrees indexed by the Arrow-Pratt coefficient of relative risk aversion, $R$. Developing country agents and farmers in industrial countries, whose risks stemming from price fluctuations are significant in relation to their net income, are assigned a value of $R = 2$. Food price risk is less significant for consumers and taxpayers in industrial countries, however, so they are assigned a value of $R = 1$. In estimating the welfare impacts of changes in price stability for each group of agents, the approach of Newbery and Stiglitz (1981) is adopted, with only minor modification (Tyers and Anderson, 1992, chap. 3).

Empirical evidence as to the levels of short- and long-run market insulation (Tyers, 1990) suggests that market insulation is generally partial in both industrialized and developing countries. The case examined is therefore that of a partial insulation that reduces the coefficient of variation of the domestic price by half. As the results presented in Table 1 demonstrate, farmers are comparatively indifferent to market insulation in the developing country but could be expected to favour it in the industrial country. This is primarily because farmers in developing countries commit a relatively large share of their income to the purchase of farm products. Their gain from revenue stabilization is largely offset by losses that stem from their relatively elastic consumer behaviour. In the industrial country, on the other hand, farmers commit little of their income to farm products and the revenue (and hence income) stabilization effects are dominant.

Table 1—Benefits from Partial (50-Percent) Insulation in
"Typical" Developing and Industrial Economies

|  | Benefits as a Percent of Average Group Income or of Government Expenditure | |
|  | Developing Economy | Industrial Economy |
| --- | --- | --- |
| Farmers | 0.2 | 4.0 |
| Workers | −0.05 | 0.02 |
| Industrial capital owners | 2.7 | 0.0[a] |
| Government revenue | 0.6 | 0.0 |

[a]Zero since, while wages in the developing country are adjusted for food price changes, they are not in the industrial country. Industrial profits are therefore not significantly affected by food price risks.

Source: Calculations drawing on parameter values from World Bank (1986) and detailed algebraically in Tyers and Anderson (1992).

Non-agricultural workers in both the developing and the industrial country are roughly indifferent to market insulation. In the developing country, this is because worker income is adjusted for food price changes through wage indexing or partial payments in kind. In the industrial country, it is because demand is inelastic and workers spend only a small share of their income on food. Food market insulation is clearly favoured by industrial capital owners in developing countries. This is because payments to labour dominate the value added in the non-agricultural sector. Fluctuations in these payments therefore result in substantial profit risk. In industrial countries, worker income tends not to be compensated for short-term changes in food prices and, in any case, the nonagricultural sector is less labour-intensive.

The government revenue effects are dominated by shifts in mean revenue due to the partial insulation policy. These revenue gains depend primarily on the elasticity of domestic consumer demand in the short run. Since this elasticity is comparatively high in developing countries, the revenue effects of partial stabilization are significant there.

In both cases, there are net gains nationally from the insulation, supporting the public interest explanation for insulating policies. The results do, however, bear out the hypothesis that the most influential group has the most to gain from market insulation in each case. The gains to industrial capital owners and to government revenue are dominant in the developing country, where industry tends to be protected at the expense of agriculture and where the cost of collecting revenue by other means is especially high. In the industrial country, on the other hand, where agriculture tends to be protected at the expense of other sectors, farmers have the dominant interest in price stabilization. In addition, since no group of agents in the domestic economy would appear to lose significantly, governments tend not to find market-insulating policies costly to sell to non-beneficiaries.

The reasoning above is readily extended to long-run insulation, leading to the drift of domestic prices away from international prices and to pure protection. Although domestic rent-seeking pressures obviously play a role, it is one that is greatly facilitated when insulating policies are present. This is because, first, such policies always separate domestic from border prices and hence distort domestic incentives, at least in the short run. And, second, because the current and future trend of international market prices is uncertain, there is no obvious and undisputed level at which domestic prices should be set in order to achieve the objective of comparatively stable domestic prices. The process by which the domestic price is set is therefore subject to lobbying by vested interests.

## Some Implications of Market Insulation for Agricultural Trade Policy Analysis

The presence of market-insulating policies makes the analysis of policy reform and its interpretation more difficult. Moreover, commonly used comparative static analysis can be badly misleading. To demonstrate this, I examine two key implications of market insulating behaviour. These are, first, that magnitudes of price distortions vary from year to year as international prices fluctuate. The results from any comparative static analysis therefore depend on which year is chosen for analysis. Second, when price distortions are measured in a single year, it is impossible to tell what part of these is due, on the one hand, to governments' commitments to keeping domestic prices above the trend of world prices (pure protection) and, on the other hand, to risk-avoiding market insulation.

To illustrate these implications for policy analysis, I draw on the Tyers/Anderson model of world trade in grains, livestock products, and sugar. This dynamic model is equipped for this purpose with endogenous policy formation and stock-holding behaviour (Tyers and Anderson, 1992, chap. 5). Its base period, 1980–82, has average international food prices roughly on the long-run trend. Simulations run from 1983 to the year 2000. Disturbances to food production provide the main source of uncertainty in the model and are introduced stochastically beyond 1987, the last year for which comprehensive quantity and price data were available at the time of writing.

The analysis begins with a reference simulation that projects a continuing downward trend in real international food prices during the 1990s, despite the temporary resurgence of real prices in the late 1980s. Protection rates in countries with insulating policies rise and remain higher than in the base period (Table 2), although they do abate slightly in the late-1990s as the dip in world prices is gradually passed through to some domestic agents. Price distortions and their associated efficiency losses peak in 1987, and their projected mean declines thereafter.

These results clearly illustrate the first of the above implications of insulating policies. Comparative static analysis would yield conclusions about price distortions and their economic cost, which would vary enormously, depending on the year chosen. In particular, studies based on statistics for 1986 or 1987 would yield global efficiency losses twice as large as for subsequent years and five times larger than they were in 1980–82. This is a major difficulty with studies such as that by Horridge, Pearce, and Walker (1990). They address distortions that appear high because of the year chosen (1986) but that are mere symptoms of more complex policies not intended to distort prices to that extent in all years.

To address the second implication, that the effects of the pure protection and the market-insulating components of trade policy are difficult to separate, I have made an additional simulation in which the insulating component of policy is removed in all countries from the base period, 1980–82 onwards. In our model, this is the equivalent of the conversion of all policies into *ad valorem* taxes or subsidies at the border as of 1982 and to the binding of their rates in that year. Thereafter, while the level of pure protection is held constant, all proportional fluctuations in international prices are fully transmitted to all domestic markets. Not surprisingly, the projected path of international prices is made more smooth by the wider spreading of price risk. More importantly, however, the decline in prices beyond the base period is substantially reduced. This is because, when domestic markets are not insulated, the increases in *ad valorem* protection rates in Table 2 no longer occur. The differences between the retained protection rates of the base period and those that occur when markets are insulated is then that part of price distortions due to the insulating component of policies. The two components of policy are separated in this way in Table 2 and their global efficiency losses compared.

Table 2—Changes in Protection and Efficiency Losses Due to Insulation since 1982

|  | 1980–82 | 1987 | 1990 | 1995 | 2000 |
|---|---|---|---|---|---|
| OECD nominal protection coefficient |  |  |  |  |  |
| Reference | 1.40 | 1.96 | 1.81 | 1.91 | 1.85 |
| No insulation |  | 1.43 | 1.39 | 1.38 | 1.37 |
| Percent of distortion due to insulation |  | 55 | 51 | 58 | 56 |
|  |  |  |  |  |  |
| Annual global net welfare cost of OECD protection,[a] 1985 US $'000 million |  |  |  |  |  |
| Reference | 16 | 83 | 45 | 46 | 50 |
| No insulation |  | 20 | 18 | 13 | 12 |
| Percent of cost due to insulation |  | 76 | 60 | 72 | 76 |

[a]The welfare measures used here are equivalent variations in income. They ignore risk benefits, assuming agents are risk neutral.

Source: Simulations using the Tyers-Anderson trade model (Tyers and Anderson, 1992).

These results suggest that the effects of insulation since 1982 are substantial. More than half of the average price distortion in the OECD is due to insulation. Note that the average rate of protection varies slightly from year to year even when there is no insulation. This is due to changes in the volume mix of food commodities in production and trade. The efficiency losses due to insulation since 1982 are also the major part of the net global cost of price distortions both in the OECD countries and in all countries. Had the GATT Round concluded in 1982 with agreement to cease market insulation but to retain existing pure protection (which held very high levels in some countries, particularly in Europe and Japan), the majority of the distortions and of the costs now being borne by the world economy would not have arisen. These are good reasons why market-insulating policies should have been given a higher profile earlier in the current round; they were not, at least in part because of a reliance on "standard" models.

# Note

[1]Australian National University.

# References

Anderson, K., Hayami, Y., et al., The Political Economy of Agricultural Protection: East Asia in International Perspective, Allen and Unwin, Sydney, Australia, 1986.

Burniaux, J.M., and Waelbroeck, J., "The Impact of the CAP on Developing Countries: A General Equilibrium Analysis," in Stevens, C., and Verloren van Themaat, J. (Eds.), Pressure Groups, Policies, and Development, Hodder and Stoughton, London, UK, 1985.

Center for Rural Affairs, "The Consequences of Free Trade: USDA Confesses," GATT Tales, Lincoln, Nebr., USA, 1990.

Grilli, E.R., and Yang, M.C., "Primary Commodity Prices, Manufactured Goods and the Terms of Trade of Developing Countries: What the Long Run Shows," World Bank Economic Review, Vol. 2, No. 1, 1988, pp. 1–48.

Hertel, T.W., "General Equilibrium Analysis of U.S. Agriculture: What does it Contribute?," Journal of Agricultural Economics Research, Vol. 42, No. 3, 1990, pp. 3–9.

Horridge, M., Pearce, D., and Walker, A., "World Agricultural Trade Reform: Implications for Australia," Economic Record, Vol. 66, No. 194, 1990, pp. 235–248.

Kindleberger, C.P., "International Public Goods without International Government," American Economic Review, Vol. 76, 1986, pp. 1–13.

Newbery, D.M.G., and Stiglitz, J.E., The Theory of Commodity Price Stabilization: A Study in the Economics of Risk, Oxford University Press, Oxford, UK, 1981.

OECD (Organization for Economic Cooperation and Development), Modelling the Effects of Agricultural Policies, Paris, France, 1990.

Parikh, K.S., Fischer, G., Frohberg, K., and Gulbrandsen, O., Towards Free Trade in Agriculture, Mortimers Nijoff, Amsterdam, Netherlands, 1988.

Rausser, G.C., and de Gorter, H., "Endogenizing Policy in Models of Agricultural Markets," in Maunder, A., and Valdés, A. (Eds.), Agriculture and Governments in an Interdependent World, Dartmouth, Aldershot, UK, 1989.

Rojko, A.S., et al., "Alternative Futures for World Food in 1985," Foreign Agricultural Economic Report, Nos. 146 and 151, Economic Research Service, US Department of Agriculture, Washington, D.C., USA, 1978.

Roningen, V.O., "A Static World Policy Simulation (SWOPSIM) Modeling Framework," Staff Report No. AGES 860625, Economic Research Service, US Department of Agriculture, Washington, D.C., USA, 1986.

Roningen, V., and Dixit, P., "Economic Implications of Agricultural Policy Reforms in Industrialized Market Economies," Staff Report No. AGES 89-363, Economic Research Service, US Department of Agriculture, Washington, D.C., USA, 1989.

Runge, C.F., von Witzke, H., and Thompson, S.J., "International Agricultural Policy: A Political Coordination Game," in von Witzke, H., Runge, C.F., and Job, B. (Eds.), Policy Coordination in World Agriculture, Wissenschaftsverlag, Kiel, Germany, 1989.

Thompson, R.L., "A Survey of Recent U.S. Developments in International Agricultural Trade Models," Bibliographies of Literature in Agriculture, No. 21, Economic Research Service, US Department of Agriculture, Washington, D.C., USA, 1981.

Tyers, R., "Agricultural Protection and Market Insulation: Model Structure and Results for the European Community," Journal of Policy Modeling, Vol. 7, No. 2, 1985, pp. 219–251.

Tyers, R., "Trade Reform and Price Risk in Domestic and International Food Markets," World Economy, Vol. 20, No. 2, 1990.

Tyers, R., and Anderson, K., Disarray in World Food Markets: A Quantitative Assessment, Cambridge University Press, Cambridge, UK, 1992.

Valdés, A., and Zeitz, J., "Agricultural Protection in OECD Countries: Its Cost to Less Developed Countries," Research Report No. 21, International Food Policy Research Institute, Washington, D.C., USA, 1980.

World Bank, World Development Report, Oxford University Press, Oxford, UK, 1986.

## Discussion Opening—*Jock R. Anderson* (World Bank)

I feel somewhat like a spiritual adviser receiving a confession—if one of the major global commodity modellers wants to whisper that all has not been wise and well in this field, I am inclined to listen carefully. Thus it was that I found this paper to be of great interest. The interest starts with the catchy title. "Searching under the Light" can, as the author suggests, sometimes be about all that one can do.

My remarks on the presented version of this paper opened by a complaint that, at least in its abbreviated form, I found the formal logic and persuasiveness to be less than optimal. Since that time, however, I have had the opportunity of reading a more complete telling of the story, and I find the general argument to be entirely reasonable.

Models are intrinsically simplifications, and, if a particular element is excluded from the model of reality, it is more or less inevitable that new information about this element or its links to other parts of the modelled system simply cannot be gleaned from the model.

Risk management, to focus on one highlighted matter seemingly excluded from most of the "bad" models, has probably been an important element of the political economy of protectionism. Thus, to the extent that models have inadequately represented the stochastic structure of production and trade and have ignored the risk aversion of all relevant agents, they will probably not be useful to policy analysts concerned with such topics. Indeed, I have argued elsewhere that the more deterministic models often used are, through such omissions, systematically wrong in estimates of even the mean or average effects.

Similar arguments surely pertain to the author's claimed excessively underplayed dynamic elements of model structures. All this remote philosophizing is one thing and, I would contend, not really debatable. Whether, however, such underspecified models have, as he argues, also variously misled, misguided, and misinformed the negotiators in the Uruguay Round seems to be quite another thing. It suggests, for instance, that negotiators really take the results of these intrinsically inadequate models seriously.

From the observational perspective of one who has not been actively involved, however, it seems that the debate in much of the Uruguay Round is, for better or worse, driven by other forces that are overwhelmingly political. If this is the case, it may be that all the formal econometric modelling of the costs and benefits of trade reforms has, to continue the metaphor of the title, been conducted in a conveniently lit part of the policy arena, and that the real action has been elsewhere. Notwithstanding this possibility, I enjoyed this paper and am firmly of the belief that trade policy analysts will be usefully informed by it.

*[Other discussion of this paper and the author's reply appear on page 190.]*

# Subsidies and Cattle Production in the Amazon: An Economic Policy Analysis

*Emily McClain, Catherine Halbrendt,
Jennifer Sherbourne, and Conrado Gempesaw*[1]

**Abstract:** Cattle production has been a major source of agricultural deforestation in Brazil's Amazon rainforest. Brazilian credit subsidies have been blamed for speeding cattle expansion and thus deforestation. A stochastic coefficients regression approach was used to quantify the effects of credit subsidies and world prices on cattle numbers in five Amazon regions for the 1963–83 period. Results show that cattle production has been positively correlated to both prices and credit. Elasticities show that these relationships strengthened over time and that production has been more responsive to credit than to world prices.

## Introduction

The international scientific community continues to strengthen the link between deforestation and global climatic change. With an increased focus on deforestation, there is a need to identify and quantify the forces leading to it. Of particular interest are those factors subject to direct control, such as government policies.

Brazil, estimated to hold one-third of the world's rainforests in its Amazon Basin, has received harsh international criticism over the acceleration of deforestation, which began in the 1970s. While it is uncertain how much of Brazil's rainforest has been cleared, estimates range from 5.1 percent (Brazilian government) to 12 percent (Browder, 1988). In either instance, a large area has been deforested.

Environmental issues present difficult challenges for policy makers, particularly in developing countries such as Brazil. Environmental protection must balance economic development, where policy options, funding, and expertise are extremely limited. During the 1980s, Brazil's economy was increasingly unstable. Inflation reached almost 1,800 percent for 1989, and Brazil closed the decade with little real improvement in living standards.

Austerity programmes in recent years have forced cuts in crucial areas such as infrastructural maintenance, social programmes, and agricultural support, making it difficult to divert funds for the environment, funds that must compete with basic services in the world's sixth most populous country. Given such financial limitations, one obvious option for "bargain" environmental protection is to reexamine and identify public policies that may contribute to irrational deforestation, especially those that could be eliminated with cost savings or little additional expenditure on the part of the Brazilian government. In the late 1980s, one of Brazil's most applauded moves was the temporary suspension of tax credits and fiscal incentives for frontier (Amazon) farming. Worldwide, agriculture is estimated to be responsible for 76 percent of all deforestation (Keipi and Valdares, 1989), and, for Brazil's Amazon, it is estimated that cattle ranching uses twice as much land as cropping (Lewandrowski and McClain, 1990). Government policies that promoted the expansion of cattle production have been widely criticized in the literature as a major contributor to Amazon deforestation (Fearnside, 1986; Browder, 1988; and Mahar, 1989).

No studies have empirically linked credit subsidies (a major source of funding) with cattle production expansion within the Amazon. Brazil's subsidized agricultural credit programme has been criticized as ineffective because funds are often spent on nonagricultural enterprises and investments. If a positive relationship can be established between subsidies and cattle numbers, more credence can be given to the argument for their permanent elimination. If the subsidies are not the overwhelming motivation for cattle expansion, it becomes important to determine what factors are driving ranching in frontier areas in order to slow expansion and deforestation.

This paper examines the relationship between government subsidies (Brazilian policies), world beef prices (other countries' policies), and cattle numbers in the Amazon. Thus, credit subsidies and world prices are indirectly linked to deforestation through their influence on cattle production.

## Study Area

The Legal Amazon was divided into five geographic regions to capture differences in producer policy responses that are due to climatic, geographic, market, and regional policy differences. Four regions are individual states that account for 90 percent of all Amazon cattle production: Maranhão, Goiás (now two states: Goiás and Tocantins), Pará, and Mato Grosso (now Mato Grosso do Sul and Mato Grosso). Region five, the rest of the Amazon (ROA), is an aggregate of five states or territories: Amazonas, Amapá, Roraima, Acre, and Rondônia.

In geographical and development terms, Brazil is usually classified into five regions: North, Northeast, Centre-West, South, and Southeast. The Legal Amazon encompasses the North, plus parts of the Northeast and Centre-West. Cattle expansion has varied throughout Brazil due to the differences in land availability, ranch sizes, and government policies that tended to favour certain interest groups or locals. Rates of cattle expansion were phenomenal in the Amazon when compared to southern Brazil, the largest cattle-producing region. During 1962–85, cattle expansion rates were 271 percent, 105 percent, and 130 percent for the North, Northeast, and Centre-West, regions, respectively, but only 35 percent for the South and Southeast (non-Amazon) regions. Government development programmes probably contributed to these rapid growth rates in the Amazon.

## Background Information

In the early 1960s, Brazil decided actively to develop the Amazon for two main reasons: to defend Brazil's borders and to exploit the region's vast wealth and agricultural potential (Mahar, 1979). Later, the Amazon became an outlet for the poor of the Northeast and a way to alleviate land-reform pressures using migration policies.

Initial attempts at development came through the construction of the 2,000 km Belém-Brasília highway, completed in 1964. After establishing this north-south road link, the government launched Operation Amazônia in 1966. The plan's goal was to attract large-scale corporate investment to the region; corporate cattle ranching was identified as a promising enterprise because the region's fragile tropical soils were best suited for pasture and also because of an FAO study that indicated that Brazil could be a world-class beef exporter.

As a result, a massive infusion of credit subsidies occurred. During 1968–70, the subsidy grew from $848,000 to $33 million (Browder, 1988). Cheap land, low labour requirements, flexible marketing, and low maintenance costs also contributed to rapid cattle expansion. Livestock producers could receive subsidized financing for production, marketing, and investment activities under the national programme for rural credit (one of the government's main agricultural support programmes).

In the early 1970s, the Brazilian government changed its Amazon development strategy, giving priority to colonization and increased migration. Funding for livestock enterprises decreased during 1970–74 by 7 percent as funds were diverted to a colonization programme of the National Colonization and Land Reform Institute. Consequently, cattle expansion rates decreased in almost every region of the Amazon during the period, compared to annual rates of increase of 10 percent during 1962–70.

The beef sector was restimulated in 1975 by Brazil's National Development Plan, prompted by the 1973 oil price shock and the need to increase foreign exchange earnings. Since beef was becoming an important export, cattle production incentives were increased. The 1975 plan prompted a subsequent plan, the Brazilian National Cattle Plan, to stimulate beef production in regions where expansion had stagnated under decreased funding and high production costs. Other programmes, the Livestock Development Programme, the National

Pasture Programme, and the Livestock Herd and Management Programme, were developed and implemented. Cattle production responded rapidly in the Amazon, with an expansion of 31 percent in the North, 20 percent in the Centre-West, and 24 percent in the Northeast during 1974–79 (IBGE).

During the 1980s, cattle expansion continued but the pace slowed, perhaps due to cuts in government funding. Much of the Amazon region showed a food deficit in the 1970s. Although several factors could have stimulated Brazil's cattle industry, this study examines the roles of the much-maligned government subsidies and world prices in promoting cattle production during the 1969–83 period, when both cattle and deforestation expanded rapidly.

## Model Development

Theory suggests that production is a function of prices and supply shifters, which in this study are government subsidies. Tax credits offered to corporate livestock operators were added to expenditures under the national agricultural subsidized credit programme to represent direct transfers to the agricultural sector by the government. Although not ideal, the aggregation was needed to reduce the numbers of regressors, given the small sample size.

The end of slaughter and export quotas in the Brazilian beef sector in 1971 meant that international prices may have influenced the evolution of ranching for most of the period. Thus, a world price variable was included in the analysis. If significant, one may infer that policies in beef exporting and importing countries that affected world prices may also have influenced cattle production and, indirectly, deforestation in the Amazon. Theoretically, the domestic price of beef and prices for commodities competing in the production of cattle should also be considered. However, competing enterprises are difficult to identify, and a reliable series of domestic beef prices could not be compiled due to high inflation rates and aggregation in reporting.

Finally, an interactive dummy variable was used to separate the oil shock period of 1974–79 from the rest of the observations. The oil shock period also coincided with domestic beef shortages that caused a draw-down of cattle inventories. The dummy variable is interactive with the aggregate credit term, changing the slope of the equation during the 1974–79 period.

The general specification of the empirical model is:

$$(1) \quad CATTLENO_{it} = \beta_0 + \beta_1 GOVTSUB_{it} + \beta_2 PRICE_{t-3} + \beta_3(DUM_{it}GOVTSUB_{it}) + e_{it}$$

Cattle numbers ($CATTLENO$) in the $i$th region in time $t$ are expressed as a function of world beef prices ($PRICE$) lagged three periods, an aggregate term representing government subsidies ($GOVTSUB$) in the $i$th region, and a dummy variable ($DUM$). The configuration of the dummy variable is $DUM = 1$ for all periods where $t > 1973$ and $< 1979$; otherwise $DUM = 0$.

The availability of tax credit data limited the study period to 1969–83, or 14 total observations. Cattle numbers and subsidized credit expenditures by region were obtained from the *Anuário Estatístico do Brasil* (IBGE). Tax credits were available from a secondary source, and both credit series were deflated to a mid-1980 base period using a general price index (IBGE). World prices were assumed to be utility beef prices in $/t, f.o.b. US Gulf. Prices were lagged and converted to domestic currency to include the effects of exchange rate policy (over- or under-valuation) on producer (exporter) incentives for cattle production.

In pre-test experiments, the un-lagged $GOVTSUB$ term was more important than lagged terms in explaining variation in cattle numbers. *A priori*, it is expected that credit under the rural credit system (production, marketing, and investment) should have some lagged effect, but this did not show up in the aggregate credit term. Some lag may be inherent in the data since credit is issued early in the year and cattle inventories made late in the year. *A priori*, positive coefficients were expected for $PRICE$ and $GOVTSUB$, while a negative coefficient was expected on *DUM* due to the contractionary impacts of the oil shocks on Brazil's economy and domestic beef shortages.

Elasticities for both price and credit were expected to be inelastic (less than 1.0), since cattle production is characterized by large initial investments. Costs of acquiring and clearing land, plus pasture establishment, are substantial, but, once this investment occurs, large reductions in price or credit are necessary to remove area from production and reduce cattle numbers. Cattle are primarily forage-fed; thus, carrying costs are low and producers respond to low prices by withholding cattle from markets.

## Estimation Method

Conventionally, fixed coefficient models with dummy variables such as ordinary least squares (OLS) are used to capture policy impacts on the structure of a sector. The large numbers of cattle development programmes instituted during the time period plus the volatility of Brazilian sectoral and macroeconomic policies during the period make conventional modelling unusually difficult and questionable. The timing and net effects of political and structural adjustment on the livestock sector would be hard to hypothesize.

Since a fixed coefficient estimation method assumes constancy in the marginal contribution of causal factors, OLS was considered to be too restrictive when evaluating the impacts of government subsidies on cattle production. Over time, investments in research and infrastructure would have changed the way producers respond to policy and policy changes. With no strong theoretical grounds for choosing linear specifications, a stochastic coefficient regression approach (SCM) developed by Swamy and Tinsley was used in this study (Swamy and Tinsley, 1980).

Reasons for choosing the SCM approach included functional form specifications, changes in technology such as breeding and pasture, and the ability to accommodate changing structural parameters. The SCM can adequately represent several forms of nonstationary processes and can respond quickly to changing economic conditions, yielding predictions that are superior to those from fixed-slope coefficient models.

## Estimation Results

Parameter estimates of the SCM model are presented in Table 1. Overall, the results are very good (significant at the 95-percent level), with the exception of the model for Mato Grosso and the coefficient on subsidies for Maranhão. One plausible explanation for the lack of success in explaining cattle expansion in Mato Grosso is that the model ignores the effects of alternative agricultural enterprises. Cattle expansion in Mato Grosso may have been driven by a migration of São Paulo's cattle industry westward as it was displaced by the expansion of sugarcane under Brazil's National Fuel Alcohol Programme in the mid-1970s (USDA).

An FAO study of agricultural credit in the Northeast found that government-subsidized credit for livestock enterprises in Maranhão was being diverted to other investments. This finding is supported by the non-significance and wrong sign of the credit coefficient in the Maranhão equation.

All but three of the credit and price terms were significant and/or positive, indicating that both government subsidies and world prices positively affected cattle expansion in the Legal Amazon during the study period. As hypothesized, all elasticities were inelastic, but in general became more elastic over time (Table 2). Estimates for subsidies in ROA were the most inelastic until 1980. This can be partially explained by the fact that production in this region may have been discouraged by the scarcity of roads and markets in the Amazon's most remote area.

All the elasticities for the oil shock period are less responsive than for the non-oil shock period (i.e., all the dummy variable coefficients were of the expected negative sign). In addition, comparisons of the variations in elasticities across the five modelled regions show a consistent pattern. This pattern supports SCM's ability to distinguish structural changes through time, as all regions faced similar changes in macroeconomic conditions.

Table 1—Stochastic Coefficients Estimation Results

| Coefficient | Goiás | Mato Grosso | Maranhão | Pará | Rest of the Amazon |
|---|---|---|---|---|---|
| Intercept | 6715.56 | 7501.48 | 2186.96 | 1176.6 | 1577.42 |
| T-ratio | (19.72) | (0.2) | (113.8) | (35.96) | (26.46) |
| Subsidy | 0.11 | 0.11 | 0.01 | 0.05 | 0.05 |
| Subsidy (1973–79) | 0.16 | 0.04 | 0.22 | 0.01 | 0.001 |
| T-ratio | (3.02) | (0.02) | (1.41) | (2.21) | (4.06) |
| World price $(t-3)$ | 241.75 | 250.83 | 23.26 | 34.7 | −68.82 |
| T-ratio | (13.37) | (0.11) | (25.5) | (6.35) | (−4.67) |
| Dummy | 0.05 | −0.08 | 0.21 | −0.04 | −0.04 |
| T-ratio | (1.87) | (0.02) | (5.88) | (−2.34) | (−5.35) |

Note: All estimates are significant at the 95-percent level, except those for Mato Grosso and the credit subsidy parameters for Maranhão.

## Implications

The elasticities confirm that government subsidies have indeed played a positive role in expanding cattle production (and implicitly, deforestation) in most regions of the Amazon. It is important to remember topographical differences within the Amazon when attaching importance to this result. Most of the true rainforest falls within the region defined as ROA, while the remaining areas are primarily palm forests, floodplains, or grassland savannas.

The SCM estimates show that cattle numbers in the ROA were highly inelastic in response to subsidies during the first part of the period, but that responsiveness increased rapidly over time, with ROA exhibiting the highest elasticity of all regions in the last year. Thus, the elimination or reduction of credit should be very effective in slowing cattle expansion in rainforest areas.

World price was shown to have a very small negative effect on cattle production in the ROA. This is probably due to the fact that the region has typically been food-deficit; the distance from markets also weakens price transmission to this region. The general increases in the responsiveness of cattle numbers to world prices suggested by the elasticity estimates for all regions is consistent with declining government intervention and loosening of trade restrictions in the beef sector over time. This means that Brazil's 1990 switch to a floating exchange rate and the subsequent devaluation to equilibrium exchange levels now occurring may serve to fuel cattle expansion and exacerbate deforestation. Since cattle numbers were generally more responsive to subsidies than to price signals, price declines would have a smaller impact on cattle production than reductions in credit, *ceteris paribus*.

## Note

[1]Clemson University, University of Delaware, University of Missouri, and University of Delaware, respectively.

Table 2—Subsidy and Price Elasticity Estimates

| Year | Subsidy Elasticities | | | | | Price Elasticities | | | | |
|---|---|---|---|---|---|---|---|---|---|---|
| | Goiás | Pará | Mato Grosso | ROA | Maranhão | Goiás | Pará | Mato Grosso | ROA | Maranhão |
| 1969 | 0.129 | 0.120 | 0.081 | 0.003 | 0.006 | 0.138 | 0.133 | 0.137 | -0.030 | 0.053 |
| 1970 | 0.198 | 0.630 | 0.143 | 0.033 | 0.014 | 0.119 | 0.116 | 0.120 | -0.027 | 0.048 |
| 1971 | 0.276 | 0.410 | 0.214 | 0.008 | -0.067 | 0.105 | 0.118 | 0.108 | -0.027 | 0.051 |
| 1972 | 0.274 | 0.530 | 0.214 | 0.025 | -0.061 | 0.111 | 0.124 | 0.112 | -0.028 | 0.053 |
| 1973 | 0.251 | 0.106 | 0.213 | 0.119 | -0.022 | 0.139 | 0.137 | 0.138 | -0.032 | 0.065 |
| 1974 | 0.187 | -0.087 | 0.175 | 0.036 | -0.360 | 0.201 | 0.233 | 0.193 | -0.052 | 0.123 |
| 1975 | 0.284 | -0.053 | 0.167 | 0.013 | -0.350 | 0.188 | 0.239 | 0.207 | -0.058 | 0.129 |
| 1976 | 0.299 | -0.051 | 0.165 | 0.080 | -0.404 | 0.216 | 0.275 | 0.244 | -0.068 | 0.157 |
| 1977 | 0.322 | 0.028 | 0.188 | 0.137 | -0.401 | 0.202 | 0.254 | 0.229 | -0.061 | 0.144 |
| 1978 | 0.399 | 0.159 | 0.277 | 0.197 | -0.342 | 0.137 | 0.165 | 0.156 | -0.039 | 0.096 |
| 1979 | 0.424 | 0.208 | 0.321 | 0.250 | -0.257 | 0.137 | 0.159 | 0.153 | -0.038 | 0.092 |
| 1980 | 0.406 | 0.326 | 0.378 | 0.448 | 0.126 | 0.186 | 0.186 | 0.185 | -0.042 | 0.104 |
| 1981 | 0.355 | 0.364 | 0.351 | 0.528 | 0.104 | 0.252 | 0.217 | 0.243 | -0.052 | 0.143 |
| 1982 | 0.196 | 0.305 | 0.212 | 0.619 | 0.054 | 0.419 | 0.327 | 0.396 | -0.078 | 0.231 |
| 1983 | 0.225 | 0.359 | 0.216 | 0.673 | 0.064 | 0.415 | 0.314 | 0.406 | -0.072 | 0.239 |
| Minimum | 0.196 | -0.087 | 0.081 | 0.003 | -0.404 | 0.105 | 0.116 | 0.108 | -0.078 | 0.048 |
| Average | 0.282 | 0.224 | 0.221 | 0.211 | -0.126 | 0.198 | 0.200 | 0.202 | -0.047 | 0.115 |
| Maximum | 0.424 | 0.630 | 0.378 | 0.673 | 0.126 | 0.419 | 0.327 | 0.406 | -0.027 | 0.239 |
| Avg. 1974–79 | 0.319 | 0.034 | 0.216 | 0.119 | -0.352 | 0.180 | 0.221 | 0.197 | -0.053 | 0.124 |

# References

Browder, J., "Public Policy and Deforestation in the Brazilian Amazon," in Repetto, R., and Gillis, M. (Eds.), *Public Policies and the Misuse of Forest Resources*, Cambridge University Press, Cambridge, UK, 1988.

Fearnside, P., *Human Carrying Capacity of the Brazilian Rainforest*, Columbia University Press, New York, N.Y., USA, 1986.

IBGE (Instituto Brasileiro de Geografia e Estatística), *Anuário Estatístico do Brasil*, Rio de Janeiro, Brazil, various years.

Keipi, K., and Valdares, T., "Deforestation: Impact and Alternatives," paper presented at a meeting sponsored by the Latin American Program of the Woodrow Wilson Center, Washington, D.C., USA, 1989.

Lewandrowski, J.K., and McClain, E., "Resource Use and Deforestation in Brazil's Amazon," *World Agriculture: Situation and Outlook Report*, WAS–60, Economic Research Service, US Department of Agriculture, Washington, D.C., USA, 1990.

Mahar, D.J., *Frontier Development Policy in Brazil: A Study of Amazônia*, Praeger, New York, N.Y., USA, 1979.

Mahar, D.J., *Government Policies and Deforestation in Brazil's Amazon Region*, World Bank, Washington, D.C., USA, 1989.

Swamy, P.A.V.B., and Tinsley, P.A., "Linear Prediction and Estimation Methods for Regression Models with Stationary Stochastic Coefficients," *Journal of Econometrics*, Vol. 12, 1980, pp. 103–142.

USDA (US Department of Agriculture), "Brazil: Agricultural Situation Report," Foreign Agricultural Service, unpublished, various years.

---

**Discussion Opening**—*Gervásio Castro de Rezende* (Instituto de Pesquisa Econômica Aplicada, Brazil)

The authors included a dummy variable, interactive with the credit term, for the 1973–79 period, based on the argument that in this period there were "beef shortages that caused a draw-down of cattle inventories," loosely attributed to the oil shocks. This does not seem plausible; a closer look at the Brazilian data on livestock herd and beef cattle slaughter reveals what really happened in this period. The "draw-down" of cattle inventories of which the authors speak actually occurred only during 1970–73 (i.e., before the first oil shock) and was due to the fact that the data were collected, up to 1970, by the Ministry of Agriculture and, from 1973, by IBGE. Because these institutions' methodologies were different, the two series are not comparable; e.g., the Ministry of Agriculture's estimate of cattle numbers in Brazil for 1970 is 98 million head, while that of IBGE for 1973 is only 90 million.

If the authors are able to circumvent this problem and revise their paper, some additional comments are appropriate. First, the data for Brazil show that the cattle numbers series contains a clear cyclical component, which conforms to the livestock cycle (about which there is not a single word in the paper). Since one cannot expect pasture area to adhere to this cycle, it is questionable whether inferences about deforestation are valid. (The same is true in regard to comparison of long-run trends, since there has been much pasture improvement in Brazil.) In this connection, the finding that the credit variable enters with no lag in the equations (contrary to theoretical expectations) may have to do with its pro-cyclical behaviour.

There was no reason to use world beef prices, since data on domestic beef cattle prices have existed at the state level since 1966.

It is not clear whether credit volumes or credit subsidies were used; since the subsidy rate varied wildly during the period, the two series present very different behaviour. In any case, the aggregation of credit volumes and tax credits does not make sense.

*[Other discussion of this paper and the authors' reply appear on the following page.]*

**General Discussion**—*Léo da Rocha Ferreira, Rapporteur* (Instituto de Pesquisa Econômica Aplicada, Brazil)

Discussion on the Herrmann *et al.* paper concentrated on the model specification. Herrmann agreed that many other factors could be considered in order to improve wider understanding of food aid and food trade. The exogenous or endogenous nature of imports was discussed. It was also argued that emergency aid is no answer to food problems and that relative prices are distorted in the developing countries.

On the Tyers paper, the author stressed that the application of "standard" but inappropriate models misinforms and misleads conclusions and policy implications and the necessity to include the combination of explicit food price risk with dynamic behaviour and market insulating policies. However, the stochastic effects were not elaborated in the paper. Other topics discussed were whether trade promotes sustainable agriculture, whether the paper's results might underestimate the welfare gains due to reducing insulation (since agents are risk averse), and whether the design of commodity programmes worldwide indicates that their purpose is to raise income rather than to reduce variability. Other comments were that models are not effective in sustainable agriculture, that risk aversion was properly included in the paper, and that farmers are more interested in income policies. Finally, it was argued that the more simple the model, the more powerful it is, and that protection is not supposed to yield welfare improvements.

The focus of the discussion on the McClain *et al.* paper was on the data used in the study, where two different methodologies resulted in two series that were not comparable (methodologies used by the data source). The role of informal credit and the sequence of causality from credit subsidies to more beef, to more pasture area, and to more deforestation were also discussed. Macroeconomic and trade policy are important in making land investment more attractive as a hedge against inflation and consequently a more powerful force for deforestation. Finally, it was argued that the authors were unable to quantify several variables and were unaware of informal credit.

Participants in the discussion included N. Devisch (Boerenbond, Belgium), M. Hartmann (Universität Frankfurt), J. Kola (Agricultural Economics Research Institute, Finland), J. Lundy (University of California), H.A. Mahran (University of Gezira), L.R. Sanint (CIAT), D. Tomić (Economic Institute, Yugoslavia), and L. Tweeten (Ohio State University).

# Textile Trade Liberalization and its Welfare Implications for US Cotton Producers

*Shangnan Shui, John C. Beghin, and Michael Wohlgenant*[1]

**Abstract:** This study analyses the impact on the US cotton industry of removal of the Multi-Fibre Arrangement (MFA) using a multimarket displacement equilibrium model. The model captures the basic linkages of textile products and cotton markets in the USA and in non-US markets. Different textile trade policy reforms are simulated. Results suggest that removal of textile trade restrictions in the OECD countries induces a decrease and structural change in the total demand for US cotton towards a larger dependency on the world market. The decrease in total demand for US cotton has negative welfare effects on the US cotton industry. However, the welfare loss depends on how non-US cotton exporters respond to changes in OECD trade policy. The largest estimated loss is about $200 million. Ignoring agricultural linkages of the textile industry in the analysis of textile trade liberalization would induce an upward bias in estimated welfare gains for the US economy. The results suggest the likely formation of a coalition of US cotton-textile-apparel producers to generate political pressure for more trade protection.

## Introduction

International trade in textiles and apparel is comprehensively regulated and managed under the Multi-Fibre Arrangement (MFA). The MFA provides the framework for the negotiation of bilateral agreements between importing and exporting countries to control textile and apparel trade among its signatories. The USA currently has bilateral restraint agreements with 43 countries and regions, covering 80 percent of textile and apparel imports from developing countries. In the EC, the MFA regulations cover about 77 percent of total EC textile and apparel imports from 27 countries.

The MFA has been under scrutiny because it conflicts with the basic principles of the GATT; i.e., the use of quotas and country-based discrimination. The MFA also brings about substantial welfare losses because of its trade distortions. The welfare implications of removing the MFA have been extensively analysed in the literature (Jenkins, 1980; Hufbauer *et al.*, 1986; Cline, 1987; and Trela and Whalley, 1990). These studies have shown that substantial welfare gains for both exporting and importing countries would be induced by the removal of MFA. Most of these studies, however, centre their analysis on manufacturing sectors, and little attention has been paid to the impact on derived demand industries, especially the cotton industry, of removal of the MFA.

The objective of this paper is to analyse the impact of removing textile and apparel trade barriers in the USA and other OECD countries on US raw cotton producers, for both domestic and export markets. US cotton is one of the basic raw materials for US textiles; it is also exported and enters textile production abroad. Since both US and non-US textile and apparel production would be considerably affected by a removal of the MFA, the US cotton sector is likely to be deeply influenced by such trade policy changes. Cotton is one of US agriculture's largest field crops, and the USA is the world's largest cotton producer and exporter.

## Textile and Apparel Industries and the MFA

Derived demand relationships link the cotton, textile, and apparel markets. Two basic fibre types—cotton and manufactured fibre—comprise the raw materials for the textile industry. Textile manufacturing involves several intermediate stages and products. These products, which are heavily traded domestically and internationally, are used to produce three types of end-goods: apparel, home furnishings, and industrial products. The demand for fibres such as cotton is a derived demand for two end-use goods (home furnishings and industrial products) and for a semi-finished good, fabrics, which enters into apparel production. All these commodities are traded internationally.

The original objectives of the MFA were to allow developing countries to increase their shares of the world market for textiles and apparel and to encourage developed countries to abandon production of non-competitive textile products. In practice, the MFA has evolved into an import-restraining device for developed countries, which use a battery of quantitative and tariff restrictions to protect their domestic textile and apparel producers.

The Uruguay Round of the GATT negotiations has been exerting pressure for removal of the MFA, since it contradicts the GATT principles (Goto, 1989). Developed countries will have to make major concessions on the MFA if they expect trade liberalization in agriculture and services of developing countries. Several schemes have been considered for textile and apparel trade liberalization: tariffication of quotas, phasing out of quotas, total removal of the MFA, etc. The major scenario considered in this paper—total removal of OECD trade barriers—is a benchmark case. The qualitative implications of this scenario remain valid for less radical reforms.

## Model

The analysis relies on a multi-market model comprising cotton, textile, fabric, and apparel markets in different country groups. Following Muth (1964), the model solves for the comparative statics of shocks in exogenous policy variables; i.e., the tariffs and quotas on textile, fabric, and apparel imports in the USA and other OECD countries. The endogenous variables are the equilibrium price and quantity in the markets mentioned above. The welfare changes for US cotton producers are approximated by producer surplus and revenue changes.

OECD countries are divided into the USA and other OECD countries. Developing countries are categorized by destination of their exports (USA, other OECD, and non-OECD) and by the structure of their cotton derived demand (users of US cotton, users of US and non-US cotton, and users of non-US cotton). To avoid double-counting problems, four-digit SIC classification was used to define the goods. Textile goods are defined as household and industrial end-use of textiles; fabrics are the semi-finished textile products entering apparel production.

In the USA, textile and fabric production use US cotton exclusively, whereas in other OECD and some developing economies, US cotton competes with non-US cotton. In all OECD countries, apparel production uses both domestic and imported fabric. Similarly, in final consumption of OECD countries, textile and apparel commodities compete with developing countries' substitutes. Market equilibrium conditions close the model. Import tariffs on textiles, fabric, and apparel enter as wedges in identities linking the price paid for these goods in the USA and other OECD countries and the price received by the non-OECD producers of these goods. All producer prices are endogenous and respond to input price changes. Import quotas are explicitly modelled as exogenous quantity constraints entering market equilibrium conditions for imported textile, fabric, and apparel goods in the USA and other OECD countries.

Additional simplifying assumptions are made for tractability of the model. All textile fabric and apparel production exhibits constant returns to scale and nonjointness in technology, and all producers are price takers. These assumptions are convenient to trace output effects on cotton derived demand and cotton price effects on textile and fabric prices. In addition, US, other OECD, and non-OECD goods are assumed to be imperfect substitutes.

The removal of quotas and tariffs in OECD countries has a direct negative impact on their textile, fabric, and apparel production. These in turn induce a decrease in domestic demand for US cotton. Conversely, non-OECD textile, fabric, and apparel producers expand their output because of export expansion. This production expansion outside the OECD countries stimulates both US and non-US cotton exports. Hence, there is a structural change in the composition of US cotton demand (a decrease in domestic use along with an expansion of exports). There are also secondary substitution effects between US and non-US cotton because of changes in their relative price. The net impact of these three effects (lower domestic use, increased export, relative price change) is ambiguous analytically.

The full model includes 54 equations and identities describing cotton, textile, fabric, and apparel production and trade for 29 countries (the USA, other OECD, and 22 developing economies). The equations are log-differentiated and show changes in endogenous variables caused by policy shocks.

## Data

Although conceptually simple, the model requires a larger number of cost and market share parameters and elasticity estimates. The existing empirical literature is used to define ranges of values for most price elasticity estimates that will serve as bound for sensitivity analysis (Duffy et al., 1987 and 1990; Anson and Simpson, 1988; Cline, 1987; and Wohlgenant, 1986). The estimates are adjusted for consistency with the underlying structure of the model. The detailed procedures for this step are explained in Shui (1990). Output elasticities of cotton demand are derived assuming constant returns to scale and non-jointness in textile, fabric, and apparel production. Most share parameters come from USDA data and Anson and Simpson (1988). Shares are average values for 1982–87, excluding 1985 because of its unusual cotton trade flows.

## Policy Simulations

Results are reported for five policy scenarios. The first two consider the removal of all quotas and tariffs in all OECD countries for short-run cotton supply response and long-run supply response for both US and other cotton growers. The third policy scenario looks at the same policy reform assuming a non-US cotton expansion of 10 percent (supply shift of 10 percent). This third simulation is motivated by the steady expansion of non-US cotton production in the last decade. The last two cases analyse the implications of tightening MFA quotas (decrease of import quotas of 5 percent in all OECD countries). These last scenarios differ in their assumptions on cotton supply price response: short-run and long-run. These two pessimistic cases reflect the attempt by US textile producers and the politicians to pass the textile trade bill. The results for the five simulations are presented in Table 1. The table shows changes in US cotton price, total demand, export demand for other OECD countries, non-US textile exporters to OECD countries, and other non-US textile producers. Table 1 also gives changes in revenue and producer surplus for US cotton production.

Strong tendencies emerge from these simulations. The policy reform effects on US total cotton demand and price are small, but a considerable structural change in the composition of cotton demand would occur with trade liberalization. US cotton growers would be much more exposed to world competitive forces because US domestic use of cotton would decrease and export demand would represent the lion's share of total demand. In the long run (assuming more elastic cotton supply responses or non-US cotton supply expansion), total demand for US cotton would decline, with a maximum welfare loss of around $110 million. However, in the short run, textile trade liberalization would increase US cotton growers' surplus because of a large substitution effect in the non-OECD demand for cotton. Because non-US cotton supply is inelastic in the short run, the output effect induced by the expansion of non-US textile production creates a strong increase in the non-US cotton price. This, in turn, induces a substitution effect against non-US cotton towards increased use of US cotton. Conversely, a tightening of MFA quotas would be beneficial to US cotton production, although such restrictions decrease non-US use of US cotton because of the smaller quotas and through substitution effects driven by higher US cotton prices.

Sensitivity analysis suggests that these results are extremely robust. Most results are driven by share parameter values and output effects in textile, fabric, and apparel production.

The cost and trade share parameters are available and are not as arguable as the choice of elasticity estimates. Hence, the sensitivity analysis centres on price elasticity estimates. The production effects on derived demands for US and non-US cotton systematically dominate price (substitution) effects, although the latter are significant. For the 10-percent supply shift scenario, the maximum welfare loss estimated for US cotton producers is around $200 million.

Table 1—Changes in US Cotton Price, Demand, Revenue, and Producer Surplus

| Variable | Policy | | | | |
|---|---|---|---|---|---|
| | Removing All Trade Restrictions | | Removing All Trade Restrictions with non-US Expansion of 10 Percent | USA Tightens Quota Restrictions (Quotas Decrease by 5 Percent) | |
| | Short-Run | Long-Run* | Supply Shock | Short-Run | Long-Run* |
| Price (percent) | 1.660 | −0.989 | −3.684 | 1.198 | 0.730 |
| Demand | | | | | |
| Total demand | 0.581 | −0.633 | −1.289 | 0.420 | 0.467 |
| US mill | −26.953 | −24.631 | −22.243 | 0.729 | 1.152 |
| Total exports | 28.559 | 23.752 | 21.003 | −1.588 | −2.112 |
| Other OECD | −7.780 | −12.686 | −16.010 | −0.627 | −0.834 |
| Textile exporters | 45.133 | 42.752 | 42.610 | −3.955 | −4.098 |
| Other importers | 4.452 | 2.225 | 1.386 | −0.255 | −0.287 |
| Revenue | | | | | |
| Percent | 2.241 | −1.622 | −4.973 | 1.618 | 1.197 |
| Value (million 1982 $) | 66.672 | −48.256 | −147.950 | 48.137 | 35.612 |
| Producer Surplus | | | | | |
| Percent | 1.665 | −0.986 | −3.660 | 1.201 | 0.732 |
| Value (million 1982 $) | 49.535 | −29.334 | −108.889 | 35.731 | 21.778 |

*Long-run cotton supply price elasticity is 0.64 for the USA and 2.36 for other cotton producers. Short-run supply price elasticity is 0.35 for the USA and 0.38 for other cotton producers.

The results imply the existence of an upward bias in estimated welfare gains such as in Cline (1987), when the backward linkages of US textile and fabric production are not incorporated into the analysis. Nevertheless, this bias is small compared to the expected consumer welfare gains from textile and apparel trade liberalization.

Another implication concerns the political economy of the MFA. The results show that US cotton farmers would lose from trade liberalization (lower producer surplus in the long run and greater exposure to world market forces). They are likely to join a grand coalition of US textile-apparel and cotton producers to put pressure on policy makers for more trade distortions. The costs of adjustment associated with the change in the composition of US cotton demand is likely to be significant while US growers work their way through new export channels. These expected adjustment costs associated with trade liberalization, and not accounted for in this paper, may reinforce the aversion of cotton producers to less distorted textile trade.

# Conclusions

In this paper, an attempt was made to analyse the implications for the US cotton market of liberalization of the textile and apparel trade. A multi-market equilibrium displacement model was used to trace the impact of exogenous changes in the import tariffs and quotas in OECD countries on the derived demand for cotton in the USA and in the rest of the world. The major impacts were the increase in export demand for US cotton and the long-run decrease in producer surplus due to a sharp decrease in textile and apparel production in OECD countries. The efforts of US cotton, textile, and apparel producers are likely to converge in putting pressure on the political body for more protection.

# Notes

[1] North Carolina State University.
[2] A companion paper that describes the model more fully is available from the authors.

# References

Anson, R., and Simpson, P., "World Textile Trade and Production Trends," Economist Intelligence Unit, London, UK, 1988.

Cline, R., *The Future of World Trade in Textiles and Apparel*, Institute for International Economics, Washington, D.C., USA, 1987.

Duffy, P.A., Richardson, J.W., and Wohlgenant, M.K., "Regional Cotton Acreage Response," *Southern Journal of Agricultural Economics*, Vol. 19, No. 1, 1987, pp. 99–109.

Duffy, P.A., Wohlgenant, M.K., and Richardson, J.W., "The Elasticity of Export Demand for U.S. Cotton," *American Journal of Agricultural Economics*, Vol. 72, No. 2, 1990, pp. 468–474.

Goto, J., "The Multi-Fiber Arrangement and its Effects on Developing Countries," *World Bank Research Observer*, Vol. 4, No. 2, 1989, pp. 203–227.

Hufbauer, G., Berlinger, D., and Elliott, K., *Trade Protection in the United States: Thirty-One Case Studies*, Institute for International Economics, Washington, D.C., USA, 1986.

Jenkins, G.P., "Costs and Consequences of the New Protectionism: The Case of Canada's Clothing Sector," North-South Institute, Ottawa, Canada, 1980.

Muth, R., "The Derived Demand for a Productive Factor and the Industry Supply Curve," *Oxford Economic Papers*, Vol. 16, 1964.

Shui, S., "Impacts on the U.S. Cotton Industry of Changes in Textile Trade Policies," unpublished Ph.D. dissertation, North Carolina State University, Raleigh, N.C., USA, 1990.

Trela, I., and Whalley, J., "Do Developing Countries Lose from the MFA?" *Economic Journal*, Vol. 100, 1990, pp. 1190–1205.

Wohlgenant, M.K., "Impact of an Export Subsidy on the Domestic Cotton Industry," Bulletin No. B–1529, Texas Agricultural Experiment Station, College Station, Tex., USA, 1986.

*[Discussion of this paper and the authors' reply appear on page 211.]*

# Effect of Imports on US Prices for Fresh Apples

*Amy L. Sparks and Boris E. Bravo-Ureta*[1]

**Abstract:** This paper presents new econometric evidence concerning the variation of fresh apple prices in the US market as a function of fluctuations in supplies from seven major US and non-US supply areas. The Rotterdam inverse demand system recently developed by Barten and Bettendorf is used in the analysis. The results show that the impact on prices stemming from a change in overall quantities of fresh apples varies widely across supply regions. The price of Washington apples would suffer a marked drop with a rise in total quantity. The analysis also suggests that Chile's entrance into the US apple market has had a negligible impact on prices.

## Introduction

Over the last several years, apple growers in the USA have encountered an adverse economic environment stemming from lower prices for their products (Sparks, 1989). A major determinant of lower farm prices has been a sharp increase in US production resulting from heavy tree plantings in the late 1970s and early 1980s. In addition, apple production in Chile, New Zealand, South Africa, and Canada and imports from these countries into the US market have increased substantially in the last two decades.[2] Another major factor contributing to lower apple prices has been a relatively constant consumption level.

The increase in apple production observed in recent years in several major supplying nations, including New Zealand, South Africa, and Chile, is expected to continue. This production growth is likely to be particularly strong in less-developed countries as they struggle to generate badly needed foreign exchange (USDA, 1989 and 1990). For example, apple production in Chile, a relatively new player on world apple markets, is expected to double in the next decade (O'Rourke, 1987). These non-US suppliers will be looking beyond their national borders for markets in which to sell their products, including in the USA.

Apple production makes a significant contribution to farm income in various regions of the USA and in several other countries. Nevertheless, there is little recent economic analysis designed to understand the interaction between apple production and price levels as supply from competing regions changes. The purpose of this paper is to present new econometric evidence concerning the variation of fresh apple prices in the US market as a function of fluctuations in supplies from major US and non-US supply areas. The recently developed Rotterdam inverse demand system (Barten and Bettendorf, 1989) is used in the analysis.

## Methodology

Econometric estimates of relationships between prices and quantities of apples using a variety of specifications have been reported in the literature.[3] Studies using an inverse demand specification include those of Carman and Kenyon (1969), Edman (1972), and Baritelle and Price (1974). These studies, however, have focused on local or regional US markets and have often been formulated on an *ad hoc* basis without much attention given to *a priori* restrictions stemming from demand theory.

The model used in this paper explicitly accounts for restrictions derived from demand theory. It starts by assuming that the total commodity bundle is weakly separable, which makes it possible to treat the commodity group of interest, fresh apples in this case, as being independent of all other groups (Phlips, 1974). Under weak separability, the (direct) market demand system for fresh apples can be written as:

(1) $D_i = f(P, E)$

where $D_i$ is the demand function for fresh apples produced in the $i$th region, $P$ is a vector of apple prices, $E = P'Q$ is total expenditure on apples, and $Q$ is a vector of apple quantities. In this formulation, fresh apples are differentiated by the region where they are produced, and they are not considered to be perfect substitutes. The inverse demand system corresponding to Equation (1) is given by:

(2)  $P_i = f^{-1}(Q, E)$

Although the direct demand system is equivalent to the inverse system from a theoretical point of view, these two approaches are not equivalent econometrically. The choice of one approach over the other in empirical work depends on which variable, price or quantity, is considered to be exogenous. Waugh (1966, p. 81) suggested that in agricultural markets "... changes in prices are generally determined by changes in quantities and changes in income— not the other way around." More recently, Salvas-Bronsard et al. (1977, p. 310) have argued that a price-dependent demand system is appropriate when "... supply is quite inelastic, or [if] ..., within a year, quantities are determined independently of prices."

Apples are perennials. The size of the apple crop in any one year, depends, to a large extent, on planting decisions made several years earlier and is not related to current market conditions. Based on these arguments, the inverse demand framework is chosen in this study to model the price formation of fresh apples in the wholesale market.

The specific model used is the price-dependent Rotterdam demand system (Barten and Bettendorf, 1989), which is the inverse equivalent of the regular Rotterdam system developed by Theil (1975 and 1976). The equation to be estimated is:

(3)  $\bar{s}_{it}\Delta \ln r_{it} = \beta_i \Delta \ln Q_t + \Sigma_j \beta_{ij} \Delta \ln q_{jt} + \varepsilon_{it}$        (i, j = 1, 2, ..., 7; t = 1969, ..., 1986)

where:

(4)  $\bar{s}_{it} = 0.5(S_{it} + s_{i,\,t-1})$

(5)  $r_{it} = P_{it}/S_t$

(6)  $Q_t = \Sigma_j \bar{s}_{jt} \Delta \ln q_{jt}$

and $s_{i,\,t}$ is the market share of fresh apples from region $i$ sold in year $t$; $P_i$ is the average wholesale price of apples from region $i$; $S_t$ is the total value of fresh apples from all regions sold in the US market in year $t$; $q_{jt}$ is the quantity of fresh apples sold in the US market from supply region $j$ in year $t$; $\beta_i$ and $\beta_{ij}$ are parameters to be estimated; and $\varepsilon_{it}$ is an error term that is assumed to be normally distributed with mean zero and constant variance. The inverse demand system consists of seven equations, one from each producing region, as discussed below.

Restrictions resulting from demand theory require that $\Sigma_i \beta_i = -1$ and $\Sigma_i \beta_{ij} = 0$, which corresponds to the adding-up property. The homogeneity property is satisfied when $\Sigma_j \beta_{ij} = 0$, and symmetry requires that $\beta_{ij} = \beta_{ji}$ (i.e., the estimated coefficients on the quantity parameters between the $i$th and $j$th supply regions are symmetric). The system is estimated using the iterative least squares option of the TSP econometric package.

Once the system of Equations (3) is estimated, it is possible to calculate the scale effect, which gives the change in the $i$th price $(P_i)$ in response to a proportionate change in total quantity $(Q)$. As shown by Barten and Bettendorf (1989), the scale elasticity $(SE)$ can be computed from Equation (3) as:

(7)  $SE_{iQ} = \dfrac{\partial \ln r_i}{\partial \ln Q} = \dfrac{\partial \ln s_{it}}{\partial \ln Q} - 1 = \dfrac{\beta_i}{s_i}$

A second relationship of interest obtained from an inverse demand system is the Antonelli substitution $(AS)$ or quantity effect, which is equivalent to the substitution effect for a direct demand formulation. The $AS$ effect is defined as the change in the normalized price of the $i$th good with respect to a marginal change in the consumption of the $j$th good, holding utility

197

constant (Anderson, 1980). The Antonelli substitution effect between the $i$th and the $j$th good, calculated from Equation (3), is equal to the coefficient $\beta_{ij}$, and the Antonelli substitution elasticity is $\beta_{ij}/s_i$. Demand theory requires that the own Antonelli substitution effects $\beta_{ii}$ be negative, while there is no *a priori* expectation on the sign of the cross effects $\beta_{ij}$ (Anderson, 1980). To implement the model, the world is divided into seven apple-producing regions: Washington State; North-Central and Northeastern USA (NC–NE), including Michigan, New York, Ohio, Pennsylvania, Vermont, Massachusetts, Maine, Connecticut, and New Hampshire; rest of the USA; Canada; New Zealand; South Africa; and rest of the world. Annual price and quantity data for the 1969–86 period were obtained from USDA and UN sources.

Table 1 presents the prices, quantities, values, and value shares of apples in the USA from each of the seven supply regions for 1969 and 1986, and for Chile for 1975 and 1986. As can be seen from the data, the quantities and values of imported apples from Canada, Chile, New Zealand, South Africa, and the rest of the world increased substantially in this time period. Washington has experienced major gains as a supplier to the US market, in terms of both quantity and value share, while other regions of the USA have suffered considerable losses. The share of non-US apples, excluding Canada, has increased markedly over the past two decades but still plays a relatively minor role in the US fresh market. Chile has shown a very impressive growth, increasing from 600 t in 1975 to over 31,000 t in 1986, which, in the latter year, was over two-thirds of US imports from the rest of the world.

Table 1—Prices, Quantities, and Values of Fresh Apples in the USA from Major Suppliers, 1969–86

| Supply Region | Year | Price ($/t) | Quantity (t) | Value ($) | Value Shares (%) |
|---|---|---|---|---|---|
| Washington | 1969 | 77.0 | 552,476 | 42,385,933 | 13.50 |
| | 1986 | 410.0 | 1,106,766 | 453,840,341 | 32.20 |
| NC–NE | 1969 | 163.0 | 498,770 | 81,409,239 | 25.80 |
| | 1986 | 457.0 | 452,413 | 206,906,561 | 14.80 |
| Rest of the USA | 1969 | 154.0 | 1,179,839 | 181,553,625 | 57.60 |
| | 1986 | 419.0 | 1,572,242 | 658,580,729 | 46.80 |
| Canada | 1969 | 211.0 | 36,738 | 7,761,000 | 2.50 |
| | 1986 | 409.0 | 44,565 | 18,213,000 | 1.30 |
| New Zealand | 1969 | 323.0 | 2221 | 717,000 | 0.20 |
| | 1986 | 1026.0 | 26,917 | 27,605,000 | 1.90 |
| South Africa | 1969 | 446.0 | 1735 | 774,000 | 0.30 |
| | 1986 | 785.0 | 14,482 | 11,371,000 | 0.80 |
| Rest of the world | 1969 | 222.0 | 1,842 | 409,000 | 0.10 |
| | 1986 | 683.0 | 45,667 | 31,207,000 | 2.20 |
| Chile* | 1975 | 275.0 | 600 | 165,000 | — |
| | 1986 | 554.0 | 31,040 | 17,202,000 | — |

*Chile is included as part of the rest of the world.

# Results

The parameter estimates for the seven-equation system are presented in Table 2. To facilitate the reading of the table, all estimated parameters and their standard errors are multiplied by 100. Given that all quantity effects $(\beta_{ij})$ are symmetric, the top portion of the parameter matrix is omitted. A total of 35 parameters are estimated, of which 10 are statistically significant at the 0.01 level, three at the 0.05 level, and two at the 0.10 level. As already mentioned, economic theory dictates that the diagonal elements of the Antonelli substitution matrix should be negative. This condition holds for all seven of the diagonal elements of the matrix shown in Table 2.

The Antonelli substitution elasticities along with their approximate standard errors are reported in Table 3. The cross-substitution elasticities reveal that Washington apples, the largest single-state source of apples in the USA, have a highly significant complementary relationship with apples from the rest of the USA. There is little relationship between the price of Washington apples and the quantities supplied by individual regions in the US market. The NC–NE region has a highly significant and fairly large complementary relationship with apples from the rest of the USA. It has a complementary relationship with Canadian and New Zealand apples, although these elasticities are quite small. Washington and New Zealand apples are complements to apples from the rest of the USA.

In contrast, apples from Canada and New Zealand are more evenly divided between substitutes and complements with respect to other suppliers to the US market. Canadian apples have a complementary relationship with those from NC–NE and South Africa, and are substitutes for New Zealand apples. NC–NE and rest-of-the-world apples are weak complements.

Of the seven scale elasticities shown in Table 3, four are statistically significant at the 5-percent level or higher. Of these four, two are negative, Washington and rest of the USA, and two are positive, New Zealand and South Africa. The negative scale elasticities suggest that a 1-percent increase in overall quantity would lead to a 2.2-percent and 0.8-percent drop in the prices of apples supplied by Washington and rest of the USA, respectively. In contrast, a similar rise in quantity would yield a gain in price of 2.0 percent for New Zealand and 1.6 percent for South African apples, both Southern Hemisphere producers.

As noted earlier, apple production in Chile has grown very rapidly, from 125,300 t in 1975 to 515,000 t in 1986. Most of this additional apple production has been for export to various non-Chilean markets, since Chilean consumption reached only 141,000 t in 1988/89. Chile began exporting apples to a significant degree in the mid- to late-1970s, and its role in international markets, which is still relatively small, is expected to increase as recently planted trees come into bearing age. There is concern that increased imports of Chilean apples into the USA have depressed the prices received by competing suppliers.

To assess the impact of imports from Chile on US fresh apple prices, the demand system was re-estimated.[4] A dummy variable with a value of zero in 1963–73 (when Chile was not a supplier) and a value of one in 1974–87 (when Chile was a supplier) was multiplied by the aggregate quantity variable. The significance of the coefficient on the dummy variable was used as a test of whether the entrance of Chile into the US apple market has had an impact on the scale elasticities of each of the major suppliers. For all suppliers, the coefficient on the dummy variable was statistically insignificant. This is not surprising if we consider that, despite the major rise in Chilean apple exports, this country remains a small participant in the US market.

Table 2—Parameter Estimates (×100) for an Inverse Rotterdam Demand Model for Fresh Apples, 1969–86

| Supply Region | Quantity Effects | | | | | | | Scale Effects |
| --- | --- | --- | --- | --- | --- | --- | --- | --- |
| | Washington | NC–NE | Rest of the USA | Canada | New Zealand | South Africa | Rest of the World | |
| Washington | -3.40 (2.20) | | | | | | | -54.06*** (8.11) |
| NC–NE | -0.68** (1.30) | -9.87*** (1.55) | | | | | | -6.15 (5.01) |
| Rest of the USA | 3.97* (1.87) | 9.59*** (2.03) | -13.26*** (3.28) | | | | | -42.49*** (5.98) |
| Canada | 0.28 (0.19) | 0.53** (0.24) | -0.29 (0.40) | -0.34 (0.26) | | | | -0.12 (0.63) |
| New Zealand | -0.09 (0.15) | 0.44** (0.20) | 0.08 (0.33) | -0.26** (0.15) | -0.40*** (0.15) | | | 1.89*** (0.49) |
| South Africa | -0.06 (0.07) | -0.08 (0.07) | 0.02 (0.12) | 0.12** (0.06) | 0.05 (0.04) | -0.09*** (0.02) | | 0.65** (0.27) |
| Rest of the world | -0.04 (0.11) | 0.06 (0.15) | -0.12 (0.24) | -0.05 (0.05) | 0.16*** (0.04) | 0.03*** (0.01) | -0.05* (0.03) | 0.28 (0.41) |

***Significant at the 1-percent level. **Significant at the 5-percent level. *Significant at the 10-percent level.

Table 3—Scale and Antonelli Substitution Elasticities Calculated from an Inverse Rotterdam Demand Model for Fresh Apples

| Supply Region | Substitution Elasticities (calculated at sample means) | | | | | | | Scale Elasticities |
|---|---|---|---|---|---|---|---|---|
| | Washington | NC–NE | Rest of the USA | Canada | New Zealand | South Africa | Rest of the World | |
| Washington | -0.14 (0.09) | -0.03 (0.05) | 0.16** (0.07) | 0.01 (0.01) | -0.00 (0.01) | -0.00 (0.00) | -0.00 (0.01) | -2.22*** (0.33) |
| NC–NE | -0.03 (0.06) | -0.47** (0.07) | 0.45*** (0.10) | 0.02** (0.01) | 0.02** (0.01) | -0.00 (0.00) | 0.00 (0.01) | -0.29 (0.23) |
| Rest of the USA | 0.08** (0.04) | 0.19*** (0.04) | -0.26*** (0.06) | -0.01 (0.01) | 0.00 (0.01) | 0.00 (0.00) | -0.00 (0.01) | -0.84*** (0.11) |
| Canada | 0.20 (0.14) | 0.37** (0.16) | -0.20 (0.27) | -0.23 (0.18) | -0.18* (0.10) | 0.08** (0.04) | -0.04 (0.03) | -0.09 (0.44) |
| New Zealand | -0.09 (0.17) | 0.48** (0.22) | 0.09 (0.36) | -0.28* (0.16) | -0.43*** (0.16) | 0.06 (0.05) | 0.18*** (0.04) | 2.04*** (0.53) |
| South Africa | -0.14 (0.19) | -0.19 (0.19) | 0.04 (0.30) | 0.30 (0.14) | 0.13 (0.11) | -0.22*** (0.05) | 0.08** (0.03) | 1.63** (0.67) |
| Rest of the world | -0.05 (0.14) | 0.08 (0.18) | -0.14 (0.29) | -0.06 (0.05) | 0.20 (0.05) | 0.04** (0.01) | -0.07 (0.04) | 0.34 (0.50) |

Note: Approximate standard errors in parentheses. ***Significant at the 1-percent level. **Significant at the 5-percent level. *Significant at the 10-percent level.

# Conclusions and Implications

While consumption of fresh apples in the USA has remained fairly constant, production has been increasing in both the USA and several other countries. This trend is expected to continue for the next several years. These conditions are placing downward pressure on the prices US farmers receive for their product, forcing adjustments in the industry.

In order to measure the effects of increasing quantities on prices of fresh apples from each of the major suppliers to the US market, an inverse Rotterdam model was constructed and estimated. The results indicate that the impact on US apple prices of increased supplies varies considerably by supply region. The price of apples from Washington state, the largest single supplier in the USA, is affected positively by increasing supplies from the rest of the USA. Changes in the quantity of apples from other individual supply regions have little effect on the price of Washington apples. Nonetheless, the scale elasticities indicate that the price of Washington apples, as well as that of the rest of the USA, is negatively affected by increases in total supplies to the US market. In contrast, empirical results indicate that the prices of apples from New Zealand and South Africa, both Southern Hemisphere suppliers, are augmented significantly by increased apple supplies in the US market. An analysis to account for Chile's entrance into the US apple market showed that, up to now, Chile has had a negligible impact on the price of apples sold in the USA.

# Notes

[1]US Department of Agriculture and University of Connecticut, respectively.

[2]For political reasons, the USA has imposed a ban on imports from South Africa. Before the ban, South Africa was a relatively large non-US supplier of fresh apples to the US market. Hence, a lifting of the import restriction would probably enable South Africa to resume its role of relative prominence in the US market.

[3]For a review of several studies see Nuckton, 1978.

[4]The results are not shown due to space limitations.

# References

Anderson, R.W., "Some Theory of Inverse Demand for Applied Demand Analysis," *European Economic Review*, Vol. 14, 1980, pp. 281–290.

Barten, A.P., and Bettendorf, L.J., "Price Formation of Fish: An Application of an Inverse Demand System," *European Economic Review*, Vol. 33, 1989, pp. 1509–1525.

Baritelle, J.L., and Price, D.W., "Supply and Marketing for Deciduous Crops," *American Journal of Agricultural Economics*, Vol. 56, 1974, pp. 245–253.

Carman, H.F., and Kenyon, D.E., *Economic Aspects of Producing and Marketing California Apples*, Giannini Foundation Research Report No. 301, University of California, Berkeley, Calif., USA, 1969.

Edman, V.G., *Retail Demand for Fresh Apples*, Market Research Report No. 952, Economic Research Service, US Department of Agriculture, Washington, D.C., USA, 1972.

Nuckton, C.F., *Demand Relationships for California True Fruits, Grapes, and Nuts: A Review of Past Studies*, Giannini Foundation Special Publication No. 3247, University of California., Berkeley, Calif., USA, 1978.

O'Rourke, A.D., *Chile: US Supplier and Competitor*, IMPACT Center Information Series No. 17, Washington State University, Pullman, Wash., USA, 1987.

Phlips, L., *Applied Consumption Analysis*, North-Holland/Elsevier, Amsterdam, Netherlands, 1974.

Salvas-Bronsard, L., LeBlanc, D., and Bronsard, C., "Estimating Demand Equations: The Converse Approach," *European Economic Review*, Vol. 9, 1977, pp. 301–321.

Sparks, A.L., "Situation and Prospects for Fresh Apple Markets," in *Fruit and Tree Nuts Situation and Outlook Report,* Economic Research Service, US Department of Agriculture, Washington, D.C., USA, 1989.

Theil, H., *Theory and Measurement of Consumer Demand,* North-Holland, Amsterdam, Netherlands, 1975 (Vol. I) and 1976 (Vol. II).

UN (United Nations), trade data, SITC code 051.4.

USDA (US Department of Agriculture), *Developing Economies: Agriculture and Trade—Situation and Outlook Series,* No. RS–90–5, Economic Research Service, Washington, D.C., USA, 1990.

USDA (US Department of Agriculture), *Fruit and Tree Nuts: Situation and Outlook Yearbook,* No. TFS–250, Economic Research Service, Washington, D.C., USA, 1989.

USDA (US Department of Agriculture), *Noncitrus Fruits and Nuts,* National Agricultural Statistics Service, Washington, D.C., USA, various issues.

Waugh, F.V., *Demand and Price Analysis,* Technical Bulletin No. 1316, Economic Research Service, US Department of Agriculture, Washington, D.C., USA, 1966.

---

## Discussion Opening—*Sibylle Scholz* (Winrock International)

The methodology used is well thought out with respect to the choice of an inverse demand system. Among agricultural goods, tree crops are probably the least influenced by short-term price fluctuations. The correct sign and significance of the estimated parameters partially confirm the appropriateness of the model.

However, the size of the parameters is in need of some explanation. Given an ever-increasing amount of substitutes for apples, one would expect the demand elasticity for the USA to be in the elastic range. The coefficients for Washington, NC–NE, and the rest of the USA are all greater than one in absolute terms, which means that demand is in the inelastic range since these coefficients are from the inverse demand function.

Given that the demand elasticities are in the inelastic range, it is correct to conclude that an increase in supply will lead to downward pressures on producers' revenues. However, it is not quite correct to use a fairly constant US consumption of fresh apples as an added condition on downward pressures of producer revenues. This latter point leads into an additional question about the unit of analysis.

Is the choice of apple produced related to final consumption? In other words, is an apple produced for fresh consumption a different apple from one produced for sale to Sara Lee to go into an apple pie? To what extent do processing industries absorb the increased supply of apples? What proportion of wholesale fresh apples is consumed as fresh apples? What proportion of total apples produced is consumed fresh?

Given the recent development in post-harvest technologies and an ever-increasing supply of prepared, premixed juices, jellies and chewables, microwaveables, individually wrapped deep-frozen "ready-to-pop" tarts, supply fluctuations, at least within the USA, might be absorbed in processing. This, coupled with a highly concentrated food processing industry, might explain why demand is in the inelastic range. I mean to imply that price-strategic behaviour by apple-processing industries might push demand into the inelastic range. This would imply that USA apple producers have more to fear from US industries than from non-US apple producers.

*[Other discussion of this paper and the authors' reply appear on page 211.]*

# Advertising Check-Off Programmes

*Hui-Shung Chang and Henry Kinnucan*[1]

**Abstract:** This paper examines the impact of a change in the advertising tax on prices, output, and welfare. Results show that a supply shift alone (i.e., advertising is ineffective and hence there is no demand shift) will result in higher retail prices, lower farm output, higher retail-farm price ratios, and losses in benefits to society. If the supply shift is accompanied by a demand shift due to effective advertising, the retail price will be higher. Farm output and the retail-farm price ratio, however, will be smaller compared to an isolated supply shift. Given the advertising elasticities found so far in empirical studies (less than 0.10), an increase in producer assessments or check-offs for the purpose of increasing demand through advertising will lead to welfare loss. Research and new product development may be better alternatives to increasing demand from a social perspective.

## Introduction

Commodity check-off programmes have been in existence in the USA since 1935 (the first was the Florida Citrus Advertising Act) and have proliferated over time. In 1989, there were 350 federal- and state-legislated promotional programmes that covered over 80 farm commodities and cost more than $530 million. Funding for these programmes comes primarily from mandated producer assessments and check-offs based on two types of legislation, research and promotion acts and marketing orders.

Under the check-off programmes, producers are required by statute to pay advertising excise taxes or assessments for each unit of output sold. The revenues are then used by commodity promotional organizations and marketing boards for market development and research (Armbruster and Frank, 1988). The dairy programme is by far the largest, with annual collection exceeding $200 million, followed by beef and pigmeat, which have annual collections exceeding $80 million and $26 million, respectively. The assessment rates are 15 cents per cwt for milk, $1.00 per head for beef, and 25 cents per $100 of pig value for pigmeat.

## Model

Let us consider a competitive food marketing industry using two types of inputs, farm-based inputs, $a$, and marketing inputs, $b$, to produce a food product, $x$, sold in the retail market (Gardner, 1975). All firms are assumed to be price takers in both the product and factor markets.

The marketing industry's production function:

(1) $x = f(a, b)$

is assumed to yield constant returns to scale. The demand function for $x$ at the retail level is:

(2) $x = D(P_x, A)$

where $P_x$ is the retail price of $x$, and $A$ is advertising for $x$.[2]

The model also includes four equations representing both the demand and supply for $a$ and $b$. Assuming profit-maximizing behaviour, the demand for the farm-based input is:

(3) $P_a = P_x f_a$

The demand for marketing inputs is:

(4) $P_b = P_x f_b$

204

where $f_a$ and $f_b$ are the marginal product of $a$ and $b$, respectively. Supply functions of farm-based and marketing inputs are represented, respectively, by:

(5) $P_a = h(a, T)$

(6) $P_b = g(b)$

where $T$ is the advertising tax or the check-off amount.

When a specific (as opposed to *ad valorem*) advertising tax is imposed, the total tax revenue, $TR$, is $Ta$. $TR$ may or may not equal $A$ in Equation (2). This depends *inter alia* on how $TR$ is allocated among market development activities and how advertising is measured. For example, $A$ may be measured in dollars, advertising goodwill (Nerlove and Waugh, 1961), advertising effort (Zufryden, 1978), or gross rating points (Heath, 1990). In the simplified case, $A$ represents the total tax collection, which is spent entirely on advertising; i.e., $A = TR$. More generally, $A$ is a function of the tax rate and farm output:

(7) $A = k(T, a)$

Partial derivatives of $A$ with respect to $T$ and $a$, $A_T$ and $A_a$, respectively, are assumed to be positive.

Equation (7) plays a significant role in the analysis. It links the supply shifter (tax rate) to the demand shifter (advertising) and makes it possible to analyse both shifts simultaneously with respect to just one exogenous variable, the tax rate.[3] The endogenous variables are $x$, $a$, $b$, $P_x$, $P_a$, and $P_b$. Changes in market equilibrium conditions are analysed by differentiating Equations (1)–(6) with respect to $T$.

This model differs from Gardner's in that, instead of considering an exogenous shift in demand or supply separately, it considers shifts in retail demand and farm supply simultaneously.[4] This is necessary because a change in the advertising tax rate affects not only the cost of production but also the budget constraint for advertising, which, in turn, affects the demand for an advertised product.

## Total Elasticities

The principal objective here is to determine the effect of a change in the advertising excise tax on retail price, farm output, and the retail-farm price ratio. Following Gardner's derivation procedures, total elasticities for these three variables with respect to the tax rate are presented as $E_{PxT}$, $E_{aT}$, and $E_{Px/PaT}$ in Equations (8)–(10).

(8) $E_{PxT} = e_T e_a S_a (e_b + \sigma) + \dfrac{-\eta_A[(e_{Aa}e_T e_a - e_{AT})(S_a e_b + \sigma) - e_{AT} S_b e_a]}{D}$

(9) $E_{aT} = e_T e_a [\eta \sigma + e_b (S_a \eta - S_b \sigma)] + \dfrac{\eta_A e_{AT} e_a (e_b + \sigma)}{D}$

(10) $E_{Px/PaT} = e_T e_a S_b (\eta - e_b) + \dfrac{-\eta_A \{e_{Aa} e_T e_a (S_a e_b + \sigma) - e_{AT}[(S_b e_a + S_a e_b + \sigma) - e_a (e_b + \sigma)(\frac{1}{\eta_a})]\}}{D}$

where $S_a = aP_a / xP_x$, the relative shares of $a$, $\sigma$ = elasticity of substitution between $a$ and $b$, $\eta$ = retail demand elasticity of $x$, $e_a$ = own-price elasticities of supply for $a$, $e_b$ = own-price elasticities of supply for $b$, $\eta_A$ = advertising elasticity of demand for $x$, $e_T$ = elasticity of $P_a$ with respect to $T$, $\eta_a$ = elasticity of derived demand for $a$, $e_{AT}$ = elasticities of advertising with respect to $T$, and $e_{Aa}$ = elasticities of advertising with respect to $a$.

The denominator of Equations (8)–(10) is:

(11) $\quad D = - \eta(S_b e_a + S_a e_b + \sigma) + e_a e_b + \sigma(S_a e_a + S_b e_b) - \eta_A e_{Aa} e_a (e_b + \sigma)^5$

The sign of $D$ can be determined by rewriting Equation (11) as:

(12) $\quad D = (1 - \eta_A e_{Aa}) e_a e_b + \sigma[S_a e_a + S_b e_b - (\eta_A e_{Aa} e_a + \eta)] - \eta(S_b e_a + S_a e_b)$

$D$ is positive if $\eta < 0$, $0 < \eta_A < 1$, $0 < e_{Aa} < 1$, and $e_a$, $e_b$, $e_T > 0$ and if $(\eta_A e_{Aa} e_a + \eta) < 0$. The latter is true if the advertising elasticity is small relative to demand elasticity in absolute value. This assumption is reasonable since most empirical advertising elasticities for food items are less than 0.10 (Hurst and Forker, 1989), whereas demand elasticities range from −0.14 to −2.63 (Brandow, 1961; and George and King, 1971).

Both the numerator and denominator in Equations (8)–(10) can be separated into two parts: the supply shift component (SSC) resulting from a change in the tax rate, and the demand shift component (DSC) resulting from a change in advertising. Note that $e_T$ and $\eta_A$ are the key parameters for the SSC and DSC, respectively.

## Analysis

If demand and supply curves have normal slopes, the SSCs represented by the first terms in Equations (8) and (9) are expected to be positive and negative, respectively, because a parallel upward shift of the farm supply curve induced by a tax increase leads to higher prices and lower output. The DSCs represented by the second term are expected to be positive because an upward (not necessary parallel) shift of the retail demand curve leads to higher prices and output. In Equation (10), the SSC is expected to be negative and the DSC positive.

If advertising is ineffective (i.e., if $\eta_A = 0$), the second terms in Equations (8)–(11) vanish; therefore, only an isolated supply shift is present, as in Gardner (1975). As expected, an increase in advertising tax in this case will increase the retail price, decrease farm supply, and decrease the retail-farm price ratio. The economic logic behind these results is that, when farm supply shifts to the left, both retail and farm price will tend to increase, causing a decrease in the quantity demanded of $x$. The decrease in the quantity demanded for $x$ releases marketing inputs. So long as $e_b > 0$, $P_b$ will fall, which reduces the cost of marketing inputs relative to farm input and hence the ratio $P_x / P_a$.[6]

On the other hand, if a tax increase does not affect farm price (i.e., if $e_T = 0$), then SSC vanishes. This happens if farm demand is perfectly elastic; therefore, a supply shift does not affect farm price. An increase in advertising tax in this case will result in increases in retail price, farm supply, and the retail-farm price ratio.

When $e_T$ and $\eta_A$ are not both zero, the signs of Equations (8)–(10) cannot be determined a priori. Hypothetical parameter values can be used for illustrative purposes. Following Gardner, $e_a$, $e_b$, $\eta$, and $S_a$ were set equal to 1.0, 2.0, −0.5, and 0.5, respectively; and $\sigma$, the elasticity of substitution between farm-based input and marketing inputs, is set alternately to 0 and 0.5. $e_T$ is set alternately to 0, 0.5, and 1.0. The long-run advertising elasticities are set between 0 and 0.35. Parameters pertaining to advertising such as $e_{Aa}$ and $e_{AT}$ are set equal to 1.0.[7]

Simulated results show that total elasticities are sensitive to changes in the advertising elasticity, $\sigma$, and $e_T$ (Table 1). For example, when the advertising elasticity becomes greater, the total elasticity of retail price continues to rise while the total elasticities of farm output and the retail-farm price ratio first decrease, reach zero, and then increase. The sign changes for the latter two imply an increase in the importance of the DSC relative to the SSC as advertising becomes more effective. On the other hand, the greater the $\sigma$, the less the retail price, farm output, and the retail-farm price ratio will change as alternative input combinations exist. Further, all three elasticities increase with $e_T$ in absolute terms, which means that the greater the impact of the tax on farm price, the greater the impact of the tax at all levels of the market.

Table 1—Economic Effects on Prices and Output of a
10-Percent Increase in the Advertising Tax

| $e_T$ | $\eta_A$ | Retail Price Elasticity | | Farm Output Elasticity | | Price Ratio Elasticity | |
|---|---|---|---|---|---|---|---|
| | | $\sigma$ | | $\sigma$ | | $\sigma$ | |
| | | 0.0 | 0.5 | 0.0 | 0.5 | 0.0 | 0.5 |
| 0.50 | 0.00 | 1.82 | 1.67 | −0.91 | −1.33 | −2.27 | −1.67 |
| 0.50 | 0.05 | 2.08 | 1.90 | −0.57 | −1.00 | −0.47 | −0.86 |
| 0.50 | 0.07** | 2.18 | 1.99 | −0.42 | −0.86 | 0.29 | −0.52 |
| 0.50 | 0.09 | 2.30 | 2.09 | −0.27 | −0.72 | 1.07 | −0.18 |
| 0.50 | 0.11** | 2.41 | 2.19 | −0.12 | −0.57 | 1.88 | 0.18 |
| 0.50 | 0.13* | 2.53 | 2.30 | 0.04 | −0.42 | 2.71 | 0.55 |
| 0.50 | 0.17 | 2.78 | 2.52 | 0.37 | −0.10 | 4.46 | 1.32 |
| 0.50 | 0.19* | 2.91 | 2.63 | 0.55 | 0.07 | 5.38 | 1.72 |
| | | | | | | | |
| 1.00 | 0.00 | 3.64 | 3.33 | −1.82 | −2.67 | −4.55 | −3.33 |
| 1.00 | 0.13 | 4.28 | 3.84 | −0.96 | −1.78 | −0.06 | −1.56 |
| 1.00 | 0.14** | 4.33 | 3.88 | −0.89 | −1.71 | 0.32 | −1.41 |
| 1.00 | 0.22 | 4.81 | 4.25 | −0.26 | −1.06 | 3.64 | −0.13 |
| 1.00 | 0.23** | 4.87 | 4.30 | −0.17 | −0.98 | 4.08 | 0.05 |
| 1.00 | 0.24 | 4.93 | 4.35 | −0.09 | −0.89 | 4.54 | 0.22 |
| 1.00 | 0.25* | 5.00 | 4.40 | 0.00 | −0.80 | 5.00 | 0.40 |
| 1.00 | 0.33 | 5.57 | 4.84 | 0.77 | −0.03 | 9.02 | 1.93 |
| 1.00 | 0.34* | 5.65 | 4.90 | 0.87 | 0.07 | 9.57 | 2.14 |

$e_T$ = elasticity of farm price with respect to advertising tax. $\eta_A$ = advertising elasticity. $\sigma$ = elasticity of substitution between farm-based input and marketing inputs. *Indicates the break-even points for farm output. **Indicates the break-even points for the retail-farm price ratio.

For illustrative purposes, let us consider the case in which $e_T = 1.0$ (the bottom half of Table 1). For $\sigma = 0$, the break-even point for farm output ($E_{aT} = 0$) occurs when the advertising elasticity is 0.25 (see Table 1, col. 5). This means that the leftward supply shift caused by an increase in advertising tax is compensated exactly by the demand shift due to advertising, leaving farm output unchanged. When the advertising elasticity is greater than 0.25, farm output exceeds the pre-tax level.

The break-even point for the retail-price ratio ($E_{Px/PaT} = 0$) occurs when the advertising elasticity is about 0.14 and becomes positive thereafter (see Table 1, col. 7).

When $\sigma = 0.5$, the break-even points, identified by the advertising elasticity, for $E_{aT}$ and $E_{Px/PaT}$ increase to 0.34 and 0.23, respectively. This means that if substitution possibilities exist between farm-base input and marketing input, more effective advertising is required to compensate for the reduced demand for the farm-based input due both to higher prices and the substitution effect.

Because $E_{P_xT}$ is positive throughout, an advertising tax increase always leads to higher retail prices. Therefore, in the case where the demand curve shifts in a parallel manner, there will be a loss in total social welfare if advertising does not shift demand sufficiently to restore farm output to the pre-tax level.[8]

Because advertising effectiveness (as measured by the magnitude of the advertising elasticity) plays a pivotal role in determining the economic impacts of check-off programmes, it is in the public interest that funds be allocated to ensure the maximum possible impact of the advertising investment. If advertising is not successful at bringing about a sufficiently large shift in the demand curve, alternatives such as research, nutrition education, or new product development may deserve greater attention.

## Concluding Remarks

Partial-equilibrium analysis of commodity check-off programmes suggests that advertising-induced demand shifts will have to be large for the programmes to compensate for the decrease in farm output due to advertising taxes.

A caveat in interpreting these conclusions is that the analysis is static and implicitly assumes perfect knowledge on the part of consumers. In a dynamic setting, the information conveyed in advertising can reduce time lags in adjusting to new equilibria and enhance competition. If the advertising results in more accurate knowledge of product characteristics, additional welfare benefits accrue to consumers. Finally, if advertising is subject to financial external economies of scale, the enhanced demand for advertising services occasioned by the recent introduction of large mandatory check-off programmes could result in economy-wide reductions in marketing cost, a welfare benefit not considered by the partial-equilibrium approach adopted in this study. As suggested by the analysis, commodity advertising taxes in general can be expected to result in higher retail prices for food and a reduction in farm output.

## Notes

[1]Auburn University.

[2]Although cross-commodity advertising may exist, its impact, like other demand shifters, is assumed to be constant and hence does not appear in Equation (2). Alternatively, $A$ can be thought of as net advertising for $x$, net of cross-commodity advertising. The second interpretation implies that the net advertising elasticity for the commodity under consideration may be smaller than the gross measure that ignores competitive advertising.

[3]The tax rate and advertising are assumed to be exogenous because the former is set, in general, by the government through referenda, and decisions about advertising expenditures usually precede price determination.

[4]Time lags may exist between the demand shift and the supply shift since producers and consumers may not respond to policy change instantaneously. This analysis, however, is restricted to a long-run equilibrium and comparative statics framework; therefore, full adjustment to a change in exogenous variables from one equilibrium to another is implied. Total elasticities derived later are subject to the same interpretation.

[5]Derivations of Equations (8)–(11) are available from the authors on request.

[6]For further discussion, see Gardner, 1975.

[7]$e_{AT}$ and $e_{Aa}$ will equal 1 if advertising is defined as $A = Ta$.

[8]Possible distributional impacts on infra-marginal consumers and producers are not considered. Moreover, this measure does not take into account the value of information and the reduced cost of entertainment to consumers due to advertising (Ekelund and Saurman, 1988; and Doyle, 1968). In the case where demand shift is not parallel, the change in welfare depends on how demand is shifted.

# References

Armbruster, W.J., and Frank, G.L., "Generic Agricultural Commodity Advertising and Promotion: Programme Funding, Structure, and Characteristics," in *Generic Agricultural Commodity Advertising and Promotion*, Report No. 88–3, Department of Agricultural Economics, Cornell University, Ithaca, N.Y., USA, 1988.

Brandow, G.E., *Interrelations among Demands for Farm Products and Implications for Control of Market Supply*, Bulletin No. 680, Pennsylvania State University, University Park, Pa., USA, 1961.

Doyle, P., "Economic Aspects of Advertising: A Review," *Economic Journal*, Vol. 68, 1968, pp. 570–602.

Ekelund, R., and Saurman, D., *Advertising and the Market Process: A Modern Economic View*, Pacific Research Institute for Public Policy, San Francisco, Calif., USA, 1988.

Gardner, B.L., "The Farm-Retail Price Spread in a Competitive Food Industry," *American Journal of Agricultural Economics*, Vol. 57, 1975, pp. 399–409.

George, P.S., and King, G., *Consumer Demand for Food Commodities in the United States, with Projections for the 1980s*, Giannini Foundation Monograph No. 26, University of California, Berkeley, Calif., USA, 1971.

Heath, F., "An Examination of Advertising Measurements in Empirical Studies," unpublished paper, Department of Agricultural Economics and Rural Sociology, Auburn University, Auburn, Ala., USA, 1990.

Hurst, S., and Forker, O.D., "Annotated Bibliography of Generic Commodity Promotion Research," Report No. 89–26, Department of Agricultural Economics, Cornell University, Ithaca, N.Y., USA, 1989.

Nerlove, M., and Waugh, F.V., "Advertising without Supply Control: Some Implications of a Study of the Advertising Orange," *Journal of Farm Economics*, Vol. 43, 1961, pp. 813–837.

Zufryden, F., "Applications of a Dynamic Advertising Response Model," in Leigh, J., and Martin, C., Jr., (Eds.), *Current Issues and Research in Advertising*, Graduate School of Business Administration, University of Michigan, Ann Arbor, Mich., USA, 1978.

---

# Discussion Opening—*Walter Armbruster* (Farm Foundation)

Chang and Kinnucan have admirably addressed a topic of increasing importance as more agricultural commodity producers seek to increase their income by product promotion. The attempts to expand demand by US producers are principally through advertising in domestic markets, although there is also some use of other market-expanding approaches, particularly in export markets.

This analysis of the effects of generic advertising programmes on various market levels provides some interesting insights. The authors' analytical model allows them to link the excise tax rate of the assessment with the advertising impact. If advertising is ineffective, the tax increases retail prices, decreases farm supply, and decreases the retail-farm price ratio.

The authors' simulated results indicate that the break-even point for farm output occurs when advertising elasticity is 0.25. Thus, economically rational producers must achieve an effective advertising elasticity of 0.25 if their goal is to increase their revenues. Since the authors report that most studies have found advertising elasticities of less than 0.10, this implies a loss in producer revenue.

If producers have estimates of effectiveness, will they abolish an ineffective programme relatively quickly? Do producers optimistically attribute all sales increases to advertising? Are producers obtaining adequate analyses of the responses to advertising expenditures to make rational decisions about continuing the programmes?

Chang and Kinnucan then turn to analysis of the policy implications reflected in welfare impacts from changes in the tax rate. What the authors conclude is that most increases in producer assessments for advertising will lead to welfare losses. Is this true for all levels of advertising? Does it imply no social welfare gain from any generic advertising programme? Or, is there a minimum or threshold amount of advertising above which the welfare losses start?

The authors indicate that welfare transfers take place among consumers as well as among producers and thus exclude marginal producers and consumers from the market. We need to ask whether agricultural economists have paid enough attention to analysing the impacts of such welfare transfers.

The authors conclude that if advertising cannot be proved successful in shifting demand, producers and society in general may be better served by the use of assessments to fund research, nutrition education, and new product development.

Identifying the advertising elasticities and then educating producers, organizational leaders, and policy officials about their implications is a major challenge. We also need to ask if agricultural economists can also offer help to producers in assessing potential effectiveness of research, education, and new product development in expanding demand as alternatives to advertising.

**General Discussion**—*Chaur Shyan Lee, Rapporteur* (National Chung-Hsing University)

On the Shui *et al.* paper, the authors were asked whether, if US producers lose from OECD trade liberalization, cotton suppliers elsewhere gain. The OECD accounts for a relatively small share of US cotton exports; e.g., about 60 percent of US cotton goes to the Pacific Rim. Why was OECD trade included? Were substitution possibilities with synthetic fibres considered in the model? Parameters of the model were both estimated and borrowed from the literature under a régime without liberalization. These parameters are then used to evaluate the effects of a policy régime that includes trade liberalization. To what extent is the Lucas critique a problem in this study?

The authors replied that the foreign cotton producers gain. Multilateral liberalization is a more reasonable assumption (all the OECD countries liberalize). The important effect comes from the USA liberalizing its textile trade and non-OECD countries expanding their textile output and demand for cotton. The paper assumes horizontal supply of manufactured fibre, so there is no substitution.

On the Sparks and Bravo-Ureta paper, the authors were asked about the theoretical and empirical justification for choosing the inverse Rotterdam demand function instead of Armington or other models, why they used total import quantity instead of income in each import demand equation, and why they did not conduct tests of theory, linear homogeneity, weak separability, etc.

The authors replied that the inverse Rotterdam demand function is ideally suited to the problems facing US apple growers, where prices are decreasing while supplies increasing. The model is also quite tractable and implementable. In contrast, the Armington model is quite cumbersome and difficult to implement. They used total quantities instead of income because the question to be answered is concerned with quantities, not income. Also, the model construction is usually done with aggregated quantities, not income, for the inverse Rotterdam model. They will be conducting tests of the theoretical restrictions as they proceed with this work. Alternative models should be estimated to evaluate which specification is more consistent with the data, statistically speaking. This applies both to the functional form and to whether the model should be price or quantity dependent.

Participants in the general discussion included D.B. Han (Korea Rural Economics Institute).

# Multi-Market Analysis of Sudan's Wheat Policies: Implications for Fiscal Deficits, Self-Sufficiency, and the External Balance

*Rashid M. Hassan,*[1] *W. Mwangi,*[1] *and B. D'Silva*[2]

**Abstract:** Highly subsidized bread prices financed partially through wheat aid and overvalued currency have stimulated rapid growth in wheat consumption in Sudan at the expense of other staple grains such as sorghum and millet. Inefficient production methods and the resultant low wheat yields have caused domestic supply to lag behind demand. Faced by serious foreign exchange shortages, severe internal and external imbalances, and reduced availability of food aid, Sudan could not sustain dependence on external sources to bridge the growing wheat gap. Given the political difficulties associated with managing demand, the government has chosen to promote local production. Research results showing high potential gains in wheat yield under improved crop management also contributed to the choice of the supply strategy. A dynamic multi-market model was developed and used to evaluate alternative supply-promoting and demand-control strategies. Competition with alternative productive uses of the country's scarce resources and substitution between wheat and other cereal grains in consumption were analysed. The impact of the various policies on net exports, food security, and the budget is measured and compared. Results of policy analysis indicate the significant contribution of production efficiency, reduced consumer subsidies, and elimination of relative price distortions to higher self-sufficiency and lower internal and external deficits.

## Introduction

A recent history of highly subsidized bread prices, financed partially through wheat aid and an overvalued currency, has trapped Sudan into a situation of rapidly growing demand for wheat at the expense of traditional food staples such as sorghum and millet. Annual consumption of wheat in Sudan has risen from around 20 kg per capita to 40 kg over the last two decades, whereas growth in domestic production has lagged far behind demand. The result has been a steady deterioration in self-sufficiency and increased wheat imports. For instance, in 1988 only 20 percent of the total wheat consumed in Sudan was locally produced compared to a self-sufficiency ratio of about 70 percent in 1971. The gap was filled by imports, of which more than 80 percent was received as food aid[3] in 1988; i.e., food aid accounted for 65 percent of total wheat consumption (Hassan, 1990). Another consequence of the aid-supported subsidies was the development of a large milling industry in Sudan, adding one more interest group to lobby for increased wheat consumption and subsidization.

With reduced availability of food aid and mounting internal and external deficits facing the country, the current wheat gap is increasingly unsustainable. Given that demand management options such as lifting consumer subsidies are politically difficult to implement, the government has chosen to rely on a crash programme to promote local wheat production in order to bridge the wheat gap. While this strategy will reduce reliance on external wheat supplies and ease the pressure on the already strained sources of foreign exchange, it will lead to greater competition with high-value crops such as cotton and faba beans for the country's agricultural resources. Moreover, the capital-intensive and highly mechanized systems of wheat production in Sudan make the foreign exchange component in local wheat production relatively high, leaving a narrow margin for potential saving of foreign resources from wheat production.

The objective of this study is to evaluate the contribution of alternative wheat production and consumption strategies to macroeconomic improvement in Sudan. The impact of various supply-promoting and demand-reducing wheat policies on net imports, food security, and the budget are analysed within a multi-market framework. This allows for substitution effects in the production and consumption of wheat and competitive products. The model developed in this study extends the static multimarket framework of Braverman and Hammer (1986) to incorporate short-run (partial) adjustment dynamics to modelling supply.

# Wheat Sector and Policies

**Production structure.** All wheat produced in Sudan receives regular irrigation during a short winter season. There are two distinct regions of wheat production representing different technologies and institutional environments, the private pump schemes of the North and the public irrigation schemes in the clay Central Plains.

Wheat is a traditional crop and the major food staple in the North. Over the last 20 years, an average yield of 2.3 t/ha has been realized in the Northern Region compared to 1.3 t/ha on the Central Plains. This yield advantage is mainly due to the relatively cooler and longer winters in the North, coupled with farmers' familiarity with the crop. Other factors critical to the horizontal (area) and vertical (yield) expansions in wheat production in both regions include high irrigation costs and water shortages, plus poor crop management practices. Land is, however, more limiting in the North.

About 75 percent of Sudan's wheat is produced on the public irrigation schemes where land allocations are determined according to a fixed crop rotation. Procurement and distribution of other critical inputs are also controlled through the schemes. Cotton is the only winter crop to compete with wheat for irrigation water. Farmers are required to deliver the produce of both crops to marketing boards at government-set prices. The other crops in the rotation (groundnuts and sorghum) are harvested before wheat planting and sold in the free market. Although wheat yields are low, there is considerable potential for improvement. A potential gain of more than 100 percent in wheat yield has been demonstrated through the adoption of a new wheat technology package, released by the Agricultural Research Corporation (ARC) of Sudan and tested by the ARC and the Sasakawa Global 2000 (a non-governmental organization) on a large number of on-farm trials and demonstrations over the past four years (Ageeb *et al.*, 1989; and Global 2000, 1990).

In the North, allocation of land, water, and other inputs is decentralized. Farmers buy all inputs from and sell their product to private traders. Faba beans are the main crop competing with wheat for land and water in the North.

**Consumption.** Sorghum, wheat, and millet are the main source of cereal calories. In spite of the rapidly growing demand for wheat, it remains largely a food of the urban population. Sorghum and millet, on the other hand, are still the basic food staples for the vast majority, particularly in rural areas where more than 70 percent of the population reside (Sudan Ministry of Finance and Economic Planning, 1988). The preference for wheat bread by urban consumers is largely due to the high bread subsidy. Wheat bread is also a convenience food that is easy to prepare and uses less time and baking energy than the popular sorghum breads such as *kisra* and *asida*. Apart from the relative price effects, accelerating rural-urban migration and increased participation of women in the urban labour force have been identified as important forces behind rising wheat consumption in Sudan (Damous, 1986; and Salih, 1985). The country is currently experimenting with the alternative of using composite flours that mix wheat with other grains for bread flour.

**Pricing policies.** Sorghum and millet prices are determined in the free market, whereas the price of wheat is regulated by the government, which controls its marketing and importation. In 1989, the producer price of wheat was set on the basis of the parallel exchange rate of S£ 12.5 per US$,[4] whereas the effective exchange rate applied to cotton exports was S£ 6.5. Both cotton and wheat, however, receive an indirect subsidy on the price of imported inputs, especially fertilizer, whose price is set at the official exchange rate.

A free-market price that is higher than the import parity cost of wheat has been reported. This price differential indicates an unsatisfied demand for wheat at the official price, reflecting the effect of a quota system on wheat imports in Sudan. Due to the discrepancy between the official producer and free market prices, farmers in the public irrigation schemes under-report their true wheat yield levels and sell the difference on the free market (Salih, 1989; Damous, 1986; and Hassan *et al.*, 1991). Consumers, on the other hand, enjoyed a high price subsidy on wheat as they paid less than 25 percent of the actual cost to the government of buying and processing wheat in 1989 (Hassan, 1990). Revenues collected from selling concessional wheat imports enabled the government to support the high subsidy on bread prices.

## A Multi-Market Model for Sudan's Wheat Economy

A proper evaluation of Sudan's wheat policy options requires a comprehensive representation of the structure of wheat production, consumption, and marketing. The multi-market approach to modelling supply-demand interactions and their macro implications is adopted in this study. This framework has been used by the World Bank to analyse the impacts of agricultural pricing and marketing policies on the level and composition of agricultural output, farmer income, government budget, and foreign trade (Braverman *et al.*, 1986 and 1987). Supply and demand decisions are modelled on the basis of the neoclassical theory of the firm and consumer behaviour. Possibilities of substitution in the production and consumption of goods competing for domestic resources and consumer budgets are allowed. The model also specifies the institutional arrangements within which agents interact and that define equilibrium conditions.[5]

**Income and final demand**. Due to data limitations on modelling agricultural factor markets, this model does not derive the functional distribution of income. Consumer demand for the three grain substitutes (sorghum, wheat, and millet) is therefore measured for aggregate spending and not classified by income groups. Total aggregate spending on cereals is accordingly allocated among the three goods using the almost ideal demand system of Deaton and Muellbauer (1980):

(1) $D_{it} = D_i (I_t, P_{it})$

where $D_i$ and $P_i$ refer to the final consumption and price of commodity $i$ (wheat, sorghum, and millet) in time $t$, respectively. $I_t$ denotes total consumer spending on cereals, which is set exogenously in this model. The consumer price of wheat is fixed by the government, whereas sorghum and millet prices are market determined.

**Output supply and factor demand**. In this model, wheat is produced in two regions, as discussed earlier. Area allocations to wheat and cotton are set by a government agency in the public irrigation schemes. Farmers decisions are therefore assumed to influence yield rather than production. Agricultural supply is consequently modelled by using area and yield response instead of output supply functions.

(2) $(A_{it}, Y_{it}) = F (ENP_t, W_{kt}, Z_t)$         $i = 1, 2, ..., n; k = 1, 2, ..., m$

Yield (Y) and area (A) of crop $i$ (except for cotton and wheat in public schemes) at period $t$ are assumed to vary with the expected net price (ENP) of all competing crops in the region, factor prices (W), and a vector of other fixed factors (Z), which includes, among others, climatic variables and diesel fuel allocations and prices.

As mentioned earlier, wheat competes with faba beans in the North and with cotton in public schemes. Sorghum is produced on the irrigated Central Plains and in the traditional and mechanized dryland farming systems of Sudan. Groundnuts under irrigation and sesame and millet in the rainfed sectors compete with sorghum for land.

Naïve price expectations are used, where the previous year's realization is used as the expected price. This specification reflects partial adjustments in domestic production. Prices are defined net of intermediate costs and indirect taxes $(t_i)$ using fixed input-output coefficients.

(3) $NP_i = P_i(1-t_i) - _j a_{ji} P_j$

Prices of all tradeables are determined in the world market:

(4) $\quad P_{it} = e_{it} P_{it}^{*}$

where the nominal exchange rate $(e_i)$ converts world price $(P_i^{*})$ into local currency prices $(P_i)$. Wheat, faba beans, and intermediate inputs are imported, while cotton, groundnuts, and sorghum are exported. The small country assumption is used where Sudan's exports face an infinitely elastic demand and imports are supplied in unlimited amounts.

The fixed coefficient technology derives demand for intermediate inputs. Demand for land is obtained from area response functions in Equation (2). Upper bounds on total acreage and regional land supply functions are specified. Demand for labour, on the other hand, is obtained from the dual representation of the specified supply structure. The nominal wage rate is fixed to define an infinite supply of labour by region. There is only one type of labour in the model, namely unskilled labour. Due to data limitations, the capital market is not modelled.

**Equilibrium conditions**. At equilibrium, total supply is equated to total demand in the product and factor markets. Total output supply is obtained by adding imports to domestic production. Exports, intermediate use, and final consumption constitute total demand. In addition to solving for equilibrium quantity and price flows, the model also computes wheat self-sufficiency ratios and traces implications for nominal macro aggregates such as the budget deficit and net foreign exchange.

Subsystem estimation is employed to generate parameter values econometrically. Supply and demand parameters are estimated separately using seemingly unrelated regressions. The Jacobian algorithm GAMS/MINOS is used to solve the model. Clearing of the product market is employed as the solution strategy for model validation and policy simulations.

## Policy Analysis and Simulation Results

Various policy scenarios are designed and evaluated as alternative avenues for improving the performance of Sudan's cereal economy. Scenario A is the present strategy, which represents the current government strategy of expanding wheat area at the expense of cotton, while maintaining a quota on imports, subsidized consumer prices for wheat, and the use of multiple exchange rates.

Improved government régimes are an advance on the present plan to attain self-sufficiency, where four scenarios are assumed. In Scenario B, present policy is adjusted for a 50-percent reduction in current consumer subsidy and 50 percent adoption of a new wheat production technology. In Scenario C, the effective exchange rate on all imports and exports is unified at the parallel rate of S£ 12.5 to eliminate internal distortions in relative prices. The wheat market is also liberalized so that the consumer subsidy and quotas on wheat are lifted; hence all wheat is traded at its import parity cost. In Scenario D, 50 percent of wheat farmers adopt the improved technology. The wheat market is partially liberalized by lifting the quota on imports as well as by removing 50 percent of the consumer subsidy. All tradeables are paid the parallel exchange rate. Finally, in Scenario E, the free market exchange rate of S£ 22 per US$ is applied to all tradeables in this experiment. The wheat sector is completely liberalized (zero subsidies and no quota), and all farmers adopt the new wheat technology. The model is solved for two consecutive years, 1988 and 1989.

While demand adjusts instantaneously in this model, the full effects of policy changes are realized a year later due to partial adjustment in supply. Table 1, therefore, reports results obtained for the second year only (1989), when full adjustment is completed. Solution values for the current government strategy were used as the basis for comparison.

Col. A of Table 1 shows that this policy could generate domestically only 41 percent of total wheat consumed. The more ambitious government policy (col. B) raised self-sufficiency in wheat to 85 percent. This is due to two forces. While wheat consumption fell by 33 percent as a result of partial lifting of the consumer subsidy, adoption of the new wheat technology increased production by 40 percent. Wheat imports consequently declined by 85 percent,

215

whereas 36 percent more inputs were imported under the new technology. This policy also saved 83 percent of the budgetary costs of the bread subsidy, or S£ 862 million under régime A. The trade balance also improved by about 8 percent under B.

Table 1—Results of Policy Simulations (1989)

| Variable | A Present Strategy (values) | B Improved Strategy (percent change)* | C Liberalization and Unified Exchange (percent change)* | D Partial Liberalization and Adoption (percent change)* | E Complete Liberalization and Adoption (percent change)* |
|---|---|---|---|---|---|
| Consumption demand (1,000 t) | | | | | |
| Wheat | 864 | −33 | −52.8 | −34.6 | −57.4 |
| Sorghum | 1,655 | 8.5 | −18.6 | −10.8 | −22.4 |
| Millet | 140 | 77 | 67.2 | 82.9 | 60.7 |
| Domestic supply (1,000 t) | | | | | |
| Wheat | 354 | 40 | −13.6 | 50.0 | 76.8 |
| Sorghum | 3,461 | 1.4 | −1.4 | 0.5 | 1.9 |
| Millet | 285 | −4.6 | 0.01 | −25 | −5.3 |
| Cotton | 87 | −6.9 | 47.2 | 49.4 | 44.8 |
| Imports | | | | | |
| Total value ($ million) | 110 | −58.2 | −60.1 | −63.7 | −60.0 |
| Inputs ($ million) | 22 | 36.4 | 13.7 | 59.1 | 100.0 |
| Wheat (1,000 t) | 510 | −83 | −80 | −93.3 | −100.0 |
| Exports | | | | | |
| Value ($ million) | 606 | −4.3 | 24.4 | 22.5 | 38.1 |
| Sorghum (1,000 t) | 1,806 | −5.1 | 14.4 | 11.0 | 24.3 |
| Millet (1,000 t) | 145 | −83.5 | −65.5 | −84.8 | 69.0 |
| Cotton (1,000 t) | 87 | −6.9 | 47.2 | 49.4 | 44.8 |
| Trade balance ($ million)** | 496 | 7.7 | 43.4 | 41.6 | 59.9 |
| Consumer wheat subsidy (S£ million) | 862 | −83.2 | −100.0 | −78.6 | −100.0 |
| Self-sufficiency ratio | 0.41 | 0.85 | 0.75 | 0.94 | 1.7 |

*Refers to percentage change relative to present government régime (col. A), except for self-sufficiency (last row), which is given as actual ratios.
**Trade balance defined as net value of exports; e.g., exports minus imports.

Relative price distortions were eliminated in experiment C as a result of the exchange rate unification, thus producing important supply and demand effects. Unification removed the wheat subsidy entirely, saving 100 percent of its budgetary burden. Wheat consumption and imports consequently dropped by 53 percent and 80 percent, respectively. Domestic supply of wheat, however, fell by 14 percent, whereas cotton production rose by 47 percent. This is because exchange rate unification removed the foreign exchange tax on cotton and revealed the comparative advantage of irrigated cotton compared to traditional wheat production practices. The trade balance then improved by 43 percent, and self-sufficiency reached 75 percent.

In addition to correcting the internal structure of incentives to producers, 50 percent of the farmers adopted the new wheat technology in Scenario D. In spite of the partial elimination of the wheat subsidy, a higher self-sufficiency ratio (94 percent) and larger decline in wheat imports were realized with this policy compared to Scenario C, in which 100 percent lifting of the consumer subsidy was adopted without improved production efficiency. This indicates the relative importance of supply-shifting policies and the contribution of the technology factor to bridging the wheat deficit in Sudan. Both scenarios produced substantial gains in net foreign exchange.

The best results, however, were obtained under complete liberalization and full adoption of improved wheat production methods (policy E). This policy could generate undesirable distributive effects that are not explored by this model. It also represents a politically sensitive option and requires considerable foreign exchange resources and critical institutional changes in the factor and product markets for Sudan's agriculture.

## Summary and Conclusions

A high subsidy on the consumer price of wheat and food aid have stimulated increased wheat consumption in Sudan. Domestic supply, on the other hand, has lagged far behind due to inefficient production practices and, hence, lower wheat yields. Faced with serious shortages in foreign exchange, severe external and internal imbalances, and reduced availability of food aid, Sudan could not sustain the high bread price subsidy and dependence on external sources to bridge the growing wheat gap. Given the political difficulties associated with demand management options, the government has chosen to promote domestic supply for higher self-sufficiency in wheat. This choice was also encouraged by research results indicating high potential gains in wheat yield under improved production methods.

A dynamic multi-market model was developed, estimated, and used to evaluate and compare alternative supply-promoting strategies and demand control options. Competition for agricultural resources with alternative crops such as cotton and faba beans as well as substitution between wheat and other cereals in consumption were analysed. Policy analysis showed that the current strategy of expanding the wheat area at the expense of cotton, while maintaining existing distortions in relative prices, consumer subsidies, and low input levels in wheat production, was out-performed by all alternative options. Much higher gains in wheat self-sufficiency, net foreign exchange, and reduced budgetary costs were realized with various combinations of lower consumer subsidies, unified exchange rates, and adoption of more efficient wheat production technologies.

## Notes

[1] Centro Internacional de Mejoramiento de Maiz y Trigo.

[2] US Department of Agriculture.

[3] Food aid here includes both donations and concessional imports.

[4] The official exchange rate in 1990 was S£ 4.5 and the free market rate S£ 22 per US$. S£ refers to Sudanese pounds.

[5] Detailed discussions of the multi-market approach are found in Singh *et al.* (1985) and Braverman and Hammer (1986). Sudan's wheat model is developed on this basis.

# References

Ageeb, O., Faki, J., El Mekki, I., Mussa, A., and El Nour, A., "Wheat Pilot Production and Demonstration Plots in Gezira," paper presented at the annual National Wheat Coordination Meeting, ARC Wad Medani, Sudan, 4–7 September 1989.

Braverman, A., and Hammer, J., "Multi-Market Analysis of Agricultural Pricing Policies in Senegal," in Singh, Inderjit, Squire, L., and Strauss, J. (Eds.), *Agricultural Household Models: Extensions, Applications, and Policy*, Johns Hopkins University Press, Baltimore, Md., USA, 1986.

Braverman, A., Hammer, J., and Young, A., "Multi-Market Analysis of Agricultural Pricing Policies in Korea," in Newberg, D., and Stern, N. (Eds.), *The Theory of Taxation for Developing Countries*, Oxford University Press, London, UK, 1987.

Damous, H.M., "Economic Analysis of Government Policies with Respect to Supply and Demand for Wheat and Wheat Production in Sudan," Ph.D. dissertation, Washington State University, Pullman, Wash., USA, 1986.

Deaton, A., and Muellbauer, J., "An Almost Ideal Demand System," *American Economic Review*, Vol. 70, 1980, pp. 312–326.

Global 2000, "Annual Report," Khartoum, Sudan, 1990.

Hassan, R.M., "Technological Options and Policy Incentives for Higher Self-Sufficiency in Wheat in Eastern and Southern Africa," paper presented at the 1990 Rockefeller Foundation Social Science Research Fellows Workshop, International Institute of Tropical Agriculture, Ibadan, Nigeria, 2–5 October 1990.

Hassan, R.M., Faki, H., and El Obeid, H., "Economic Policy and Technology Determinants of the Comparative Advantage of Wheat Production in Sudan," CIMMYT Economics Paper, Centro International de Mejoramiento de Maiz y Trigo, México, D.F., Mexico, 1991.

Salih, S.A., "Consumption Effects of Eliminating Bread Subsidies in Sudan," US Agency for International Development, Khartoum, Sudan, 1985.

Salih, S.A., "A Cheating Model of Wheat Production in Gezira," paper presented at the American Agricultural Economics Association meeting, Baton Rouge, La., USA, 3–7 August 1989.

Singh, Inderjit, Squire, L., and Kirchner, J., "Agricultural Pricing and Marketing Policies in Malawi: A Multi-Market Analysis," CPD Paper No. 1985–75, World Bank, Washington, D.C., USA, 1985.

Sudan Ministry of Finance and Economic Planning, "Economic Survey," Khartoum, Sudan, 1988.

---

## Discussion Opening—*Mesfin Bezuneh* (Clark Atlanta University)

The issue addressed in the paper is relevant and timely. It uses a straightforward household-firm model, recognizing the linkages of production to consumption in a multi-market environment. This is a highly relevant analytical approach, which captures the essential tradeoffs between competing government strategies. However, the paper did not take advantage of the model to explore explicitly the impact of food aid.

The driving force of their paper is the increasing gap between food supply (wheat in particular) from domestic production and demand for wheat due to highly subsidized bread prices. Food aid (or wheat aid) is the major source of government revenue for subsidizing bread prices. As a result, the authors assert that "annual consumption of wheat in Sudan has risen from around 20 kg per capita to 40 kg over the last two decades, whereas growth in domestic production has lagged far behind demand," and that 80 percent of the supply-demand gap was fulfilled by food aid.

The paper could have focused first on analysing the effects of food aid on production and consumption in Sudan and then evaluating the government's alternative production and policy strategies. Clearly two types of food aid linkages need to be incorporated into analysis of this type: the production linkage, which is felt through the domestic prices of wheat and other grains, such as millet and sorghum; and the consumption/demand linkage, which is felt through its income effect. Relevant questions in this regard include whether food aid results in production disincentives and whether it generates additional net income that could partially or fully offset the disincentive effects and/or exacerbate the production consumption balance by increasing demand.

It is possible to hypothesize that food aid might have disincentive effects on the production not only of wheat but also sorghum and millet, since their prices are determined in the free market. Thus, food aid not only increased demand for wheat but might also have caused a reduction in domestic cereal production, thereby exacerbating the gap.

Given the frequency of drought and the existence of other disincentive production factors in Sudan, food aid will remain an important means of filling the gap. Reliance on food aid will continue to be an increasing risk in the 1990s and beyond as the food aid donors move closer to reducing their own farm subsidies (under a GATT agreement) and the number of countries seeking food aid increases (e.g., Eastern Europe).

What is needed is thus a clear understanding of the total effects of food aid so that decision makers in the recipient countries can make the necessary policy changes in developing both food aid alternatives and efficient food aid management strategies. Kenya, for example, is formulating a desirable national food strategy by determining the appropriate levels of food aid in order to meet local, regional, and national nutritional and development needs.

Sudan has been consistently receiving over 300,000 t of grain food aid (mostly wheat) annually for the last 20 years and over 671,000 t (ranging from 330,000 t to 890,000 t) annually for the last six years (as well as about 10,000 t non-cereal food aid), making it one of the highest food aid recipient countries in Africa. The impact of such a massive infusion of food aid needs to be understood.

*[Other discussion of this paper and the authors' reply appear on page 236.]*

# Does Africa Really Lack International Competitiveness? Comparisons between Africa and Asia

*Frédéric Martin, Peter H. Calkins, and Sylvain Larivière*[1]

**Abstract:** The frequent claim that Asian development and trade strategies are superior to those of African nations is evaluated objectively. Two separate samples (one for econometric and nonparametric analyses, the other for country- and commodity-specific analyses) are used to test for significant differences between Asian and African success factors in explaining growth in GNP per capita and export values. In addition, Spearman rank correlations are performed to evaluate the ability of Asian and African nations to respond to international price, quantity, and value signals. The results show that for both continents, exports are a significant determinant of GNP growth, agricultural products are particularly determinate of such export growth, and export promotion strategies are based on the evolution of market shares rather than of prices. Extension of the World Bank's MADIA study to an Asian sample further belies the presence of an African "doom" factor. Finally, the results point to the feasibility of better education and economic policy management as a strategy for those African economies that have performed less well than others.

## Conceptual Framework

It is often asserted not only that Asian countries have had a superior record of economic development than their African counterparts, but that a large measure of this success is due to the highly successful foreign trade strategies of Asian nations. Additional arguments used to support this view include the presumption of a greater level of "industriousness" on the part of Asian nations, more efficient family investment networks, a higher rate of savings, a lower level of ethnic and social diversity, and a more benign and less prolonged period of colonial domination. More generally, the impression is often given that African nations could benefit from adopting the "Asian model" of economic development. To test the general validity of this view, this paper attempts objectively to portray, explain, and compare the development and trade performance of a similar set of countries from Africa and Asia.

According to standard international trade theory, foreign trade can contribute significantly to economic growth by promoting international specialization in production according to comparative advantage. Empirically, linkages between economic growth and international trade are significant (Petit and Knudsen, 1989). However, Helleiner (1984) notes that this linkage is weak for low-income countries. Michaely (1977) and Tyler (1981) suggest that a minimum level of development is required to observe a strong linkage.

Inversely, export performance depends partly upon domestic policy incentives. The new theory of international trade initially proposed by Helpman and Krugman (1985 and 1989) considers international trade as a strategic dynamic game and stresses the role of the state in helping the private sector to take advantage of market imperfections. This theory suggests that a potentially successful export strategy would not only respond to world prices, but attempt to increase market share to be in a position to benefit from scale economies and exercise market power.

Specifically regarding export strategy, Whee Rhee *et al.* (1984) suggest that entry into world markets can be enhanced by strategic choice of high-growth markets and products and by providing an economic policy and institutional environment conducive to export growth. This is all the more critical for low-income LDCs, since they tend to export a limited number of commodities, mainly agricultural, into strongly competitive, low-growth, and increasingly protected markets.

While Asia is often pointed to as an example of success and Africa as an example of mediocrity in terms of export and economic growth performance, few studies have actually been conducted to make consistent comparisons of the impact of agricultural trade and development strategies upon economic growth for the two continents. This paper attempts to fill part of this gap.

The above conceptual framework suggests three hypotheses for testing. First, a major determinant of the level of GNP per capita is the level of exports:

(1)  *GNP per capita = f(Asia, export value, other variables)*

Second, export value, in turn, depends critically upon the share of agricultural products within exports, as well as upon internal development policy parameters:

(2)  *Export value = f(Asia, Ag percent exports, internal development variables)*

Third, within such agricultural products, Asia and Africa have chosen their product emphases based on the evolution of world prices:

(3)  *Rank world product price = rank export quantities (by continent)*

For each of these three hypotheses, the sub-hypothesis that the performance of Asia is significantly different from that of Africa is also tested. Two separate samples are used: one for econometric and non-parametric analyses by country group, the other for country- and commodity-specific economic analyses.

## Data and Hypothesis Testing

The evaluation of like entities is critical to objective comparison. For the econometric and nonparametric analyses, the analysis is limited to non-socialist, market, coastal economies from each continent. The reasons for retaining such criteria are to avoid the inclusion of Marxist régimes such as China and Bénin, for whom international trade and other social indicators are rarely conditioned by functioning input and output markets, and to avoid penalizing Africa for its greater percentage of land-locked economies. Nevertheless, even with these criteria, lack of consistent data on key indicators prevented the inclusion of such economies as Taiwan and Guinea within the data set. A final sample (Sample 1) of 7 Asian countries (Hong Kong, Indonesia, Malaysia, Philippines, Singapore, South Korea, and Thailand) and 10 African countries (Cameroon, Côte d'Ivoire, Gabon, Ghana, Kenya, Liberia, Nigeria, Senegal, Sierra Leone, and Togo) was selected.[2]

For the country- and commodity-specific economic analysis, all 6 African countries chosen for the World Bank MADIA study were retained: Cameroon, Kenya, Malawi, Nigeria, Senegal, and Tanzania. To these were added 5 Asian economies for which trade and development case studies were available: Indonesia, Philippines, South Korea, Taiwan, and Thailand. Together, these 11 economies (Sample 2) reflect a variety of situations in terms of agricultural growth. The analysis of the factors affecting agricultural growth is first extended to all 11 economies following the methodology of the MADIA study. The factors affecting trade performance for two tropical products exported by African and Asian countries, palm oil and cocoa, are then analysed and compared.

Both samples intentionally include countries from both Anglophone and Francophone Africa. The analysis begins in the first half of the 1960s, by the end of which period all the countries of both samples had gained political independence.

The data from Sample 2 (Tables 1 and 2) show that, during 1965–88, Asian countries enjoyed a much higher economic growth rate (4.72 percent per year) than their African counterparts (1.05 percent). Agricultural production also grew at a higher rate in Asia (3.57 percent) than in Africa (2.9 percent) although the growth gap between the two continents was smaller. This difference between overall economic growth and agricultural growth arises from the much faster growth of industry in Asia. This process of economic transformation is reflected in the share of agriculture in GDP, which shrank from 36 percent to 17 percent in Asia while it decreased only slightly in Africa, from 41 percent to 36 percent.

Table 1—Economic and Agricultural Growth of Selected Asian Countries, Percent per Year

| | Taiwan | South Korea | Thailand | Indonesia | Philippines | Average |
|---|---|---|---|---|---|---|
| Growth rate of GNP per capita, 1965–88 | 6.90 | 6.80 | 4.00 | 4.30 | 1.60 | 4.40 |
| Annual growth rate of agricultural production, 1965–88 | 2.80 | 3.25 | 4.29 | 3.88 | 3.62 | 2.84 |
| Share of agriculture in GDP | | | | | | |
| 1965 | 27 | 38 | 32 | 56 | 26 | 31 |
| 1988 | 8 | 11 | 17 | 24 | 23 | 12 |

Sources: For Taiwan, Lau (1986); and for other countries, World Bank (1990).

Table 3 represents regional aggregations by period and continent for Sample 1. For each country, basic data were generated on the levels of GNP, as well as indicators of external trade and internal macroeconomic policies, for the 1961–63 and 1985–87 periods.[3] These data suggest that although Asia's overall GNP grew more rapidly than that of Africa between the two periods, the coefficient of variation was also higher in Asia in the 1960s and similar in the 1980s; thus, the two samples may not be significantly different. Second, export values soared in both continents, while the importance of agriculture as a percentage of exports fell; the slack was mostly taken up by the other sectors, which grew by varying amounts. Notably, however, the coefficient of variation in export value among African nations was higher, and only in Asia did exports increase as a percentage of GNP. Third, while savings and investment percentages in Asia tended to be higher than in Africa, African savings rates equalled those in Asia in the 1960s. The intercountry coefficient of variation for savings in Africa in the 1980s was significantly greater than for the Asian countries. Fourth, although human capital (defined as secondary schooling) tended to grow proportionally more rapidly in Africa, food production actually declined. Meanwhile, the increase in life expectancy and the percentage decline in agricultural labour tended to be similar across the two continents. Nevertheless, Asia was much less homogeneous in terms of the percentage of agricultural labour than Africa in both periods. Finally, the external debt went up about twenty-fold for both continents, but with a final coefficient of variation that was much higher in Africa than in Asia. Meanwhile, the current accounts balance was more favourable in Africa than in Asia in the 1960s, and less favourable in the 1980s, with slightly more variability within the African sample.

The data in Table 3 suggest that at least as many factors were similar (or cancelled each other out) in Asia and Africa than were different, even though mean levels for most factors were more favourable for the Asian sample. Also, high inter-country coefficients of variation within each sample could result in non-significant differences by continent.

Rigorous statistical tests of the three hypotheses are thus needed. The variables to be used are those defined in Table 3. There were 31 observations because of the pooling of two periods for each of 17 countries, less three observations for which data were missing. For this sample, equation (1a) presents the best econometric results of the test of hypothesis 1:

(1a)  Log GNP per capita = 0.075 Asia + 0.524 log export value*** + 0.356 log saving GNP*
            (0.28)         (3.59)                    (2.11)

            + 1.75 log year 1980*** – 0.538 log external debt***
            (5.21)                    (–4.08)

            $(*p < = 0.05, **p < = 0.01, ***p < = 0.001), \bar{R}^2 = 0.75, F = 19, N = 31$

Table 2—Economic and Agricultural Growth of Selected African Countries, Percent per Year

| | Kenya | Cameroon | Malawi | Nigeria | Senegal | Tanzania | Average |
|---|---|---|---|---|---|---|---|
| Growth rate of GNP per capita, 1965–88 | 1.90 | 3.70 | 1.10 | 0.90 | -0.80 | -0.80 | -0.50 |
| Annual growth rate of agricultural production, 1965–88 | 4.34 | 3.57 | 3.61 | 1.46 | 1.97 | 2.44 | 2.49 |
| Share of agriculture in GDP | | | | | | | |
| 1965 | 35 | 33 | 50 | 54 | 25 | 46 | 33 |
| 1988 | 31 | 26 | 37 | 34 | 22 | 66 | 25 |

Table 3—Summary of Trade and Macroeconomic Indicators by Continent and Period

| Variables | Asia, 1961–63 | | Africa, 1961–63 | | Asia, 1985–87 | | Africa, 1985–87 | |
|---|---|---|---|---|---|---|---|---|
| | Mean | CV* | Mean | CV* | Mean | CV* | Mean | CV* |
| GNP per capita (US$) | 322.86 | 0.73 | 202.00 | 0.53 | 3,225.70 | 0.98 | 648.18 | 1.08 |
| Export value (million US$) | 903.57 | 0.42 | 219.00 | 0.78 | 25,295.40 | 0.62 | 1,703.20 | 1.22 |
| Exports as percent of GNP | 40.43 | 0.99 | 28.90 | 0.38 | 52.00 | 0.67 | 27.40 | 0.38 |
| Agriculture as percent of exports | 51.43 | 0.50 | 61.50 | 0.44 | 20.70 | 0.68 | 42.80 | 0.53 |
| Other primary products as percent of exports | 19.71 | 0.68 | 28.20 | 0.75 | 16.60 | 1.05 | 43.70 | 0.57 |
| Machinery as percent of exports | 3.57 | 1.01 | 0.90 | 0.89 | 20.90 | 0.65 | 1.70 | 0.94 |
| Other manufactures as percent of exports | 22.00 | 1.29 | 9.40 | 1.81 | 41.70 | 0.46 | 11.40 | 1.41 |
| Investment as percent of GNP | 20.29 | 0.38 | 17.50 | 0.33 | 26.10 | 0.25 | 15.30 | 0.47 |
| Savings as percent of GNP | 17.00 | 0.46 | 17.40 | 0.55 | 31.00 | 0.25 | 15.20 | 0.57 |
| Human capital (percent secondary school enrolment) | 33.86 | 0.35 | 8.55 | 0.27 | 60.30 | 0.33 | 21.91 | 0.33 |
| External debt (million US$) | 958.33 | 0.90 | 190.45 | 0.83 | 20,629.50 | 0.57 | 4,672.91 | 1.50 |
| Life expectancy (years) | 58.57 | 0.12 | 42.73 | 0.11 | 67.30 | 0.09 | 50.73 | 0.11 |
| Food production index (1979–81=100) | 82.50 | 0.22 | 117.27 | 0.22 | 103.00 | 0.13 | 100.27 | 0.11 |
| Agricultural labour as percent of total labour | 46.71 | 0.60 | 72.55 | 0.09 | 39.50 | 0.53 | 70.14 | 0.12 |
| Current accounts (million US$) | -224.29 | 1.28 | -60.10 | 1.74 | 1,617.00 | 2.23 | -2,658.27 | 2.76 |

*CV = coefficient of variation. Sources: Table 2—World Bank (1990); Table 3—Lele (1989), World Bank (1989), and UN (1989).

The figures within parentheses represent $t$-values of the regression coefficients. Equation (1a), of Cobb-Douglas form, indicates that the level of GNP per capita is a positive function of export value. We may thus accept hypothesis 1. In addition, other policy and possibly cultural parameters enter significantly into the determination of GNP. For example, the level of external debt is strongly and negatively correlated with GNP, while national savings propensity is positively correlated. These results are consistent with neoclassical and trade theory. GNP levels in the 1980s were also higher than in the 1960s. But, significantly, Asian countries (Asia in Equation (1a)) are not statistically different from African countries in terms of their export, savings, and debt situation.[4] We thus reject the Asian sub-hypothesis to hypothesis 1.

Equation (2a), of linear form, indicates that we may accept hypothesis 2:

(2a)  Export value = 1,276.9 Asia + 536.3 human capital***
            (0.51)              (7.33)

+ 126.4 percent agricultural exports* – 8.26 percent agricultural exports × human capital
    (2.56)                                    (–4.88)

$$(*p < = 0.05, **p < = 0.01, ***p < = 0.001), \overline{R}^2 = 0.75, F = 24, N = 31$$

This is because the share of agricultural products (defined by the World Bank as non-fuel, non-mineral primary commodities) within total exports is a significant determinant of overall export value; more so than for other primary commodities, machinery and transport, and other manufactures. But the results also point to human capital formation as a second key element in any successful programme of increasing exports. Human capital is defined here as the percentage of young adults receiving a secondary school education. Interestingly, this long-term strategy of human capital formation does not conflict with the short-term strategy of increasing agricultural exports, for the sign on the interaction term between agricultural products and human capital is negative. It is concluded that agricultural products require a lower level of education than the other categories of exports.

It should be noted in addition that the human capital variable in Equation (2a) actually reflects a whole host of internal macro-level policy parameters, which were excluded from the estimation *a priori* because of a high level of multicollinearity with human capital. These parameters include life expectancy and the decline in the percentage of the labour force in the agricultural sector.

The analysis has shown both that trade has been a key determinant of GNP (hypothesis 1) and that agricultural commodities have played a privileged role within that trade (hypothesis 2). We now ask whether the choice of commodities within agriculture has respected world commodity price rankings for the 1961–63 and 1984–86 periods (hypothesis 3). The 16 commodities selected for evaluation represent the joint set of the 10 most important commodities in terms of export volume for each of the Asian and African sub-samples in each of these two periods. Table 4 gives the results of non-parametric tests of the rankings of these 16 exported commodities in each of three regions: the world, 7 Asian countries, and 10 African countries.

Surprisingly, export quantities do not seem to respond to price signals at any geographical level, except for the world in 1984–86. We must thus reject hypothesis 3. Rather, within the general category of agricultural exports, it is relative market share and not price that determines the strategy of commodity export choice. This market share can be expressed in either quantitative or value terms. Between these two, quantity signals seem to be significant more often than value signals; and wherever value signals are significant, so are quantity signals. This result is more consistent with Helpman and Krugman's (1985 and 1989) new theory of international trade than with neoclassical trade theory. Thus, we may revise hypothesis 3, and say that it is market share, rather than price signals, that tends to explain commodity rankings by geographical area. Mathematically:

(3b)  Rank world product quantity = rank export quantities (by continent)

A second conclusion one may draw from Table 4 is that in the early 1960s Asia already had its export quantities and values well aligned on world quantities and values, while Africa did not, largely because of the legacy of colonial policies.  During 1961–86, while Asia improved its quantity alignment on world quantities, Africa dramatically improved its value alignment by staying less wedded to its own 1961 value rankings.  Thus, the 1961–86 rankings of relative change rates for both Asia and Africa in terms of value growth were similar; and African quantity and value growth became tightly aligned on Asian quantity and value growth rankings.  As a result, even though Africa remains less synchronized with world values than Asia, this fact may be mainly explained by Africa's far less favourable starting point.  Africa is thus not demonstrably less responsive than Asia to international market share signals, and we may reject the Asian sub-hypothesis to the revised form of hypothesis 3 (Relation (3b)).

Table 4—Spearman Rank Order Correlations of Selected Export Strategy Parameters

| Signal | Response | Africa | Asia | World |
|---|---|---|---|---|
| | | Significance Level | | |
| **1961–63** | **1961–63** | | | |
| World quantity | Export quantity | 0.42 | 0.03** | — |
| World price | Export quantity | 0.88 | 0.88 | 0.83 |
| World value | Export value | 0.29 | 0.03* | — |
| Quantity Asia | Quantity Africa | 0.75 | | — |
| Value Asia | Value Africa | 0.38 | | — |
| **1984–86** | **1984–86** | | | |
| World quantity | Export quantity | 0.73 | | — |
| World price | Export quantity | 0.42 | 0.003** | 0.04** |
| World value | Export value | 0.15 | 0.14 | — |
| Quantity Asia | Quantity Africa | 0.80 | 0.21 | — |
| Value Asia | Value Africa | 0.13 | | — |
| **1984–86** | **1984–86** | | | |
| World quantity | Export quantity | 0.50 | 0.02* | 0.002*** |
| World price | Export quantity | — | — | 0.03** |
| World value | Export value | 0.37 | 0.36 | 0.003*** |
| Continent quantity | Continent quantity | 0.03** | 0.002*** | — |
| Continent value | Continent value | 0.09* | 0.009*** | — |
| **Δ 1961/63–1984/86** | **Δ 1961/63–1984/86** | | | |
| World price increase | Export quantity growth | 0.65 | 0.19 | 0.44 |
| World price increase | Export quantity 1986 | 0.14 | 0.19 | 0.41 |
| World quantity growth | Continent quantity growth | 0.18 | 0.09* | — |
| World price increase | Continent quantity 1986 | 0.14 | 0.19 | — |
| World value growth | Export value growth | 0.01** | 0.01** | — |
| Asia quantity growth | Africa quantity growth | 0.01** | | — |
| Asia value growth | Africa value growth | 0.01** | | — |

Notes:  *$p$ < = 0.10, **$p$ < = 0.05, ***$p$ < 0.01; "—" means irrelevant, therefore not estimated; Δ = growth rate between the two periods.

Source of raw data: FAO, 1988.

These econometric and nonparametric analyses of blocs of countries and commodity rankings have demonstrated that the Asian "success" factor and the African "doom" factor do not exist in either economic and agricultural development or trade performance.

Palm oil is a good example of a tropical product where Asia replaced Africa as the major world supplier in less than 25 years. The major Asian exporters (Malaysia and Indonesia) are much more competitive than their African counterparts (Côte d'Ivoire and Cameroon). In 1987, the cost of production of palm oil in Asia was a third of the African cost and Asian oil could be sold on European markets (the major export market for African oils) at two-thirds of the cost of production of African oil (Solagral, 1989).

There is no exclusively Asian secret recipe for Asian competitiveness; some of the contributory factors are: young, well-selected, and well-kept palm trees resulting in Asian yields being twice the African yields; planning planting to ensure regular supply to the refining industry; cost control; reliance on the private sector; good infrastructure provided by the public sector; world price signals transmitted to the producer; reasonable taxes on oil exports to remain competitive; higher taxes on raw than on refined oil, olein, and stearin to favour more processing; reasonable taxes on imported inputs such as fertilizers and crop protection products to avoid increasing production costs; and keeping the exchange rate close to the equilibrium level, etc.

The cocoa case is similar in many regards to that of palm oil, although the Asian breakthrough on this market is much more recent and African exporters still keep a major share of the market. The experience of cocoa in Ghana is particularly interesting because it illustrates the influence of economic policies and institutions on production and trade performance. Cocoa production in Ghana decreased by 72 percent during 1965–85 as a result of high export taxation, reduced producer prices, currency overvaluation, labour shortages, inadequate marketing infrastructure, and limited supply of agricultural inputs (Okyere, 1989). However, changes in economic policies institutions in the 1980s (new improved plantations, reduced overvaluation and export taxation, improved efficiency of the cocoa marketing board, privatization of input marketing, and improved infrastructure) have resulted in a substantial increase in Ghanaian cocoa production.

# Conclusion

The econometric and nonparametric analyses of Sample 1 have led us to accept two of the three hypotheses outlined at the beginning of this paper, but to reject for all three that Asia has outperformed Africa. This last point implies that there is more diversity within each continental sample than there is between Asia and Africa. Specifically, it is shown for the entire sample that: a critical determinant of the level of GNP per capita is the level of exports; export value in turn depends critically upon the share of agricultural products within exports, as well as upon internal development policy parameters; and within such agricultural products, both Asia and Africa have chosen their product emphases based on the evolution of the ranking of world commodity market shares.

Similarly, the country-specific analyses of Sample 2 and the commodity specific analyses have confirmed hypotheses 1 and 2, while rejecting the sub-hypothesis that factors affecting agricultural and trade performance differed for Asia. There is much diversity in the performance of the selected countries within each continent, arguing against a simplistic perspective of Asian success and African failure. If Asian overall performance is better than African performance, this is mainly the result of better economic policies and institutions and to a lesser extent of better initial conditions. This emphasizes the need to support provision of better education on economic policy management in Africa. Generally, the findings of the paper can be considered encouraging for long-run development in Africa.

## Notes

[1]Université Laval.

[2]In addition to the three Asian economies on which consistent econometric data were available (Singapore, South Korea, and Hong Kong), other Asian countries are included that have had lower economic growth rates.

[3]The sources of the data were the World Bank (1989 and 1990), UN (1989), and FAO (1988). In the case of the first source, the data used did not fall exactly within the periods indicated; extrapolations were made where necessary.

[4]To confirm these findings, separate linear equations were run with slope and intercept dummies for Asia; these variables also proved insignificant.

## References

FAO (Food and Agriculture Organization), Agrostat data file, 1988.

Helleiner, G.K., *Outward Orientation, Import Instability, and African Economic Growth: An Empirical Investigation*, Working Paper No. A–13, Department of Economics, University of Toronto, Toronto, Ont., Canada, 1984.

Helpman, E., and Krugman, P.R., *Market Structure and Foreign Trade*, MIT Press, Cambridge, Mass., USA, 1985.

Helpman, E., and Krugman, P.R., *Trade Policy and Market Structure*, MIT Press, Cambridge, Mass., USA, 1989.

Lau, L.J., *Models of Development: A Comparative Study of Economic Growth in South Korea and Taiwan*, ICS Press, San Francisco, Calif., USA, 1986.

Lele, U., *Agricultural Growth, Domestic Policies, the External Environment, and Assistance to Africa: Lessons of a Quarter Century*, MADIA Discussion Paper No. 1, World Bank, Washington, D.C., USA, 1989.

Michaely, M., "Exports and Growth: An Empirical Investigation," *Journal of Development Economics*, Vol. 4, No. 1, 1977.

Okyere, A., *The Effects of Domestic Policies on Exportable Primary Commodities: The Case of Ghana and Cocoa*, African Rural Social Science Series Research Report No. 1, Winrock International, Morrilton, Ark., USA, 1989.

Petit, M.J., and Knudsen, O, "Trade and Development Linkages," paper presented at the annual meeting of the Canadian Agricultural Economics and Farm Management Association, Montreal, Qué., Canada, 1989.

Solagral, "Huile de Palme: Le Péril Jaune," *Lettre de Solagral*, No. 86, 1989.

Tyler, W.G., "Growth and Export Expansion in Developing Countries: Some Empirical Evidence," *Journal of Development Economics*, Vol. 9, No. 1, 1981.

UN (United Nations), *Handbook of International Trade and Development Statistics*, New York, N.Y., USA, 1989.

Whee Rhee, Y., Ross-Larson, B., and Pursell, G., *Korea's Competitive Edge: Managing the Entry into World Markets*, Johns Hopkins University Press, Baltimore, Md., USA, 1984.

World Bank, *World Tables, 1988–89 Edition*, Washington, D.C., USA, 1989.

World Bank, *World Development Report 1990*, Washington, D.C., USA, 1990.

## Discussion Opening—*Mary E. Burfisher* (US Department of Agriculture)

The paper compares African and Asian development and export performance and, in doing so, makes an important contribution by criticizing the popular notion of a successful "Asian" model versus the strategies pursued by many African countries. The paper reaches the important conclusion that there is more diversity in economic performance among countries within Asia and Africa than between the two continents.

My main comments relate to the implications of the paper's findings for a developing country that is attempting to define a development strategy. The paper concludes that the level of exports is a major determinant of per-capita GNP and that agricultural exports have played a privileged role in that trade. These findings imply that a developing country ought to pursue export growth and should attempt to expand the share of agriculture in its exports.

First, let us examine the role of exports as a determinant of economic growth. Much research has confirmed the view that trade is a "handmaiden" of growth; i.e., trade tends to accompany growth but does not necessarily cause it. Supply conditions have been found to be critical in explaining why some countries with export growth achieve development and some do not. These supply conditions include factors such as government policies that enhance supply responsiveness and improve a country's ability to exploit export opportunities. Also, export growth may be associated with increased factor productivity, not just increased output. A related supply factor is the possible existence of beneficial externalities, so that economic growth is spurred not just by growth in exports, but also by enlarging more efficient sectors.

Countries that decide to pursue an export-led development strategy need to think about the mechanisms through which export growth is expected to stimulate economic growth and development. Policies and institutions that foster competitiveness and enhance responsiveness of the export sector to foreign market conditions are necessary elements of an export growth strategy. This argument, that countries must pursue a particular type of export strategy rather than simply export growth, is consistent with the finding of this paper that more open economic policies and institutions account for the better performance of some countries.

In relation to the question of whether countries should expand the share of agriculture in their exports, the authors accept the hypothesis that an increased share of agriculture in total exports leads to a higher export value, which in turn is estimated to be the most important determinant of per-capita GNP. The paper is not internally consistent in reaching this conclusion, however. In their time-series regression analysis, using annual averages from 1961–63 and 1985–87, the authors find a significant and positive relationship during export value and the share of agriculture in exports. Other data in the paper indicate that during 1965–88 export values on both continents soared while the importance of agriculture as a percentage of exports fell. It is difficult to determine what may account for these apparently contradictory findings.

Even if we accept the paper's conclusion that an increase in the share of agriculture in exports leads to an increase in export value, it is of interest to discuss what is encompassed in the term "agricultural exports." The paper focuses on two traditional crops: palm oil and cocoa. But expansion of agricultural trade does not necessarily imply that LDCs should focus on traditional tropical crops. In fact, the fastest growing segment of LDC agricultural exports has been high-value and processed commodities. Farm-reared shrimps and fresh cut flowers are two examples.

Finally, in this paper, and in much recent research, emphasis has been placed on creating supply conditions that are favourable to export growth. This is crucial. But it is useful to remember that an objective in improving supply conditions is to improve market responsiveness to foreign demand conditions. Foreign market opportunities are shaped by foreign incomes and taste, as well as by their trade policies. These create an environment that opens opportunities for some products but limits them for others. Both domestic supply and foreign demand conditions combine to explain successful export performance.

*[Other discussion of this paper and the authors' reply appear on page 236.]*

# Problems of Agricultural Restructuring in South Africa: Lessons from the Hungarian Experience

*Ferenc Fekete, Tamas I. Fènyes, and Jan A. Groenewald*[1]

**Abstract:** In the wake of current moves to dismantle apartheid in South Africa, the agricultural sector is seeking ways of restructuring its dualistic nature. Restructuring may affect both equity and productivity. Hungarian and other experiences can provide useful guidelines. Hungary's restructuring has virtually come full circle from individual units to collectivization to individualization. Individual units have performed better than collective units. In its restructuring, South Africa must avoid the mistakes of others, striking a balance between equity and productivity.

## Introduction

South African agriculture is on the eve of serious restructuring, which will, as elsewhere, be part of wider-ranging political and socioeconomic restructuring. Wisdom dictates that it is essential to benefit from the experience of others to forestall costly errors and unnecessary hardship.

Agricultural restructuring has traditionally had two main objectives: equity and productivity. The equity objective, which is closely allied to political egalitarian objectives, has for a long time often occupied centre stage. Egalitarian motives were often regarded as so important that any resulting disruption of agricultural production could be ignored. Examples include post-revolutionary restructuring in the USA, Western European reforms after the French Revolution, Eastern European restructuring after World War I, Latin American reforms after 1910, and Japanese, Korean, and Taiwanese reforms after World War II (Ruttan, 1969).

Some other conditions clearly also cause the productivity objective to be vitally important. They include high rural population/land ratios, high rural/urban population ratios, and a high degree of dependence on agricultural exports, all of which certainly apply to South Africa.

Equity and productivity can be but are not necessarily conflicting objectives. Warriner (1964) compared Bulgaria, which restructured agriculture in the early twentieth century, with Hungary, which had not done so before 1940. By 1940, Bulgarian farmers' living standards were much more equitable but, on the average, hardly better than at the turn of the century. The Hungarian distribution was more unequal, and the peasants were poorer than the average Bulgarian farmer. But Hungarian productivity had improved much more and landowners had invested in industrial development, which drew some poor peasants off the land to other occupations.

If a process of restructuring will improve equity and simultaneously induce incentives to produce more efficiently and effectively, then both the equity and productivity goals will be served. Such restructuring must be accompanied by other needed agricultural support activities.

The concept of a break with the past can be a powerful political incentive for a nation. But "... in agriculture there can be no break with the past. Continuity is the essence of its growth. For the world's agricultural countries, the maintenance and increase of agricultural production are now quite literally a matter of life and death. If governments and people wish to break with the past, they must find ways of doing so which will increase the incentives to produce more and to invest more in the land" (Warriner, 1964).

## The South African Agricultural Scene

European colonization of South Africa started in the 17th century when the Dutch East India Company established a settlement in the South (Cape Province) with the aim of supplying fresh products for seafarers sailing between Europe and Asia. This settlement became a colony, which expanded and was later annexed by Britain. In the 19th century, some settlers who were dissatisfied with the colonial rulers trekked northward, established independent republics, but lost this independence following the Boer War. The Union of South Africa was formed in 1910, gained independence in 1932, and became a republic in 1960.

Over almost three and a half centuries, the European settlers expanded their land area, often at the expense of the indigenous African population who were ill-equipped militarily to check this expansion. This part of South African history has marked similarities with that of some other countries settled by Europeans since the seventeenth century. By the end of the nineteenth century, European farmers had occupied most of the land and Africans had largely retreated to areas today known as reserves, locations, or "homelands." Some took employment on farms operated by European farmers, mostly because of two factors: impoverishment after inter-tribal and inter-racial wars and overpopulation of people and animals in the reserves, resulting in further poverty and the increased attractiveness of selling their labour to European farmers (Grosskopf, 1933).

Rapid mining development following the discovery of diamonds (1866) and gold (1885) led to commercialization and development of agriculture among European farmers, but not in the reserves, which were geographically removed from the new markets and railways. In addition, and more important in the long run, a series of acts were passed which severely restricted African farmers' ability to compete. The Glen Grey Act of 1894, for example, enforced the "one person, one plot" (approximately 4 ha) principle in eastern Cape Province. More legislation and other measures either discriminated against indigenous land ownership or favoured the mainly European commercial farmers with respect to infrastructure. Some other measures entrenched outmoded tenure systems in the reserves (Louw and Kendall, 1986; Davenport, 1990; Leseme et al., 1980; and Kassier and Groenewald, 1990).

Historical developments gave rise to a distinct dualistic agricultural structure comprising a commercial sector (mainly European farmers) and a subsistence sector (mainly African farmers in the reserves). The commercial sector comprises some 65,000 farmers and a total land area of approximately 86 Mha, employing 1.37 million workers. It is somewhat similar to the farming sectors of the developed world; it produces surpluses and uses considerable amounts of purchased inputs. It also shares many problems of First-World agriculture. To these are added problems induced by inflation at higher rates than those encountered by most competitors and trading partners. Owner-operated farms predominate and are well supported by infrastructure.

The subsistence sector is rather similar to the agricultural sectors in less-developed countries of Africa and shares most of their problems. While the two sectors involve roughly the same number of people, the commercial sector occupies roughly 6 times as much land as the subsistence sector and produces more than 20 times its output per capita (Cobbett, 1987). Differences in output per person and per acre have been growing consistently and have been caused largely by differences in technology, capital, marketing infrastructure, and public and private support institutions (e.g., credit, research, and extension).

Commercial agriculture is facing many problems. Double-digit inflation has caused input prices to outstrip output prices. The resultant continuous decrease in its terms of trade has eroded its competitiveness on export markets, led to large increases in indebtedness (13.4 percent per year), and caused many insolvencies. Managerial problems and drought aggravated this situation (van Zyl et al., 1987; and Janse van Rensburg and Groenewald, 1987).

At the same time, the subsistence sector, being largely dependent on remittances from family members in urban employment, has sunk deeper into poverty. A sluggish economy reduced urban employment opportunities, depressed wages, and increased costs of purchased production and consumption goods. The subsistence farmers were very vulnerable to the same droughts as the commercial farmers.

South Africa is now on the threshold of important political and economic restructuring. The apartheid system has been discredited and is being dismantled. Laws relating to racial division of land were revoked in 1991. Some institutions, such as the statutory Land Bank, have been ordered to remove discrimination. At the same time, however, the indigenous farming population still has serious handicaps in its potential ability to compete with the European farmers. Backlogs in provision of education have left them with fewer farming and managerial skills, their poverty has left them with very little capital, and the traditional communal tenure system in the reserves has left them with a lack of experience in individual entrepreneurial action.

The challenge is to find an orderly process of restructuring that will achieve, simultaneously, a more equitable distribution of resources and returns, improvement or at least maintenance of productivity, and stability. Possible guidelines are sought in the Hungarian experience, augmented by experiences from elsewhere.

Why choose Hungary? Besides the first two authors' familiarity with Hungarian agriculture, the choice was also influenced by some similarities in the two countries' agricultural resource base and performance (Table 1).

Table 1—Comparative Data: Hungary and South Africa

|  | GNP/Capita (US$) | Percentage Share of Top 10 Percent | Population (millions) | Life expectancy (years) |
|---|---|---|---|---|
| Hungary | 2,240 | 21 | 10.6 | 70 |
| South Africa | 1,890 | 50 | 33.1 | 60 |

The Hungarian population is, on the average, better off than the South African population, and the welfare is more evenly spread. The total geographical area of Hungary is 9.3 Mha, of which 6.5 Mha are suitable for farming; of that, 4.7 Mha are arable. By comparison, South Africa consists of 119.5 Mha, of which 99.2 Mha are available for farming; of that, only 18.3 Mha are arable and only 4 Mha are considered to be high potential arable land—less than in Hungary. Both Hungary and South Africa consistently produce surpluses for export.

## Hungarian Experience In Restructuring Agriculture

Before the second World War, Hungary's agriculture was dominated by large, privately owned estates. Six percent of the owners owned 68 percent of the land. Restructuring started immediately after the war. An aggressive land reform programme affected more than one third of the land. On average, each "new" farmer (previously farm labourer) received a grant of 2.9 ha; 400,000 new farms were established.

In the second phase of restructuring (1949–61), these new farmers were forced into cooperative farms. The cooperatives use land partly owned by the cooperatives themselves and partly by their members who receive rent for this privately owned but collectively cultivated land. The privately owned land is inheritable, but beneficiaries who are not members of the cooperative are obliged to sell the inherited land to the cooperative; the land then becomes collective property. In this limited "land market," the selling price is determined by the monopsonistic buyer, much below what it could have been in a free market. During the last decade, some state-owned land cultivated by the cooperatives has also become collective property. Parallel with the collectivization, state farms were established on socialized (confiscated) feudal estates, with the main aim of applying modern technology and farming methods and assisting the fledgling cooperative movement.

A further stage in the restructuring process was the establishment of household plots for cooperative members, especially in the late 1960s, to provide incentives for private initiative. By the late 1970s, well over 95 percent of agricultural land was employed in the socialist sector. A restructuring in the opposite direction also occurred, however, especially after 1968.

The aim of this counter-reform was to increase the efficiency of the socialist system through the liberalization of planning, greater independence for enterprises, the acceptance of profit as the main indicator of economic performance, strengthening of material incentives to labour, price reforms (whereby a larger proportion of prices can be determined by the market forces of supply and demand), a greater role for finance and credit (by more flexible use of interest rates, credit, and taxes), a closer link between production and distribution (mostly by basing profit calculations on quantities sold rather than mere quantities produced), and a stronger orientation towards foreign trade outside CMEA.

A partially hidden element of these reform approaches is a recognition of the scarcity of non-labour resources and an implied marginal analysis. These are in conflict with the labour theory of value, a central theme of Marxist economics.

The process of restructuring is far from complete. With the collapse of one-party socialism and establishment of a multi-party democracy, privatization and deregulation of the Hungarian agricultural economy have gained new momentum. It is not certain what form new structures will take. In general, the different parties favour more individualization. One partner in the governing coalition, the Smallholder Party, demands the re-establishment of the pre-1947 landed relationships, with certain limitations on farm size.

The whole process of restructuring of Hungarian agriculture has virtually come full circle, involving a starting point with huge inequality of privately owned landholdings, nationalization including confiscation, creation of a new smallholder class, elimination of the smallholder class and formation of cooperative and state enterprises, inefficiency of production, establishment of household plots as a first step away from Marxist dogma (with marginal results), further liberalization of the economy, and movement towards the elimination of socialist structures in land ownership and agricultural production.

Various authors have analysed agricultural productivity in Hungary since the start of restructuring (Donáth, 1980; Fekete et al., 1976; Fekete, 1989; and Fekete, 1990). Such analyses may promote an objective view of the alternative farming models. Relevant data appear in Table 2.

Table 2—Comparative Data for Hungarian Farming Structures, 1989

| | Cooperative farms (large-scale section) | State farms | Individual Units |
|---|---|---|---|
| | Percentage share | | |
| Land ownership | 76.2 | 14.3 | 9.5 |
| Farm assets | 64.6 | 21.2 | 14.2 |
| Output (GDP) | 47.1 | 13.6 | 36.7 |
| Value added | 40.1 | 8.9 | 44.4 |

Source: Fekete, 1990.

A salient feature is the relatively high productivity obtained by the individual units. This productivity performance has been effectively supported, particularly on cooperative members' domestic plots, by inputs from large units—feed, soil cultivation, transport, and services. There have been important differences in activity structures and in factor combinations. For example, small private farms have concentrated largely on labour-intensive crop production and grain-fed livestock, and state farms have enjoyed investment priorities, enabling them to develop plantations and dairy and poultry activities with high capital requirements.

In terms of national average yield, the state farms performed best in wheat and sunflower production, while the cooperatives excelled in maize and potato production, and the individual farmers were superior in sugar beet and tobacco. Data on GDP and value added per hectare reveal that state farms achieved double the yields of cooperatives. Individual units' output varied between double and four times the levels achieved by cooperatives (Fekete, 1989).

Differences in capital requirements and efficiency can be illustrated as follows: for one unit of GDP, the cooperatives employ 23 percent less capital than state farms; and the private farms employ only approximately one quarter of the capital used by large units to achieve the same output. However, organization of new family and/or market-oriented farms in the recent transition period has required large amounts of capital.

In summary, relative productivity performances dictate the following future strategies: decrease the role of state and cooperative sectors; increase the role and share of small and medium-sized private farms in harmony with the supply of land, financial, physical and human resources; decentralize the organizational structures on cooperative farms; decrease the average size of large-scale units; and strengthen tenure arrangements, inter-farm cooperation, and activity (first-stage marketing) linkages in the mainstream development process towards market-oriented mixed (property-based) farming.

## Lessons for South Africa

Restructuring of South African agriculture starts from a situation of inequity, inequality, and non-sustainability in both subsistence (Vink, 1986) and commercial agriculture (Fényes et al., 1988). The pattern of restructuring will depend partially on which political grouping predominates. The Land Commission of the African National Congress (ANC, 1990) recently issued a statement on their present stance. Some of their recommendations resemble the early stages of the Hungarian experiment. They propose to start with measures almost identical to those of the immediate postwar period in Hungary. According to the statement, the state should play the principal role in redistributing land. It sees an urgent need for a programme of affirmative action involving the acquisition of land for African people and in support of aspiring African producers.

In this sense, the compilers of the report paid insufficient attention to Hungarian and other experience. In Hungary, state farms were used for the same purpose, but with poor results. Zimbabwe has also experienced problems with similar programmes. One problem is that success cannot be hoped for unless the settlers are selected according to definite criteria (Lewis, 1964), a point well-proved by Zimbabwean disappointments (Eicher and Rukuni, 1990). Another problem is that rapid population growth can turn such a programme into a demographic treadmill. This consideration should overshadow the idealistic dream of "one man, one farm." It also leads to resource degradation. A third problem is that planning, servicing, and staffing of resettlement programmes require very high human resource input (Eicher and Rukuni, 1990).

However, the ANC does not necessarily regard nationalization as the only redistributive instrument; selective nationalization according to land use need is mentioned as a possible choice. There appears to be an understanding that, after nationalization, the government will return the redistributed land to the people—indicating further similarities with Hungary.

The possibility of state and/or cooperative farming has also been mentioned in some circles. Once again, experience in Hungary and other countries can provide guidelines. Such units can reduce incentive and productivity, smallholders become mere wage earners, and managerial problems can occur. In Zambia, cooperative farming units were a failure—largely because of managerial problems—and have virtually disappeared (Watts, 1990); nor have state farms had a happy history in Africa.

Agricultural restructuring will inevitably involve tenurial changes. Even if the state does not use expropriation as a redistributive tool, it has some land available in trust. The Land Bank (a statutory finance institution) and private banks have taken over land because of debt delinquencies. Such land can be used for redistributive purposes.

Customary communal tenure in the subsistence sector (i.e., the reserves) has been a source of low productivity but not necessarily of inequitable distribution. In Africa, as in Hungary (Fekete, 1990), it is clear that absence of a land market hinders productivity. Recent analysis (Lyne, 1990) shows that even if a land market could only be developed in the form of allowing people to rent one another's use rights, productivity would improve to the benefit of both parties.

In the final analysis, all considerations regarding agricultural restructuring in South Africa should strike a balance between equity and productivity. The challenge is one of increasing equity while simultaneously maintaining and improving productivity. South Africa does not need to repeat mistakes made elsewhere; the experience of Hungary over the last 40 years, and other examples, are very relevant.

# Note

[1]Budapest University of Economic Sciences, Vista University, and University of Pretoria, respectively.

# References

ANC (African National Congress), "Report on a Workshop on Land Distribution," ANC Land Commission, Johannesburg, South Africa, 1990.

Cobbett, M.J., "The Land Question in Southern Africa: A Preliminary Assessment," *South African Journal of Economics*, Vol. 55, 1987, pp. 63–77.

Davenport, T.R.H., "Land Legislation Determining the Present Racial Allocation of Land," *Development Southern Africa*, Vol. 7, 1990, pp. 431–440.

Donáth, F., *Reform and Revolution: Transformation of Hungary's Agriculture, 1945–1970*, Corvina Kiadó, Budapest, Hungary, 1980.

Eicher, C.K., and Rukuni, M., "Namibia: Restructuring and Strengthening the Institutional Base of Agricultural Development," IAAE–AGRECONA Inter-Conference Symposium, Swakopmund, Namibia, 1990.

Fekete, F., "Ownership Structures and Nationalization Experiences in the Land Economy of Hungary," IAAE–AGRECONA Inter-Conference Symposium, Swakopmund, Namibia, 1990.

Fekete, F., "Progress and Problems in Hungarian Agriculture," in Clarke, R.A. (Ed.), *Hungary: The Second Decade of Economic Reform*, Longman, Colchester, UK, 1989.

Fekete, F., Heady, E.O., and Holdren, B.R., *Economics of Cooperative Farming: Objectives and Optima in Hungary*, A.W. Sijthoff, Leiden, Netherlands, and Akadémiai Kiadó, Budapest, Hungary, 1976.

Fényes, T.I., van Zyl, J., and Vink, N., "Structural Imbalances in South African Agriculture," *South African Journal of Economics*, Vol. 56, Nos. 2/3, 1988, pp. 181–195.

Grosskopf, J.F.W., "Vestiging en Trek van die Suid-Afrikaanse Naturellebevolking na Nuwere Ekonomiese Voorwaardes," *South African Journal of Economics*, Vol. 1, 1933, pp. 261–280.

Janse van Rensburg, B.D.T., and Groenewald, J.A., "The Distribution of Financial Results and Financial Ratios among Farmers during a Period of Financial Setbacks: Grain Farmers in Western Transvaal, 1981/82," *Agrekon*, Vol. 26, No. 1, 1987, pp. 1–7.

Kassier, W.E., and Groenewald, J.A., "The Agricultural Economy of South Africa," IAAE–AGRECONA Inter-Conference Symposium, Swakopmund, Namibia, 1990.

Leseme, R.M., Fényes, T.I., and Groenewald, J.A., "Traditional and Legal Aspects Influencing Agricultural Development in Lebowa," *Development Southern Africa*, Vol. 2, 1980, pp. 171–195.

Lewis, W.A., "Thoughts on Land Settlement," in Eicher, C., and Witt, L. (Eds.), *Agriculture in Economic Development*, McGraw-Hill, New York, N.Y., USA, 1964.

Louw, L., and Kendall, F., *South Africa: The Solution*, Amagi Publications, Bisho, South Africa, 1986.

Lyne, M.C., "Distortion of Incentives for Farm Households in Kwazulu," Ph.D. dissertation, University of Natal, Natal, South Africa, 1990.

Ruttan, V.W., "Equity and Productivity Issues in Modern Agrarian Reform Legislation," in Papi, U., and Nunn, C. (Eds.), *Economic Problems of Agriculture in Industrial Societies*, Macmillan, London, UK, 1969.

van Zyl, J., van der Vyver, A., and Groenewald, J.A., "The Influence of Drought and General Economic Effects on Agriculture," *Agrekon*, Vol. 26, No. 1, 1987, pp. 2–12.

Vink, N., "An Institutional Approach to Livestock Development in South Africa," Ph.D. dissertation, University of Stellenbosch, Stellenbosch, South Africa, 1986.

Warriner, D., "Land Reform and Economic Development," in Eicher, C., and Witt, L. (Eds.), *Agriculture in Economic Development*, McGraw-Hill, New York, N.Y., USA, 1964.

Watts, R., "Zambia's Experience of Agricultural Restructuring," IAAE–AGRECONA Inter-Conference Symposium, Swakopmund, Namibia, 1990.

---

## Discussion Opening—*E. Wesley F. Peterson* (University of Nebraska)

The authors of this paper believe that the important social changes currently taking place in South Africa will not leave South African agriculture unaffected. They expect to see a "restructuring" of agriculture and argue that South Africa would do well to heed the lessons from recent Hungarian experience with changing patterns of land ownership. The main point of the paper is that collectivized agriculture was not a success in Hungary and probably would fail in South Africa as well.

There are two sets of problems with the arguments developed in the paper. First, the relevance of the Hungarian experience for South African agriculture is unclear. The authors justify their choice of Hungary as a model for South Africa by noting their familiarity with Hungarian agriculture and pointing to similarities in the "agricultural resource base and performance" of the two countries. The fact that two of the authors are familiar with both countries is irrelevant. It does not provide support for the argument that the Hungarian experience contains lessons for the future of South Africa. The figures in Table 1 as well as those cited in the text reveal more dissimilarities than resemblances between the two countries. In addition, the comparison includes nothing to account for important social, political, and historical variables that would, I suspect, lead most to conclude that there are very few similarities between the two countries.

It seems to me that the argument against collectivized agriculture could be made by considering a wide range of experiences, not only in Europe but in other parts of Africa, Asia, and Latin America. In fact, South Africa could probably learn a great deal more about appropriate structural changes in agriculture from neighbouring Zimbabwe, for example. Zimbabwean resettlement programmes, combined with the development of marketing infrastructure, smallholder credit programmes, and price policies, have led to a dramatic expansion of food production by smallholder farmers. This point is related to the second set of problems in the paper, the excessive focus on land tenure arrangements. While rules to govern land ownership are undoubtedly of great importance for both efficiency and equity, the prosperity of a national agricultural sector depends on much more. For example, allowing individual land ownership in Eastern Europe has not solved problems associated with inefficient marketing systems. Again, the experience of Zimbabwe in providing marketing, credit, technical, and policy support for resettled farmers is probably of greater relevance to South Africa than Hungary's present anguish over who owns land that was nationalized after the second World War.

*[Other discussion of this paper and the authors' reply appear on the following page.]*

# General Discussion—*Johan van Zyl, Rapporteur* (University of Pretoria)

Three issues were raised with respect to Sudan's wheat policies: to what extent political instability has contributed towards widening the supply-demand gap, to what extent the subsidy on wheat is a subsidy on bread consumption, and why Sudan would want to expand wheat production in the first place, given the comparative advantage of sorghum and millet. In reply, Hassan stressed that the model used assumes that Sudan is a price taker with respect to wheat (small-country assumption) and that food aid is only one issue aimed at bridging the gap between supply of and demand for wheat. Managing food aid is the most important point. However, Sudan has not done this adequately.

The paper on the relative international competitiveness of Africa and Asia received favourable comments. However, one speaker felt that the paper asks many new questions without addressing old ones. In particular, the role of size of a country in determining its trade is not clear. In their response, the authors emphasized that the results may be questionable due to data problems. However, the three different analyses used to overcome this shortcoming all yielded similar results. Size of a country and trade were also controlled through sample selection, which improves the reliability of the results. The case studies are valuable in this regard as they imply that countries should find a specific niche where they have a comparative advantage.

In relation to the third paper, almost all participants agreed with the discussion opener that Hungary cannot really be compared with South Africa. South Africa can learn much more from the experiences of Kenya and Zimbabwe with respect to land reform. Other issues raised included the relative weight of productivity and equity in a new strategy as well as different policy options, what Hungary can learn from South Africa, and what can be learned with respect to other prime movers of development such as technology, marketing services, and credit. The authors replied that access, not only to land, is critical in a new strategy. In this respect, South Africa can learn much from Kenya and Zimbabwe. However, a critical issue is that the supporting structure must be in place in order to facilitate development. Land tenure reform is a necessary condition but not a sufficient condition to ensure development. It is not possible to put weights to productivity and equity in a new strategy. The authors elaborated further on the Hungarian experience of land reform and argued that while other countries are relevant to South Africa, Hungary provides a different or even outside perspective.

Participants in the discussion included H. Alfons (Federal Institute of Agriculture Economics, Austria), K. Daubner (Budapest University of Economic Sciences), G.T. Jones (University of Oxford), H.A. Mahran (University of Gezira), A.W. Mukhebi (ILRAD, Nairobi), J. van Rooyen (Development Bank of Southern Africa), and L.D. Smith (University of Glasgow).

# Appraising Rice Production Efficiency in Taiwan under the Contract Cultivation System

*Ming-Che Lo and Tsorng-Chyi Hwang*[1]

**Abstract:** The popularly accepted rice production system in Taiwan is contract cultivation. This paper investigates the production efficiency of rice farming under the contract cultivation system. Data obtained from the financial accounts of 60 rice-farming families are used and estimated in a translog cost function. The efficiency indicators are the elasticities of input demand combinations with respect to own- and cross-input prices and the average incremental cost. The results show that contract labour demand is the most rigid input for the adjustment of rice production efficiency. The low elasticities of input substitution and of per-unit output variable cost further prove the limited ability of rice farmers to improve efficiency except via deregulation and land policy. Some policy implications are also presented.

## Introduction

Over the past three decades, several major systems of farm operation have been introduced in Taiwan's agriculture to expand the scale of farming. These are joint, contract, and cooperative operations and contract cultivation. Each system is designed to do what its name implies. The underlying policy objectives were to promote modernization and industrialization of small farms and effectively to solve family farming problems in the long run. In order to improve the efficiency of application of farm resources, programmes to expand the scale of farming were advocated and introduced. However, the performance of the various systems have shown great discrepancies. Recent reports showed that only the contract cultivation method is acceptable to general farmers.

The popularity of rice contract cultivation has been influenced by the supply of inputs and equipment from seedling centres. These centres were established after 1979 with the aim of encouraging joint planting and use of farming equipment. The major objective was to increase productivity. As a result of outflow of farm labour from the agricultural sector, continuous technology transfer, and the subsidized purchase of farm machinery, some specialized farmers have bought machines to substitute for labour. However, most small farmers have relied on contract cultivation in the form of soil management, transplanting, harvesting, and drying. As a result, contract cultivation has become the major operating system of the rice industry. Under the regulated rice supply policy, the rice price is at a comparatively low level, which focuses rice production on efficiency of resource distribution and minimum costs.

To study the performance of rice production under the contract cultivation system, a cross-sectional multi-input translog cost function was specified. A survey was carried out in September 1990 involving 60 rice farms in central Taiwan. Data on annual cost structure and output were collected to estimate the model. Neoclassical duality theory provides an approach for computing various pairwise elasticities of substitution between inputs, own-, and cross-price elasticities of demand for inputs and for computing product-specific elasticities in the short run.

## Theoretical Model

The 1-output-$n$-input translog cost function is specified as:

$$(1) \quad \ln C_v = \alpha_0 + \alpha_y \ln Y + \sum_i \beta_i \ln P_i + \sum_i \sum_j \beta_{ij} \ln P_i \ln P_j + \sum_i \delta_i \ln Y \ln P_i$$

$C_v$ is total variable cost, $Y$ is rice output, and $P_i$ is the price of input $i$. Technology transfer is excluded from the cross-sectional approach. Neoclassical theory further suggests a symmetric property of $\beta_{ij} = \beta_{ji}$. Any sensitive cost function must enable inputs to be homogeneous of one degree (Ray, 1982). It is assumed that the input prices of (1) are linearly homogeneous of degree one. Thus, the quadratic approximation of the cost function implies the following $i + 1$ conditions:

(2) $\sum \beta_i = 1, \sum \beta_{ij} = 0, \sum \delta_i = 0$

By differentiating the translog cost function applying Shephard's lemma, the following variable cost share functions are obtained:

(3) $S_i = \dfrac{\partial \ln C_v}{\partial \ln P_i} = \beta_i + \sum_j \beta_{ij} \ln P_j + \delta_i \ln Y$

To coordinate with the linearly independent requirement, one of the cost share functions must be excluded from the estimation of the equation system (Christensen and Greene, 1976). The price of the excluded cost share would become the numeraire price of other inputs. Since other variable costs are minimal, they are substituted by one as a proxy price.

The estimation involves six major steps. First, survey data are transformed into variable costs, output, variable input prices, fixed costs, and cost shares, while at the same time all variables are deflated by their geometric means. Second, a system of joint variable cost and cost share functions is jointly estimated using the iterated seemingly unrelated regression technique with the imposition of linear restrictions. Third, the estimated parameters are further investigated with tests on heteroscedasticity as well as input separability (Berndt and Christensen, 1973). Fourth, the estimated parameters are used to calculate Allen $(_aE_{ij})$, Morishima $(_mE_{ij})$, and McFadden shadow $(_sE_{ij})$ input substitution elasticities:

(4) $_aE_{ij} = \dfrac{1}{S_i S_j} \beta_{ij} + 1 = \dfrac{\beta_{ij}}{S_i^2}(\beta_{ij} + S_i^2 - S_i)$

(5) $_mE_{ij} = S_i \left(_aE_{ij} - _aE_{jj}\right)$

(6) $_sE_{ij} = \dfrac{S_i S_j}{S_i + S_j}(2 _aE_{ij} - _aE_{it} - _aE_{jt})$

Finally, the elasticity of the average incremental cost (AIC) is calculated:

(7) $AIC(Y) = \dfrac{\partial \ln \dfrac{C_v}{Y}}{\partial \ln Y}$

The establishment and estimation of the model were also assisted by reference to Hanoch (1975) and Debertin and Paagoulatos (1985).

## Cost Structure and Statistical Results

The cost structure of rice production in Taiwan can be explained through the production procedure. Inputs can be categorized as the fixed inputs of land and long-term capital as well as the variable inputs of contract labour, own labour, and variable capital. Long-term capital includes the discount values of farm buildings, machinery, and equipment. Land includes own- and rented-land values. Labour costs comprise the expenses resulting from soil management, transplanting, weeding, fertilizing, harvesting, and drying. Farmers are free to choose a combination of contract and own cultivation. Short-run capital involves expenditure on seeds, fertilizers, pesticides, and insecticides.

The descriptive statistics are shown in Table 1. The largest share of long-term capital $(F_1)$ implies the intensive capitalization of rice production. The large share of the variable cost $(C_v)$ indicates possible improvement of the cost structure. Other statistics further show that existing data have included a wide range of farm sizes. Among the variable input prices $(P_1, P_2,$ and $P_3)$, the mean price of own labour is the highest. This implies that most farmers put a lot of time into rice production. The research thus represents not only part-time farming but rice-specialized farming.

Table 1—Descriptive Statistics of Farms under the Contract Cultivation System

| Variable | Sample Mean | Standard Error | Minimum Value | Maximum Value | Geometric Mean |
|---|---|---|---|---|---|
| $C_v$ | 299,161.7 | 153,344.4 | 130,788 | 929,442 | 270,866.10 |
| $Y$ | 19,532.25 | 15,579.79 | 1,900 | 89,100 | 14,823.87 |
| $P_1$ | 28,618.96 | 12,641.64 | 5,600 | 62,636.36 | 25,656.18 |
| $P_2$ | 51,030.42 | 36,183.52 | 6,893.94 | 204,545.45 | 40,078.57 |
| $P_3$ | 25,264.04 | 6,447.43 | 17,210 | 48,553.33 | 24,580.64 |
| $F_1$ | 935,535.70 | 1,305,423 | 20,016 | 9,382,500 | 540,660.5 |
| $F_2$ | 56,999.53 | 48,761.42 | 5,557.50 | 235,188.50 | 40,309.21 |
| $S_1$ | 28.54 | 9.90 | 11.06 | 54.62 | 26.76 |
| $S_2$ | 44.37 | 13.94 | 11.63 | 68.89 | 41.80 |
| $S_3$ | 27.08 | 8.75 | 10.11 | 55.00 | 25.64 |

$C_v$ = variable costs (NT\$/year); $Y$ = output (kg/year); $P_1$ = contract labour price (NT\$/acre); $P_2$ = own labour price (NT\$/acre); $P_3$ = short-run capital price (NT\$/acre); $F_1$ = long-run capital (NT\$); $F_2$ = land (NT\$); $S_1$ = contract labour price (percent); $S_2$ = own labour share (percent); and $S_3$ = short-run capital share (percent).

The estimated results are shown in Tables 2 and 3. The exponential of the intercept in the cost function represents total variable costs at its geometric mean. Using the estimated coefficients and linear restrictions, the coefficients of the third cost share equation are calculated.

Tests for heteroscedasticity and separability are also performed. The Glejser and Breusch-Pagan tests are applied for the former. The resulting $\chi^2$ values are large enough for the rejection of heteroscedasticity. The test for separability is performed by taking some cost elements such as interest payments, material input costs, and meal subsidies out of the original cost categories and making them the fourth cost category. Using the $\chi^2$ test on the results of three variable inputs and four variable inputs, the three-input case is accepted as reasonable. The estimated results are then used to calculate related elasticities.

Table 4 shows Allen, Morishima, and McFadden own- and cross-price elasticities as well as those for factor demand. Factor demand own-price elasticities are inelastic and negative in sign. Short-run capital input has the greatest variability among the three variable inputs. Own-price factor demand elasticity is also high as compared to short-run capital input. However, the price elasticity of demand for contract labour is close to zero. It represents existing conditions in the prevailing contract cultivation system. The number of aged farmers and the farm labour shortage are the main reasons for the decision of most farms to adopt contract cultivation. Such farms also have to pay for contract labourers using suitable machines in order to complete certain rice production processes.

Own labour has very little substitution effect on the contract labour price, with a cross elasticity of only 0.0967; i.e., a 0.967-percent increase in own labour demand is expected in response to a 10-percent increase in the contract labour price. When the own labour price is increased, farmers may start thinking of some substitution away from contract labour and short-run capital. If own-labour price increases by 10 percent, the demands for contract labour and short-run capital are anticipated to increase by 1.504 and 3.405 percent, respectively. Short-run capital is the preferred choice in such a situation. Short-run capital may be partially substitutable for contract input, and vice versa. In the short run, both contract and own labour may be substitutable for capital. As a result, farmers may choose variable substitutes while factor prices, except for contract labour, change, but would not take such action in the short run.

Table 2—Estimated Parameters of the Translog Variable Cost Function

| Variable | Estimated Coefficient | $t$-value |
|---|---|---|
| Constant | −0.057755** | −2.91 |
| $\ln Y$ | 0.743668** | 12.35 |
| $\ln P_1$ | 0.285571** | 172.83 |
| $\ln P_2$ | 0.443337** | 217.98 |
| $\ln P_3$ | 0.271093** | 140.91 |
| $\ln P_1 \ln P_1$ | 0.091472** | 50.79 |
| $\ln P_1 \ln P_2$ | −0.083706** | −26.15 |
| $\ln P_1 \ln P_3$ | −0.007766* | −2.51 |
| $\ln P_2 \ln P_2$ | 0.111653** | 36.15 |
| $\ln P_2 \ln P_3$ | −0.027947** | −7.61 |
| $\ln P_3 \ln P_3$ | −0.023121 | −1.45 |
| $\ln F_1$ | −0.015458 | −0.84 |
| $\ln F_2$ | 0.224017** | 4.41 |
| $\ln F_1 \ln F_1$ | 0.030887* | 1.98 |
| $\ln F_1 \ln F_2$ | 0.143247* | 2.11 |
| $\ln F_2 \ln F_2$ | −0.220992* | −2.35 |
| $\ln Y \ln P_1$ | 0.038127** | 12.29 |
| $\ln Y \ln P_2$ | 0.019036** | 3.47 |
| $\ln Y \ln P_3$ | −0.057163** | −10.39 |
| $\ln Y \ln F_1$ | −0.058386 | −0.81 |
| $\ln Y \ln F_2$ | 0.171398 | 1.58 |
| $\ln P_1 \ln F_1$ | 0.079078 | 1.70 |
| $\ln P_1 \ln F_2$ | −0.120905* | −2.43 |
| $\ln P_2 \ln F_1$ | 0.187154* | 2.06 |
| $\ln P_2 \ln F_2$ | −0.149429 | −1.83 |
| $\ln P_3 \ln F_1$ | 0.005382 | 0.08 |
| $\ln P_3 \ln F_2$ | 0.242092* | 2.41 |

$R^2 = 0.953$, $DW = 1.638$, $D.F. = 33$
*Significance level is 0.025.
**Significance level is 0.005.

Table 3—Estimated Coefficients of the Share Functions

| | Estimated Coefficient | $t$-value |
|---|---|---|
| $S_1$: | | |
| Constant | 0.2855706 | 172.83 |
| $P_1$ | 0.1829436 | 50.79 |
| $P_2$ | −0.08370554 | −26.15 |
| $P_3$ | −0.007766269 | −2.51 |
| $Y$ | 0.03812664 | 12.29 |
| $S_2$: | | |
| Constant | 0.4433368 | 217.98 |
| $P_1$ | −0.08370554 | −26.15 |
| $P_2$ | 0.2233054 | 36.15 |
| $P_3$ | −0.02794714 | −7.60 |
| $Y$ | 0.01903598 | 3.47 |
| $S_2$: | | |
| Constant | 0.2710926 | |
| $P_1$ | −0.007766269 | |
| $P_2$ | −0.02794714 | |
| $P_3$ | −0.0462416 | |
| $Y$ | −0.05716262 | |

The Allen elasticity considers the effects of relative factor price movement on factor substitution. The own-price elasticities of own labour and short-run capital are elastic and negative in sign. The own-price elasticity of contract labour input is also very inelastic. For cross-price elasticity in the short run, own labour and contract labour possess very inelastic substitutability. However, short-run capital is considered as a near perfect substitute for both contract and own labour.

The Morishima elasticity takes own elasticity into account weighted by own cost share. Since the own-price elasticities of own labour and short-run capital are high in value, the Morishima elasticities are mostly near one for these two inputs. Here, contract labour shows greater substitutability for own labour than that in Allen's. The substitutability of own labour and short-run capital for contract labour is very inelastic. The low own-price Allen's elasticity of contract labour depresses the substitutability of short-run capital for contract labour. In the short run, these results better capture the reality.

Table 4—Own- and Cross-Price Elasticities

| Inputs | $P_1$ | $P_2$ | $P_3$ |
|---|---|---|---|
| Factor demand: | | | |
| Contract labour | −0.0735923 | 0.1504082 | 0.2435884 |
| Own labour | 0.0967467 | −0.7449533 | 0.2078136 |
| Short–run capital | 0.2567213 | 0.3404981 | −0.8145803 |
| Allen elasticity: | | | |
| Contract labour | −0.2578567 | 0.3389862 | 0.899514 |
| Own labour | | −1.6789571 | 0.7674062 |
| Short–run capital | | | −3.0080513 |
| Morishima elasticity: | | | |
| Contract labour | 0 | 0.8953614 | 1.0581686 |
| Own labour | 0.1703389 | 0 | 1.0223938 |
| Short–run capital | 0.3303135 | 1.0854513 | 0 |
| Shadow (McFadden) elasticity: | | | |
| Contract labour | 0 | 0.4541423 | 0.7037936 |
| Own labour | | 0 | 1.0462924 |
| Short–run capital | | | 0 |

$P_1$ = per acre price of contract cultivation; $P_2$ = per acre price of own cultivation; and $P_3$ = per acre price of short-run capital input.

The McFadden elasticity considers own-price elasticity and relative cost share more than the Morishima elasticity and becomes symmetrical. As a result, the cross relationships among the three short-run inputs are similar to Allen's except for the substitution between own labour and short-run capital. Own labour and short-run capital turn out to have elastic substitution while the variables remain inelastic. This results from the high values of both own elasticities.

The result of average incremental cost ($AIC$) is 10.39, which indicates an inefficient production system. As rice output increases by 1 percent, the per-unit variable cost increases nearly tenfold. Under conditions of rising labour wages and other input costs, a low productivity growth rate would be detrimental to rice production efficiency. Such an inefficient production situation could be highly correlated to government programmes on land ownership, mechanized farming, and crop conversion.

## Conclusion

This research uses 1990 sample cost data from 60 rice farmers to analyse cost structure and production efficiency in Taiwanese rice production. The translog cost function associated with cost share equations is estimated using the seemingly unrelated regression technique. The estimated results are collected to calculate short-run own- and cross-input price elasticities and the average incremental variable cost. The calculated elasticities are then analysed to evaluate of rice production efficiency in Taiwan.

Own- and cross-price elasticities and the average incremental costs well capture rice farmers' production characteristics in Taiwan. Farmers are concerned about own labour and short-run capital price changes but not about contract labour. The rigidity of adjusting contract cultivation under conditions of labour shortage is evident in rice production. It turns out that only own labour and short-run capital can be substituted elastically. Some farmers may still be able to substitute short-run capital for contract labour while substitution between own labour and contract labour is limited. As a result, rice farmers have a very limited ability to perform short-run cost minimization and thus improve production efficiency. Moreover, the elasticity of per unit output variable cost gives little hope of improving rice production efficiency without deregulation.

The results may provide some policy implications for the rice programme in Taiwan. The variability of short-run capital input prices must be maintained within a certain range under the contract cultivation system for the possible improvement of the cost structure and thus production efficiency. Since there is little substitutability of contract cultivation for other variable inputs, the extension programme to organize contract cultivation for efficient operation may be important for the improvement of overall production efficiency. Finally, the high cost share of own labour may imply failure of programmes to mechanize rice production and need re-evaluation.

## Note

[1]National Chung-Hsing University.

## References

Berndt, E.R., and Christensen, L.R., "The Internal Structure of Functional Relationships, Separability, Substitution, and Aggregation," *Review of Economic Studies*, Vol. 40, 1973, pp. 403–410.

Christensen, R.L., and Greene, W.H., "Economies of Scale in U.S. Electric Power Generation," *Journal of Political Economy*, Vol. 84, 1976, pp. 655–676.

Debertin, D.L., and Paagoulatos, A., *Contemporary Production Theory, Duality, Elasticity of Substitution: The Translog Production Function, and Agricultural Research*, Agricultural Research Report No. 40, Department of Agricultural Economics, University of Kentucky, Lexington, Ky., USA, 1985.

Hanoch, G., "The Elasticity of Scale and the Shape of Average Costs," *American Economic Review*, Vol. 65, 1975, pp. 492–497.

Ray, C.S., "A Translog Cost Function Analysis of U.S. Agriculture, 1939–1977," *American Journal of Agricultural Economics*, Vol. 64, No. 3, 1982, pp. 491–498.

---

**Discussion Opening**—*Bradley J. McDonald* (US Department of Agriculture)

Lo and Hwang use a translog cost function approach to estimate econometrically input relationships for rice production in Taiwan. The study uses data from a cross-section of 60 Taiwanese rice farms, apparently all using a system of contract cultivation. The parameter estimates from the translog cost function are used to draw conclusions about the production efficiency of the contract cultivation system for rice in Taiwan and for Taiwanese rice policy.

My first concern with the paper is that its stated objective is to investigate production efficiency, but this is not actually done. Instead, the econometric estimation in the paper relates to the measurement of optimal input relationships.

The type of exercise conducted in this paper is, nevertheless, valuable. Among many potential applications is the use of the estimated parameters in calibration studies, such as applied general equilibrium models. The authors of this paper are to be congratulated for the thorough reporting of results. For example, they not only present the Allen elasticity of substitution matrix resulting from their estimation but also the Morishima and McFadden substitution elasticities and the price elasticities. They point out important differences between these measures and, although these could be derived by the reader, it is convenient to have them reported in the paper.

A second concern is that, unfortunately, the integrity of the results can be called into question by the apparent failure of some parameter restrictions, suggested by microeconomic theory, to hold. For example, in Equation (2), it is noted that the sum over $i$ of $\beta_{ij}$ should equal zero. The restriction seems to have been applied for the first two inputs but not for the third (see Table 2). The Euler condition on the Allen elasticity of substitution matrix, which requires the row sum of Allen elasticities, when weighted by cost shares, to equal zero, is therefore violated (see Table 4 and the cost shares in Table 1).

These two concerns lead me to believe that the conclusions drawn in the paper regarding efficiency of rice production in Taiwan may not be valid.

If the authors' interest lies in the comparison of production efficiency under the contract cultivation system with that of other systems in Taiwan (such as joint management or cooperative operation), it may be desirable to use another approach.

# Economic Implications of Taxing Agricultural Exports: The Case of Pakistan's Basmati Rice

*Anwar F. Chishti and Stephen C. Schmidt*[1]

**Abstract:** The impact of an *ad valorem* export tax on Pakistan's Basmati rice trade is analysed in a partial equilibrium framework. Results suggest that producers of Basmati rice lost a considerable amount of their producer surplus while consumers gained in terms of their consumer surplus. The national treasury received tax revenues as well as some positive increases in foreign exchange earnings. The nation as a whole thus gained. However, the tax pushes up international prices, which may encourage other producers to increase their production and compete for their shares. A gradual decrease in the level of the export tax and an increase in producer price may take care of the adverse effects of the export tax.

## Introduction

It is not always necessary that a country's major export trade policy goal should be the expansion of its export sales. Governments often have programmes that aim at reducing exportable quantities of particular commodities to protect domestic consumers and/or to raise revenues for the national treasury (Houck, 1986, pp. 120–131). A frequently used programme for the achievement of these goals is the taxing of exports. In particular, a large country with some market power has the option to maximize its national welfare by applying an export tax (McCalla and Josling, 1985, pp. 133–141).

Imposition of an export tax may have different implications for different sectors of the national economy. This paper attempts to evaluate such implications and to quantify their impact, particularly on the producers, consumers, government revenue, and foreign exchange earnings of the exporting country. The analysis is confined to an estimation of the effects of taxes placed on Pakistan's Basmati rice exports in a partial equilibrium framework.

## Theoretical Framework

The impact of an export tax can perhaps best be illustrated in graphical form (Figures 1 and 2). Figure 1 represents the case of a large country that has some market power to influence foreign demand for its exports, and therefore faces a downward-sloping excess-demand curve $(ED)$ as compared to a small country (Figure 2) that faces horizontal $ED$ from the rest of the world.

In the absence of an export tax (free trade), world prices transmit fully to the domestic economy, so that domestic prices $(P_d)$ are equal to world prices $(P_w)$. Consequently, the exporting country's total supply $(Q_s)$ is greater than its total domestic demand $(Q_d)$. This creates an excess supply $(ES)$ for the rest of the world. The difference between $Q_s$ and $Q_d$ is exported, and is equal to the quantity traded $(Q_t)$ in panel (b) in Figures 1 and 2.

After an *ad valorem* tax is imposed, the exporting country's $ES$ shifts to $ES^*$, reducing $Q_t$ to $Q_t^*$. This leftward shift of ES has different impacts on $P_w$ and $P_d$. In the case of a large country, there is an upward pressure on ED from rest of the world, and consequently $P_w$ rises to $P_w^*$ (Figure 1). In the case of a small exporting country, $P_w$ remains intact since such a country faces a horizontal ED for its exports (Figure 2).

Imposition of a tax reduces the domestic price by the amount of the tax, and $P_d$ falls to $P_d^*$. This causes downward pressure on domestic production and upward pressure on domestic consumption; producers respond by moving from $a$ to $b$ along their supply curve $(S_d)$, while consumers move from $c$ to $d$ along their demand curve $(D_d)$. As a consequence, the quantity supplied reduces to $Q_s^*$, and the quantity demanded increases to $Q_d^*$. The country now has less to export $(Q_t^*)$ than when it had no tax on its exports.

The imposition of an export tax thus has the following impacts. First, it reduces the producer surplus $(PS)$ by the amount equal to area $P_d a b P_d^*$. Second, it increases consumer surplus (CS) by the amount equal to area $P_dcdP_d^*$. Third, the treasury of the exporting country receives export tax earnings equal to area *efbd* or *jklm*. Fourth, the social costs of imposing the tax are equal to the sum of the changes in *PS*, *CS*, and tax revenue to the treasury. These costs are clearly positive in the case of a small country, and are equal to areas *ced* and *afb* in Figure 2. In the case of a large country, the magnitude and sign of the social cost depend upon the nature of the ED of the rest of the world and how effectively the exporting country can exploit its market power to raise extra tax revenues to offset the "deadweight" losses, namely the areas *cid* and *ahb* in Figure 1. Lastly, whether an export tax has a positive or negative impact on foreign exchange earnings depends on the difference between the areas $caQ_sQ_d$ and $efQ_s^*Q_d^*$ in both cases; but it will be clearly negative in the case of a small exporting country and will depend upon the magnitude of the two areas in the case of a large country.

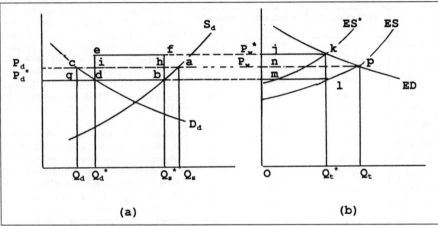

Figure 1—Effect of an Export Tax: Large-Country Case

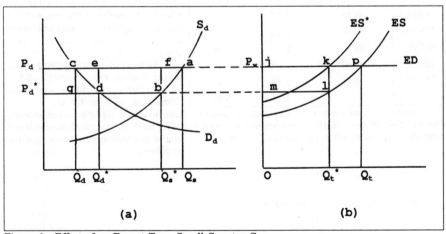

Figure 2—Effect of an Export Tax: Small-Country Case

245

# Model Specification and Estimation Techniques

While Basmati rice is traded freely on the domestic market in Pakistan, its export is monopolized by the state-run Rice Export Corporation of Pakistan. The RECP procures rice at the procurement price announced by the government each year, and then sells it on the world market (Chishti, 1990). A comparison of the procurement and export prices of Basmati rice, given in Table 1, indicates that the latter prices were much higher than the former throughout the 1968–88 period; export prices were more than twice as high as procurement prices during the latter half of the reference period. This is a clear indication of the fact that Pakistan has been taxing its rice exports.

Table 1—Rice Price Ratios

| Year | Procurement/ Export | Basmati/ Thai | Basmati/ US |
|------|------|------|------|
| 1968/69 | 0.82 | 1.25 | 1.33 |
| 1969/70 | 0.87 | 1.20 | 1.19 |
| 1970/71 | 0.86 | 1.44 | 1.08 |
| 1971/72 | 0.69 | 2.34 | 1.59 |
| 1972/73 | 0.32 | 2.35 | 1.63 |
| 1973/74 | 0.42 | 1.34 | 1.00 |
| 1974/75 | 0.31 | 1.43 | 1.39 |
| 1975/76 | 0.46 | 1.44 | 1.25 |
| 1976/77 | 0.79 | 1.35 | 1.11 |
| 1977/78 | 0.60 | 1.65 | 1.35 |
| 1978/79 | 0.40 | 2.01 | 1.86 |
| 1979/80 | 0.41 | 2.13 | 1.87 |
| 1980/81 | 0.48 | 1.64 | 1.43 |
| 1981/82 | 0.52 | 1.49 | 1.27 |
| 1982/83 | 0.46 | 2.17 | 1.74 |
| 1983/84 | 0.48 | 2.18 | 1.59 |
| 1984/85 | 0.41 | 2.49 | 1.58 |
| 1985/86 | 0.40 | 3.08 | 1.75 |
| 1986/87 | 0.41 | 3.42 | 2.10 |
| 1987/88 | 0.49 | 3.14 | 2.24 |

Sources: Government of Pakistan, *Economic Survey 1989–90*, and IMF, *International Financial Statistics Yearbook*, 1989, pp. 182–183.

In the world market, Pakistan's Basmati rice enjoys a special preference among its customers (Ali and Flinn, 1989; and Slayton, 1984). The fine cooking quality and distinct aroma of this rice variety have given Pakistan a special advantage in the world rice market, obtaining much higher prices than the two leading rice exporters, Thailand and the USA (Table 1).

In the light of these considerations, the theoretical framework applicable to a large country, as shown in Figure 1, seems an appropriate methodology for analysing the effects and implications of Pakistan's Basmati rice policies. In terms of the symbols used in Figure 1, the procurement and export prices of Basmati rice are represented by $P_d^*$ and $P_w^*$, respectively, and the difference between the two indicates the tax that goes to the national treasury. Without the imposition of this tax, world and domestic prices would have been the same at $P_w$.

As has already been pointed out, taxing Basmati rice exports has differential effects on producers and consumers as well as for the national treasury and foreign exchange earnings. Such impacts and the social cost of the programme can be quantified as follows:

$$(1) \quad \Delta PS = -P_d abP_d^* = -\int_{P_d^*}^{P_d} S(P)dP < 0$$

$$(2) \quad \Delta CS = P_d cdP_d^* = -\int_{P_d^*}^{P_d} D(P)dP > 0$$

(3) $G = efbd = (P_w^* - P_d^*)(Q_s^* - Q_d^*) = jklm = TQ_t^* > 0$

(4) $NSC = \Delta PS + \Delta CS + G < = > 0$

(5) $\Delta FE = np Q_t O - jk Q_t^* O = P_w Q_t - P_w^* Q_t^* < = > 0$

where $\Delta PS$, $\Delta CS$, and $\Delta FE$ denote changes in producer surplus, consumer surplus, and foreign exchange earnings, respectively; $S(P)$ and $D(P)$ are the total supply and domestic demand functions; $G$ is export tax revenue; and $NSC$ is net social cost. The definition of all other symbols coincides with those used in Figures 1 and 2.

Equations (1) and (2) estimate changes in the producer and consumer surpluses caused by the export tax; Equation (3) estimates revenue accruing to the national treasury; Equation (4) works out the net social cost to the economy resulting from the imposition of the export tax; and Equation (5) provides estimates of whether the foreign exchange earnings from rice exports have been affected by this tax and, if so, in what direction.

Among the variables involved, $P_d^*$, $P_w^*$, $Q_s^*$, and $Q_d^*$ are already observable in the form of the country's domestic procurement and export prices and quantities produced and consumed on the domestic market. Data have to be generated for other variables such as $P_w$, $P_d$, $Q_s$, and $Q_d$, which requires the estimation of related functions including the export demand $(ED)$ and export supply $(ES)$ functions as well as the domestic demand, $D_d$, and total supply, $S_d$, functions.

A 2SLS econometric estimation technique with data for 20 years (1968–88) gives the following estimates of demand and supply functions:

(6) $S_d = 427.3568 + 0.10532\ PD$

(7) $D_d = 661.8715 - 0.056696\ PD$

(8) $ED = 642.932 - 0.76625\ PE$

where $PD$ and $PE$ denote the domestic and export prices, respectively. The fourth function, $ES$, can be calculated from (6) and (7), as follows:

Equating $S_d$ and $D_d$ gives the value of $PD$ (1447.4725) for which quantity exported becomes zero. For the mean value of $PD$ (2880.3), the quantity exported $(ES)$ is 232.13. Substituting these values in the export function, $ES = d + (dES/dPD)PD$, we obtain:

(9) $ES = -234.5016 + 0.162008\ PD$

The values of the intercepts in Equations (6)–(9) are valid only for the mean values of the quantities and their respective prices. For other values, the equations would be:

(10) $S_d = a + 0.10532\ PD$

(11) $D_d = b - 0.056696\ PD$

(12) $ED = c - 0.76625\ PE$

(13) $ES = d + 0.162008\ PD$

where intercepts $a$–$d$ vary with values of prices and quantities for each year.

Equations (12) and (13) represent the export demand and export supply functions. The relationship between $PE$ and $PD$ is given by:

(14) $PD + T = PE \times EXR, \quad T > 0$

247

(15) $PD = PE \times EXR, \ T = 0$

where $EXR$ = exchange rate and $T$ = export tax.

Equations (12), (13), and (14) give $P_w^*$, $P_d^*$ and $Q_t^*$ while Equations (12), (13), and (15) estimate $P_w$ and $Q_t$ of Figure 1. In the latter case, estimates for $PE$ and $PD$ are needed first; the resultant equations are then equated in the form given in (15) and solved for the tax-free equilibrium quantity traded $(Q_t)$, which is given by:

(16) $Q_t = \dfrac{(0.162008c \ ^* \ EXR) - 0.76625d}{0.76625 + 0.162008 \ EXR}$

Substituting the value of $Q_t$ into (12) gives the tax-free equilibrium export price, $P_w$, which is given by:

(17) $P_w = \dfrac{c - Q_t}{0.76625}$

## Empirical Findings and Conclusions

The estimated results given in Table 2 indicate that the export tax imposed on Pakistan's Basmati rice has had a positive social welfare effect for the domestic economy. The tax gave an annual average net social gain of Rs 83.86 million,[2] Rs 284.47 million, Rs 624.94 million, and Rs 946.60 million during the 1968–73, 1973–78, 1978–83, and 1983–88 periods. Producers of Basmati rice lost, on average, Rs 112.41 million and Rs 1314.83 million in producer surplus while consumers, on the other hand, gained Rs 53.11 million and Rs 831.32 million in consumer surplus during the 1968–73 and 1983–88 periods. The national treasury, in addition, received Rs 143.16 million and Rs 1,430.11 million as tax revenue during the reported periods. The nation as a whole thus benefited. Foreign exchange earnings, on average, decreased by $1.93 million per year in the initial period but increased subsequently by $10.51 million, $16.55 million, and $14.48 million during the 1973–78, 1978–83, and 1983–88 periods.

Results suggest that Pakistan has been able to obtain a positive net social gain from the export tax on Basmati rice. The tax, however, has some adverse effects. First, it hurts producers and causes a considerable decrease in their producer surplus. Second, it drives up international prices, which may encourage other producers to increase their production and compete for their shares. A gradual decrease in the level of the export tax and an increase in the producer price may take care of these adverse effects.

## Notes

[1]Peshawar Agricultural University and University of Illinois, respectively.
[2]The average exchange rate was Rs 10.585 per US$ during the period under study.

## References

Ali, M., and Flinn, J.C., "Profit Efficiency among Basmati Rice Producers in Pakistan Punjab," *American Journal of Agricultural Economics*, Vol. 71, No. 2, 1989, pp. 303–310.

Chishti, A.F., *Demand for and Supply of the Pakistan Basmati Rice*, master's thesis, University of Illinois, Urbana, Ill., USA, 1990.

Houck, J.P., *Elements of Agricultural Trade Policies*, Macmillan Publishing Company, London, UK, 1986.

Table 2—Implications of Taxing Basmati Rice Exports (1968/69–1987/88)

| Year | $\Delta PS$ | $\Delta CS$ | G | NSC | $\Delta FE$ |
|------|------|------|------|------|------|
| 1968/69 | −62.67 | 50.49 | 22.65 | 10.47 | 1.72 |
| 1969/70 | −39.55 | 33.79 | 10.88 | 5.12 | 1.10 |
| 1970/71 | −27.21 | 16.28 | 21.31 | 10.38 | −0.20 |
| 1971/72 | −88.03 | 44.12 | 80.30 | 36.40 | 0.89 |
| 1972/73 | −344.59 | 120.85 | 580.67 | 356.93 | −13.15 |
| Average | −112.41 | 53.11 | 143.16 | 83.86 | −1.93 |
|  |  |  |  |  |  |
| 1973/74 | −387.80 | 189.09 | 477.79 | 279.07 | −3.63 |
| 1974/75 | −1182.74 | 683.91 | 810.97 | 312.15 | 59.81 |
| 1975/76 | −620.47 | 342.13 | 658.12 | 379.79 | 3.76 |
| 1976/77 | −151.56 | 56.72 | 280.58 | 185.74 | −8.30 |
| 1977/78 | −332.99 | 167.95 | 430.63 | 265.59 | 0.91 |
| Average | −535.11 | 287.96 | 531.62 | 284.47 | 10.51 |
|  |  |  |  |  |  |
| 1978/79 | −1358.25 | 936.01 | 796.80 | 374.57 | 47.55 |
| 1979/80 | −1277.20 | 712.57 | 1300.08 | 735.45 | 4.66 |
| 1980/81 | −1223.12 | 630.47 | 1487.79 | 895.14 | −13.34 |
| 1981/82 | −1219.75 | 835.39 | 887.75 | 503.39 | 26.26 |
| 1982/83 | −1256.76 | 877.48 | 995.43 | 616.15 | 17.62 |
| Average | −1267.02 | 798.38 | 1093.57 | 624.94 | 16.55 |
|  |  |  |  |  |  |
| 1983/84 | −1121.36 | 579.05 | 1715.01 | 1172.70 | −28.02 |
| 1984/85 | −1369.09 | 993.87 | 983.91 | 608.70 | 25.88 |
| 1985/86 | −1320.50 | 770.98 | 1658.54 | 1109.01 | 5.68 |
| 1986/87 | −1377.06 | 880.04 | 1375.11 | 878.09 | 36.22 |
| 1987/88 | −1386.15 | 932.66 | 1418.00 | 964.51 | 32.63 |
| Average | −1314.83 | 831.32 | 1430.11 | 946.60 | 14.48 |

Note: Changes in producer and consumer surpluses ($\Delta PS$ and $\Delta CS$), government revenue (G), and net social loss/gain (NSC) are in Rs million and changes in foreign exchange earnings ($\Delta FE$) are in \$million.

McCalla, A.F., and Josling, T.E., *Agricultural Policies and World Markets*, Macmillan Publishing Company, London, UK, 1985.

Slayton, T.M., "Some Pieces of the World Rice Puzzle," *Rice: Outlook and Situation Report*, No. RS–43, Economic Research Service, US Department of Agriculture, Washington, D.C., USA, 1984, pp. 11–13.

## Discussion Opening—*Farman Ali* (University of East Anglia)

The paper by Chishti and Schmidt provides an interesting time-series 2SLS estimation of the net social gain resulting from the imposition of an export tax on Pakistani Basmati rice. It is particularly comforting to know that the authors found a positive net social gain. However, the paper has some deficiencies.

The difference between the procurement price and the export price of Basmati rice does not necessarily mean the existence of an export tax. This difference could be due to handling and transport charges and could widen with an increase in fuel cost. The ratios in Table 1 make it clear that the figure for 1972/73 is less than half of the previous year's value because of the oil shock and the subsequent rise in transport cost.

I am not convinced by the view that Pakistan's behaviour as a Basmati rice exporter may be viewed as that of a large country because Pakistan is not the sole producer of Basmati rice and, since there are other varieties of rice, consumers may switch from consumption of Basmati to its substitutes if there is an increase in the price of Basmati. It is thus hard to believe that Pakistan can affect or influence the world rice market.

The only explanatory variable used in the demand and supply functions is the domestic price of Basmati rice. The inclusion of other variables such as consumer income, rainfall, export tax, and price of other varieties of rice could also be considered.

The basic diagnostic statistics such as $R^2$, $t$-ratios, etc., are not provided, so that we do not know how reliable the results are. For example, if the goodness of fit is poor, then there is no need to proceed any further. Similarly, if the coefficients are not significant, then there is no point in relying on them.

*[Other discussion of this paper and the authors' reply appear on page 258.]*

# Cross-Country Comparison of Agricultural Performance Using Different Data Sets

*Asatoshi Maeshiro and Jerome C. Wells*[1]

**Abstract:** This paper examines the consistency of two series of aggregate output data in comparing the agricultural performance of 60 developing countries. Employing two approaches that take into account the fact that deepening of intermediate goods takes place in agriculture as development proceeds, it is found that, in less than half the country cases tested, the data on gross agricultural output compiled by the US Department of Agriculture and the series on value-added in agriculture produced by the World Bank are consistent with each other and with the expectation of deepening of intermediate goods. The World Bank's aggregates provide the more optimistic reading of performance of the two series, but neither series indicates that a majority of developing countries have performed adequately in terms of the test of agricultural performance suggested by Johnston and Mellor in 1961.

## Introduction

Some 30 years ago, in their classic statement on the role of the agricultural sector in economic development, Johnston and Mellor (1961) set forth a simple formula for estimating the growth of domestic demand for agricultural production as development proceeds. Their formula is:

(1) $\quad D^* = P^* + e_y Y^*$

where $D^*$ = the growth rate of domestic demand, $P^*$ = the growth rate of population, $e_y$ = the income-elasticity of demand for agricultural products, and $Y^*$ = the growth rate of per-capita income.[2] Equation (1) provides a means of estimating demand for agricultural production across developing economies and also suggests a test of adequate agricultural performance. If the growth rate of domestic agricultural output ($A^*$) equals the Johnston-Mellor growth rate of demand (i.e., if $A^* = D^*$), then domestic production growth will be sufficient to leave a country's external agricultural orientation (the degree to which it is a net importer or exporter of agricultural produce) the same. This condition, which is considerably more general than the notion of self-sufficiency,[3] is denoted as Johnston-Mellor adequacy. When the Johnston-Mellor adequacy test is used to examine the agricultural performance of 84 developing economies since 1950, the results suggest that the "typical" country has failed to achieve such adequacy (Wells, 1989).

This finding depends on the accuracy of the measures of aggregate agricultural production by country used for comparison. Such measures of aggregate agricultural output are compiled by several different agencies, including the US Department of Agriculture (USDA), FAO, and, as part of its coverage of national accounts aggregates, the World Bank.

Although the sets of agricultural aggregates provided by these agencies are compiled independently, each depends ultimately on field surveys of crop and livestock activities conducted by agricultural ministries or offices of statistics of the individual developing countries. Each agency adjusts the basic data in different ways to prepare aggregate measures of agricultural output.

Questions regarding the reliability of the different series have been raised for some time and deal with the methods by which aggregates are prepared (Farnsworth, 1961), discrepancies among the major reporting sources (Paulino and Tseng, 1980; and Wells, 1988), and with the interpretation of the tenuous record of Africa's agricultural performance in the past two decades (e.g., World Bank, 1981; and Berry, 1984). Although these problems have led to some caution in the use of the data,[4] there is little alternative for those dealing with the overall record of agricultural growth in developing countries.

Recognizing these problems, the purpose in this paper is to examine the consistency of agricultural production aggregates derived from two major data sources: the USDA series and that produced by the World Bank as part of its comparative survey of national accounting aggregates. To do this, several tests are employed that recognize a fundamental conceptual difference between the two data sources: the World Bank's series records, in constant prices, the value added in the agricultural sector while the USDA series records gross agricultural output; i.e., the total value of agricultural output without deduction from intra- or inter-industry inputs.[5] The tests derive from the differences expected in estimates of both long-run growth and year-to-year fluctuations in series measuring the value added in agriculture as opposed to gross agricultural output.[6]

## Deepening of Intermediate Goods and Estimates of Output over Time

The approach used derives from the well-known proposition that, as economic development proceeds, inputs to agriculture from other sectors of the economy (including agriculture) increase as a share of total output. This deepening of the intermediate goods component implies that measures of the growth of value-added in agriculture ($V^*$) will differ from measures of the growth of gross agricultural output ($A^*$).

Let $V_t$ = value added in agriculture, in constant prices, in time $t$; and $A_t$ = the value of gross agricultural output in time $t$, also at constant prices.[7]

Next let $v_t = V_t/A_t$. The expectation is that $v_t$ declines over time with economic development as the weight of intermediate inputs into agriculture increases. Denoting the exponential growth rate of a variable (estimated over the time series available) with an appended ($^*$), it is therefore concluded that:

(2) $\quad v^* = V^* - A^* < 0$, for all cases of developing economies where $A^* > 0$

A second and more complex test employing the same concept of deepening of intermediate goods can be developed using the same notation as above, and, defining $\beta_t$ as the share of intermediate inputs in $A_t$, gives:

(3) $\quad V_t = A_t - \beta_t A_t + \varepsilon_t = (1-\beta_t)A_t + \varepsilon_t$, where $\varepsilon_t$ is an error term

Reducing these to indices implies that:

(4) $\quad V_0 \dfrac{V_t}{V_0} = (1-\beta_t)A_0\dfrac{A_t}{A_0} + \varepsilon_t$

(5) $\quad \dfrac{V_t}{V_0} = (1-\beta_t)\dfrac{A_0}{A_0}\dfrac{A_t}{A_0} + \varepsilon'_t$

Thus:

(6) $\quad \dfrac{V_t}{V_0} = \dfrac{A_0}{V_0}\dfrac{A_t}{A_0} - \beta_t\dfrac{A_0}{V_0}\dfrac{A_t}{A_0} + \varepsilon'_t$

Now, let $h(t)$ be some generalized but positive and increasing function of time and reformulate (6) as:

(7) $\quad \dfrac{V_t}{V_0} = \gamma\dfrac{A_t}{A_0} - \eta h(t)\dfrac{A_t}{A_0} + \varepsilon'_t$

The expectation is that $\gamma > 1$ and that $\eta < 0$. We are experimenting with different forms of $h(t)$ and here use the value $h(t) = (t^\lambda - 1)\lambda$ to estimate Equation (7).

## Year-to-Year Fluctuations

The third means of comparing the USDA and World Bank data involves the consistency of year-to-year movements in output in each of the 60-country sample of LDCs. Here, it is assumed that the observed changes in output from one year to the next reflect a predicted component ($d\hat{V}_t$ or $d\hat{A}_t$) arising from returns normally expected to inputs and a variable component $\delta_t$ reflecting disturbances in the production function due to weather and other factors.

The expected change in value-added ($d\hat{V}_t$) is related to the expected change in gross agricultural output, $d\hat{A}_t$, by:

(8) $\quad d\hat{V}_t = (1-\beta_t)d\hat{A}_t - d\beta_t\hat{A}_t$

where $d\beta_t > 0$ by deepening of intermediate goods, and the actual change in value-added, $dV_t$, is given by:

(9) $\quad dV_t = d\hat{V}_t + \delta_t = (1-\beta_t)d\hat{A}_t - d\beta_t\hat{A}_t + \delta_t$

The actual change in gross agricultural output, $dA_t$, is given by:

(10) $\quad dA_t = d_t + \delta_t > dV_t$, because $d\hat{A}_t > d\hat{V}_t$

Hence, we predict that when $dV_t > 0$, $dA_t$ should also be $> 0$.

## Preliminary Results

Selected results are presented in Tables 1 and 2 and indicate that for most of the countries, the USDA and World Bank data are inconsistent with each other given the postulate of deepening of intermediate goods in agriculture.

In Table 1, "consistent" reflects cases that meet the first and second tests and essentially reflect deepening of intermediate goods taking place. Col. 1 shows the number of country cases where Equation (7) holds at least 80 percent of the time.[8] Employing the 80-percent measure for the third (year-to-year) test, 12 of the sample of 60 pass all three tests of consistency. Four additional countries pass at least one of the long-run consistency tests—i.e., $\eta$ from Equation (7) is negative and significant or $v^*$ is negative—as well as the short-run tests, but for at least 44 countries in the sample, the World Bank and USDA estimates exhibit neither long- nor short-run patterns of agricultural growth consistent with being taken from the same data series in a case where deepening of intermediate goods is occurring.

Although the results of the first and second tests are highly correlated—only 4 of the 60 cases show discrepancies between these two tests—the short- and long-run tests act quite independently. About half the countries (25–33) pass the long-run test, and 31 countries pass the short-run test. The long- and short-run results vary by country so that overall only about a quarter of the sample passes both short- and long-run tests that the series are similar.

Table 2 provides an appraisal of the impact of differences in USDA and World Bank measures of agricultural performance by comparing the numbers and proportions of countries that meet various yardsticks of Johnston-Mellor adequacy used in previous study (Wells, 1989).

For the 60-country sample, the level of performance implied by the World Bank series is considerably more optimistic than that implied by the USDA series.[9] A similar result is found in another test of estimated growth rates using the USDA, FAO, and World Bank series for 13 African economies (Wells, 1988).

Table 1—Distribution of 60 Developing Countries

| | $dV > 0$ and $dA > 0$ | | Subtotal |
|---|---|---|---|
| | $\geq 80\%$ | $< 80\%$ | |
| Low-income countries: | | | |
| $v^* < 0$ and coefficient $b$: | | | |
| Consistent[1] | (a) 4 | (b) 5 | 9 |
| Unclear | | (c) 5 | 5 |
| Not consistent | (d) 3 | (e) 4 | 7 |
| Subtotal | 7 | 14 | 21 |
| Lower-middle-income countries: | | | |
| $v^* < 0$ and coefficient $b$: | | | |
| Consistent | (f) 3 | (g) 5 | 8 |
| Unclear | (h) 2 | | 2 |
| Not consistent | (i) 6 | (j) 7 | 13 |
| Subtotal | 11 | 12 | 23 |
| Upper-middle-income countries: | | | |
| $v^* < 0$ and coefficient $b$: | | | |
| Consistent | (k) 5 | (l) 2 | 7 |
| Unclear | (m) 2 | | 2 |
| Not consistent | (n) 6 | (o) 1 | 7 |
| Subtotal | 13 | 3 | 16 |
| Combined total: | | | |
| $v^* < 0$ and coefficient $b$: | | | |
| Consistent | 12 | 13 | 25 |
| Unclear | 4 | 4 | 8 |
| Not consistent | 15 | 12 | 27 |
| Subtotal | 31 | 29 | 60 |

Notes: The two series appear consistent when $v^* < 0$ and $\eta$ (Equation 7) is significant. If one of these conditions does not hold, the country is coded "unclear," and if neither holds, the country is coded "not consistent." (a) India, Niger, Malawi, and Pakistan; (b) China, Bénin, Burkina Faso, Senegal, Tanzania, and Zaire; (c) Bangladesh, Ethiopia, Sudan, and Uganda; (d) Burundi, Kenya, and Sri Lanka; (e) Myanmar, Ghana, Sierra Leone, and Togo; (f) Jordan, Tunisia, and Zimbabwe; (g) Côte d'Ivoire, Ecuador, El Salvador, Nigeria, and Zambia; (h) Indonesia and Morocco; (i) Egypt, Paraguay, Peru, Philippines, Thailand, and Turkey; (j) Bolivia, Cameroon, Colombia, Costa Rica, Dominican Republic, Jamaica, and Nicaragua; (k) Barbados, Greece, South Korea, Panama, South Africa, and Venezuela; (l) Argentina; (m) Malaysia and Mexico; (n) Algeria, Brazil, Chile, Cyprus, Uruguay, and Yugoslavia; and (o) Trinidad and Tobago.

Table 2—Measures of Agricultural Performance: USDA versus World Bank Data

| Per-Capita Growth Rates | Number of Countries | | Percent | |
|---|---|---|---|---|
| | USDA | World Bank | USDA | World Bank |
| Low-income countries: | | | | |
| < 0 | 13 | 14 | 61.9 | 66.7 |
| 0–0.008 | 7 | 4 | 33.3 | 19.0 |
| 0.008–0.012 | 0 | 1 | 0.0 | 4.8 |
| > 0.012 | 1 | 2 | 4.8 | 9.5 |
| Total | 21 | 21 | 100.0 | 100.0 |
| Lower-middle-income countries: | | | | |
| < 0 (a) | 8 | 10 | 34.8 | 43.5 |
| 0–0.008 | 10 | 4 | 43.5 | 17.4 |
| 0.008–0.012 | 2 | 2 | 8.7 | 8.7 |
| > 0.012 | 3 | 7 | 13.0 | 30.4 |
| Total | 23 | 23 | 100.0 | 100.0 |
| Upper-middle-income countries: | | | | |
| < 0 (a) | 6 | 3 | 37.5 | 18.8 |
| 0–0.008 | 4 | 6 | 25.0 | 37.5 |
| 0.008–0.012 | 0 | 0 | 0.0 | 0.0 |
| > 0.012 | 6 | 7 | 37.5 | 43.8 |
| Total | 16 | 16 | 100.0 | 100.0 |
| Total 60 developing countries: | | | | |
| < 0 (a) | 27 | 27 | 45.0 | 45.0 |
| 0–0.008 | 21 | 14 | 35.0 | 23.3 |
| 0.008–0.012 | 2 | 3 | 3.3 | 5.0 |
| > 0.012 | 10 | 16 | 16.7 | 26.7 |
| Total | 60 | 60 | 100.0 | 100.0 |

# Conclusions

The results are not encouraging to those who would like to use the international record of agricultural production for precise comparisons of country performance over time. The discrepancies found between the USDA and World Bank series appear to be very little associated with conceptual differences in the measures being used, and they are so numerous that the method used here does not give much promise of identifying a few country cases where the most serious inconsistencies appear. With discrepancies of one sort or another occurring in over half the country cases, little can be done to identify the source of these disparities on the basis of a few crucial country studies.

Finally, it must be noted that even results indicating far more consistency between the World Bank and USDA estimates would not necessarily imply that these estimates were correct. As budgets for the type of field work that underlies good agricultural reporting are cut in Third-World countries and in the international agencies that have in the past supported development of agricultural statistics, the developers of the international statistical base must rely increasingly on indirect estimates and attempts to develop consensus estimates. That indirect methods are inadequate has long been recognized (Farnsworth, 1961), but the types of effort needed to improve the international statistical base have not received the priority in funding necessary to develop a fully reliable and useful data base.

## Notes

[1]University of Pittsburgh.

[2]Throughout this paper, ($^*$) indicates the (exponential) growth rate of a variable. Here also, $e_y$ is presumably a pattern variable ranging (per Johnston and Mellor's initial estimates) from 0.8–0.9 in low-income countries to 0.2–0.3 in developed economies.

[3]Self-sufficiency implies that $A = D$, thus presuming that each country has no starting point in development as a net importer or exporter of agricultural product (Wells, 1989, pp. 167–169).

[4]"Needless to say, agricultural production estimates for less developed countries need to be treated with reserve, although we have confidence in the broad trends that they reveal" (Mellor and Johnston, 1984, p. 538).

[5]The FAO series also measures gross agricultural output.

[6]An alternative approach, disaggregating the production indices into their component parts and prices, is possible when comparing the USDA and FAO series (Wells, 1988) but cannot be used with World Bank data because an aggregate amount for the value added in agriculture is all that is reported.

[7]For comparability between the USDA and World Bank series, indices of $V_t/V_O$ and $A_t/A_O$ are used, based on $t = 0$ for 1980.

[8]That is, for 80 percent of the years where $dV_t > 0$, $dA_t$ is also $> 0$. The 80-percent standard is arbitrary and can be replaced by a formal test of $H_O: d\hat{A} - d\hat{V} \leq 0$ versus $H_A: d\hat{A} - d\hat{V} > 0$.

[9]The test of whether per-capita agricultural output grows at 1.2 percent corresponds to matching the growth of demand where per-capita output is seen to be growing at 2 percent and $e_y$, is 0.6. The 0.8-percent per-capita growth corresponds to expected demand growth if per-capita income is growing at 1 percent in a country where $e_y$ is 0.8. The assumption of deepening of intermediate goods would imply that $A^*$ is expectedly greater than $V^*$, but we have no basic estimates of $v^*$ to allow us to convert the World Bank's $V^*$s into corresponding $A^*$s.

## References

Berry, S.S., "The Food Crisis and Agrarian Change in Africa," *African Studies Review*, Vol. 27, No. 2, 1984, pp. 59–112.

Farnsworth, H.C., "Defects, Uses, and Abuses of National Food Supply and Consumption Data," *Food Research Institute Studies*, Vol. 2, No. 3, 1961, pp. 179–203.

Johnston, B.F., and Mellor, J.W., "The Role of Agriculture in Economic Development," *American Economic Review*, Vol. 51, No. 4, 1961, pp. 566–593.

Mellor, J.W., and Johnston, B.F., "The World Food Equation: Interrelations among Development, Employment, and Food Consumption," *Journal of Economic Literature*, Vol. 22, No. 2, 1984, pp. 531–574.

Paulino, L., and Tseng, S.S., "A Comparative Study of the FAO and USDA Data on Production, Area, and Trade of Major Food Staples," Research Report No. 19, International Food Policy Research Institute, Washington, D.C., USA, 1980.

Wells, J.C., "On the Agricultural Performance of Developing Nations," *Food Research Institute Studies*, Vol. 21, No. 2, 1989, pp. 165–191.

Wells, J.C., "On the Measurement and Interpretation of Agricultural Performance in 13 African Economies, 1950–85," Working Paper No. 236, Department of Economics, University of Pittsburgh, Pittsburgh, Pa., USA, 1988.

World Bank, *Accelerated Development in Africa: An Agenda for Action*, Washington, D.C., USA, 1981.

---

## Discussion Opening—*Habibullah Khan* (National University of Singapore)

Maeshiro and Wells illustrate the age-old problem of inconsistency in economic data. Although the main focus of the paper is on agricultural statistics, various issues pertaining to economic data limitations are relevant to the discussion.

Ever since empirical research in economics began, there has been concern about the lack of precision in economic statistics and the consequences for the validity of econometric forecasts. About 40 years ago, Morgenstern raised serious doubts about the quality of many economic data series and asked whether such data were good enough for the purposes for which economists and econometricians were using them. He viewed errors in economic statistics as an expression of imperfection and of incompleteness in description and claimed that such errors might come from at least eight different sources.

There has been very little coherent response to Morgenstern's criticisms. Only recently, the Harvard econometrician Griliches made a thorough examination of economic data problems and gave four responses to Morgenstern's criticisms: (1) The data are not as bad as suggested. There has been significant progress both in the quality and quantity of the available data in the past decades. (2) The data are poor, but it does not matter. Empirical economists have over generations adopted the attitude that having bad data is better than having no data at all, that their task is to learn as much as possible about how the world works from these available poor data. (3) The data are bad, but we have learned how to live with them and adjust for their foibles. (4) The data available are all there are.

I am inclined to adopt a similar attitude towards economic data as Griliches. The paper under review has unnecessarily expressed pessimism about the quality of agricultural data and their future improvement. The finding that, in less than half the country cases, the two sets of output data (USDA versus World Bank) are consistent with each other is not surprising at all. First of all, the result is based upon the so-called "deepening of intermediate goods" hypothesis, which is yet to be firmly established. Because of complexities in the nature of the development process, any generalization in economic development is bound to be tentative. Secondly, the agricultural data are likely to contain relatively more errors due to well-known factors such as presence of large-scale subsistence farming, illiteracy of rural population, fluctuations caused by climatic or natural factors, and so on.

Finally, in order to reduce the sensitivity of the results to data errors, multiple indicators of agricultural performance should be used, rather than a single output indicator. This would make the estimated results much more stable.

*[Other discussion of this paper and the authors' reply appear on the following page.]*

**General Discussion**—*Sophia Wu Huang, Rapporteur* (US Department of Agriculture)

In response to the opener's question about the use of the difference between procurement price and export price to represent an *ad valorem* export tax, Chishti and Schmidt replied that the average ratio between procurement and export prices has been 0.47, which means that 53 percent of the export price has been going to the government while only 47 percent to the farmers. This is an export tax and not just handling and transport charges; according to rough estimates, handling and transport charges may range from 5 to 8 percent of the export tax amount. Regarding the validity of large country assumption, Basmati rice is a distinct rice variety with special aroma and good cooking quality. Basmati rice is usually sold at a price three times that of IRRI rice, while its export price has been more than three times of and twice that of Thai rice (5 percent broken) and US long-grain rice (Zenith No. 2), respectively. Thus, Basmati rice has a special edge over the other rice varieties; and it would be inappropriate if a small country assumption were used for its export demand. The opener's impression of using price as the only explanatory variable is incorrect. The demand and supply functions were fully specified, but the detailed specification could not be given in the paper due to the page limitation. The basic diagnostic statistics could not be given for the same reason. Regarding a question about the significance of the coefficient of export price of world demand for Pakistani Basmati rice, the coefficient is significant at the 5-percent level.

In response to the opener, Maeshiro and Wells agreed that he correctly invokes Morgenstern's curse and the antidote for it from Griliches to defend them from the charge that economic data are meaningless. Their intent is not to express Morgenstern-like shock as much as to ask where discrepancies in the data make a serious difference to the reading of performance. Their standards are fairly forgiving. They are dealing with directions of change over a 25-year period; so the 50 percent or so of inconsistences do indicate problems. They feel that deepening of intermediate goods is virtually received doctrine—it amounts to saying little more than that the proportion of fertilizer, chemical, etc. inputs to output increases over time. In reply to comments on the inconsistency between FAO and World Bank data series, they reply that the inconsistency is caused by factors such as different national offices collecting data, different handling of subsistence estimates, and different data coverage. As budgets for data collection are reduced, the agencies are forced to resolve inconsistencies via consensus estimates. Even perfect consistency would not allow the assumption of truly reliable data.

Participants in the discussion included N. Alexandratos (FAO), P. Dixit (US Department of Agriculture), H. Tsujii (Kyoto University), and C.L.J. van der Meer (Agricultural Research Council, Netherlands).

# Technical Efficiency of Melon Farms under the Marketing Strategy of Agricultural Cooperatives

## Takafusa Shimizu[1]

**Abstract:** The purpose of this paper is to measure the technical efficiency of farms on the basis of data envelopment analysis (DEA) and to present an analytical method of farm diagnosis using measurements obtained for melon farms in Japan. The paper first describes the objectives of melon farms operating under the marketing strategy of agricultural cooperatives and explains DEA, which measures the technical efficiency of such farms. It then uses this analytical method to measure technical efficiency for each farm and demonstrates how technical efficiency is dependent not only on the management performance of production of melons and other crops but also on the number of full-time farm workers. In addition, the effect of management performance on technical efficiency is analysed quantitatively for each size group of full-time farm workers. The paper also illustrates how DEA can be used for farm diagnosis when the management performance of inefficient melon farms is compared with that of efficient farms. DEA is able to indicate some efficient farms as the optimum solution that should be treated as the model for the improved management of inefficient farms. Data are drawn from a sample of 91 melon farms in Choshi, Chiba Prefecture.

## Introduction

Fruit have tended to be in a state of overproduction in Japan since 1980, due to factors that include a slump in demand, the conversion from rice cultivation to other crops, and an increase in yield. The oversupply situation of melons had begun to ease, thanks to an increase in the disposable income of consumers and high price elasticity, but recently the dramatic increase in supply has resulted in a levelling-off of the price received by producers. As a result, melon producers in different producing areas have now started to compete strongly with one another in order to achieve higher prices.

Agricultural cooperatives in melon-growing regions sell most of the fruit and vegetables produced under their control to wholesalers on behalf of producers. In the case of melons, the cooperatives devise a marketing strategy to maximize the total income of melon producers and advise growers to produce and market according to this strategy. The melon marketing strategy of agricultural cooperatives calls for the regular supply to wholesale markets of a large volume of melons that are of the quality and size required to meet buyers' expectations. Consequently, producers are subject to the constraints of the marketing strategy of agricultural cooperatives and must set a farm management goal of maximizing farm income while determining the timing of melon cultivation and the marketing of superior quality produce.

Melon farms are typically mixed farms that also produce other crops, such as rice and vegetables, so that other crops compete with melon production in the use of farm resources. Income from melon production in mixed farms must be maximized subject to the constraints of increasing income from other crops while giving due regard to the competitive relationship with these crops.

Melon farms have multiple goals in that they are subject to both the marketing strategy of agricultural cooperatives and the constraints of the production of other crops. As a result, the technical efficiency of melon farms must be judged in accordance with standards that comprehensively satisfy the multiple output indices for representing differing characteristics. These types of judgement standards are particularly useful for agricultural cooperatives in farming regions that manage agriculture for the region while seeking to improve the efficiency of individual farmers.

The objective of this paper is to indicate an appropriate management diagnosis method for this situation, using data obtained from the measurement of technical efficiency on the basis of data envelopment analysis (DEA), by focusing on the objective of melon farms, which is to maximize multiple outputs under the marketing strategy of agricultural cooperatives. Agricultural cooperatives will be able to apply this analytical method to various farming types, including melon growing, to support farmers' decision making.

The paper looks specifically at farms under the control of the Choshi Agricultural Cooperative in Chiba Prefecture that produce the *Ams* type of melon. This cooperative grades the melons strictly by size and quality to distinguish them from those grown in other areas, and also promotes melon cultivation under plastic to extend the shipment period to the market, which at present is concentrated in July. The data analysed consist of records for 1988 obtained from 91 farms for which this cooperative ships the melons.

## Estimation of Technical Efficiency Based on DEA

The traditional concept of technical efficiency can be applied to farms only when they aim at a single goal (Farrell, 1957; and Timmer, 1970) and cannot be measured when the farm seeks to satisfy multiple goals. Banker, Charnes, and Cooper developed the DEA method that measures the technical efficiency of the decision-making unit when there are multiple outputs and inputs (Banker, 1984; and Banker, Charnes, and Cooper, 1984). Their analytical methods are explained here using the example of melon farms in Choshi.

Mixed farms whose melon-marketing strategy is planned by agricultural cooperatives allocate their cultivated lands between planted areas of melon $(X_1)$ and planted areas of other crops $(X_2)$, use these inputs and the labour of full-time farm workers $(X_3)$, and adopt the behaviour that maximizes the total outputs of melons $(Y_1)$, the total outputs of other crops $(Y_2)$, the percentage of high quality quantity of total marketed melons $(Y_3)$, and the percentage of early quantity of total marketed melons $(Y_4)$.

Capital goods are not included in the inputs described above. It is assumed that capital goods are in proportion to each of the outputs. Under this assumption, which approximately expresses the actual situation, the effects of other inputs on each of the melon outputs are not dependent on the inputs of capital goods. The behaviour of melon farms according to this assumption can be formulated as the fractional programming of maximizing $h_{j0}$ in objective function (1), subject to the constraints of (2), (3), and (4):

$$(1) \quad h_{j0} = \frac{\sum_{r=1}^{4} u_r Y_{rj0}}{\sum_{i=1}^{3} v_i X_{ij0}}$$

$$(2) \quad \frac{\sum_{r=1}^{4} u_r Y_{rj}}{\sum_{i=1}^{3} v_i X_{ij}} \leq 1 \quad (j=1, 2, ..., 91)$$

$$(3) \quad u_r > 0 \ (r=1, 2, 3, 4)$$

$$(4) \quad v_i > 0 \ (i=1, 2, 3)$$

The fractional programming is converted into linear programming which maximizes (5), subject to the constraints of (6), (7), (3), and (4):

$$(5) \quad Z_{j0} = \sum_{r=1}^{4} u_r Y_{rj0}$$

$$(6) \quad \sum_{i=1}^{3} v_i X_{ij0} = 1$$

(7) $\sum_{r=1}^{4} u_r Y_{rj} - \sum_{i=1}^{3} v_i X_{ij} \leq 0$  $(j=1, 2, ..., 91)$

If the optimum solution $Z_{j0}^*$ of the linear programming is $(u_r^*, v_i^*)$, it equals the optimum solution of fractional programming $h_{j0}^*$ and represents technical efficiency. Consequently, farming is most efficient when $Z_{j0}^* = 1$ and is inefficient the closer $Z_{j0}^*$ is to 0.

## Measurement of Technical Efficiency in Melon Farms

Measures of the technical efficiency of the 91 melon farms can be obtained from the optimum objective function of the linear programming described in the previous section. The second column of Table 1 shows the level of technical efficiency of the 91 farms. This is shown as an average of 0.81, centred on the range 0.7–0.8 and 1.0 but covering a wide distribution from 0.5 to 1.0.

Table 1—Choshi Melon Farms Measured in Terms of Technical Efficiency

| | Total | Number of Full-Time Workers | | |
|---|---|---|---|---|
| | | 1 | 2 | 3 or more |
| Number of samples | 91 | 28 | 34 | 29 |
| Grouped by measure of technical efficiency | *Percent* | | | |
| 0.5–0.6 | 8.8 | 3.6 | – | 3.4 |
| 0.6–0.7 | 17.6 | 3.6 | – | 10.4 |
| 0.7–0.8 | 26.4 | 28.6 | 8.8 | 10.4 |
| 0.8–0.9 | 13.2 | 10.7 | 17.7 | 27.6 |
| 0.9–1.0 | 7.7 | 10.7 | 14.7 | 3.4 |
| 1.0 | 26.4 | 42.8 | 58.8 | 44.8 |
| Average | 0.8076 | 0.8802 | 0.9442 | 0.8727 |
| Standard deviation | 0.1503 | 0.1312 | 0.0870 | 0.1411 |

Notes: Based on total sample farms in Choshi. Number of full-time workers based on farms grouped by full-time workers.

The technical efficiency of farms is affected by management performance in production of melons and other crops and by the number of full-time farm workers. If technical efficiency $T$ is represented as a dependent variable, the high quality of total marketed melons as $Q$ percent, the total quantity of early melons marketed as $E$ percent, melon yield per are as $M$ kg, the yield of other crops per are as $P$ ¥1,000, and the number of full-time workers $N$ as independent variables, the multiple regression equation is computed as follows (figures in parentheses represent $t$-values):

(8)  $T = 0.0029Q + 0.0042E + 0.0249M + 0.0008P - 0.0311N + 0.4032$
       (4.0790)    (5.2914)    (4.5465)    (2.6640)   (−2.2651)            $R^2 = 0.5491$

The coefficient of determination is 0.55, which is not at all large, but all the regression coefficients are significant at the 5-percent level. The effect of the independent variables on the measures of technical efficiency in terms of elasticity coefficients is greatest in the "melon yield per are" category (0.30) and smallest in the "percentage of early melons marketed" category (0.06).

Of particular interest is the fact that the higher the number of full-time workers, the lower the level of technical efficiency. There are two reasons for this. One is that on melon farms, as the number of full-time workers increases, the number of part-time workers falls or remains unchanged. Nor does the ratio of fixed equipment increase. The second reason is that the percentages of high quality and early crop melons in total melons marketed obtained to measure technical efficiency do not increase in proportion to the number of full-time workers.

The measurements of technical efficiency for full-time workers are shown in Table 1 (Cols. 3–5). In all groups, the number of efficient farms with a technical efficiency of 1.0 is increasing and the number of inefficient farms with a technical efficiency of 0.7 or less is decreasing. This is due not to measurement in terms of absolute efficiency on farms but to measurement in terms of relative efficiency within each group. This indicates that differences in technical efficiency are small among farms in the group comprising two full-time workers and large in the groups comprising one or three-or-more full-time workers. The results of measurements of the multiple regression equations, which make technical efficiency a dependent variable and (excluding the number of full-time workers) make other factors independent variables, are shown in Table 2 in order to clarify the factors determining technical efficiency in terms of number of full-time workers.

Table 2—Regression Relationships between the
Measure of Technical Efficiency and Its Factors

| Number of Full-Time Workers | Linear Regression Coefficients (t-values) | | | | $R^2$ |
|---|---|---|---|---|---|
| | Q | E | M | P | |
| 1 | 0.0048** (3.6690) | 0.0033* (2.5977) | 0.0072** (3.1201) | 0.0005 (1.0696) | 0.6217 |
| 2 | 0.0027** (3.6888) | 0.0016 (1.7016) | −0.0012 (−1.0025) | 0.0006 (2.0098) | 0.3967 |
| 3 or more | 0.0015 (1.1730) | 0.0038** (3.0455) | 0.0028 (1.6225) | 0.0016* (2.1435) | 0.5378 |
| | Elasticity at Average Samples | | | | |
| 1 | 0.175 | 0.032 | 0.361 | 0.042 | |
| 2 | 0.095 | 0.021 | 0.058 | 0.053 | |
| 3 or more | 0.054 | 0.048 | 0.143 | 0.141 | |

Note: $Q$ = percent of high quality melons marketed; $E$ = percent of early melons marketed; $M$ = Melon yield per are; and $P$ = Yield of other crops per are. *Significant at 5-percent level. **Significant at 1-percent level.

A comparison of the elasticity coefficients by size group, with particular attention to the significance of the regression coefficients, shows that, in the size group of one full-time worker, the effect of melon yield per are is at the maximum, followed by the effect of the high quality of total melons marketed, and the effect of the total quantity of early melons marketed is at the minimum. The yield per are of other crops displays no significant effect. In the size group of three-or-more full-time workers, the effect of the yield per are of other crops is at the maximum, followed by the effect of the total quantity of early melons marketed; melon yield per are and the high quality of total melons marketed do not display a significant figure. In groups of two full-time workers, the difference in technical efficiency among farms is small, so that the coefficient of determination is small and only the effect of the high quality of total melons marketed is significant, which is at the maximum. The effect of this group by factors shows a middle character between the group of one full-time worker and the group of three-or-more full-time workers.

The means of raising technical efficiency based on this analysis differ among groups with different numbers of full-time workers. In the group with one full-time worker, labour is relatively scarce so that technical efficiency is raised by means of increasing the yield per are of melons grown in tunnels and marketed in July and by improving the quality of the melons. However, this has little effect on increasing the yield per are of other crops. In comparison, in groups with three-or-more full-time workers, labour is relatively abundant, so that technical efficiency is raised by means of increasing the quantity marketed of early melons cultivated in plastic houses and the yield of other crops, but the effects of increasing the yield per are and high quality of melons are insignificant. Agricultural cooperatives must change the method of inducing improvements in farm management in relation to the number of full-time workers.

## Diagnosis of Farm Management Performance

The technical efficiency of melon farms shows a wide difference even among farms with the same number of full-time workers. The method of improving technical efficiency differs among groups. The procedure and method for diagnosing management performance are illustrated using the example of farm G from among the class with the lowest efficiency.

The inputs and outputs of inefficient farms are changed into those of a theoretical optimum farm by computing the weighted sum of inputs and outputs of some efficient farms based on the optimum solution of DEA (Tone, 1988). Table 3 shows the inputs and outputs of farm G and three efficient farms, A, B, and C, composing its solution and a theoretical optimum farm G'. The inputs and outputs in G' are the figures arrived at by the sum of the products in A, B, and C multiplied by 0.00816, 0.53243, and 0.43456, respectively. The diagnosed farm, according to those measurements, applies approximately the same inputs as in an actual situation and can be made more efficient by raising all the outputs of melon and other crops, the percentages of high quality, and total quantity of early melons marketed.

Table 3—Inputs and Outputs of the Diagnosed Farm G and the Efficient Farms Composing an Optimum Solution on that Farm

| | Diagnosed Farm | Efficient Farms Composing Optimum Solution on the Diagnosed Farm | | | Theoretical Optimum Farm |
|---|---|---|---|---|---|
| | G(0.668) | A(1.000) | B(1.000) | C(1.000) | G'(1.000) |
| Weight | | 0.00816 | 0.53243 | 0.43456 | |
| $X_1$ (a) | 60 | 70 | 30 | 100 | 60 |
| $X_2$ (a) | 110 | 65 | 155 | 62 | 110 |
| $X_3$(worker) | 3.0 | 3.0 | 3.0 | 3.0 | 2.9 |
| $Y_1$(million ¥) | 4.05 | 7.11 | 3.65 | 9.34 | 6.06 |
| $Y_2$(million ¥) | 8.67 | 7.57 | 16.41 | 9.63 | 12.98 |
| $Y_3$(percent) | 12.1 | 59.1 | 68.8 | 43.6 | 56.1 |
| $Y_4$(percent) | – | 38.0 | 1.0 | 32.8 | 15.1 |

Note: Figures in parentheses represent the measure of technical efficiency.

It is difficult to perceive precise methods of improving the diagnosed farm from the theoretical optimum farm in Table 3. The method of improvement can be clarified by investigating the technology of efficient farms B and C with a higher weight and surveying the record of the types of crops they grow and their melon production management. Table 4 shows a comparison of the actual results of types of crops and melon production in the diagnosed farm G with those of efficient farms B and C. Diagnosed farm G, for winter cropping, whether it uses B or C efficient farms as the model, is forced to avoid competition over labour when harvesting cabbage and planting melons by eliminating mixed cabbage cropping. For summer cropping, several differences occur in cropping types depending on whether B or C is used as the model. When B is used as the model, the diagnosed farm reduces the areas of melon under extensive tunnel cultivation and converts part of this to intensive cultivation in plastic houses to bring the shipment time forward while increasing area planted to the more labour-extensive watermelon to make more effective use of uplands. However, when C is used as the model, the farm stops growing watermelon and expands the planted area of melons, combining tunnel cultivation with plastic house cultivation to avoid competition over use of labour. The diagnosed farm can raise technical efficiency by selecting either of cropping systems B or C or a combination of both. The choice of which cropping system is to be adopted depends on decision making by the farmers themselves.

Table 4—Use of Cultivated Land and Management Performance of
Melon Crops in the Diagnosed Farm and Efficient Farms

| | Diagnosed Farm | Efficient Farms | | | | |
|---|---|---|---|---|---|---|
| | G (0.668) | B (1.000) | | C (1.000) | | |
| Size of cultivated land (a) | 170 | 185 | | 162 | | |
| Planted area of: | | | | | | |
| Melon (a) | 60 | 30 | | 100 | | |
| Watermelon (a) | 20 | 100 | | – | | |
| Cabbage (a) | 120 | 100 | | 25 | | |
| Radish (a) | 120 | 70 | | 70 | | |
| Other crops (a) | 50 | 15 | | 95 | | |
| Total (a) | 370 | 315 | | 290 | | |
| Melon crops: | | | | | | |
| Average yields per are (kg) | 235 | 331 | | 309 | | |
| Average prices sold (¥/kg) | 287 | 368 | | 302 | | |
| Field practice: | | | | | | |
| Planted area (a) | 22 | 18 | 20 | 10 | 28 | 25 |
| Month and day of seeding | 2/10 | 3/10 | 2/10 | 2/20 | 2/25 | 3/24 |
| Period from pollination to harvesting (days) | 58 | 49 | 58 | 56 | 56 | 51 |
| Number of seedlings per are | 91 | 67 | 69 | 72 | 69 | 56 |

Note: Under Field practice, only those fields seeded within the 10 February–31 March period are presented because of space limitations.

Problems exist in the management of melon production on diagnosed farm G in the case of both plastic house cultivation (seeded in February) and tunnel cultivation (seeded in March). In the former case, the seedlings are planted too densely. Consequently, it is necessary to bring forward the shipment time by planting seedlings sparsely and promoting their growth. In the latter case, the melon harvest date is forced forward by impeding the growth process, as can be estimated from the fact that the time from pollination is short.

The DEA can thus be seen to be useful in measuring technical efficiency and diagnosing management performance of farming systems that have multiple objectives. Farm diagnosis can obtain the desired effect by analysing technical efficiency based on DEA in combination with other information that represents technology indices and the management performance of crop production. The analytical method used in this study can be of practical use in supporting the decision making of individual farms and various other agricultural production organizations.

## Note

[1]Chiba University.

## References

Banker, R.D., "Estimating Most Productive Scale Size using Data Envelopment Analysis," *European Journal of Operations Research*, 1984, pp. 35–44.

Banker, R.D., Charnes, A., and Cooper, W.W., "Some Models for Estimating Technical and Scale Inefficiencies in Data Envelopment Analysis," *Management Science*, Vol. 30, 1984, pp. 1078–1092.

Farrell, M.J., "The Measurement of Productive Efficiency," *Journal of the Royal Statistical Society*, Series A, Vol. 120, 1957, pp. 253–290.

Timmer, C.P., "Using a Probabilistic Frontier Production Function to Measure Technical Efficiency," *Journal of Political Economy*, Vol. 79, 1970, pp. 776–794.

Tone, K., "Analytical Method of Firm Efficiency," *Communications of the Operations Research Society of Japan*, Vol. 33, 1988, pp. 191–198.

---

**Discussion Opening**—*Mária Sebestyén Kostyál* (Budapest University of Economic Sciences)

The paper provides a useful method for commercial cooperatives to help less efficient farms improve their competitiveness. Comments are related to three main questions: the method selected, some of the conclusions, and the role of agricultural cooperatives.

Data envelopment analysis (DEA) has been developed as a tool for analysing production data. It is an interesting initiative to use this method in the case of mixed farms, but great care has to be taken when using DEA in the case of agricultural production units. This is because of the high degree of uncertainty (as with the weather) influencing the results of agricultural production, and is also why the method is applied mainly to industrial production units.

The next concern is the time horizon used, since the analysis was done on the basis of one year. It would have been useful if the final analysis had used at least a three-year data base.

Since the method handles the inputs in a highly simplified way (land, without any differentiation, and labour are represented in the model), the results obtained must be used very carefully when resource allocation is the main question. It can be assumed that the interest of the producers is to gain the highest margin, under the given conditions and constraints, between the cost of the inputs used and the market values of products. To set up the best production strategy, more factors have to be taken into consideration than the method presented can handle. It would be most interesting to learn what kind of additional methods the author may suggest in order to provide a more comprehensive set of tools for farm-level decisions.

The conclusions include the interesting finding that the higher the number of full-time workers, the lower the level of technical efficiency. This statement needs careful analysis. The two reasons given in the paper are not totally convincing. First, if the number of full-time workers increases and the number of part-time farmers decreases, technical efficiency can decrease only in one case; when the performance of the full-time workers is much lower than that of the part-time ones. Second, it seems natural that, after employing more full-time workers, the ratio of fixed assets is not bound to rise. On the contrary, it might even fall if output is not expected to grow.

Since the category of melon farms is not defined, it is hard to see whether the paper deals only with those mixed farms where the land is used for melon production above a certain ratio (e.g., 40 percent) or on every farm where melons are grown at all. Fairly big differences can be envisaged in technical efficiency measures between farms where melon can be considered as a main product and those where melon production has a very low share.

A fuller description of the role of cooperatives is needed, especially their marketing strategy and how they make cooperatives follow it. Is the interest of the cooperative always the same as that of every single farm? The topic also offers an opportunity to discuss the types of activities that cooperatives should carry out to assist member farms in order to make them more competitive; for instance, the tools used by cooperatives to maintain production at the level that the market needs.

*[The author's reply appears on page 282.]*

# Cereal Import Demand in Developing Countries

*Terrence S. Veeman, Maxine Sudol,*
*Michele M. Veeman, and Xiao-Yuan Dong*[1]

**Abstract:** The major determinants of cereal import demand in 74 less-developed countries (LDCs) were analysed using an econometric cross-sectional model. Key explanatory factors included the level of income and degree of urbanization, financial capacity proxies, and domestic grain supply variables. A major innovation involved the analysis of the impact of income distribution on LDC cereal import demand in 1986 and 1987 for a more restricted sample of 23 nations. These developing countries exhibit a greater than proportional increase in cereal imports due to an increase in the income share of the poorest 40 percent of their populations. The inclusion of regional slope and intercept dummies in the cereal import demand model also provides improved results. High levels of government debt appear to have inhibited cereal imports in nations in South America but not in Asia and Africa. In all three continental regions, particularly Africa, there is a positive relationship between food aid and cereal imports. The model predicts cereal imports more satisfactorily for nations in Asia and South America than for those in Africa. Finally, the results support the view that improvements in income distribution in developing nations would considerably stimulate cereal imports.

## Introduction

There have been dramatic changes in the structure of the international grain trade in recent decades. Not only has the volume of grain trade increased, particularly in the 1970s, but also the import shares of the different socioeconomic regions have changed. Less-developed countries (LDCs) became the fastest growing import market segment, while developed country import markets declined significantly. Cereal imports into the LDCs increased by 5.6 percent per year between the early 1960s and the early 1980s, the LDC share of world cereal imports increasing from 36 to 46 percent in the process (Mellor, 1988). In the 1980s, however, there were concerns that slower economic growth and high levels of debt, which constrain the financial capacity of many LDCs, may have limited LDC grain imports. The relative importance of various import demand factors is assessed in this analysis through the development and testing of a cross-sectional model of import demand for cereals. This analysis includes two notable improvements over previous research (Morrison, 1984): the incorporation of dummy variables and an investigation into the effects of income distribution on cereal import demand.

## Model and Data

The factors affecting cereal import demand can be broadly categorized into four groups: development variables, that attempt to quantify the level, growth, and distribution of income and the degree of urbanization in a country; financial capacity variables, that measure a country's ability to afford imports; potential and actual domestic cereal supply, that measure the gap between demand and supply; and socioeconomic dummy variables, that quantify structural differences in import demand across countries. These four categories are included in the following single-equation import demand model:

$$(1) \quad CM = f(X_1, X_2, X_3, X_4)$$

where:
$CM$ = cereal imports
$X_1$ = vector of development variables ($GNP$, $rGDP$, and $URB$)
$X_2$ = vector of financial capacity variables ($LRES$, $AID$, $LDBT$, $TDS$, $X86$, $EXP$, and $LACN$)
$X_3$ = vector of domestic grain supply variables ($CP$, $FLUC$, and $DENS$)
$X_4$ = vector of intercept and slope dummy variables

Price variables are omitted because the analysis is cross sectional and prices are assumed to be fixed for the year (Christiansen, 1987, p. 5; and Morrison, 1984, p. 21). Table 1 contains a summary of the definitions and data sources of the various alternative proxy variables. The data are for the year 1986, with all lagged variables being from 1985. Per capita values are used in order to eliminate the influence of different country sizes from the data set. For the initial analysis, 74 LDCs are chosen from three continents (South America, Asia, and Africa), and from all income levels (low, medium, and high). All net cereal importers are included in the sample, with the exception of high-income oil exporters (Saudi Arabia, Kuwait, and United Arab Emirates), which are excluded as being atypical developing nations.

Table 1—Variable Definitions and Data Sources, LDCs

| Variable | Definition | Source[1] |
|---|---|---|
| POP | 1986 population | A (1988) |
| GNP | 1986 GNP per capita, $US/capita | A (1988) |
| rGDP | Average annual growth rate of GDP, 1980–86 | A (1988) |
| URB | 1985 percent urban population of total population | A (1988) |
| AID | Quantity of cereal food aid, kg/capita | A (1988) |
| LRES | 1985 gross international reserves, $US/capita | A (1987) |
| LACN | 1985 current account balance, $US/capita | A (1987) |
| LDBT | 1985 external public debt, outstanding and disbursed, $US/capita | A (1987) |
| TDS | 1986 total debt service on government debt, $US/capita | D (1987) |
| EXP | Average annual growth rate of merchandise exports, 1980–86 | A (1988) |
| X86 | 1986 value of merchandise exports, $US/capita | A (1988) |
| CM | 1986 gross quantity of cereal imports, kg/capita (SITC 041–046) | B (1987) |
| CP | 1985 quantity of cereal production, kg/capita | C (1987) |
| FLUC | Difference between 1985 and 1986 cereal production, kg/capita | C (1987) |
| DENS | 1986 population density on arable land, 1,000 persons/ha | C (1987) |
| DSA | Dummy variable for 20 South American countries | |
| DAS | Dummy variable for 18 Asian and Mid-Eastern countries | |
| DAF | Dummy variable for 36 African countries | |

[1]A: World Bank, *World Development Report*; B: FAO, *Trade Yearbook*; C: FAO, *Production Yearbook*; and D: World Bank, *World Debt Tables*, Vol. 2.

Data for the dependent variable, cereal imports, include concessional food aid imports as well as commercial cereal imports (Huddleston, 1984, pp. 13–14). Since food aid enters the regression as an independent variable, the preferred procedure would be to express cereal imports net of food aid. Unfortunately, cereal imports are measured on a calendar year basis, while food aid data are measured on a crop year basis (July–June). Therefore, the dependent variable, cereal imports, cannot be expressed net of food aid, which limits the explanatory power of the food aid variable (*AID*).

The intercept dummy variables *DSA*, *DAS*, and *DAF* divide the sample set on the basis of geography to account for factors such as general weather patterns, resource endowments, and cultural differences that may influence tastes and preferences across nations. In addition to these intercept dummies, slope dummy variables were also included in the analysis once the preliminary set of significant variables was identified.

## Results of the Cereal Import Demand Model

Equation 2 presents the final results of the preliminary model, which was estimated with a linear functional form using the statistical package SHAZAM, version 6.1. $t$-statistics appear in brackets; $t$-critical (2-tailed, $\alpha = 0.05$, 60 d.f.) = 2.000, and $t$-critical (2-tailed, $\alpha = 0.01$, 60 d.f.) = 2.660.

$$CM = -31 - 24DAF - 86DSA + 0.03GNP + 0.89URB + 1.11AID - 0.15CP +$$
$$\quad\ (2.04)\ \ (2.29)\qquad (7.26)\qquad (6.84)\qquad (3.65)\qquad (5.27)\qquad (3.76)$$

(2)

$$0.03LDBT - 0.24FLUC \qquad \text{adj. } R^2 = 0.83$$
$$(2.58)\qquad\ (2.15)$$

Of the two alternative income variables, $GNP$ was a significant explanatory factor in cereal imports, but average annual growth in income was not and is therefore omitted from the regression. Two of the financial capacity variables were significant, $AID$ (food aid) and $LDBT$ (lagged government debt). Contrary to expectations, the coefficient on the lagged debt variable is positive; i.e., countries with heavier loads of debt per capita tend to import more cereals. This factor is further explored in the next section of the paper. The alternative debt variable, total debt service ($TDS$), was also significant in separate regressions, but $LDBT$ explains more variation in cereal imports than does $TDS$.

It was initially surprising that $LRES$, the foreign exchange variable, is insignificant in the regression. Further investigation revealed that $LRES$ is significant, but only when the variable $GNP$ is omitted. When both $GNP$ and $LRES$ appear in the same regression, the coefficient on $LRES$ is insignificantly different from zero and has a counter-intuitive sign. This result is the consequence of strong, destructive collinearity between these two variables (discovered through testing using the procedure outlined by Belsley, Kuh, and Welsch, 1980). Since the variable $LRES$ is more adversely affected by the collinearity than is $GNP$, $LRES$ was dropped from the regression. The same destructive collinearity with $GNP$ also applies to $X86$, the value of merchandise exports; like $LRES$, $X86$ was dropped from the regression due to this. The other two finance variables, $LACN$ and $EXP$, are simply insignificant and were also dropped. The cereal production variables, lagged cereal production and production fluctuations, were significant explanatory factors in cereal imports; population density on arable land was not.

The geographical intercept dummies indicated that there are significant differences in the level of cereal imports by Asian, African, and South American countries.[2] Slope dummy variables were then introduced to test for significant regional differences in import response as measured by the independent variables. These included: $FLUC.AF$ (cereal production fluctuations in Africa), $FLUC.SA$ (production fluctuations in South America), $CP.AF$ (cereal production in Africa), $CP.SA$ (cereal production in South America), $AID.AF$ (food aid in Africa), $LDBT.AF$ (lagged debt in Africa), and $LDBT.SA$ (lagged debt in South America).

Slope dummies for the cereal production variables ($FLUC$ and $CP$) are tested because there may be regional production and, therefore, import differences in different regions due to factors such as resource endowments and continental weather patterns. The slope dummy for food aid in Africa is included because African countries rely more on food aid as a source of cereal imports than do Asian or South American countries (Huddleston, 1984, p. 25), and aid may therefore have a differential impact on African cereal import demand. Finally, the dummy variables for government debt are included to test whether differences in cereal imports are associated with regional differences in different levels of debt or different reactions to external debt. While most LDCs face major debt problems, these have been particularly severe in South America (Holley, 1987, p. 9; and Kuczynski, 1988, p. 1). A government debt slope dummy variable is also included for Africa.

The seven slope dummy variables were entered into the regression in various combinations and $F$-tests were applied to assess which combination of variables was significant. The results are presented in Equation (3), which represents the best set of tested explanatory variables for cereal import demand in LDCs. Testing the model indicates that there is no significant heteroscedasticity in the regression at the 5-percent level ($\chi^2$ = 15.61, 10 d.f., with $\chi^2$ critical = 18.302).

$$
(3) \quad CM = 42 - 41DSA - 55DAF + 0.023GNP + 0.689URB + 0.729AID - 0.190CP
$$
$$
\quad\quad (2.72)\ (2.56) \quad (3.67) \quad\quad (7.31) \quad\quad\quad (3.09) \quad\quad\quad (3.19) \quad\quad\quad (4.39)
$$

$$
+ 0.040LDBT + 0.134CP.AF + 1.353AID.AF - 0.051LDBT.SA \quad\quad \text{adj. } R^2 = 0.867
$$
$$
(4.17) \quad\quad\quad (1.98) \quad\quad\quad\ (3.20) \quad\quad\quad (2.97)
$$

The significant negative slope dummy variable for government debt in South America indicates that cereal imports in that region are adversely affected by the level of government debt. For South America, the value of the coefficient on *LDBT* is –0.011 (derived by adding the coefficients for *LDBT* and *LDBT.SA*). In contrast, the implication from the positive coefficient on *LDBT*, that government debt did not act as a dampening agent on cereal imports in 1986, reflects the lower levels of debt in Asia and Africa relative to South America and the possibility that cereals are given a very high import priority in these two regions.

The slope dummy variable for food aid in Africa has a coefficient value of 2.082 as opposed to 0.729 for South America and Asia (2.082 is derived from the sum of the coefficients for *AID* and the African *AID* slope dummy). The higher value for Africa suggests that, as expected, African countries do indeed have a higher dependence on food aid as a form of cereal imports than the other two regions. For all three regions, the positive sign on the *AID* variable coefficient suggests that cereal food aid and cereal imports are complementary, rather than competitive, goods.

The only cereal production slope dummy that is significant is that for Africa. For the entire sample, the coefficient on *CP* is –0.190, while for Africa this value is –0.056. In all regions, domestic cereal production acts as a substitute for cereal imports, but more so in Asia and South America than in Africa. Addition of the slope dummy variables caused the variable *FLUC* (cereal production fluctuations) to become insignificant in Equation (3) (*FLUC* was a significant variable in the preliminary regression results given in Equation (2)). It appears that the level of cereal production is a more important determinant of cereal imports than production fluctuations. The variables *GNP* and *URB* (percentage of urbanization) both have the same effect on cereal imports across all countries: cereal imports increase as GNP levels increase and as urbanization increases. While cereal imports of countries in Asia (and South America) are predicted quite well by the model, cereal imports for certain African countries are not predicted as well. There is no evident unifying geographical or income characteristic among the African countries to suggest a reason for the relatively poorer predictive ability of the model for that continent.

Table 2 contains the estimated cereal import elasticities of demand from the results in Equation (3). All are relatively inelastic. For example, a 1-percent increase in per capita national income, *GNP*, causes only a 0.5-percent increase in cereal imports. The elasticities of import demand with respect to the variables *AID* (food aid), *LDBT* (government debt), and *CP* (domestic cereal production) differ among regions.

Cereal imports are slightly more elastic with respect to food aid (*AID*) for Africa than for Asia or South America. This may reflect Africa's high level of cereal food aid in cereal imports relative to the other two regions. The responses in cereal imports to changes

Table 2—Elasticities of Import Demand for Cereals in LDCs

| Variable | Elasticity |
|---|---|
| GNP | 0.477 |
| URB | 0.407 |
| AID | |
| Africa | 0.232 |
| Asia | 0.123 |
| South America | 0.123 |
| LDBT | |
| Africa | 0.306 |
| Asia | 0.306 |
| South America | –0.037 |
| CP | |
| Africa | –0.314 |
| Asia | –0.449 |
| South America | –0.449 |

in government debt for both Africa and Asia show positive elasticities, while South American countries exhibit a negative and very inelastic response in cereal imports to government debt. The cereal import elasticities with respect to cereal production (*CP*) reveal that Africa reduces cereal imports less for each unit of domestic production increase than do either Asia or South America. This may result from Africa's relatively high cereal deficit compared to Asia and South America.

## Income Distribution and Cereal Imports

It has long been argued that income inequality is one of the principal causes of the food problems in LDCs. According to Yotopoulos (1985), income distribution influences both the quantity and composition of cereal import demand and the total supply of cereal available for consumption through direct and indirect (i.e., animal product) means. However, the issue of income distribution is often overlooked in the study of cereal import demand in LDCs, despite cereals being a major component of the human diet and LDCs being the fastest growing market segment for cereal imports. Since income distribution influences both the quantity and composition (food or feed grains) of cereal import demand, an empirical investigation of the impact of income distribution on the demand for cereal imports will improve our understanding of the world food economy.

The cereal import demand in Equation (3) is re-estimated with the addition of two kinds of income variables. One variable is *SH*, the share of income of the poorest 40 percent of the population. This variable is a measure of the income distribution within a single country. An alternative measure of relative inequality in the distribution of income, the Gini coefficient, was also used but proved to be a weaker explainer of cereal imports than the income share of the poorest 40 percent. The second kind of additional income variable is a set of slope dummy variables that divides the sample between countries on the basis of low, middle, and high GNP. These GNP dummies are initialized using the World Bank definitions of low, middle, and high income: *DL* = 1 for 7 countries with per capita GNP < US$350, otherwise = 0; *DM* = 1 for 9 countries with US$450 < per capita GNP < US$1,800, otherwise = 0; and *DH* = 1 for 7 countries with per capita GNP > US$1,800.

These variables enter the regression as slope dummies for low- and middle-income countries on the GNP variables (*GNP.DL* and *GNP.DM*) and for low- and middle-income countries on the income distribution variables (*SH.DL* and *SH.DM*). The divisions are based on 1987 data and then imposed on 1986 data.

The results of the regressions appear in Table 3. The model is estimated for data from 1986 and 1987 to determine the stability of the results over time. The sample sizes are limited to 23 countries due to the availability of data for the variable *SH*.[3]

Table 3—Results of the Cereal Import Demand Regressions Involving Income Distribution

| Year | Constant | DSA | GNP | GNP.DM | AID | AID.AF | CP | SH | SH.DL | SH.DM |
|------|----------|-----|-----|--------|-----|--------|-----|-----|-------|-------|
| 1986 | 29 (1.45) | −64 (5.03) | 0.03 (3.25) | 0.08 (4.19) | 1.37 (2.49) | 0.04 (3.24) | −0.17 (3.92) | 7.35 (3.40) | −7.17 (4.52) | −10.226 (4.49) |
| 1987 | −27 (1.15) | −57 (3.90) | 0.04 (3.40) | 0.06 (3.88) | 0.59 (1.04) | 2.56 (4.05) | −0.19 (4.49) | 10.19 (3.80) | −6.66 (3.07) | −8.09 (3.10) |

Notes: For the *t*-statistics: *t*-critical (2-tailed, $\alpha$ = 0.05, 14 d.f.) = 2.145, and *t*-critical (2-tailed, $\alpha$ = 0.01, 14 d.f.) = 2.977. Adjusted $R^2$s are 0.91 and 0.93 for 1986 and 1987, respectively.

Two tests are run on each model, the Breusch-Pagan (BP) test for heteroscedasticity and the Ramsey RESET test for misspecification. The BP test indicates that there is no significant heteroscedasticity in any regression. The RESET tests (not presented here) indicate that the linear functional form is appropriate and that there is probably no misspecification error.

The results, given in Table 3, reveal that the coefficient estimates for most of the variables were fairly stable in the two years considered. With the exception of government debt, the explanatory variables from Equation (3) are still significant when income distribution variables are added to the basic cereal import demand model. The remainder of the income distribution discussion centres on the 1987 equation since the 1986 equation is only presented to assess the stability of the regression results.

The 1987 regression results reveal that all of the variables, with the exception of the constant and food aid, are significant at the 95-percent confidence level. The insignificant food aid variable, *AID*, can be interpreted to mean that cereal food aid in Asia and South America did not influence cereal imports very much. The significant food aid dummy variable for Africa, *AID.AF*, means that, as found previously in Equation (3), Africa is relatively more reliant on cereal food aid than are the other two regions. In contrast to the results noted earlier from the larger sample, there is a significant slope dummy variable on income for countries in the middle-income category, *GNP.DM*, which suggests that these middle-income countries display a different import demand behaviour than do either low- or high-income developing countries. It appears that among the 23 countries in the smaller cross section, middle-income countries tend to import more cereals for a given increase in per capita income than do either the low- or high-income countries. The dummy variable on income for low-income countries, *GNP.DL*, was dropped from the regression as it was insignificant.

The income distribution variables reveal some interesting within-country and between-countries import demand behaviour. First, the significant share variables indicate that the income distribution within a country does have an impact on cereal import demand. Improving the equity of income distribution within a country, increasing the share of income of the poorest 40 percent and thereby reducing the income share of the richer 60 percent, has a large, positive impact on the demand for cereal imports. This result conforms with evidence that income elasticities of demand for food by the poor in developing countries are relatively high (Mellor, 1988). Second, the significant share slope dummy variables for low- and middle-income countries reveal that, between countries, improvement in the equity of income distribution can be expected to have different impacts on cereal imports that depend on the level of per capita income the countries have attained. Specifically, an increase in the income distribution equity of the 7 countries with a national per capita income greater than US$1,800 has a relatively larger impact on cereal imports than the same increase in equity of the 7 countries with national per capita income less than US$450 or the 9 countries with national per capita income between $450 and $1,800. It may be that this differential impact on cereal import demand is a result of the poorest 40 percent of the population in high-income developing countries having a relatively higher level of income and therefore different cereal demand pattern than the poorest 40 percent in middle- and low-income countries. Admittedly, the sample of 7 high income LDCs is relatively small, contains several nations with high degrees of inequality, and thus may not be fully representative.

Table 4 contains the income and income distribution elasticities for the 1987 regression. The income elasticities reveal that a 1-percent increase in GNP in middle-income countries causes a greater than proportional increase in cereal import demand, probably due to an increase in

Table 4—Cereal Import Elasticities with Respect to Income and Income Distribution Variables

| Variable | Low GNP Countries | Middle GNP Countries | High GNP Countries |
|---|---|---|---|
| GNP | 0.76 | 1.12 | 0.76 |
| SH | 1.39 | 1.25 | 1.80 |

feed grain and meat consumption. This impact on cereal import demand of an increase in income is less than proportional in low- and high-income LDCs.

The elasticities of cereal imports with respect to the income distribution variable, *SH*, reveal that at all three income levels, developing countries exhibit a greater than proportional increase in cereal imports due to an increase in the income share of the poorest 40 percent of their populations. This increase is greatest for the high-income developing countries and smallest for the middle-income LDCs in the sample.

# Conclusions

The results of the estimation of the import demand for cereals in LDCs reveal that cereal imports are determined by such factors as the geographical location of an individual country, the level of development as measured by income and the degree of urbanization, and domestic cereal production. Cereal food aid appears to be a complementary rather than competitive goal to cereal imports (although this is clouded by the data on cereal imports, which are not net of food aid). The relationship between cereal imports and variables postulated to reflect financial capacity was tested. Lagged foreign exchange reserve levels and value of exports were expected to be significantly positively associated with cereal imports. This was the case, although the destructive collinearity that exists between these variables and GNP led to deletion of both financial capacity variables from the model. Lagged levels of government debt were expected to be significantly negatively associated with cereal imports. This was the case for South American countries but not for Asian and African countries. Indeed, the final results, for the sample of 74 countries, suggest that for African and Asian countries, lagged government debt levels have not been a deterrent to cereal imports, at least in cross section.

The investigation into the impact of income distribution on cereal import demand for a sample of 23 countries reveals that income distribution is an important determinant of the demand for cereal imports in developing countries and that improving distributive equity has a positive effect on cereal imports. The results of incorporating slope dummy variables for GNP and the income distribution proxy, *SH*, on the basis of different development levels indicate that cereal import response differs across nations with different levels of income. More extensive work on the impact of income distribution needs to be undertaken when data on income distribution in more LDCs become available.

In this study, the importance of including income distribution as an important determinant of cereal import demand in developing nations, the difficulties of analysing financial capacity constraints on LDC import demand, and the importance of considering regional and socioeconomic differences in cereal import demand are all illustrated. The analysis lends strong support to Mellor's (1988) contention that the fortunes of the developed and developing nations are closely intertwined in the world food economy. The pace at which poor nations can develop, both through increasing income levels and improving income distribution, significantly influences their cereal imports and, concomitantly, cereal exports, largely from rich nations.

# Notes

[1] University of Alberta.

[2] An alternative set of dummy variables based on income level are not significant in this regression.

[3] Low-income countries: Bangladesh, Tanzania, India, Kenya, Zambia, Sri Lanka, and Indonesia. Middle-income countries: Philippines, Egypt, Côte d'Ivoire, El Salvador, Turkey, Chile, Peru, Mauritius, and Costa Rica. High-income countries: Malaysia, Mexico, Brazil, Panama, South Korea, Venezuela, and Trinidad and Tobago.

# References

Belsley, D.A., Kuh, E., and Welsch, R.E., *Regression Diagnostics*, John Wiley and Sons, New York, N.Y., USA, 1980.

Christiansen, R.E., *The Impact of Economic Development on Agricultural Trade Patterns*, Staff Report No. AGES–861118, Economic Research Service, US Department of Agriculture, Washington, D.C., USA, 1987.

Holley, H.A., *Developing Country Debt: The Role of the Commercial Banks*, Routledge and Kegan Paul Ltd., London, UK, 1987.

Huddleston, B., *Closing the Cereals Gap with Trade and Food Aid*, Research Report No. 43, International Food Policy Research Institute, Washington, D.C., USA, 1984.

Kuczynski, P.P., *Latin American Debt*, Johns Hopkins University Press, Baltimore, Md., USA, 1988.

Mellor, J.W., "Food Demand in Developing Countries and the Transition of World Agriculture," *European Review of Agricultural Economics*, Vol. 15, 1988, pp. 419–436.

Morrison, T.K., "Cereal Imports by Developing Countries," *Food Policy*, Vol. 9, 1984, pp. 13–26.

Yotopoulos, P.A., "Middle-Income Classes and Food Crises: The New Food-Feed Competition," *Economic Development and Cultural Change*, Vol. 3, 1985, pp. 463–483.

---

## Discussion Opening—*John Dyck* (US Department of Agriculture)

This paper raises important issues about trade flows and global interdependence. It provides some interesting empirical results. The paper reports that each equation was estimated several times, with different specifications, in order to see which variables' coefficients were significant. Such respecification means that the final estimated equation is tailor made for the particular data set and the coefficients may not be as robust as the $t$-test suggests. The estimations do suggest that GNP/person is positively correlated with cereal import demand, as is a lessening of income inequality. However, the study regresses import demand for all kinds of cereals on explanatory variables. Cereal imports behave quite differently if they are for food or for feed. Cereal imports for food are expected to rise at very low levels of income. But at some point, cereal food consumption/person needs are satisfied, and that may happen at relatively low income levels. At higher levels, as in East Asia, such consumption may even decline.

However, the demand for livestock products grows throughout the development process, and it is hard to see where it will stop. Livestock-derived foods require more grain than do cereal foods. Thus there are two distinct demands for cereals, and mixing the two can be misleading. The income elasticity of cereal import demand calculated in the paper is unlikely to hold in other situations because the relative importance of food and feed imports will have changed. It would be preferable to use data that allow calculation of cereal imports for food and feed and estimate the effect of income on each type of import. Feed imports should measure both direct feed grain imports and the feed equivalent of imported livestock products.

Cereal imports by LDCs are important to developed country agriculture, and a correlation of those imports with income growth in the past seems clear, although not well quantified in the current study. It is true that rising incomes in the Third World may lead to continued rising cereal imports, which could alleviate the problem of excess cereal production in the developed countries. However, surpluses in the developed countries have been created by policies that can be changed. Also, food supply growth in a developing country can initially lag behind food demand growth, but, given favourable policies and technological progress, can catch up. Thus, rising, income-led cereal consumption by LDCs in the 1990s and beyond may not translate into rising cereal import demand as it has in the past if supply curves in LDCs shift out because of technical change and in developed countries shift in because of policy reforms. An interesting question to discuss is whether the current fit between excess cereal demand in developing countries and excess supply in developed countries will continue, especially in light of policy reforms in agriculture and evidence of how much technology can boost supplies.

*[Other discussion of this paper and the authors' reply appear on page 282.]*

# Food Grain Marketing Reform in Ethiopia

## Abebe Teferi[1]

**Abstract:** During the past 15 years, there has been a high level of direct government intervention in food grain marketing in Ethiopia. To maintain its strong hold on the grain market, the government established a compulsory delivery quota system for both producers and traders, with geographically uniform fixed prices with no seasonal variation. Licenced traders were banned in surplus-producing regions, and grain movements between surplus and deficit regions were restricted. As a result of the government's direct intervention in the grain market, markets became disrupted geographically. The integrated market systems were cut off, giving way to the creation of segmented markets. The prices in these segmented markets were uncorrelated and created distorted prices. The distorted prices in turn led to artificial shortages of food crops. Recognizing the social and economic problems associated with the government's marketing and pricing policies during the late 1970s and the 1980s, a free market system was adopted in February 1990. Since then, the market has experienced price fluctuations related to supply and demand.

## Introduction

Agricultural practices in Ethiopia are primitive, and production is limited to subsistence mixed farming. The average annual growth rate of agriculture was only 1.2 percent in 1973–83, and annual agricultural GDP declined to –3.9 percent in 1980–86 (World Bank, 1988). The rate of growth of food crop production has also been stagnant and in some cases decreasing, with the population growing at the rate of 2.9 percent per year. During recent years, agriculture has been unable to provide sufficient food relative to demand. Food grain supply has been in deficit in all urban centres and in many rural areas and even more scarce in areas of frequent drought. Because of bad marketing and pricing policies, the grain marketing system is thought to have been a major contributor to food shortage problems. As shown in Table 1, nearly 10 percent of the population is affected by food shortages every year. Marketing as an institution has contributed little to the reduction of those shortages, especially in famine-stricken areas.

Table 1—Population Affected by Food Shortages during 1979–88, by Administrative Region (in thousands)

| Region | 1979 | 1980 | 1981 | 1982 | 1983 | 1984 | 1985 | 1986 | 1987 | 1988 |
|---|---|---|---|---|---|---|---|---|---|---|
| Shewa | 14 | 176 | 239 | 533 | 195 | 204 | 852 | 709 | 330 | 511 |
| Welo | 996 | 950 | 450 | 592 | 1,069 | 1,821 | 2,587 | 1,547 | 334 | 1,017 |
| Tigre | 919 | 1,213 | 500 | 600 | 1,000 | 1,332 | 1,429 | 1,000 | 358 | 1,036 |
| Eritrea | 1,000 | 218 | 650 | 713 | 842 | 518 | 827 | 650 | 399 | 1,050 |
| Gonder | 317 | 109 | 67 | 202 | 493 | 470 | 363 | 341 | 221 | 292 |
| Gojam | – | – | 99 | 20 | 35 | 76 | – | – | – | – |
| Illubabor | 22 | – | – | 6 | – | – | – | 102 | 209 | – |
| Welega | 12 | – | 28 | 97 | – | – | – | 116 | 118 | – |
| Bale | 527 | 368 | 379 | 109 | 35 | 58 | 84 | 99 | 30 | 30 |
| Sidamo | 203 | 153 | 310 | 333 | 145 | 186 | 533 | 442 | 221 | 320 |
| Gemu Gefa | 14 | 81 | 232 | 108 | – | 80 | 106 | 153 | 27 | 33 |
| Arusi | – | – | 185 | 220 | 60 | 21 | 82 | 20 | 66 | – |
| Harer | 287 | 1,177 | 420 | 777 | 285 | 329 | 875 | 1,517 | 576 | 925 |
| Kefa | – | 39 | 13 | 26 | – | – | – | 90 | 39 | – |
| Ethiopia | 4,308 | 4,484 | 3,473 | 4,415 | 4,144 | 5,054 | 7,814 | 6,786 | 2,928 | 5,214 |

Source: Alemayehu (1990).

For the last 15 years, the grain market has been disrupted in terms of both crops and geographical differentiation. The direction of trade was also disrupted. Market competitiveness was affected by the creation of barriers to free entry into the market. At the same time, since prices were fixed by the state rather than by market forces, the free market was unable to perform its function on the basis of supply and demand. In this process of pricing and marketing activities, the producer has been very disadvantaged. Because the level of food insecurity is chronic and there are many vulnerable groups, it is assumed that efficient marketing would help to minimize the food insecure groups and reduce the level of relief food aid. On the other hand, the level of food insecure groups will increase when the market is disrupted. Bigman (1982) argues that, according to World Bank estimates, as much as 40–60 percent of the population in developing countries is undernourished, and the number may increase even further as the food problem becomes especially acute in times of market disruption. With the current state of grain marketing, Ethiopia may not be an exception to this rule.

The level of government intervention in the food grain marketing system over the past 15 years has been very high. During this period, the role of the private sector in food grain marketing was minimized, and in some regions private traders were totally banned. Abbott (1987) writes that during the 1970s and 1980s the combination of Marxist political views and the interests of development planners encouraged governments to take an increasing role in agricultural marketing. The potential role of indigenous private enterprise was largely ignored.

Ethiopia established, as did many other African countries, government marketing institutions: the Ethiopian Grain Board and the Ethiopian Grain Council, in 1950 and 1960, respectively. The first acted as a regulatory body while the latter acted as a price stabilizer by actively participating in the market by holding stocks. However, these institutions were not able to regulate the grain marketing system, so that low prices for producers and high prices for consumers existed until the onset of the revolution. A study by Stanford University's Food Research Institute (Thodey, 1969) recommended that the integration of the marketing systems should be improved. But before any improvement could be made, the government was overthrown. During the period in power of the new government, the grain marketing system developed a new feature with the creation of the Agricultural Marketing Corporation (AMC). Unlike similar institutions in many underdeveloped countries, the AMC is socialist in concept, which led to the banning of private traders and creation of fixed prices with a compulsory delivery quota system.

Grain marketing in Ethiopia was problematic in the past but has become more so in recent years. The current grain marketing problems are often associated with government intervention in the market through the AMC. The AMC, as a government marketing institution, controlled the grain marketing sector through the imposition of compulsory delivery quotas for both producers and traders, and application of fixed prices and restriction of free movement of grain among regions have been practised for more than a decade. During this time, licenced traders were banned in most of the surplus-producing regions, and nearly all traditional market structures that had positive integration and that used to be competitive were abolished. Free entry into the market was also restricted to give more monopoly power to the government marketing agency. A uniform pricing system was applied in all parts of the country and at all times of year, ignoring transport costs and seasonal price variations.

## Objectives of Government Intervention in Grain Marketing

The socialization of the distribution system through government intervention was believed to be the major characteristic of a socialist planned economic management system. Thus, controlling prices and the market as an integral part of a socialist economic management system was a major objective of government intervention in the market. In relation to this, Franzel et al. (1989) assert that the driving force for attaining control over marketing is both ideological and pragmatic. On the ideological side, there is a strong belief that merchants and other intermediaries exploit the peasantry and consumers and that state intervention is required to curtail exploitation. The pragmatic reasons are associated with the post-revolutionary land reform. The end of share tenancy in grain surplus areas led to increased on-farm consumption; thus, the share of peasant production marketed declined from 25 percent to 10 percent during 1974–78. This, they claim, had resulted in higher urban prices that promoted the establishment of the AMC in 1976. This again led to a situation where the activities of private traders were sharply curtailed. It was these conditions and general thinking prevailing at the time that led the government to intervene in the market.

To implement government intervention in the market, many government agencies were established with regard to grain marketing, of which the AMC was the principal one. The AMC was actually initiated as a project during the previous government, with the assistance of the World Bank. Since the project proposal was found to be suitable for the new system, the project was easily accepted by the new government with some modifications. At the same time, while the World Bank was very opposed to the activities of the AMC, it has continued its financial assistance. Under the framework of the formation of a socialist economic system, the AMC's objectives were to ensure stable producer and consumer prices, maintain adequate producer incentives, reduce marketing margins through greater efficiency and reduced risks and profits, and ensure adequate food supply in all parts of the country through a policy of low food prices.

To attain active government participation in the grain market and to achieve these objectives, a committee was set up and it decided to fix prices for 14 types of grains on 6 markets. A compulsory quota delivery system for both producers and traders was introduced with the new pricing system. The minimum grain quota for each peasant association was 100 quintals and, for licenced grain wholesalers, 30 percent of their purchase, to be delivered to the AMC. In 1980/81, these quotas were raised to 150 quintals for the peasant associations and 50 percent of purchases for wholesale traders. State farms and producer cooperatives were obliged to deliver all their marketed output to the AMC.

In order to ensure grain delivery, 1,768 grain collection centres were established by the AMC. Every year, before each harvest, fixed quotas were allocated to each region by the AMC with the approval of the national grain purchase task force and of the Office of the National Committee for Central Planning. The quotas were allocated to each region on the basis of purchase demand rather than production structures. The national grain purchase task force, operating through a series of grain purchase task forces, was usually responsible for allocating regional quotas down to the *woreda* level, and eventually quotas were set for service cooperatives and individual farmers. The fulfilment of quotas by service cooperatives and individual farmers was strictly enforced.

Prices paid by the AMC were established by the Council of Ministers for the farm gate, wholesale markets, and state farms. The ranging of prices and price differentiation based on distances and storage costs were abandoned in 1980/81. Instead, prices became geographically uniform throughout the country and all-year-round, regardless of differences in transport and storage costs and demand. The uniform pricing policy lasted up to beginning of 1990 when the new market liberalization policy became effective.

## Results of Government Intervention

To facilitate the role of the AMC, improved infrastructure, storage facilities, and transport equipment were set up and operated by a series of crop purchasing committees, task forces, and service cooperatives. These facilities have been expanding as a result of greater government attention at both national and regional levels. For example, when the storage capacity increased from 1.6 million quintals in 1979 to 6.4 million quintals in 1989, the total work force also increased, from 2,019 in 1979 to 4,191 in 1989. Despite all these measures taken at both the policy and execution levels, grain marketing and pricing activity remained a major problem. Limited marketable surplus, rising food prices, increasing food imports and international food aid, low level of agricultural growth, and unavailability of staple food items are the major characteristics of the grain production and marketing systems in Ethiopia today.

These problems are further aggravated by critical shortages of basic consumer goods, with rising prices and an increasing share of black market activities in major cities and towns and in rural areas. These problems are manifestations of contradictory situations in the grain production and distribution systems. On the one hand, agricultural production is by-and-large organized under small-scale production, with stagnant agricultural development and decreasing agricultural production; while on the other hand, there is a huge government food grain marketing agency, an expanding compulsory delivery quota system, and highly organized government intervention in the market.

While it is very difficult accurately to assess the market system, two features are clear. First, two market and pricing systems have been operating, one distorted and the other controlled, and, second, the government marketing system has had a negative impact on the distribution of commodities.

To give more purchasing power to the AMC, grain movements among regions were stopped. Checkpoints to stop grain movement were established in key places, so that the traditional trade connections between small town traders and large wholesalers in large urban centres were very much minimized, and marketing chains were broken off and became virtually non-existent. The traditional trade connections between surplus and deficit regions, in particular, came to a complete stop. The integrated market system, that was moving towards full market integration, was thus disrupted, and created several segmented markets.

Initially, the procurement and handling capacity of the AMC was substantially less than the marketable surplus. At the same time, private traders were allowed to operate so long as they submitted 30 percent of their purchase to the AMC. Then the government felt the need to strengthen the AMC, and free entry into the market was restricted, not only through central government measures but mainly by the regional administration. The regional administrations in Arusi and Gojam actually banned all private grain traders, while elsewhere traders operated on the basis of area-specific regulations. Such restrictions paved the way for the creation of a single monopolistic government marketing agency and disrupted or destroyed the competitiveness of the grain market, contributing to the creation of market inefficiencies. Since private grain traders were banned in two heavy surplus regions, Arusi and Gojam, and allowed in other parts of the country, government intervention created geographical distortions of the grain market. With the start of fixed prices and a compulsory quota delivery system, two market systems and two pricing systems were operating within a single economy. While the controlled market system played a dominated role, the so-called free or open-market system became a residual market system. Distortion of the market and prices meant replacement of competitive traditional market structures by a system that is superimposed by government intervention involving deliberate stoppage of movement of commodities among markets, a compulsory quota delivery system, and fixing of market prices. As a result of such intervention, the market and prices of food grains during the 15 years became highly distorted. The formation of prices in the residual market did not follow the rules of supply and demand; nor did the controlled market system give appropriate signals for the allocation of resources or any indication of shortages and abundance of commodities.

Geographically, the AMC's major concentration for purchases of cereals was on the Arusi, Gojam, and Shewa regions. The AMC's purchase of cereals in different regions was not based on production structures but emphasized buying preferred crops in a given region and leaving other crops for the open market system. The AMC's preference for one crop over another created market disruption on the basis of crop differentiation. For example, teff and maize represented 36 and 24 percent, respectively, of cereal production in Gojam during 1982/83– 1987/88. In Arusi Province, the average percentage of production of wheat and maize for the same period was 38 and 11 percent, respectively. During the same period, the average percentage of the AMC purchases of teff and maize in Gojam was 71 and 6 percent, and for wheat and maize in Arusi 67 and 3 percent, respectively. So some crops in some regions, such as maize in Gojam, where private traders were banned, were left with no buyer. This again created a market distortion on the basis of crop differentiation linked to geographical market distortion.

In the pre-revolution period, grain was moved first from a smaller market to a larger market. Later, mainly during the off-season, grain was moved from a larger market such as the Addis Ababa market to smaller markets such as Nekemte and Debre Markos. Since markets were segmented due to the start of the AMC's operation, the normal links between surplus and deficit regions were disputed, which also changed the direction of trade. Geographical disruption of the market and the change of direction of trade resulted in extreme differences among commodity prices in surplus and deficit regions. For example, the producer price of black teff in Gojam between February and April 1985 was 52 birr, while it was 209 birr in Welo. When the producer price of wheat in Arusi between November 1984 and January 1985 was 61 birr, it was 141 birr in Welo. The situation indicates that the prices of commodities in these segmented markets were uncorrelated. In some cases, the prices of some commodities were moving in opposite directions. On the other hand, since the AMC's prices were geographically uniform throughout the year, the price correlation coefficient of a single commodity between any two markets was always 1.0.

The negative impact of government intervention in the food grain marketing system on short- and long-run agricultural development in Ethiopia was believed to be fully recognized by both the government and international agencies such as the World Bank. Many consultancy reports were written on the poor performance of the food grain marketing sector. But the system that led the country into the continuing food crisis continued to operate with its full-scale and original mandate until the 1990 main harvest season.

Finally, with due recognition of the social and economic problems facing the country, the 11th plenary session of the Workers Party of Ethiopia announced major economic policy changes. One of these involves the food grain marketing sector. The new policy states that there will be no prices set by the government. Farmers as well as traders will no longer be required to supply compulsory delivery quotas. Prices will be determined by market forces. All checkpoints have been removed so that grains can move in any direction. The new policy clearly states that the AMC will enter the market as buyer and seller so long as it makes a profit. Generally, the food grain marketing system will operate as a free market, although this will take some time to become fully operational countrywide. It is still uncertain how the free market will function during the transitional period. Another very important question is the actual role of the AMC during the transitional period. Since the very basic conditions for the existence of a competitive market, such as free entry into the market, were restricted, and, since there are few licenced traders, it is still uncertain how fast the market will become competitive and how fast the traditional or new market structures will become functional.

However, regardless of these many uncertainties, the grain market started to react to the new economic policy changes immediately. The prices of major food grains, especially teff, went down by about 30–40 percent. The AMC's price data, collected in the free market in many grain marketing places, indicate that the prices of white teff went down to 105 birr per quintal in Addis Ababa. But from the middle of May 1990, the prices started to go up, and the prices of red teff seem to be stabilizing at about 30 percent below, and the prices of white teff at about 15–20 percent below, the old prices. Three possible reasons why the prices at first went down, started to go up later, and then seemed to stabilize, can be put forward.

First, the psychological factor, that people are happy that the controlled market system is abolished makes them want the new system to be seen as a challenge to the old system. Second, merchants with large illegal stocks were afraid that prices might decrease, so they dumped their stocks on the market, which unexpectedly increased supply and brought the prices down. And third, since the stability of prices was initially unreliable, merchants were unable to move their stock to deficit areas.

However, the prices of food grains started to increase even though they were still below the old prices. Four possible reasons may account for this. First, the AMC, which was supplying grain to the public through the *kebeles* (urban dweller associations) decreased its activities, which shifted demand from the AMC to the open market, and pushed prices upwards. Second, since the traditional market structures did not recommence normal operation, grains did not start to flow to the big terminal markets. Supply of grains to big markets such as the Addis Ababa market through the marketing chains was thus unable to catch up with market demand. Third, since licencing of traders in all urban centres, but mainly in the surplus-producing regions, will take some time, the open market is not as competitive as it should be. The presence of only a few merchants influenced prices to go up. And fourth, since May and June are the planting time for farmers, the supply of grain in the market in these months may have started to decrease and pushed up prices in the market.

## Food Grain Marketing in the 1990s

The most important question that has to be answered is the role of AMC in the future, and whether there will be any government activity to stabilize the market. The Ethiopian smallholder agricultural sector accounts for about 92 percent all food grain production. Food grain production is also the single largest source of income for the majority of the rural population. The future organization of the grain marketing sector should be directed primarily to helping smallholders increase their sources of income and to speeding up the growth of the agricultural sector.

During the past 11 years, or since the beginning of the economic campaign, there has been a great need to give more emphasis to agricultural development. The central objectives of agricultural development policy during this period were to attain food self-sufficiency, to expand the agricultural export base, and to produce enough raw materials for the small but growing industrial sector.

At the policy level, even though it was quite different in practice, the main objective of the grain marketing sector was to contribute towards the attainment of these objectives through increasing smallholder agricultural production.

At a more aggregate level, food crop production is concentrated in three surplus regions, Arusi, Gojam, and Shewa, followed by Gonder and Wolega. During 1982/83–1988/89, 79 percent of the AMC's average grain purchases came from the three surplus regions. In the past, the public grain marketing sector has marginalized the smallholder, particularly in the surplus-producing regions, which had a greater negative impact on the development of the country's agricultural sector.

When the new grain marketing policy becomes fully implemented, the contribution of the grain marketing sector to the country's agricultural development and in meeting the three policy objectives will be undoubtedly very important. The main role of the grain market is to contribute to economic growth, to attain nutritional well-being and equitable income distribution, and to achieve food security. These objectives can at best be realized only when the market is operating efficiently to a point where it is able to give signals of scarcity and abundance to buyers and sellers as well as to policy makers about food shortages and food insecure groups. It is hoped that the grain market in Ethiopia in the 1990s will be relatively free and that it will give the appropriate signals to producers and consumers, buyers and sellers, as well as for policy makers and investors. However, it is not clear whether the AMC will play a market and price stabilization role or not.

## Note

[1]Office of the National Committee for Central Planning, Ethiopia.

## References

Abbott, J.C., "Alternative Agricultural Marketing Institutions," in Elz, D. (Ed.), *Agricultural Marketing Strategies and Pricing Policy*, World Bank, Washington, D.C., USA, 1987.

Alemayehu, L., "The Assessment of Food Systems and Food Security in Ethiopia," draft report submitted to IGGAD, Vol. 2, 1990, p. 110.

Bigman, D., *Coping with Hunger: Towards a System of Food Security and Price Stabilization*, Bollinger Publishing Company, New York, N.Y., USA, 1982.

Franzel, S., *et al.*, *Grain Marketing and Peasant Production*, IAR Research Report No. 5, Institute of Agricultural Research, Addis Ababa, Ethiopia, 1989.

Thodey, O., *Marketing of Grains and Pulses in Ethiopia*, Report No. 16, prepared for the Ethiopian Government, Food Research Institute, Stanford, Calif., USA, 1969.

World Bank, *World Development Report*, Washington, D.C., USA, 1988.

---

## Discussion Opening—*William Grisley* (Centro Internacional de Agricultura Tropical)

Teferi's paper addresses a topic of major importance not only for Ethiopia, but also for many other developing countries. In his study of state participation in and price regulation of selected grain markets in Ethiopia, he concludes that the state's action contributed to both chronic food scarcity and periodic shortages and has dimmed the prospects for economic development in rural areas. This conclusion is anything but surprising. A similar conclusion could be drawn in many other developing countries, especially those in sub-Saharan Africa. Why is this the case and what are the factors responsible for its occurrence?

The wider issue in development is the roles of the state and the private sectors and the relationship of the public to these sectors. Many developing countries have elected to extend the role of the state sector beyond its traditional functions of security provision, infrastructure development, and general economic regulation to direct involvement in the production, financing, and marketing of consumer goods, services, and commodities. The crux of the problem is not necessarily the involvement of the state in these activities but the fact that the state often gives itself a monopoly position in important economic sectors. The resulting economic inefficiencies and distortions in resource allocation are thus not necessarily due to state participation but to the monopoly position of the state entity. Managers of state firms that hold monopoly positions have no incentives to become more efficient. A monopoly position held by a private firm would in theory result in similar lack of incentives and resulting inefficiencies.

If the monopoly position and not the ownership of the firm is the problem, why then have governments in many developing countries—and formerly the countries of Eastern Europe—protected the monopoly position of state or private entities? The reason lies in the absence of a public that is allowed to influence state policies. Governments that do not regularly face the whims of voters have largely ignored the wishes of the public because they have not found it costly to do otherwise. A necessary condition for effective public participation in the development of state policies is political democracy. Governments that are subject to public pressures cannot long afford to allow state or private firms to enjoy monopoly positions in non-public goods sectors and sectors in which strong externalities do not exist.

With the demise of socialist régimes in Eastern Europe, the trend in developing countries is towards a more open political system. When state firms' monopoly position is eliminated, they will have to become competitive in order to survive. In agriculture, the flexibilities required in production and marketing will make continued participation by unprotected state firms difficult and many will fail. Over the longer term, both producers and consumers will benefit from this political trend.

*[Other discussion of this paper and the author's reply appear on the following page.]*

# General Discussion—*L.P. Apedaile, Rapporteur* (University of Alberta)

Responding to the opener's remarks, Shimizu noted that melon yields are stable from year to year, allowing the DEA method to be applied to measuring management performance. He further noted that sustainable farming necessarily involves multiple objectives to which the DEA technique is well suited. He agreed that a longer data base would be useful; in such a longer data series he will introduce farm growth as one of the output indices. Responding to criticism of the conclusion that the presence of more full-time workers is associated with lower technical efficiency, he referred to two problems. First, the exclusion of part-time workers leads to overestimation of technical efficiency for one-worker farms and underestimation for three-worker farms. Second, dropping the fixed asset variable had the same effect. This estimation problem is related to the fixed upper limit of 100 percent for quality and quantity in the output index, although the estimates improve as the number of full-time workers increases. Finally, he noted that the measures of technical efficiency are independent of the degree of specialization in melons on the farms because of the low degree of correlation between melon output and total farm output.

Veeman *et al.* were asked why a price variable was excluded, noting that even with cross-sectional data, border prices for cereal imports would vary across nations. The response was that a price series is not included in the World Bank report used as the data base, and that the limited availability of consistent price data to match that base would have restricted the sample size. Another problem raised was the specification of the model. The size of a country could affect the level of imports. Larger countries could avoid imports by interregional transfers of cereals. Veeman accepted the point and observed that a size dummy might work. Veeman further admitted the absence of a strong theoretical base to the model, emphasizing the absence of literature on the effects of income distribution on cereal import demand. Disaggregation of the cereals variable to enable a focus on food grains would be a good idea. The results of their work could not be construed to mean that imports by LDCs would increase further, observing that yield barriers and poverty are the most important long-run determinants of cereal imports.

Relating to Abebe's paper, he was asked if there had been economic growth in Ethiopia during the period under study and whether a surge in cereal output could be expected now that the Marxist policies were being changed. Abebe responded that the growth rate had declined from 1.2 percent is in the early 1980s to negative values in the late 1980s up to the overthrow of the government in early 1991. He felt that it was too early to tell whether cereal output would respond dramatically to the new policy environment. Another participant listed a number of features of the Ethiopian situation and asked if the almost 50-percent commitment of treasury funds to the military would now be redirected to finance a food policy. He wondered about the inflationary consequences of consumer food subsidies. A question was also asked about the likelihood of land redistribution under the new government. Abebe said that all land belongs to the state and that there are no plans to privatize ownership. He noted, however, that there is now secure access to land under long-term leases, which is a major improvement over the previous policy. In reply to the opener's remarks, Abebe reaffirmed that the quotas had to be abolished, especially for the food surplus areas, because they are the major impediment to production increases. The Agricultural Marketing Corporation (AMC) would probably be needed to procure grain to feed the army. However, in Abebe's view, the AMC should be abolished along with the delivery quota system. The only possible future role for the AMC would be as a sort of buffer stock manager to stabilize prices.

Participants in the discussion included D. Belshaw (University of East Anglia), J. Benet (Hungarian Academy of Sciences), R. Herrmann (Universität Giessen), P.J. Lund (Ministry of Agriculture, Fisheries, and Food, UK), and H. Shinoura (Agricultural Research Institute, Japan).

# Structural Adjustment, Agriculture, and the Poor: A General Equilibrium Analysis of the Kenyan Economy

*Godfrey J. Tyler and Oludele A. Akinboade*[1]

**Abstract:** A computable general equilibrium model based on a social accounting matrix for Kenya is used to simulate the effects of 10-percent devaluation, 10-percent increased investment, and 10-percent agricultural productivity improvement on the macroeconomy and on the real incomes of the poor. For each policy simulation, two specifications for the labour markets are adopted, the first assuming unlimited supplies of labour at given nominal wages and the second assuming fixed supplies so that wages are determined endogenously. These crucially affect the results. Under the first assumption, devaluation provides a 10-percent boost to real GDP and has highly favourable effects on agricultural production, exports, the current account deficit, employment, and poverty. Under the second assumption, it has a largely inflationary impact, with attenuated effects on real GDP and other variables and no effect on the current-account deficit. Agricultural productivity improvement is less affected by the different specifications and compares favourably with devaluation except for its smaller impact on GDP. The increased investment policy is found to be inferior on most counts. All three policies decrease poverty, though income distribution remains stable.

## Introduction

There has been controversy for some time among economists and international agencies about the appropriateness of structural adjustment and stabilization programmes for the economic problems of less-developed countries (LDCs), particularly those in sub-Saharan Africa. Concern has also been expressed in some quarters that such programmes, even if judged successful at the macroeconomic level, may have had deleterious effects on the incomes of the poorest in those countries, especially in the short to medium term.[2] Recently, increasing emphasis has been placed by the IMF and World Bank on "supply-side" policies to stimulate economic growth while at the same time correcting external and internal imbalances.[3] The present paper is a modest contribution to the analysis and debate on such concerns. It focuses on Kenya and attempts to trace the impact of a limited number of policies on the macro-economy and on the real incomes of the poorest sections of the Kenyan population. The options considered are devaluation, investment, and agricultural productivity improvement.

## Model and Data

It is now increasingly being accepted that such policy concerns as expressed earlier are best analysed by means of economy-wide multi-sector models that are able to incorporate the interrelationships between productive activities, factors of production, households, government, and the rest of the world in a general equilibrium framework.[4] Hence, we have used a computable general equilibrium (CGE) model based on a social accounting matrix (SAM) for the Kenyan economy. The model is medium term in outlook, short enough to ignore population growth but long enough to allow productivity changes and some factor mobility.

The CGE model consists of a large set of structural equations linking producing sectors, factor markets, households, government, and the rest of the world, with market prices as well as production, employment, consumption, savings, trade, etc. being endogenously determined. As it is based on a SAM, the accounting framework ensures internal consistency; thus, the usual national accounting identities hold.

The SAM was based on 1976, the most recent data set available, but the general pattern of the results from the simulations using this data set have relevance to the present situation in Kenya. A certain amount of aggregation, disaggregation, and other adjustments were made to the original SAM to produce a 69 × 69 matrix.[5]

**Production accounts.** There are three production activities representing the three sectors of agriculture, manufacturing, and services. Each activity is assumed to combine labour of different categories, capital stock, and "operating surplus" in a constant-returns-to-scale Cobb-Douglas production function to produce value added. "Operating surplus" is a catch-all category for the input of, and returns to, entrepreneurship, management, and, in the case of agriculture, land. Factors are combined optimally, such that their marginal value products are equated to factor prices. Value added is then combined in fixed Leontief fashion with purchased intermediate inputs to produce gross output, at producer prices.

**Commodity accounts.** For each production sector, there are four commodity accounts: domestic, imported, exported, and composite commodities. Where applicable, indirect domestic taxes and import taxes are added to domestic production at producer prices and imports at c.i.f. prices, respectively, to produce domestic commodities and imports at market prices. They are combined as a constant elasticity of substitution (CES) aggregate (with elasticities of substitution of 1.0 for agriculture and 0.5 for the other two sectors) to produce composite commodities at minimum cost. These composite commodities are then available to meet total demand; i.e., for household consumption, government consumption, intermediate use, and investment.

**Factor accounts.** There are 14 labour, 3 capital, and 3 operating surplus categories. The labour categories comprise unskilled, skilled, professionals, and self-employed. Factor incomes from each of the categories are distributed to the institutions of households, companies, the government, and the rest of the world. It is assumed that, in value terms, the distribution is in the same proportions as the original SAM, which is the same as assuming that factor ownership by institutions does not change. In the urban unskilled labour market, labour moves freely between the industrial and services sector. Skilled labour and self-employed labour move freely across all sectors but are specific to either rural or urban areas. Professionals are assumed to be specific to each sector and to each area. Capital and "operating surplus" are assumed to be specific to each sector.

**Institution accounts.** There are four household groups in urban areas and four in the rural areas, differentiated according to income class. Other institutions are the government and companies. Incomes accruing to the various household groups are allocated to consumption, savings, direct taxes, transfers to companies, and remittance transfers to other households in fixed value shares. These shares are as in the original SAM except for remittances, which have been mapped according to other evidence.[6] Total consumption expenditure is then allocated to consumption of composite commodities in fixed proportions in quantity terms. This implies unitary income elasticity and zero price elasticity of household demand. The government account is modelled similarly.

**Capital accounts.** Total savings in the economy, by households, companies, the government, and foreign savings are distributed to investment in the three sectors (by destination) in fixed value shares. Within each sector, this investment is then used to purchase the three composite commodities in fixed value shares (investment by sector of origin). Foreign savings is specified as a residual. It meets the gap between total savings and total investment. By definition it is identically equal to the current balance-of-payments deficit.

**Rest-of-the-world account.** The rest of the world pays the Kenyan economy in foreign currency for exports and receives foreign currency for imports. There are also transfers in both directions, in fixed foreign currency, with factors and institutions. Imports of commodities are assumed to be available in perfectly elastic supply to Kenya at fixed world prices (the small country assumption). However, exports of domestic production are not assumed to face perfectly elastic demand. The demand for exports is assumed to depend on the price of exports relative to the world market price of comparable goods, with elasticities of demand of 3.0, 1.5, and 1.0 for agriculture, industry, and services, respectively. Producers are assumed to be indifferent between sales to the export and domestic markets, as they sell at the same (producer) price. Finally, the exchange rate is fixed as numeraire. However, in the devaluation experiment, it is varied exogenously.

**Closure of the model.** Closure of the model is essentially concerned with ensuring that the model has the same number of endogenous variables as equations.[7] As said previously, product and factor market prices are in general determined endogenously, with, in the present model, only world prices of imports and exports in foreign currency and the exchange rate taken as exogenous. However, in the case of the labour markets, because of the uncertainty surrounding these markets in Kenya,[8] two alternative specifications are used in each simulation. The first, the so-called Keynesian closure, assumes an unlimited supply of labour at a fixed nominal wage, so that the level of employment is determined endogenously by the demand for labour. The second, the so-called neoclassical closure, assumes a fixed supply of each labour category, so that the market wage is determined endogenously in a situation of full employment. Investment is, in general, endogenously determined, but, in some of our simulations, investment is exogenously increased.

## Results of Policy Simulations

As intimated in the introduction, the three policy experiments reported here are: 10-percent nominal devaluation, 10-percent increase in total real investment, and 10-percent improvement in agricultural efficiency (through adoption of new technology, improved methods, etc.). The figure of 10 percent was arbitrary; it was chosen to provide a basis of comparison between the various policy options and was considered as representing an order of change that was reasonable and attainable. No claim is made that the ease or difficulty of achieving these objectives is equal.

Table 1 presents the results. Only the important indicators are reported here. The first column contains the actual levels of the major aggregates in the base year in millions of Kenyan pounds. As all prices (including wages) are taken as unity in the base year solution, the physical quantity of (for example) employment of each category of labour is given in money terms equal to the total of wages in the base year.

Because a major concern is the impact of policy on the real incomes of the poor, we have included employment and wages of the urban unskilled and skilled labour and of rural self-employed labour, these being major sources of income for the urban and rural poor.

The results of the three policy experiments are given, as percentage changes from the base solution, in two blocks. Block A assumes an unlimited supply of labour at fixed wages. Block B assumes fixed supplies of labour. Under the latter assumption, real GDP, being the aggregate of the returns to the primary factors at constant factor prices, is inherently constrained. It can only increase under the first two policy experiments if the capital stocks or entrepreneurial inputs increase. This explains the very much smaller changes in GDP in Block B.

**Devaluation.** The vastly different outcomes of devaluation under the A and B assumptions highlight the critical importance of the labour market situation. With labour available in perfectly elastic supply, there is a 10-percent increase in real GDP, a 15-percent increase in exports leading to a 38-percent reduction in the current account deficit, increases in production, especially in agriculture, of 15 percent, increases in employment ranging from 8 percent in urban unskilled to 13 percent in rural self-employment and, importantly, increases in real consumption of poor households of between 6 and 9 percent. Market prices only rise by between 1.4 and 5.2 percent. There is virtually no change in the GDP deflator.

On the other hand, with fixed supplies of labour, the nominal devaluation has a large inflationary impact. Wages rise approximately in line with the devaluation, domestic market prices rise by 8 percent, and the overall GDP deflator by a similar amount. There is only a 2.5-percent rise in real GDP, production increases of, at most, 3 percent, virtually no change in the current account deficit (even though the volume of exports increases by 3.6 percent, this is outweighed by the increased cost of imports in domestic currency), and the consumption of the poor only increases by about 2 percent.

Though the magnitudes of the effects differ markedly as between the two assumptions about the labour market, the response of real GDP is positive, the elasticity of response with respect to the exchange rate varying between 0.25 and 1.0. This is in stark contrast to the estimate by Branson (1985) for Kenya, which was negative.

Table 1—Policy Simulations

| | Base Solution (K£M) | Percentage Change from Base Solution | | | | | |
| | | A* | | | B** | | |
| | | 10% Devaluation | 10% Boost to Investment | 10% Rise in Agricultural Efficiency | 10% Devaluation | 10% Boost to Investment | 10% Rise in Agricultural Efficiency |
| --- | --- | --- | --- | --- | --- | --- | --- |
| GDP at factor cost constant | 1,296.1 | 10.2 | 3.2 | 3.4 | 2.5 | 1.2 | 0.8 |
| GDP deflator price index | | 0.7 | -0.5 | 0.3 | 7.7 | 1.6 | 2.7 |
| Exports at constant prices*** | 478.1 | 15.7 | 0.8 | 12.1 | 3.6 | -1.8 | 8.4 |
| Imports at constant prices*** | 461.6 | 2.7 | 5.0 | 0.3 | 0.6 | 4.7 | 0.3 |
| Current account deficit | 51.9 | -38.3 | 42.4 | -48.7 | -1.2 | 50.5 | -35.6 |
| Net indirect taxes | 181.9 | 13.7 | 3.1 | 3.4 | 10.6 | 2.5 | 2.5 |
| Government expenditure | 215.2 | 12.2 | 2.7 | 3.4 | 10.4 | 2.5 | 2.9 |
| Total real investment | 294.2 | -2.1 | 10.0 | -4.9 | -0.5 | 10.0 | -4.2 |
| Urban unskilled—employment | 65.0 | 8.2 | 2.9 | 3.3 | — | — | — |
| —wages | | — | — | — | 9.7 | 3.7 | 3.8 |
| Urban skilled—employment | 162.2 | 8.3 | 2.9 | 3.2 | — | — | — |
| —wages | | — | — | — | 9.7 | 3.7 | 3.8 |
| Rural self-employment | 334.3 | 13.3 | 2.5 | 4.1 | — | — | — |
| Household real consumption: | | | | | | | |
| Urban poor | 89.2 | 6.5 | 3.1 | 5.1 | 1.6 | 1.9 | 3.3 |
| Rural very poor | 42.2 | 8.9 | 3.0 | 7.9 | 2.2 | 1.5 | 5.5 |
| Rural poor | 62.3 | 9.2 | 2.9 | 8.3 | 2.3 | 1.4 | 5.9 |
| Agricultural production | 513.9 | 15.0 | 2.7 | 14.7 | 3.4 | 0.4 | 10.8 |
| Agricultural producer prices | | 1.1 | -0.4 | -9.2 | 7.9 | 0.9 | -7.4 |
| Industrial production | 599.8 | 9.2 | 2.9 | 4.2 | 2.6 | 1.1 | 1.8 |
| Industrial producer prices | | 3.0 | -0.6 | -2.0 | 7.6 | 0.7 | -0.4 |
| Service production | 1,114.5 | 7.1 | 3.6 | 2.6 | 1.6 | 1.8 | 0.5 |
| Service producer prices | | 1.8 | -0.4 | -0.2 | 8.3 | 1.9 | 2.1 |
| Agricultural market prices | | 1.4 | -0.4 | -8.9 | 8.0 | 0.8 | -7.2 |
| Industrial market prices | | 5.2 | -0.4 | -1.3 | 8.4 | 0.5 | -0.3 |
| Service market prices | | 2.3 | -0.3 | -0.2 | 8.4 | 1.7 | 2.0 |

*Simulations A assume unlimited supply of labour. **Simulations B assume fixed supply of labour. ***Excludes factor payments and transfers.

The importance of the labour market situation in a particular country for the outcome of devaluation is clearly demonstrated. The model has shown that very different results can occur under general equilibrium and competitive assumptions in the labour and product markets. There is no need to call on non-competitive assumptions, mark-up pricing, institutional rigidities, or political power to explain such differences, and, in particular, the possible inflationary consequences of devaluation.

The simple analytics of devaluation with traded and non-traded sectors usually concludes that real wages have to decline, suggesting a real decline in living standards. The model shows that under Assumption A, where nominal wages remain constant and there are increases in both producer and market prices, real wages clearly decline. However, there is a large increase in employment leading to a significant rise in real household incomes and consumption. Under Assumption B, there is by definition no change in overall employment, but the rise in money wages is somewhat greater than the rise in both producer and market prices. There is in this case a small rise in real wages, which consequently feeds through to a small rise in real consumption.

**Investment.** In the case of increased investment, there is a very much less marked difference between the results under the two alternative assumptions about the labour market. Real GDP increases by 3.2 percent and 1.2 percent under Assumptions A and B, respectively. Because of the importance in Kenya of capital imports for investment purposes, there is an increase of 5 percent in the volume of imports. With little change in exports, the balance of payments deteriorates markedly. This, however, provides increased foreign savings, which help to finance the increased investment. Domestic savings increase only marginally. Increased wage income of between 2.2 and 3.7 percent comes from either increased employment or increased wages. With market prices virtually unchanged, these translate into real increases in household incomes and consumption of the poor of about 3 percent under Assumption A and 1.4–1.9 percent under Assumption B.

The estimates of the elasticity of GDP with respect to investment of between 0.32 and 0.12 under Assumptions A and B, respectively, imply incremental capital-output ratios for the whole economy of between 3 and 8, respectively. Though these appear to be wide differences, they are in the range of estimates given in Godfrey (1986).

**Productivity improvement in agriculture.** As would be expected, a 10-percent improvement in agricultural productivity results in a large boost to agricultural production, 14.7 percent and 10.8 percent under simulations A and B, respectively. This is accompanied by decreases in agricultural producer prices of 9.2 and 7.4 percent, respectively, insufficient to offset the rise in output. Agricultural exports increase significantly, leading to overall exports rising by 12.1 and 8.4 percent under the two simulations, respectively. Imports are virtually unchanged, so that there is a dramatic decrease in the current account deficit. However, there are only minor increases in production from the industrial and service sectors, so that overall GDP at factor cost improves by 3.4 percent under the assumption of unlimited supplies of labour and by only 0.8 percent under fixed supplies of labour. Indirect taxes and government expenditure both increase by about 3 percent, but there is a reduction in investment of 4 to 5 percent, allied to a drop in foreign savings. Either employment or wages of workers increases by 3 to 4 percent. In conjunction with reductions in market prices, particularly those of agricultural products, the real consumption of the poor rises significantly, especially that of the rural poor, where it is as much as 8.3 percent under simulation A and 5.9 percent under simulation B.

## Conclusions

A major conclusion is that the results of the policy simulations depend crucially on whether labour is assumed to be available in unlimited supplies or fixed within the time period considered. This is clearly particularly true for the effect on real GDP but also holds for most other variables of interest. The importance of having knowledge of the labour market situation in the particular country is thus highlighted. There has been controversy over this question for the Kenyan situation in the past, but our understanding of the present reality is that it is nearer the first assumption (Godfrey, 1986).

If that is the case, then, comparing the three policy options, a devaluation of the order of 10 percent appears very favourable. Only the productivity improvement option gives similar effects on exports, the balance of payments, agricultural production, and the real incomes of the poor. As would be expected, it does not give anything like the boost to overall GDP. We must, however, reiterate the proviso that we have no way of comparing the ease or difficulty of implementing identical proportionate changes in the three policy instruments.

Another important conclusion stems from the general equilibrium nature of the model. In discussing the impact of, say, devaluation on agricultural production, we are operating in the world of *mutatis mutandis*, not *ceteris paribus*. Under simulation A, export and producer prices for agricultural commodities only increase by 1.1 percent. Yet there is a massive 15 percent increase in production. This comes about through a large increase in employment and the use of intermediate inputs and a small reduction in the capital stock and is accompanied by a shift in demand through higher real incomes of all households. Under simulation B, agricultural export and producer prices rise by 7.9 percent, almost in line with the devaluation, yet there is only a 3.4 percent increase in production. This is principally because there is no change in employment and the rise in wages is greater than that of producer prices. There is only a small shift in demand. The great merit of the general equilibrium approach is that the results take into account simultaneously all the various linkages and feedback mechanisms between the sectors, between the product and factor markets, and between these and households, and so on, which no amount of juggling with partial equilibrium estimates can match.

Our final conclusion concerns income distribution. Though we have not dwelt on this in the paper, our results support the conclusions of many CGE analyses; i.e., that the overall household distribution of incomes remains remarkably stable in face of quite significant changes in policy variables. However, our concern was not primarily with income distribution but with the absolute incomes of the poor. We are heartened by the fact that, although the aggregative nature of our model has to be borne in mind, all three policy options point to improvements in their real incomes and consumption.

## Notes

[1]University of Oxford and United Nations Development Programme, Gambia, respectively.

[2]See, for example, on the first point, Rose (1985), Helleiner (1986), Smith (1988), and Commander (1989), and, on the second point, Cornia et al. (1987).

[3]See, for instance, Corbo et al. (1987).

[4]See, for example, Dervis et al. (1982), Helleiner (1987), and Demery and Addison (1987).

[5]This is explained fully in Akinboade (1990).

[6]A review of the evidence on remittances in Kenya is given in Akinboade (1990).

[7]A good discussion of this problem appears in Drud et al. (1985).

[8]See, for instance, the divergent views of ILO (1972), Collier and Lal (1986), and Godfrey (1986).

## References

Akinboade, O.A., *Agriculture, Income Distribution and Policy in Kenya: A SAM Based General Equilibrium Analysis*, unpublished Ph.D. thesis, University of Oxford, Oxford, UK, 1990.

Branson, W.H., *Stabilization, Stagflation and Investment Incentives: The Case of Kenya, 1979–80*, Discussion Papers in Economics No. 89, Woodrow Wilson School of Public and International Affairs, Princeton, N.J., USA, 1985.

Collier, P., and Lal, D., *Labour and Poverty in Kenya, 1900–80*, Clarendon Press, Oxford, UK, 1986.

Commander, S., *Structural Adjustment and Agriculture*, James Currie for Overseas Development Institute, London, UK, 1989.

Corbo, V., Goldstein, M., and Khan, M. (Eds.), *Growth-Oriented Adjustment Programs*, International Monetary Fund and World Bank, Washington, D.C., USA, 1987.

Cornia, G.A., Jolly, R., and Stewart, F., *Adjustment with a Human Face*, Clarendon Press for UNICEF, Oxford, UK, 1987.

Demery, L., and Addison, T., "Stabilization Policy and Income Distribution in Developing Countries," *World Development*, Vol. 15, No. 12, 1987, pp. 1483–1498.

Dervis, K., de Melo, J., and Robinson, S., *General Equilibrium Models for Development Policy*, Cambridge University Press for the World Bank, Cambridge, UK, 1982.

Drud, A., Grais, W., and Pyatt, G., *An Approach to Macroeconomic Model Building Based on Social Accounting Principles*, Discussion Paper Report No. DRD150, Development Research Department, World Bank, Washington, D.C., USA, 1985.

Godfrey, M., *Kenya to 1990: Prospects for Growth*, Special Report No. 1052, Economist Intelligence Unit, London, UK, 1986.

Helleiner, G.K. (Ed.), *Africa and the International Monetary Fund*, International Monetary Fund, Washington, D.C., USA, 1986.

Helleiner, G.K., "Stabilization, Adjustment and the Poor," *World Development*, Vol. 15, No. 12, 1987, pp. 1499–1513.

ILO (International Labour Office), *Employment, Incomes and Equality*, Geneva, Switzerland, 1972.

Rose, T. (Ed.), *Crisis and Recovery in Sub-Saharan Africa*, Organization for Economic Cooperation and Development, Paris, France, 1985.

Smith, L.D., *Structural Adjustment, Prices and the Agricultural Sector in Sub-Saharan Africa*, Occasional Paper No. 2, Centre for Development Studies, University of Glasgow, Glasgow, UK, 1988.

---

# Discussion Opening—*James V. Stout* (US Department of Agriculture)

The paper describes the results of a model of the Kenyan economy under two assumptions on labour (unlimited supplies of labour at a fixed wage and fixed labour supply with wages determined endogenously) and for three scenarios (devaluation of the Kenyan pound, improvement in agricultural productivity, and a scenario which is described as an exogenous increase in investment).

The authors of this paper have not given us enough information to allow us to make full use of their model and results. One of the major conclusions highlighted is that the results of the model depend crucially on whether labour is assumed to be fixed or available in unlimited supply at a fixed wage. This result is qualitatively not surprising. Has this model captured some particular characteristic or characteristics of the Kenyan economy or can the results of this model can be generalized to other developing nations facing similar circumstances and similar sorts of policy alternatives? We would also benefit from more information on parameter values used in the model and a more detailed description of the model structure.

In the scenario described as an exogenous increase in investment, since the investment comes first (i.e., it spurs the economy rather than resulting from an acceleration of the economy), the economic stimulus could be better described as an exogenous increase in government spending financed by deficits. If this is the case, the fact that this scenario involves a transfer of purchasing power from the future is an important reason why it shows significant welfare benefits in the static general equilibrium framework of this model.

*[Other discussion of this paper and the authors' reply appear on page 306.]*

# Macroeconomic Adjustment and the Performance of the Agricultural Sector: A Comparative Study of Cameroon and Kenya

*William A. Amponsah and Leroy J. Hushak*[1]

**Abstract:** The interaction between macroeconomic and trade issues and agriculture is not well understood. In sub-Saharan Africa, this relation becomes critical in recent structural adjustment programmes, especially in countries where agriculture is the mainstay of the economy. This paper examines empirically examples of a non-liberalized economy and a gradually liberalizing economy in the region. It demonstrates that a sufficiently monolithic and inefficient agricultural marketing structure results in the slow transmission of foreign exchange gains to farm gate prices that constrain potential supply response. Monetary policy impacts seem to be a function of greater macroeconomic liberalization than non-liberalization, especially when, in the latter case, domestic currencies are pegged directly to foreign currencies.

## Introduction

There is a general need adequately to treat the interaction between macroeconomic and trade issues and agriculture. This link is particularly critical for many sub-Saharan African economies, but is often either neglected or not sufficiently understood. The designs of recent structural adjustment facilities in sub-Saharan Africa may not adequately incorporate the implications arising from the nature of the prevailing agricultural production and marketing régimes, which may seriously hamper the long-run success of those adjustment programmes, especially when a country is very dependent on agricultural exports.

The outcome of macroeconomic adjustment programmes may depend critically on the policy framework in the agricultural sector; implying that some agricultural sector régimes may be consistent with successful macroeconomic adjustment, while others may not. An evolving agricultural policy framework subsequently becomes a necessary integral part of such adjustment. For simplicity of exposition, the World Bank's taxonomy of agricultural régimes (non-liberalized, gradually liberalizing, and fully liberalized) is used.

The primary characteristics of a non-liberalized régime are found in the major government and parastatal roles in a single-channel marketing structure, a small or non-existing private sector role in marketing, and rigid structures of controlled prices for consumers and producers. In a fully liberalized régime, there is a reduced government role, a multiplicity of marketing channels with strong private sector participation, and market-based pricing mechanisms. A gradually liberalizing régime is an intermediate level, a gradually evolving agricultural framework from a non-liberalized to a fully liberalized régime.

The focus of macroeconomic adjustment is usually exchange rate reforms, production incentives, monetary stability, and trade reforms. In sub-Saharan Africa, exchange rate reforms tend to be synonymous with currency devaluations. When complemented by reduced industrial protection, it is believed to result in a shift in internal terms of trade towards agricultural production.

A potentially perverse effect on internal terms of trade often arises under non-liberalized or gradually liberalizing régimes when the agricultural marketing structure is sufficiently monolithic and inefficient. It results in the slow transmission of exchange rate changes to farm gate prices even as imported input prices adjust rapidly to the devalued currency. The incidence of taxation on farmers poses potential barriers to production response. In sub-Saharan Africa, this raises immense challenges in slow-growing economies, especially when exacerbated by monetary instability.

Trade liberalization must be intimately related to the liberalization of agricultural marketing and prices. Yet sometimes an important industrial subsector may be massively protected under an overvalued and quantitatively restricted external trade régime. The level of protection may be reduced under trade liberalization, yet the agricultural marketing scenario remains non-liberalized, whereby protected parastatal monopolies function inefficiently.

Of the 22 sub-Saharan African countries that have been undertaking structural adjustment since 1980 through World Bank and/or International Monetary Fund auspices, Kenya fits the above profile as a gradually liberalizing economy, while Cameroon is an example of a non-liberalized economy.

## Analytical Approach

A large proportion of agricultural production in sub-Saharan Africa is traded to earn sorely needed foreign exchange. Thus, the link between the external sectors and agriculture takes place through the real exchange rate (measured in units of domestic currency per US$ foreign price index/domestic price index). The effect of commercial policy on the structure of incentives in the agricultural sector is analysed using the traded-to-non-traded goods model and abstracting from the elasticity approach to balance-of-payments disequilibria.

The analytical approach of the study of the performance of the goods, money, and financial markets follows from Garcia Garcia and Montes Llamas (1988). Whenever increases in government expenditure are financed through taxes, an excess demand for non-traded goods develops if the marginal propensity of the government to spend on non-traded goods is larger than the corresponding marginal propensity of the private sector. Thus, the prices of non-traded goods rise.

These occurrences reduce incentives to produce in other tradeable activities such as agriculture and industry, which may explain the economic performance of the selected countries. So the supply responses to the agricultural ($AG^s$), industrial ($NA^s$), and service ($NT^s$) sectors are determined to capture the various influences in terms of the relative prices of agriculture and industry to services. It is possible that increases in the world interest rate, $r^*$ (measured by the London Interbank Offer Rate), may have transmitted to the lean capital markets of these countries to contribute to recessions and, especially, high fluctuations in their monetary reserves.

Fluctuations in the monetary reserves ($dR/M$) are usually attributed to the effects of changes in the current and capital accounts. The analysis isolates the real per capita GDP ($PGDP$), the per capita value of exports ($PVEX$), or the per capita value of agricultural exports ($PAGVEX$)—including coffee exports ($COVEX$) in Kenya—and the per capita balance of payments ($PBOP$) as the determinants of the current account. The domestic-to-foreign interest rate differential, $r/r^*(1+e)$, and the difference between per capita debt and debt service ($PDBT$) are the major determinants of the capital account. This is consistent with Fleming (1962) and Mundell (1962 and 1968).

The monetary ($M^d/P$) specification follows Dornbusch's demand-for-money equation to determine the response to interest rate changes in a semi-open economy (Dornbusch, 1980, p. 176). The impact of certain monetary instruments on the real exchange rate is also measured, abstracting from the portfolio balance approach of the asset market and defining the real exchange rate as a logarithmic function of the real market exchange rate ($LEXCH$), lagged real domestic interest rate ($r_{-1}$), $PGDP$, and $PBOP$. Consistency with the terms of trade ($TOT$) definition is achieved by defining the real exchange rate as the ratio of the price of tradeables to the price of non-tradeables ($LPCNA$). The size of government expenditure relative to GDP ($EGDP$) and $PGDP$ are also included.

A general equilibrium framework similar to Garcia Garcia (1981), Oyejide (1986), and Tsibaka (1986) is also used in arriving at an equation that allows measurement of the incidence of governments' trade and exchange rate policy taxation on agriculture. The underlying static equilibrium approach follows Dornbusch (1974) and Sjaastad (1980) in that it demonstrates that exchange rate policies sometimes have global economic repercussions quite different from those intended by policy makers. It is also consistent with Balassa (1981), Little *et al.* (1970), and Krueger *et al.* (1981), whose studies dealt with how incentive systems and resource flows in developing countries have concentrated on the degree of protection for competing manufacturing activities by trade and exchange rate policies.

# Results

The estimated equations for Cameroon and Kenya are presented in Tables 1–7. They cover the 1970–88 period, and the data are derived from World Bank, FAO (*Production* and *Trade Yearbooks*), and IMF (*International Financial Statistics*) sources. The *t*-values are presented in parentheses below their respective coefficients.

The static supply response equations (Table 1) for the real sector did not provide the expected signs and significance of price coefficients. However, modifications of Nerlove's (1956) partial adjustment model provided more consistent results. The own prices for Cameroon were all positive as hypothesized. However, for Kenya, the own prices for agriculture and industry were all negative, which probably suggests the potential perverse effect of agriculture in an economy that is gradually shifting its dependency towards the industrial and service sectors. However, a generally higher supply response to the previous year's relative prices in Kenya was observed. The elasticity of acreage response (*XC*) expectation to the previous year's agricultural prices (*PAG*) was higher for Kenya than for Cameroon, and conformed to the non-zero results of Grilliches (1959). Farmers in both countries, therefore, respond rationally to price incentives. For both countries, all sectors exhibited positive trend variability, even though Kenya's was more significant.

Table 1—Supply Response: Partial Adjustment

| | | Constant | $PAG_{-1}$ | $PNA_{-1}$ | $AG^s_{-1}$ | $XC_{-1}$ | $T$ | $R^2$ | $DW$ |
|---|---|---|---|---|---|---|---|---|---|
| $AG^s$ | C | 242.05 (2.96) | 174.36 (0.95) | −161.08 (1.21) | 0.03 (0.10) | − | 15.26 (2.29) | 0.68 | 2.04 |
| | K | 100.93 (18.11) | −1,005.30 (0.81) | 290.23 (0.24) | 0.17 (1.09) | − | 527.41 (5.14) | 0.98 | 1.82 |
| $NT^s$ | C | 268.31 (3.47) | −25.27 (0.17) | −16.02 (0.14) | − | − | 41.86 (8.84) | 0.92 | 1.18 |
| | K | 1,1096.0 (10.87) | −4,359.0 (2.44) | 3,957.7 (2.41) | − | − | 1,364.4 (24.97) | 0.98 | 0.66 |
| $NA^s$ | C | 15.61 (0.15) | 7.98 (0.04) | 82.43 (0.47) | − | − | 51.71 (7.31) | 0.89 | 0.36 |
| | K | 3,915.00 (6.41) | 2,001.20 (1.87) | 32.13 (0.33) | − | − | 467.93 (14.32) | 0.96 | 1.32 |
| $XC$ | C | 553.46 (12.12) | 111.13 (1.11) | −240.18 (2.59) | − | 0.73 (5.43) | 5.76 (1.18) | 0.94 | 1.59 |
| | K | 4,7366.0 (112.88) | 2,201.40 (0.20) | 38,336.0 (2.48) | − | 0.10 (2.03) | 1,306.90 (5.67) | 0.72 | 1.56 |

Note:  C = Cameroon and K = Kenya.  Absolute values of *t*-statistics are in parentheses.

The potential impact of agricultural price incentives becomes very apparent in both non-liberalized and gradually liberalizing examples.  An increase in agricultural price should reduce the supply of non-agricultural tradeables in the short-run as resources (including human capital) are transferred to agriculture. This resource movement effect, however, is expected to create an excess demand for all tradeables (since industrial tradeables depend on agricultural inputs in both countries) and, hence, should reduce the price of non-traded goods. On the demand side, increases in agricultural prices also increase income and subsequently expenditures, which drive up the demand for non-agricultural tradeables. Thus, the prices of industrial tradeables go up, thereby driving down the prices of non-traded goods in the long run. Such scenarios have strong implications in narrowing the rural/urban terms of trade—a necessary expectation in any adjustment scheme.

In Cameroon, increases in both agricultural and industrial tradeable prices do indeed reduce the supply of non-tradeable goods. In Kenya, however, agricultural prices reduce the supply of non-tradeables more significantly, but industrial prices significantly increase non-tradeable goods supply.

The demand for money equation (Table 2) provided consistent results as expected for both countries, except for the negative sign on per capita income in Cameroon. For Kenya, the marginal propensity to spend out of increased income is 0.002, which is very small compared to that obtained for the USA by Alexander (1952). Thus, locally induced investment is very small in Kenya. For Cameroon, the negative marginal propensity probably suggests the potential existence of monetary quirks—an adjunct to the Cameroon-Francophone monetary alliance.

Table 2—Demand for Money

| | | Constant | $PI$ | $r$ | $(M^d/P)_{-1}$ | $TOT$ | $R^2$ | $DW$ |
|---|---|---|---|---|---|---|---|---|
| $M^d/P$ | C | 0.47 (0.60) | −0.001 (0.71) | 0.30 (2.53) | 0.52 (3.83) | 0.006 (0.75) | 0.88 | 1.35 |
| | K | −46.48 (0.69) | 0.002 (0.08) | 4.67 (2.49) | 0.56 (2.90) | 0.46 (1.67) | 0.75 | 1.93 |

The interest rate equation (Table 3) shows the expected signs for all variables, except for the rate of devaluation in Cameroon. While it is possible that Kenya's domestic interest rate follows the parity rate (and, thus, allows for flexibility in capital mobility), the negative value of the rate of devaluation ($e$) and the non-significant parity rate in the second real interest rate equation for Cameroon seem to suggest that the domestic interest rate may depend on external factors but not on internal factors. Domestic monetary policy may, therefore, have no impact on Cameroon's economy; however, it does have an impact on Kenya's economy.

Table 3—Interest Rate

| | | Constant | $r^*$ | $e$ | $PGDP$ | $PBOP$ | $r^*(1+e)$ | $R^2$ | $DW$ |
|---|---|---|---|---|---|---|---|---|---|
| $r$ | C | 1.33 (1.63) | 0.02 (0.33) | −1.10 (1.38) | 0.04 (5.15) | 0.03 (0.74) | − | 0.87 | 1.92 |
| | K | −0.45 (0.10) | 0.41 (2.70) | 5.40 (8.17) | −0.005 (0.40) | 0.009 (0.18) | − | 0.75 | 1.93 |
| $r$ | C | 0.96 (1.45) | − | − | 0.03 (5.07) | 0.002 (0.06) | 0.02 (0.64) | 0.84 | 1.35 |
| | K | 2.67 (0.36) | − | − | −0.004 (0.18) | 0.008 (1.12) | 0.37 (4.03) | 0.57 | 1.04 |

The changes in international reserves equation (Table 4) provides the customary mixed results for small countries. For example, the signs on *PVEX* and *PCOVEX* were negative for Kenya. For Cameroon, *PAGVEX* was negative, demonstrating agriculture's weak role as an export sector in the economy's reserve accumulation. Also, *PDBT* was positive for Cameroon but negative for Kenya. Generally, Kenya's relatively greater debt-to-debt service ratio negatively affects its reserves.

For both countries, the parameters on the interest rate differential were positive but insignificant, which suggests some international mobility of capital into the countries. However, real income and the value of exports play weak roles in the current account portion of the balance of government reserves.

Table 4—Changes in International Reserves

| | | Constant | PGDP | PBDP | $r/r^*(1+e)$ | PVEX | AGVEX | PCOVEX | PDBT | PBOP | $R^2$ | DW |
|---|---|---|---|---|---|---|---|---|---|---|---|---|
| dR/M | C | 0.04 (0.16) | -0.001 (0.45) | – | 0.23 (0.66) | 0.005 (0.08) | – | – | 0.009 (0.38) | -0.03 (0.08) | 0.29 | 2.21 |
| | K | -0.09 (0.27) | – | -0.008 (0.79) | 0.22 (0.80) | -0.002 (0.08) | – | – | -0.001 (0.18) | 0.01 (4.98) | 0.77 | 1.82 |
| dR/M | C | 0.30 (3.07) | 0.09 (3.08) | – | 0.28 (1.12) | – | -0.09 (1.97) | – | – | 0.004 (1.00) | 0.48 | 2.95 |
| | K | -0.12 (0.34) | – | -0.004 (0.44) | -0.26 (0.87) | -0.003 (0.3) | – | -0.001 (0.25) | -0.003 (0.4) | -0.02 (4.58) | 0.77 | 1.86 |

The real exchange rate equation (Table 5) provides consistent results for Kenya, except for PGDP. However, the results of the second equation provide a better fit for Kenya. As a result of its pegged exchange rate with the French franc, it is possible that the level of real output and prices of tradeables do not positively affect real exchange rate movements in Cameroon. In other words, an improvement in terms of trade towards tradeables and an increase in government expenditure would tend towards an increase in the relative price of non-traded goods. The negative effects of government expenditure on the relative price of tradeables also suggest that Cameroon's propensity to spend on non-traded commodities is higher than that for traded commodities.

Table 5–Exchange Rate

| | | Constant | $r_{-1}$ | ln PGDP | ln PNAG | ln PNA | GBOP | ln EGDP | $R^2$ | DW |
|---|---|---|---|---|---|---|---|---|---|---|
| LEXCH | C | 5.64 (23.2) | 0.16 (2.5) | -0.39 (1.8) | -0.26 (0.63) | 0.93 (1.2) | -0.51 (0.23) | – | 0.48 | 1.43 |
| | K | 2.69 (22.7) | 0.04 (4.1) | -0.11 (2.3) | 0.10 (0.3) | 0.54 (1.6) | 0.57 (0.5) | – | 0.63 | 0.85 |
| LPCNA | C | 0.72 (16.2) | 0.02 (1.5) | -0.10 (2.7) | -0.42 (5.6) | 0.33 (2.8) | – | -0.007 (0.28) | 0.82 | 1.65 |
| | K | 0.82 (26.7) | 0.005 (1.2) | 0.04 (3.2) | 0.46 (5.6) | 0.14 (1.4) | – | -0.02 (0.8) | 0.76 | 2.52 |

Note: All variables are lagged.

A key observation for both countries is that the interest rate has a positive effect on the price of traded goods. So policies that increase real interest rates lead to an increase in the real exchange rate, *ceteris paribus*. This relationship is in agreement with the presumption that increases in the interest rate also generally increase relative prices of tradeables. Nevertheless, the relative prices of agricultural and non-traded outputs equation (Table 6) seems to suggest that increases in the interest rate decrease the relative price of agriculture (logarithmically) in Kenya but not in Cameroon. An improvement in the terms of trade of agriculture and industry, real income, and government expenditure (biased towards agriculture) should, therefore, increase the relative price of agriculture in Kenya. Indeed, this is a cardinal imperative in any gradually liberalizing marketing régime.

The commercial policy equations (Table 7) yield excellent results for the incidence of taxation on the agricultural sector. Highly significant levels of taxation of 64 percent and 83 percent, respectively, were obtained for agriculture in Cameroon and Kenya. These results are similar to earlier results obtained for Nigeria (Oyejide, 1986) and for Colombia (Garcia Garcia, 1981). The results confirm the slow transmission mechanism of price incentives to agricultural producers, especially in a gradually liberalizing economy such as Kenya. Unfortunately, data unavailability poses a binding constraint on the examination of sector-specific policies, in terms of the direct and indirect effects on the main agricultural export sectors (coffee and tea) in Kenya.

Table 6—Relative Prices of Agricultural and Non-Traded Output

|   |   | Constant | $\ln (PGDP)_{-1}$ | $r_{-1}$ | $\ln TOT_{-1}$ | $BGDP_{-1}$ | $R^2$ | DW |
|---|---|---|---|---|---|---|---|---|
| $\ln$ PAG/ PNT | C | −0.25 (1.81) | 0.15 (0.68) | 0.03 (0.65) | −0.19 (1.04) | 1.60 (0.84) | 0.48 | 2.10 |
|  | K | −0.19 (2.16) | 0.27 (2.73) | −0.01 (1.21) | −0.37 (2.54) | 2.13 (2.26) | 0.54 | 1.47 |

Table 7—Commercial Policy Impacts

|   |   | Constant | $\ln (PNA/PAG)$ | $\ln PGDP$ | PBOT | $R^2$ | DW |
|---|---|---|---|---|---|---|---|
| $\ln PNT/PAG$ | C | 0.27 (0.52) | 0.64 (6.23) | −0.40 (0.40) | −0.005 (1.61) | 0.89 | 0.98 |
|  | K | −4.13 (2.38) | 0.83 (8.39) | 0.50 (2.36) | 0.0004 (0.65) | 0.91 | 2.08 |

## Conclusion

The possibility that a successful macroeconomic adjustment process in a sub-Saharan African country may not necessarily translate into a good performance by the agricultural sector is provided by the Kenyan example. In the medium to long term, there may be negative repercussions on the economy-wide adjustment process, especially when the slow transmission mechanism of relative prices does not allow better price incentives to farmers. Farmers in both countries respond rationally to price incentives. The high incidence of taxation on agriculture seem to suggest that, in the gradually liberalizing régime, the agricultural sector is not liberalized enough. In any case, the gradually liberalizing régime demonstrated a higher propensity for production growth than the non-liberalized régime.

Monetary policy has no impact on the Cameroon economy, but does on the Kenyan economy. But the monetary sector's performance in Cameroon may reflect not the non-liberalized nature of its régime but rather its monetary relationship with France. Capital flows into both countries are not very significant, even though flows into Cameroon may be constrained by its relationship with France. Increases in interest rate in the medium run translate into increases in the prices of tradeables, including agricultural goods for Kenya. However, government policy seems to delay price incentives to farm producers. Faced with a volatile international commodities market, it is imperative that a gradually liberalizing economy that is dependent on agriculture reduce the incidence of taxation on agriculture to allow for rapid domestic production response.

The study also demonstrates the possibility that, for an non-liberalized economy whose export sector does not depend heavily on agriculture, formal macroeconomic structural adjustment may not be a necessary requirement in effecting efficient performance in the agricultural sector. Furthermore, it proves that a more diversified export sector may offer other alternatives that reduce the high incidence of taxation on agriculture.

## Note

[1]Ohio State University.

## References

Alexander, S.S., *The Effects of Devaluation on a Trade Balance*, Staff Paper No. 2, International Monetary Fund, Washington, D.C., USA, 1952, pp. 263–278.

Balassa, B., *The Structure of Protection in Developing Countries*, Johns Hopkins University Press, Baltimore, Md., USA, 1981.

Dornbusch, R., "Tariffs and Non-Traded Goods," *Journal of International Economics*, 1974, pp. 177–185.

Dornbusch, R., *Open Economy Macroeconomics*, Basic Books, Inc., New York, N.Y., USA, 1980.

Fleming, M., *Domestic Financial Policies Under Fixed and Under Floating Exchange Rates*, Staff Paper No. 9, International Monetary Fund, Washington, D.C., USA, 1962.

Garcia Garcia, J., *The Effects of Exchange Rates and Commercial Policy on Agricultural Incentives in Colombia: 1953–78*, Research Report No. 24, International Food Policy Research Institute, Washington, D.C., USA, 1981.

Garcia Garcia, J., and Montes Llamas, G., *Coffee Boom, Government Expenditure, and Agricultural Prices: The Colombian Experience*, Research Report No. 68, International Food Policy Research Institute, Washington, D.C., USA, 1988.

Grilliches, Z., "The Demand for Inputs in Agriculture and a Derived Supply Elasticity," *Journal of Farm Economics*, Vol. 38, 1959, pp. 309–322.

Krueger, A.O., et al. (Eds.), *Trade and Employment in Developing Countries*, University of Chicago Press for the National Bureau of Economic Research, Chicago, Ill., USA, 1981.

Little, I.M.D., Scitovsky, T., and Scott, M., *Industry and Trade in Developing Countries*, Oxford University Press, Oxford, UK, 1970.

Mundell, R.A., *The Appropriate Use of Monetary and Fiscal Policy for Internal and External Stability*, Staff Paper No. 9, International Monetary Fund, 1962, pp. 70–77.

Mundell, R.A., "Capital Mobility and Stabilization Policy Under Fixed and Flexible Exchange Rates," *Canadian Journal of Economics*, Vol. 29, pp. 475-485. Reprinted in: Caves, R.E., and Johnson, H. (Eds.), *Readings in International Economics*, Irving, Homewood, Ill., USA, 1968, pp. 487–499.

Nerlove, M., "Estimates of the Elasticities of Supply of Selected Agricultural Commodities," *Journal of Farm Economics*, Vol. 38, 1956, pp. 496–509.

Oyejide, A.T., *The Effects of Trade and Exchange Rate Policies on Agriculture in Nigeria*, Research Report No. 55, International Food Policy Research Institute, Washington, D.C., USA, 1986.

Sjaastad, L.A., "Commercial Policy, True Tariffs, and Relative Prices," in Black, J., and Hindley, B. (Eds.), *Current Issues in Commercial Policy and Diplomacy*, St. Martin's Press, New York, N.Y., USA, 1980.

Tsibaka, T.B., *The Effects of Trade and Exchange Rate Policies on Agriculture in Zaire*, Research Report No. 56, International Food Policy Research Institute, Washington, D.C., USA, 1986.

## Discussion Opening—*Timothy O. Williams* (International Livestock Centre for Africa)

Economic theory and experience both suggest that in sub-Saharan African countries, where agriculture accounts for a substantial share of GNP and trade, macroeconomic adjustment programmes will impinge on the agricultural sector. Indeed, the success of such adjustment programmes depends largely on improved performance of the agricultural sector. By assessing the impact of adjustment programmes on the performance of the agricultural sector in two sub-Saharan African countries, this paper addresses a topic that is of relevance not only to the countries studied but also for other developing countries currently undertaking structural adjustment programmes.

However, in assessing agricultural sector performance, the paper uses an eclectic framework that requires further elaboration in places. First, in estimating the response of agricultural production to adjustment policies, the approach used ignores the importance of non-price variables in determining supply response. There is ample evidence that weather, infrastructure, and extension services together affect aggregate agricultural output more than prices alone. Incorporation of these structural variables into the framework employed would produce different comparative results, given that Kenya has relatively better infrastructure and extension services than Cameroon.

Second, the framework used concentrated entirely on the performance of agricultural exports. To the extent that macroeconomic adjustment programmes are executed to promote growth and efficiency in the agricultural sector as a whole, the performance of other aggregates (such as domestic food production) needs to be evaluated as well. Available evidence suggests that net exports respond quickly to higher prices induced by adjustment policies, but domestic food supply may not increase as rapidly. Further empirical evidence on the response of these two aggregates is needed in order to improve the design of structural adjustment programmes.

Another point to note is that there is little discussion of the effects of agricultural sector-specific reforms on production incentives in the two countries. Yet the conclusions reached in this paper and the results of other studies have shown that macroeconomic adjustment policies (involving devaluation, fiscal austerity, and reduction in industrial protection) alone are unlikely to have a significant impact on aggregate agricultural output unless accompanied by a number of sectoral reforms, including elimination of export taxation, reduction of subsidies to agricultural marketing boards, and improvement in services.

On a more general note, a careful interpretation of some of the estimated parameters reported in the paper produces results that are difficult to reconcile with available data on the countries studied.

Nonetheless, I share the authors' concern for a better understanding of the links between macroeconomic policy changes and agricultural incentives. I consider the approach used in this paper as a useful first step which, with elaboration, can produce results that could be used to improve the design of structural adjustment programmes.

*[Other discussion of this paper and the authors' reply appear on page 306.]*

# A Computable Household Model Approach to Policy Analysis:
# Structural Adjustment and the Peasantry in Morocco

*A. de Janvry, M. Fafchamps, M. Raki, and E. Sadoulet*[1]

**Abstract:** This paper opens a new field in quantitative policy analysis by developing a methodology for the construction and simulation of computable non-separable household models. Non-separability originates in market failures and in a binding credit constraint, both of which transform the products and factors affected into non-tradeables. The methodology is applied to a simulation of the impact on Moroccan peasant households of the new pricing rules for cereals introduced by structural adjustment. The results show that the elasticity of supply of tradeables is hampered by the presence of nontradeable factors in the household; that small farmers are pushed on to the labour market as a strategy to relax a credit constraint; and that technological change should be directed at enhancing the productivity of non-tradeables to increase the elasticity of supply response of tradeables. Further, a rising price for cereals shifts the farm economy from animals to crops. Nevertheless, as the price of tradeable animal forage rises, child labour used for herding in the commons is substituted for this forage, and children's work load increases. The expected consequence is increased school absenteeism and more overgrazing in the commons, two historical curses of Moroccan underdevelopment.

## Structural Adjustment and Moroccan Peasants

As in many other countries, structural adjustment in Morocco has led to the definition of new rules for price formation that bring the domestic prices of tradeables closer to international prices. With a long history of price discrimination against the cereals sector, these new rules have the potential significantly to raise the prices of hard and soft wheat, barley, and maize. Through multimarket and general equilibrium effects, these price changes also affect all other prices in the economy, including wages and the exchange rate. Because, in Morocco, smallholders in dryland areas are the main producers of cereals, and because it is among these households that the greatest incidence of absolute poverty is found, the new prices have the potential not only to induce import substitution in cereals if these producers are able to respond to price incentives but also to help reduce poverty among this segment of Moroccan society.

The smallholder economy is, however, highly heterogeneous as it relies on a complex portfolio of activities that include not only a variety of crops but also, very prominently, livestock and wage earnings. Smallholders are also important buyers of many of the commodities whose prices are rising. Further, their economy is characterized by numerous market failures (in particular for child labour, which is essential for the herding of animals in the commons) and by credit constraints that limit their adaptability to the new set of costs and incentives. As a result, it is not clear whether these households will benefit or not from the price adjustments or how different members of the household will be affected, women and children in particular.

This paper develops the methodology of computable non-separable household models so as to allow measurement and simulation of the complex implications on different types of households and different household members of price changes brought about by structural adjustment. It is applied to small and medium farm households in the Chaouia, a dryland area with extensive poverty.

## Household Model with Market Failures and Credit Constraint

The household produces ($q \geq 0$) hard wheat, soft wheat, coarse grains, other crops (legumes, fruit, and vegetables), forage, milk, and meat. Nonagricultural sources of income include handicrafts and services and the sale of labour. Production factors ($q \leq 0$) used are machinery and fertilizers; coarse grains and forage; male, female, and child labour; and the depreciation of fixed factors. Products and factors are related through the production technology $G(q, Z)$, where $Z$ is a vector of structural characteristics of the farm household and fixed factors (land, livestock, and capital). The household consumes ($c \geq 0$) hard wheat, soft wheat, coarse grains, other crops, milk, meat, nonagricultural goods and services, and leisure time (male, female, and child) and also saves.

The household has initial endowments ($T \geq 0$) in total time (male, female, and child) and receives net transfers $S$. Expenditure on machinery, fertilizers, forage, and hired labour has to be incurred ahead of harvest, and this requires financial liquidity at that time of year. For this, the household has access to credit in an exogenous amount $K$ (including transfers received ahead of harvest) and to cash income from wage earnings if it is a net seller of labour. Since all farms have a surplus of coarse grains, they carry over stocks of these grains with the result that they do not enter the liquidity constraint. According to the net amount of these entries and outlays that occur before harvest, the credit constraint may or may not be binding.

The household may be a net seller or net buyer of any product and factor. It is a price taker ($\bar{p}$) for all products and factors for which markets exist (or more exactly for which the subjective equilibrium price falls outside a price band between risk equivalent sale and purchase prices). For milk and child labour, market failure (or a subjective equilibrium within the effective price band) implies that an internal equilibrium must obtain between the supply ($q + T$) and demand ($c$) of these nontradeable commodities.

The household maximizes a utility function, $U(c, z)$, where $z$ denotes exogenous household characteristics, with respect to production and consumption decisions subject to a cash constraint, a credit constraint, a technology constraint, and equilibrium conditions for tradeables and non-tradeables. Goods are decomposed into three subsets: tradeables that are not subject to a credit constraint, $TNC$; tradeables subject to a credit constraint, $TC$ (jointly, these two subsets of tradeables are also indexed as $T$); and non-tradeables, $NT$.

The household's problem is thus to:

$$MaxU(c, z)$$
$$c, q$$

subject to:

$$\sum_{i \in T} P_i(q_i + T_i - c_i) + S \geq 0 \qquad \text{(cash constraint)}$$

$$\sum_{i \in TC} P_i(q_i + T_i - c_i) + K \geq 0 \qquad \text{(credit constraint)}$$

$$G(q, Z) = 0 \quad \text{(production technology)}$$

(1a) $p_i = \bar{p}_i, \ i \in T$ (exogenous market price for tradeables)

(1b) $q_i + T_i = c_i, i \in NT$ (equilibrium for non-tradeables)

The Lagrangean length scale associated with the constrained maximization problem is written as:

$$L = U(c, z) + \lambda \left[ \sum_{i \in T} \bar{P}_i(q_i + T_i - c_i) + S \right] + \eta \left[ \sum_{i \in TC} \bar{P}_i(q_i + T_i - c_i) + K \right] + \phi G(q, Z) + \sum_{i \in NT} \theta_i(q_i + T_i - c_i)$$

The three types of goods can be treated symmetrically in the first-order conditions by defining endogenous decision prices as follows:

$$P_i^* = \bar{P}_i, i \in TNC$$

$$P_i^* = \bar{P}_i(1+\lambda_c), \quad \lambda_c = \eta/\lambda, \quad i \in TC$$

$$P_i^* = \bar{P}_i \; \theta_i/\lambda, \quad i \in NT$$

In the reduced form, production decisions are represented by a system of supply and factor demand functions in the endogenous decision prices $p^*$ that derive from maximizing a generalized profit function $\Pi$ for all tradeables and non-tradeables:

(2a) $\Pi^* = \sum P_i^* q_i$

(2b) $q = q(P_i^*, Z)$

Consumption decisions in terms of the $p^*$ prices are represented by:

(2c) $c = c(p^*, Y^*)$

subject to the credit-extended full income constraint:

(2d) $Y^* = \sum_i P_i^* c_i = \Pi^* + \sum_i P_i^* T_i + S + \lambda_c K$

The Kuhn-Tucker condition on the credit constraint can be written using a slack variable $K_{net}$ in the credit constraint as:

(2e) $K_{net}\lambda_c = 0, \; K_{net} = K + \sum_{i \in TC} \bar{P}_i(q_i+T_i-c_i) \geq 0, \; \lambda_c \geq 0$

In these equations, either the credit constraint is effective, in which case $K_{net} = 0$ and $\lambda_c > 0$ or it is ineffective, in which case $K_{net} \geq 0$ and $\lambda_c = 0$.

The household model with non-tradeables and credit-constrained tradeables thus contains three sets of prices: decision prices, prices of tradeables, and prices of non-tradeables. Decision prices $p^*$ affect how production and consumption decisions are taken to accommodate the credit constraint. The endogenous markup $\lambda_c$ on the price of the credit-constrained tradeables serves to raise the decision price of the credit-constrained tradeable products and factors with a positive marketed surplus (in particular, labour on the small farms). Even though these goods are transacted at the market price $\bar{p}$, their supply increases and their use by the household falls, since $p^* > \bar{p}$, reflecting the fact that higher exports of these goods and factors help ease the credit constraint. Similarly, the endogenous markup $\lambda_c$ raises the decision price of the credit-constrained tradeables of which the household is a net buyer, such as forage on all farms and labour on the medium farms, inducing it to produce for import substitution and to use less in production. Even though the transaction occurs at the market price $\bar{p} < p^*$, imports of these goods and factors are reduced to accommodate the credit constraint.

The model thus consists of production decisions (2a) and (2b), consumption decisions (2c) and (2d), a credit constraint (2e), and equilibrium conditions (1a) and (1b). Because of the existence of both a credit constraint that transforms the prices of credit-constrained tradeables into endogenous prices and of endogenous non-tradeables prices, production and consumption decisions are consequently not separable. This system of equations consequently needs to be solved simultaneously. Since this is analytically intractable, a computable version of this model is set up by specifying a generalized Leontief for the profit function and a translog for the indirect utility function.

To determine the values of the parameters of these two functions, we start from "best guess" elasticities for the medium farms derived, on the production side, from the multi-market model for Morocco developed by Aloui, Dethier, and Houmy (1989) and on the demand side from Laraki (1989). These elasticities are then calibrated to satisfy the constraints imposed by the chosen functional forms. An algorithm is used that minimizes the sum of the squares of the discrepancies between this initial set of elasticities and a set of new elasticities that satisfy all these constraints, keeping untouched the diagonal values in which the greatest confidence can be placed. For the small farms, these elasticities are scaled to correspond to their levels of fixed factors.

## Simulation Results

The new pricing rules for cereals introduced by the agricultural structural adjustment programme (ASAP), together with the secondary effects they induce in other prices and wages—predicted for the medium run using a CGE for Morocco adapted from Mateus (1988) and the multi-market model developed by Aloui, Dethier, and Houmy (1989)—result in the following percentage changes in prices: hard wheat, 17.8; soft wheat, 14.4; coarse grains, 27.8; fruit and vegetables, 8.7; forage, 24; milk, 8.3; meat, 12.8; manufactured goods, 6.1; machinery and fertilizers, 1.5; other consumption goods, 5; and agricultural wages, 6.7. Since the changes are very large, the possibility arises of a credit constraint on the ability to respond to price incentives; consequently, the effects of ASAP when this constraint is alternatively present and relaxed are simulated.

As the base-run data in Table 1 show, a key distinguishing feature between small and medium farms is that the former are net buyers of soft wheat and coarse grains while the latter are net sellers. Both have a marketed surplus of meat, but it is a much more important source of income for the small than for the medium farms. Finally, the small farms are net sellers of male and female labour while the medium are net buyers.

The effects of ASAP reported in Table 1 are clearly different between small and medium farms, with the medium farmers deriving significant welfare gains, while the gains are much more modest for the small farmers as they are caught, on the consumption side, by rising prices of food, of which they are important buyers. Rising cereal prices distort the farm economy towards crops and away from livestock. In this response, the medium farms are more constrained by credit needs, as the shadow price of credit rises by 2.9 percent on these farms as opposed to 0.4 percent on the small ones. This is due to the fact that the small farms are engaged on the labour market as important net sellers and consequently find in wage incomes an important source of credit. While all labour income is not available at the time when credit is needed, it nevertheless provides important liquidity. The effect of the credit constraint on the medium farmers is sharply to reduce their ability to hire labour and to buy machinery and fertilizers. As a result, even though cereal prices rise, the hiring of female labour and the use of machinery and fertilizers fall to accommodate the hiring of more male labour. Relaxing this constraint, by contrast, allows them to hire more labour and use purchased inputs, significantly increasing their aggregate elasticity of supply response.

The credit constraint prevents households on small farms from reducing the sale of labour in spite of rising farm prices and the incentive to substitute imports. This is because the labour market is their source of access to credit. When this constraint is relaxed, the sale of labour falls sharply (−4.7 percent for men and −59.1 percent for women) and the elasticity of supply response increases. Eliminating the credit market failure thus increases the elasticity of supply response of the traded goods that make use of credit in production.

The paradoxical result of ASAP is that, in spite of shifting the farm economy from livestock to crops, resulting in a falling production of milk and meat, rising forage prices induce a substitution in meat production from the use of forage to the use of grazing in the commons and hence intensified need for child labour. As a result, the use of children in production increases, their shadow price rises sharply, and their leisure time falls. Market failure for child labour and access to commons reduce the negative effect of ASAP on the livestock economy. The long-run consequence is increased school absenteeism and increased overgrazing in the commons, two of the curses of Moroccan underdevelopment.

Table 1—Simulation of Household Behaviour: ASAP Responses*

| Experiment | Base Run (in 1,000 dirham) | | ASAP Credit Constraint | | ASAP No Credit Constraint | |
|---|---|---|---|---|---|---|
| Farm size | Small | Medium | Small | Medium | Small | Medium |
| Utility (per 1,000 change) | 26.28 | 42.98 | 9.8 | 35.4 | 10.3 | 37.9 |
| Credit: | | | | | | |
| Credit deficit | | | 0.0 | 0.0 | 0.4 | 2.9 |
| Shadow price | | | 8.4 | 16.6 | 0.0 | 0.0 |
| Consumption: | | | | | | |
| Total consumption | 12.25 | 23.47 | 1.8 | 9.8 | −0.1 | 5.4 |
| Leisure/men | 2.95 | 7.90 | 1.4 | 6.1 | 2.6 | 8.4 |
| Leisure/women | 1.60 | 5.61 | −5.4 | −9.7 | 10.3 | 14.4 |
| Leisure/children | 1.78 | 3.22 | −0.9 | −1.9 | −0.9 | −2.8 |
| Savings | 1.33 | 4.28 | 13.7 | 27.8 | 10.7 | 20.0 |
| Production: | | | | | | |
| Hard wheat | 1.99 | 8.56 | 1.6 | 1.8 | 2.0 | 1.8 |
| Soft wheat | 0.42 | 6.73 | 2.1 | −0.7 | 8.5 | 2.3 |
| Coarse grains** | 0.17 | 6.67 | 82.5 | 8.1 | 98.6 | 11.5 |
| Forage** | −0.98 | −1.83 | −2.6 | −8.3 | −1.5 | −3.3 |
| Total crops | 3.24 | 24.54 | 4.4 | 1.8 | 6.5 | 3.8 |
| Total livestock | 9.31 | 15.67 | −1.0 | −4.1 | −1.0 | −1.8 |
| Machinery and fertilizer: | −0.90 | −8.44 | 3.1 | −2.0 | 7.1 | 4.0 |
| Labour/men | −3.55 | −6.60 | −0.5 | −5.0 | 1.0 | 2.2 |
| Labour/women | −2.53 | −2.55 | 0.1 | −0.4 | 0.7 | 5.5 |
| Labour/children | −1.76 | −1.91 | 0.9 | 3.1 | 0.9 | 4.7 |
| Shadow prices: | | | | | | |
| Labour children | 1.06 | 1.02 | 12.7 | 17.1 | 11.2 | 13.2 |
| Wage labour: | | | | | | |
| Men | 2.36 | −1.66 | −1.0 | 9.1 | −4.7 | 48.7 |
| Women | 0.31 | −1.74 | 27.5 | −31.8 | −59.1 | 54.4 |
| Marketed surplus: | | | | | | |
| Hard wheat | 1.14 | 6.17 | 3.6 | −0.5 | 4.9 | 1.4 |
| Soft wheat | −1.00 | 5.01 | 2.7 | −1.2 | −2.1 | 0.5 |
| Meat | 6.85 | 10.05 | −1.4 | −11.2 | −0.6 | −4.4 |

*Results in percentage changes over base run unless otherwise indicated.
**Net of intermediate use.

The price adjustments brought about by ASAP, which are typical of the effects of foreign sector crises on agriculture, thus have a highly positive welfare effect for medium farmers who have important marketed surpluses, particularly if the ASAP is accompanied by credit availability that allows them to incur the higher cash expenditures necessary to hire labour and purchase modern inputs. For small farmers, rising prices are a mixed blessing as they are important buyers of cereals for consumption and feed. And relaxation of the credit constraint only brings small relief as they can, in any case, use the labour market as a source of cash at a small efficiency cost. As such, the effects of ASAP are regressive on the distribution of income on agriculture. In all types of farms, the effects on the welfare of children, and thus indirectly on literacy and the environment, are negative unless the productivity of forage production is enhanced, suggesting as well complementary types of interventions to the price effects of ASAP if these negative consequences are to be avoided.

## Conclusion

This paper opens up the field of computable non-separable household (CNH) modelling as a micro-level instrument of policy analysis. This can be thought of by analogy with the computable general equilibrium (Johansen, 1964; and Adelman and Robinson, 1978), the multi-market (Quizon and Binswanger, 1986; and Braverman and Hammer, 1986), and the integrated multi-market CGE (Sadoulet and de Janvry, 1990) approaches, which offer macro-level and sectoral instruments of policy analysis that are now widely used. Indeed, the household and macro-sectoral modelling approaches have much in common, from use of the concepts of tradeables and non-tradeables to use of the same computational algorithms. A clear advantage of the CNH approach is that survey data are available for the whole model as opposed to the CGE and multi-market situations, making estimation of the model possible, clearly the next step in developing this approach. Another part of the next step is to add behaviour towards risk, which must be introduced in the computable form of the model at the level of the indirect utility function.

The usefulness in policy terms of these models lies in the systematic lack of comparable data over time that would allow separation *ex post* of the impact of policy instruments. For this reason, recourse is made to simulation of policy impacts in such models, either to retrace historical effects in duly-calibrated models or to explore alternative policy scenarios. As exemplified here, CNH models allow the exploration of the effects of adjustment programmes at the household level, an important policy question of the moment, in a considerable degree of detail that could not be achieved with available historical data.

### Results Based on Structural Features of the Model

The elasticity of supply response of tradeables is reduced by the presence of non-tradeables among products that either are consumed or factors of production. The lower the levels of substitution between these goods and tradeable alternatives and the larger the shares of these goods in production or consumption, the lower the elasticity of supply response of tradeables. To increase the elasticity of supply response of tradeables, technological change needs to be directed at the non-tradeable products and factors to ease the constraint that their production or availability imposes on the production of tradeables.

Accommodating a credit constraint imposes an endogenous markup on credit dependent tradeables. This distorts the household's allocation of resources towards import substitution and greater exports of the credit-constrained products and factors. Small farmers were thus seen to remain heavily on the labour market, in spite of rising cereals prices, because wages give them a way of escaping the credit constraint at a relatively low efficiency cost. Such low-cost escape is not available to the medium farmers. Their elasticity of supply response and their ability to benefit from the higher prices offered by ASAP depend on relaxation of this credit constraint, confirming the fundamental importance of credit components in ASAP loans.

For the small farmers, vigorous programmes of rural development must be put into place to allow them to become net sellers of the commodities whose prices have increased. The productivity of their land must consequently be raised, and the new ASAP pricing rules should be seen as an historic opportunity to mount a massive complementary programme of rural development.

## Results Based on Particular Elasticity Values

While ASAP displaces the farm economy from livestock to crops, pressures on the use of children for herding and overgrazing in the commons will not be relaxed if the price of forage increases with that of cereals due to competition in production. Avoiding this secondary effect thus requires focusing on the technology of forage production to lower its production costs. While the price adjustment occurs in cereals, technological change is needed in the forage-livestock economy. Also, infrastructure investments and institutional arrangements need to be sought that can reduce and ultimately eliminate the need for child labour in animal production. These include the enclosure of fields and innovations in contracts for herding that achieve economies of scale (and thus raise the productivity of labour in herding, making it a remunerative activity for adults) while avoiding the problems of moral hazards that maintain this ancestral practice of child use as economically rational.

Structural adjustment and the new pricing rules for cereals, by eliminating an historical bias in agricultural price formation, have the potential to benefit the peasantry in dryland areas of Morocco, the poorest segment of society, but only if accompanied by these complementary structural and policy interventions.

## Note

[1]University of California, Stanford University, Institut Agronomique et Vétérinaire Hassan II, Morocco, and University of California, respectively.

## References

Adelman, I., and Robinson, S., *Income Distribution Policy in Developing Countries: A Case Study of Korea,* Stanford University Press, Stanford, Calif., USA, 1978.

Aloui, O., Dethier, J.J., and Houmy, A., "L'Impact de la Politique d'Ajustement sur les Secteurs des Céréales et de l'Élevage au Maroc," Agriculture and Rural Development Department, World Bank, Washington, D.C., USA, 1989.

Braverman, A., and Hammer, J., "Multimarket Analysis of Agricultural Pricing Policies in Senegal," in Singh, Inderjit, Squire, L., and Strauss, J. (Eds.), *Agricultural Household Models,* Johns Hopkins University Press, Baltimore, Md., USA, 1986.

Johansen, L., A *Multi-Sectoral Study of Economic Growth,* North-Holland Publishing Company, Amsterdam, Netherlands, 1964.

Laraki, K., "Ending Food Subsidies: Nutritional, Welfare, and Budgetary Effects," *World Bank Economic Review,* Vol. 3, No. 3, 1989, pp. 395–408.

Mateus, A., "A Multisector Framework for Analysis of Stabilization and Structural Adjustment Policies: The Case of Morocco," *World Bank Discussion Paper,* No. 29, World Bank, Washington, D.C., USA, 1988.

Quizon, J., and Binswanger, H., "Modeling the Impact of Agricultural Growth and Government Policy on Income Distribution in India," *World Bank Economic Review,* Vol. 1, No. 1, 1986, pp. 103–148.

Sadoulet, E., and de Janvry, A., "Implications of GATT for the Poor African and Asian Countries: A General Equilibrium-Multimarket Approach," Department of Agricultural and Resource Economics, University of California, Berkeley, Calif., USA, 1990.

**Discussion Opening**—*Teruaki Nanseki* (National Agricultural Research Centre, Japan)

The paper makes several methodological contributions to the growing body of research on quantitative agricultural policy analysis. It presents computable non-separable household (CNH) models that are useful in micro-level analysis of the effects of alternative policy scenarios. CNH models allow the complex implications on different types of household and different household members of price changes caused by structural adjustment to be measured. Based on the application, the paper concludes that CNH models permit exploration of the effects of adjustment programmes at the household level in a considerable degree of detail that could not be achieved with available historical data. I am convinced that a combination of CNH and CGE models opens a new field in policy analysis. Furthermore, the simulation results and conclusion are generally understandable.

Several methodological questions also arise, however, since few explanations on the applied model are given in the paper. For example, no overview of the applied model, including equations and lists of variables is given (I understand this might be partly due to space limitations). It is nearly impossible to give a complete picture of the model and the results in the limited space and time. Nevertheless, since the main purpose of the paper is to develop a micro-level instrument of policy analysis, more clarification on the methodology would be useful to improve understanding of the approach.

What is the assumption on substitution among male, female, and children labour in both the technological constraint and the utility function? What is the main difference between "best guess" or initial elasticities and the calibrated elasticities used in the simulation? How are the results of the base run under actual conditions similar to the actual household survey data in terms of values of the related variables of both production and consumption? How robust is the solution of the CNH model in the face of changes in the values of the parameters?

The attitude to risk of farmers and households plays an important role in agricultural analysis. The paper also states that part of the next step is to add behaviour towards risk. The question therefore arises of why attitude to risk is ignored in the present model and whether its results are still reliable.

*[Other discussion of this paper and the authors' reply appear on the following page.]*

# General Discussion—*W.L. Nieuwoudt, Rapporteur* (University of Natal)

In relation to Tyler and Akinboade's paper, the view was expressed that CES and Cobb-Douglas production functions may not be appropriate for the economy studied, and the author was asked if the simulations were tested for Leontief technology. The perfectly elastic supply of labour scenario was not considered realistic, given that rural unemployment rates are low (about 1 percent). Another participant commented that, during the 1980s, a significant share (25–30 percent) of Kenya's officially recorded coffee exports originated in neighbouring states, with consumer good flows in the opposite direction; how could this phenomenon be incorporated in this model? Another participant questioned whether a positive sloping supply curve for labour could be incorporated in the model and to what extent the substitution of labour for other inputs allowed for in the model predetermines results. The authors were also asked if the model was consistent, given that the Cobb-Douglas production function cannot deal with optimization.

Tyler replied that the model was intended to be indicative of the likely general response of the Kenyan economy to certain important policy changes. For actual use in advising policy makers, a closer specification of the model would be required (e.g., including disaggregation of the agricultural sector and incorporation of the monetary side of the economy). The conclusion that results depend crucially on the specification of the labour markets has wider implications for CGE models in other LDCs. The use of a Cobb-Douglas production function for agriculture was based on empirical analysis of Kenya by other authors. It allows capital-labour substitution if relative prices change, and this occurred in some simulations.

In relation to the Amponsah and Hushak paper, the confidence that can be placed in the results, given the number of non-significant variables and the evidence of serial correlation was questioned. If about 25–30 percent of Kenya's coffee exports are derived from Uganda and Tanzania and the performance of the stronger economy is affected by changes in the domestic terms of trade in neighbouring states, how can this be incorporated in the model and what is the reliability of the data base? The authors were also asked about the choice criteria and characteristics of the three groups of farmers.

In reply, Amponsah stated that the objective of the study was macro-oriented and thus it does not consider the other social indicators. One of the bases for structural adjustment is to correct structural imbalances that impede the growth of the most productive sectors through the external terms of trade between exports and imports. The key argument is to improve supply response in agriculture (especially in exports) to generate needed foreign exchange. As far as the agricultural sector is concerned, a microeconomic approach could provide more in-depth information. The agricultural sector plays a weak role in both countries. Due to Kenya's arable land constraint, the country has had to depend more on tourism and other ancillary industries for foreign exchange. In the case of Cameroon, since discovery of petroleum in 1978, its agricultural exports have been stagnant.

The comment was made on the de Janvry *et al.* paper that it should be viewed as illustrative of an interesting technique and not as a reflection of what the actual experience might be, since the referenced elasticities have not been estimated and also the wheat price is now and has been for years higher than the world price and not lower as suggested in the paper. The extent to which the treatment of labour predetermines the outcome was questioned; does the model assume a fixed labour-machinery technology as is often the case in models of this nature (operations research) or can the model select a different labour-machinery technology if labour becomes more scarce? What is the derived demand for labour as incorporated in the model?

Sadoulet, in reply, indicated that labour substitutes in some areas but not in others. No labour-machinery substitution was specifically incorporated. Risk was not incorporated due to the difficulty of including risk, as price also appears in the consumption function.

Participants in the discussion included D. Belshaw (University of East Anglia), S. Hosomi (Institute of Developing Economies, Japan), H. Mahran (University of Gezira), S. Maruta (Ibaraki University), L. Nieuwoudt (University of Natal), C. Short (MPND, Kenya), and W. Tyner (Purdue University).

# Debt Peonage and Over-Deforestation in the Amazon Frontier of Brazil

## Anna Luiza Ozorio de Almeida[1]

**Abstract:** This paper proposes that under the conditions typical of the Amazon frontier of Brazil—absence of a wage labour market and abundant supply of land—farm-family labour is a positive function of debt. This labour supply response, known as "debt peonage," provides for indirect management of farm labour by local merchants via the crop lien mechanism. A microeconomic model of the family farm shows that debt-labour leads to labour-intensive farming but land-extensive farming leads to over-deforestation. Correct colonization policy should then provide for market structures that reduce indebtedness and deforestation. This would also reduce the environmental consequences of settlement in the Amazon.

## Introduction

The Amazon frontier of Brazil is being settled by a multitude of small farmers who arrive before the rest of the economy reaches each location.[2] Farmers are followed rapidly by merchants, who establish the so called "crop-lien" mechanism: advancing subsistence goods between harvests in exchange for future agricultural product.[3] Frontier merchants thus link several markets—consumer goods, agricultural product, and credit—in the manner typical of the functioning of traditional "usury-mercantile capital."[4] In a land-abundant-labour-scarce frontier economy, and in the absence of a wage-labour market,[5] merchants appropriate surplus and monitor family labour indirectly via farmer indebtedness. Different systems of "debt-peonage" exist all over the Amazon, as they have in other times and places, wherever agricultural markets are interlinked and highly concentrated.[6]

This paper calls attention to one especially grave environmental consequence of Amazonian debt-peonage: its impact in raising the intensity of family labour supply and in increasing the extent of deforestation per family.[7] The socioeconomic conditions of Amazon settlement thus provoke over-deforestation relative to what would obtain for the same population if frontier markets were more competitive. Given increasing worldwide concern for conserving the Amazon, this analysis contributes to an understanding of one of the main causes of its devastation.

## Model of Indebtedness and Labour Supply

The microeconomic model in this section combines some of the key variables that influence colonist behaviour in a frontier situation, such as chronic farmer indebtedness and the cost of access to land. Settling on virgin soil imposes abrupt discontinuities on frontier agriculture. The hazards of clearing one of the most dense and inhospitable forests in the world mean that deforestation is an especially heavy burden in the case of the Amazon frontier of Brazil. Additional modelling considerations, such as risk, contractual arrangements, household economics, subjective equilibrium, as well as many others, had to be disregarded due to space limitations.

Total annual expenditure ($E$) comprises the costs of land ($aA$), labour ($wL$), and other current costs ($cC$). Given the seasonality of agricultural production, expenditure is made throughout the year, but receipts occur only at the of the period. To finance production costs, then, producers must have on hand at the end of each year enough resources to last until the end of the harvest. The budget constraint is:

(1) $E_t = wL_t + cC_t + aA_{ti}$, where $t$ refers to the current agricultural year

Agricultural production during one agricultural year occurs according to a continuous, twice-differentiable function of land, labour, and other inputs:

(2)  $Q = Q\ (L,\ C,\ A);\ Q_i < 0;\ Q_{ij} > 0$ for $i,\ j = L,\ C,\ A$

All income is assumed to derive from agriculture, whose product is sold (all or in part) at a price $p$. Non-marketed output is valued at the same market price, so that the value of total income is $pQ$.

Incoming receipts from current loans $(D_t)$ supplement current income $(pQ_t)$. Obligations on previous debts, $(D_t)X(i+1)$, where $i$ is the interest rate, add to current expenditure $(E)$. Assuming no investment or dissaving, net income $(Y)$ is:

(3)  $Y = PQ_t - E_t$, where:

(4)  $Y - D_t - D_{t-1}\ (i+1) - 0$  exactly equals net indebtedness

When $Y < 0$, own funds $(pQ_t)$ are insufficient to cover current expenses $(E)$ plus debt repayment $(D_{t-1}(i+1))$; when $Y > 0$, own funds more than cover current expenses and contribute to debt repayment; and when $Y = 0$, current incoming loans just cover previous debt repayment.

The farmer's objective is to maximize current income $(pQt)$ subject to budget and debt constraints $(E_t+Y)$ during each agricultural year. First-order conditions for a maximum are that:

(5)  $P_i Q_{ti} - \dfrac{dY}{di} = 0$

where $P_i = w,\ c,\ a$, $Q_{ti}$ is the respective marginal product of a factor $i$ in given year $t$, and $Y = 0$.

In equilibrium, (5) is exactly satisfied. As $dy/di = 0$, this means that (5) collapses to the usual condition $(Q_{ti} = P_i/p)$, whereby each factor of production $(L,\ C,\ A)$ is used up to the point where its marginal productivity is equal to its opportunity cost; i.e., its real market price $(w/p,\ c/p,\ a/p)$. If second-order conditions for a maximum are satisfied, these equations jointly determine demand for labour, inputs, and land as well as the supply of agricultural product. The corresponding land demand and product supply equations were directly submitted to statistical testing elsewhere, where land productivity $(Q_A)$ turned out to be statistically sensitive to relative prices of land and agricultural product $(a/p)$.[8]

When the system is in disequilibrium, the respective inequalities are equalized by slack variables, which tend to move it back towards equilibrium. In this model, indebtedness $(D_t,\ D_{t-1})$ and investment are the slack variables that balance the budget constraint.

## Over-Deforestation as a Consequence of Indebtedness

If $y \neq 0$, then $dY/di \neq 0$ holds entirely. Rearranging (5) then yields:

(6)  $\dfrac{di}{dY} = \dfrac{1}{pQ_i - p_i} > 0 \Rightarrow Q_i > \dfrac{P_i}{p}$ for $i = L,\ C,\ A$

This means that the response to indebtedness is positive, in terms of labour supply $(dL/dY)$, land area $(dA/dY)$, and other inputs $(dC/dY)$ as long as marginal productivities are greater than respective real costs. In a new frontier there may yet be no land, labour, or input markets $(p_i=0)$, and only the product market may be operative $(p>0)$. The indebtedness response will then be positive as long as marginal productivities are positive $(Q_i>0)$. The initial phases of a frontier, then, should be labour-intensive family farming, as shown in Figure 1 by the horizontal tangent to the production curve $A_1$ at $L_2$, where $Q_L = 0$.

The horizontal axis indicates the quantity of total family labour supply. The vertical axis indicates the quantity of output. The slope of the ray from the origin indicates the real market wage rate $(w/p)$. Farmers clear as much land $(A_1,)$ and work as much $(L_1)$ as necessary to equate marginal productivity ($Q_L$) to the real wage rate ($w/p$), producing $Q_1$ along a production function with initially increasing and then decreasing marginal productivities.

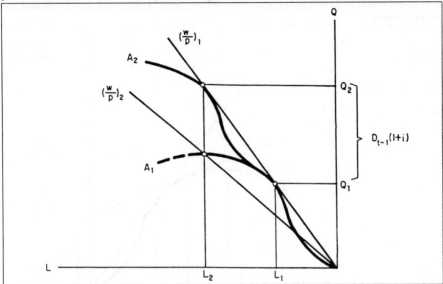

Figure 1—Debt Repayment, Deforestation, and Labour Supply

The initial section of the production curve displays increasing marginal productivity of labour, which is justified by the condition of land abundance in a frontier area. The supply of land, however, comes only from deforestation, as fallow plots are never reused in Amazonian agriculture.[9] So Equation (6) is also an equation for small farmers' demand for deforestation. In the Amazon, newly cleared land is much more fertile than "old" land, which enhances initial marginal productivities.

The higher the market wage rate ($w$) and the lower product prices ($p$), the more hiring out of labour and the less family farming there will be. The lower the market wage rate and the higher are product prices, the more family farming and the less hiring out. At the limit, when real wages rise above $(w/p)_1$, where $Q_L = Q/L$, there is no family farming. In a frontier area, where the wage-labour market is as yet non-existent, there is only family farming.

In the case of previous indebtedness ($D_{t-1}$), requirements that repayment be in kind imply that more than $Q_1$ be produced. Given the low technological level of most frontier farming, additional labour ($L_2$) along $A_1$ would reduce marginal productivity to zero, which contradicts Equation (6) whenever $(w/p) = 0$. More land is therefore deforested along $A_2$, adding to previous productive land and expanding production to $Q_2$. This discontinuous sequence of land additions implies that, along the way, itinerant farmers continuously reap the benefits of the high fertility of recently cleared land. This is why $A_2$ is "grafted" on to $A_1$, as shown in Figure 1. In order to appropriate a given level of income ($pQ_1$), then, farm families must work more and clear more forest off the land than before becoming indebted. The higher the interest rate, the greater the impact of previous debt on labour supply and deforestation.

Where the farmer cannot reach $Q_2$, current indebtedness ($D_t > 0$) can be resorted to as an income supplement, as shown in Figure 2. This reduces the reservation wage from $(w/p)_1$ to $(w/p)_3$ and increases labour intensity on each plot of land; i.e., family farming persists even at marginal productivities below the real market wage rate.

Equation (6) is of particular interest, as it refers to the labour supply response to indebtedness. This relationship is fundamental to the operation of usury-mercantile capital. Although this relationship is not tested statistically here, it is supported elsewhere by field data from small farmers in the Brazilian Amazon. Those most deeply caught in consumer indebtedness farmed their plots most extensively and cleared more additional plots.[10] Thus, debt-peonage may be an important cause of itinerant frontier farming and of over-deforestation in the Amazon.

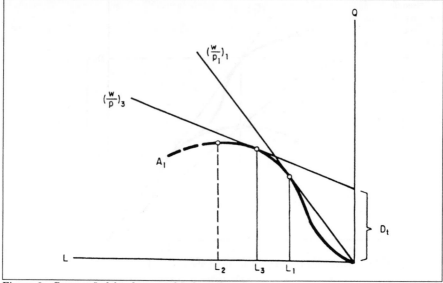

Figure 2—Current Indebtedness and Labour Supply

In summary, under the stated assumptions, recurrent indebtedness ($D_{t-1}>0$; $D_t>0$) increases deforestation, labour supply, and labour intensity in family farming. Although the model, for simplicity of exposition, lumps former debt ($D_{t-1}$) and current debt ($D_t$) together into $Y$ in Equations (3) and (4), Figures 1 and 2 show that their impacts differ.

## Conclusions and Policy Implications

Under typical Amazon frontier conditions, the lack of well structured market systems leads to farmer indebtedness. Debt, in its turn, increases the intensity of labour supply and the extent of deforestation per family on the frontier.

The main policy implication of the model is that Amazon colonization should entail more than the usual measures aimed at physical, institutional, and legal settlement; e.g., the mere provision of land, roads, and titles. It must also include measures aimed at forming a market structure consistent with agricultural production and income appropriation by colonists. If this is not done, perverse effects ensue, such as over-deforestation, already well-known in the Amazon frontier of Brazil. Such a reaction, which represents microeconomically rational behaviour under local market disequilibria, can lead to the failure of colonization policy as a whole. The effectiveness of a social policy of land distribution to small farmers in the Amazon, therefore, requires market structures that reduce farmer indebtedness and curb over-deforestation.

# Notes

[1]Instituto de Pesquisa Econômica Aplicada, Brazil. I am grateful for comments by William Savedoff, though errors and omissions are my own responsibility.

[2]See de Almeida (1991) for a measurement of the extent of Brazilian Amazon colonization in terms of number of settlers and occupied area.

[3]See Musumeci (1988) and de Almeida (1992) for analyses and bibliographies on the role of frontier merchants in the occupation of the Brazilian Amazon.

[4]The interlinking of rural markets has been studied by several authors, such as Bardhan and Rudra (1978) and Braverman (1981). For the operation of different types of capital in Brazilian agriculture, including usury-mercantile capital, see Goodman and Redclift (1977).

[5]The absence of a wage-labour market under conditions of land abundance has general consequences studied by Domar (1970); the present paper draws specific implications for deforestation.

[6]For the post-slavery South of the USA, see Ransom and Sutch (1972). For the Indian case, see Bhaduri (1973). For Caribbean plantation economies, see Harris (1988).

[7]Ransom and Sutch (1975) have already called attention to the over-production consequences of debt-peonage. This paper extends its implications to a frontier context, where over-production implies over-deforestation.

[8]See de Almeida (1992), panel 19.1.

[9]Empirical evidence and a bibliography to this effect are in de Almeida (1991, Chap. 19).

[10]Empirical evidence and a bibliography to this effect are in de Almeida (1991, Chaps. 16 and 20). The debt-peonage effect on family labour supply also yielded significant and interesting econometric results for data from the rural Brazilian Northeast (de Almeida, 1977).

# References

Bardhan, P.K., and Rudra, A., *On the Interlinkage of Land, Labour and Credit Relations in Agriculture: An Analysis of Village Survey Data in West India*, Socio-Economic Research Institute, Calcutta, India, 1978.

Bhaduri, A., "Agricultural Backwardness under Semi-Feudalism," *Economic Journal*, Vol. 83, No. 329, 1973, pp. 120–137.

Braverman, A., "Sharecropping and Interlinking of Agrarian Markets," *Landlords, Tenants and Technological Innovations*, Development Research Centre Discussion Paper No. 31, World Bank, Washington, D.C., USA, 1981.

de Almeida, A.L.O., "Parceria e Endividamento no Nordeste Brasileiro, in *Seminário Sobre Desenvolvimento, Planejamento e Políticas Agrícolas*, EIAP/FGV, Rio de Janeiro, Brazil, 1977, pp. 15–19.

de Almeida, A.L.O., *Colonização Dirigida na Amazônia*, University of Texas Press, Austin, Tex., USA, 1992.

Domar, E., "The Causes of Slavery and Serfdom: A Hypothesis," *Journal of Economic History*, Vol. 30, No. 1, 1970, pp. 18–21.

Goodman, D., and Redclift, M., "The Boias-Frias: Rural Proletarization and Urban Marginality in Brazil," *International Journal of Urban and Regional Research*, Vol. 1, No. 2, 1977, pp. 348–364.

Harris, D.J., "The Circuit of Capital and the Labor Problem in Capitalism," *Social and Economic Studies*, Vol. 37, Nos. 1–2, 1988, pp. 15–31.

Musumeci, L., *O Mito da Terra Liberta: Colonização "Espontânea," Campesinato e Patronagem na Amazônia Oriental*, ANPOCS/Vértice, São Paulo, Brazil, 1988.

Ransom, R.L., and Sutch, R., "Debt Peonage in the Cotton South after the Civil War," *American Economic Review*, Vol. 62, No. 1, 1972, pp. 77–86.

Ransom, R.L., and Sutch, R., "The Lock-In Mechanism and Over-Production of Cotton in the Post-Bellum South," *Agricultural History*, Vol. 59, No. 2, 1975, pp. 405–425.

## Discussion Opening—*Jorge A. Torres-Zorrilla* (Instituto Interamericano de Cooperación para la Agricultura)

My first comment refers to the price of the land factor in the Amazon frontier zone. It is stated in the paper that, in equilibrium, the land factor is used up to the point where the value of its marginal product equals the market price. It is also advanced that land productivity has been shown to be statistically sensitive to relative prices of land and agricultural product.

However, in the Amazon frontier, we would really not find a market for land nor a price of land since the frontier is being settled by many small farmers in an informal manner. The model should therefore perhaps use not the price of land but the cost of access to new land, a cost which is highly variable depending on the location and the particular characteristics of the local forests or the actual terrain.

Furthermore, in the Amazon, newly cleared land is much more fertile than "old" land, which will increase the dispersion of costs and values. The model and the analysis of the paper should be refined to incorporate the fact that land markets are non-existent in the Amazon frontier. Should we assume that land prices are zero? Or should we propose that the cost of deforestation and access to new land be used as the price for land?

The situation is different for the other inputs (labour, fertilizer, etc.) because there will always be a market even in a frontier area, and prices will always be positive. Labour has an alternative cost, and other inputs always have to be purchased in the modern sector of the economy and transported to the frontier, and will bear a much higher price.

The model is useful in showing that, if land markets fail to be present, marginal productivities are greater than real costs and more inputs will be used, including land. The other consequence is that debt will increase and that farm families must work more and clear more land than before. The final result will be itinerant frontier farming and a pattern of over-deforestation.

The policy implications of the paper include that development of a well structured market system could help to control the perverse effects of farmers' debt and over-deforestation in the Amazon frontier. But there is always the concern that even competitive markets may fail to produce appropriate prices for land resources. The market may imply private prices for land and forests well below the social price levels. Here, resource evaluation methods and the optimal control theory of renewable resources may give better estimates of the appropriate valuation of these ever-decreasing scarce resources.

Finally, I cannot refrain from making a comparison between these poor farmers of the Brazilian frontier and those in other poor countries in Latin America. These countries have also accumulated a large external debt that they have to repay now, making it necessary to overuse their natural resources. This also contributes to deforestation through a great expansion of agricultural exports.

# Agricultural Policies and Tropical Forests

## Béatrice Knerr[1]

**Abstract:** During the past decade, the degradation of the tropical forests has become a major topic in the worldwide discussion of the factors that threaten the long-term survival of the global ecological system. Undisputed in this context is the fact that agricultural production is among the major causes of the continuous reduction of the tree cover. As a consequence, efforts to stop deforestation have to accord high priority to agricultural policies that have played a prominent role in this outcome. This paper is concerned with these issues. It explains the role of agriculture in the process of deforestation in different regions and sub-regions and evaluates ways out of the dilemmas under different economic, social, and political conditions.

## Introduction

The worldwide destruction and degradation of tropical forests is among the most important challenges to the global environment.[2] Over the last few years, it has been the topic of an increasing number of research projects, international conferences, and government hearings that have stressed its implications for climatic changes, genetic impoverishment, and long-term economic decline (Deutscher Bundestag, 1990; and Esser, 1989). Nevertheless, the question of how the process could in practice be stopped or at least slowed down has remained open. Undisputed in this context is the fact that agriculture is among the major causes of the continuous reduction of the tree cover. Yet the situation in most tropical countries is one of deep conflict between the short-term interests of those who farm the land (and often do not know of any acceptable alternatives to earn an income other than continually to take new land under cultivation) and the long-term objectives of the society that depends on forest conservation as an economic and ecological resource of survival. While the forests disappear at an ever faster rate, lack of foreign exchange in tropical countries severely restricts commercial imports of food and agricultural inputs and makes the production of agricultural export goods of particular relevance. These conditions decisively diminish the room for manoeuvre of agricultural policies that have played a prominent role in the process of forest destruction. Nevertheless, efforts to stop deforestation have to accord high priority to changes in the agricultural sector.[3] While the inevitable conflict produced by the competition between agricultural production and forest reserves for the same resources constitutes a common problem in all tropical regions, there are large inter-regional differences in the structure of interdependencies and in the economic, social, and ecological conditions.[4] Global suggestions, therefore, are far from being appropriate and cannot serve as a basis for promising solutions. The transfer of recommendations requires careful analysis of the respective clusters of circumstances, causes, and consequences. The present paper is concerned with these issues. It explains the role of agricultural policies in the process of deforestation in different regions and evaluates ways out of the dilemmas under different economic, social, and political conditions.

## Tropical Africa

The conflicts in Africa between food production for subsistence survival and the preservation of the forest resource have become the most dramatic in the world. Over the last 40 years alone, Africa has lost about one quarter of its tree cover (World Resources Institute, 1989). The physical destruction of Africa's tropical forests is largely effected by shifting cultivators who act within a socioeconomic and policy framework that leaves them few other alternatives for survival than regularly to clear new forest plots for subsistence production. The forests are threatened by two kinds of settlers. The first is the traditional cultivators who are used to living in the forest areas and who apply farming systems developed over centuries on the basis of shifting cultivation. Such systems were sustainable as long as 1 km[2] was supporting no more than 4–5 persons and as long as a patch of land could be left fallow for 10 years or more. Nowadays, however, due to growing populations and retreating forest area, a multitude of people are living on a given expanse in many parts of Africa. These traditional forest cultivators have been increasingly joined by a second group of subsistence peasants who, due to lack of land elsewhere, are moving into forests where they adopt a slash-and-burn style of agriculture that leaves even less scope for forest regeneration. Possessing little cultural adaption to forest environments, they "tend to advance upon the natural forest in waves: they operate as 'pioneer fronts' pushing even deeper into forest tracts, leaving behind them a mosaic of degraded croplands and brush growth where there is no prospect of a natural forest re-establishing itself, even in impoverished secondary form." (Myers, 1982, p. 75). The most seriously affected region is West Africa where the rate of deforestation on closed forests is six times higher than in Central Africa, which still holds vast reserves of largely untouched forests (Myers, 1989).

A number of causes have interacted to bring West Africa's rate of deforestation to the highest level of all tropical regions: high population growth, unfavourable land tenure systems, the persistence of slash-and-burn practices and land-intensive technologies, the catalytic role of timber harvesting in forest opening, lack of employment opportunities outside the agricultural sector, and the urgent need for foreign exchange. These conditions are most apparent in Côte d'Ivoire. What can be observed there has happened to a smaller or larger extent in most of the West African countries. The agricultural policy followed in Côte d'Ivoire demonstrates two strategies which in the African context have decisively paved the way for extensive deforestation by agricultural activities: first, the attempt to achieve economic development by following an export-led growth model, which for many years made Côte d'Ivoire an example of development for African countries; and second, the consequences of large-scale timber exploitation as a source of foreign exchange.

Similar developments as in Côte d'Ivoire have been observed in Liberia, where shifting cultivators penetrate with slash-and-burn practices into logged-over forests, following the many logging roads and hauling tracks, which since 1963 have made once remote areas easily accessible (Repetto, 1988); in Ghana, where over many decades the spread of shifting cultivation, which was promoted by large-scale logging operations, and the transformation of closed woodland into permanent tree crops (mainly cocoa) have destroyed nearly all the tree cover of the once forest-rich country (Sutlive et al., 1981); and in other West African countries (Knerr, 1991).

A number of policy measures that might stop forest destruction in West Africa, and prevent other African regions from taking the same route, have been proposed. Prohibitions against forest clearing for purposes of shifting cultivation are hardly helpful in the socio-political context of most African states, as their practical enforcement poses unsurmountable difficulties, as has been exemplified in Liberia and elsewhere. Such efforts might even have reverse effects, as is demonstrated in Guinea-Bissau, where the prohibition of fire clearing caused farmers immediately to leave the area after setting it on fire, instead of controlling the fire as they did before. As a consequence, much more forest land is burned by the shifting cultivators than is intended (Knerr, 1989).

An often-proposed way to stop the deforestation is the establishment of private land ownership to increase the users' interest in higher land productivity and long-term fertility conservation, and prevent people from clearing ever new plots that are treated as a free resource. In practice, such policies have prompted a number of unwanted drawbacks. Adverse effects of land privatization leading to large-scale forest clearing—sometimes followed by the establishment of plantations of perennial crops—have been observed in a number of African countries where governments were apparently planning land registration and privatization of ownership. When, for example, in Côte d'Ivoire, a law was drafted shortly after independence saying that all cultivated land should be registered in favour of the actual cultivator, half the country's forests were immediately set on fire by slash-and-burn practices (Ley, 1982). Another often-observed reaction to increasing de facto privatization is the cultivation of permanent tree crops on large areas. This might be the cause of a significant share of Guinea-Bissau's recently established cashew plantations (Knerr, 1989).

The replacement of tribal groups' traditional control of usufruct rights by other forms of property rights, particularly by government ownership, is increasingly considered an additional threat to the forests. The perception grows that these communities had a greater interest in conserving their natural environment than private individuals or state governments. An evident example of this has occurred in Ghana, where the increasing assumption of state control over the forests made them even more vulnerable to the "tragedy of the commons" (Repetto, 1988).

The IMF and World Bank structural adjustment programmes, which West African countries apply in order to escape from their permanent debt crises, might still increase the pressure on forest land as (for lack of other realistic alternatives) they suggest the expansion of export crop production, implying the enlargement of cultivated land and the curtailment of subsidies for agricultural inputs (World Bank, 1989; Khan and Knight, 1985; and Zulu and Nsouli, 1985), causing a further drop in the price relation between land and capital inputs, which leads to more land-intensive production and decreasing land productivity.

In Central Africa, population densities are still low enough to allow a sustained agricultural use of the forest by shifting cultivators on a subsistence level. Pressure on land by farmers and spreading plantations is much lower than in other parts of the continent, and the forests are still hardly affected by timber exploitation. Zaire, in particular, still possesses vast areas of untapped but valuable forest resources (Kio, 1983). However, as the reserves of exportable timber in other regions of the world are fast being depleted, it is expected that the penetration by timber exploiters will very soon change this situation.

Large and growing cattle herds have caused important land degradation and deforestation in some tsetse-fly-free African regions such as Burundi by overgrazing and by urging crop cultivators to clear new forest land to meet their subsistence needs. The occupation of vast areas of land by such systems of extensive cattle breeding, which often contribute little to human nutrition, is often promoted by an agricultural policy that gives priority to the particular interests of the cattle herders for political or social reasons (Knerr, 1991)

Over recent decades, food production in almost all African countries has been increased mainly by enlargement of the cropped area, and to a much lesser degree by increasing land productivity. Productivity-enhancing technical progress has played a minor role. However, Africa's deforestation process can only be stopped if the productivity of the already-cultivated land increases and farmers are prevented from further land clearing. For example, fertilizer could be applied to replace soil nutrients, but this is not a practical solution in general. Fertilizers are too expensive for most African smallholders who often have no significant source of cash income, and mineral fertilizers are often not available due to weak infrastructure. The most important limitation in the long run seems to be the fact that a large share of the tropical soils can retain nutrients for only a short time; they are often leached immediately (Ellenberg, 1985). It sometimes helps to apply the fertilizer successively in very small portions and in organic form, but this is very labour-intensive and time-consuming and is therefore usually not accepted by peasant farmers.

The introduction of measures that could slow down or even reverse forest destruction would require a determined government policy that puts emphasis on long-term resource conservation. Yet governments, urged on by balance-of-payment problems and social unrest of the population, are used to deciding things on a day-to-day basis.

Lack of foreign exchange and high levels of indebtedness force African countries increasingly to provide the food needed for the growing population by domestic production. At the same time, it restricts the capacity to import inputs like fertilizers or pesticides that might contribute to higher land productivity. Moreover, the urgent need for foreign exchange implies an increasing necessity to produce exportable crops. Against this background, the present economic situation in most African countries leaves little hope for the preservation of the continent's forest reserves without major assistance from richer countries.

Putting all the arguments together, the only way to stop Africa's deforestation seems to be the introduction of sustainable agricultural production systems that both significantly increase land productivity and need a low level of external inputs. Technically mature systems, which imply a low level of external inputs, like agroforestry, have been developed and

315

are available (Adelhelm *et al.*, 1986; Haffner, 1982; and Dressler, 1984). They are, however, often not accepted by the peasants due to high input costs, particularly in the form of labour (Ellenberg, 1985).

# Latin America

In Latin America, the forest area destroyed by expanding agriculture is estimated to be around 50,000 km$^2$ each year (Myers, 1982; and Denevan, 1978), or 0.43 percent of the region's forest area. The major cause of this deforestation has been land clearing for commercial purposes. Growing domestic and international demand for beef, combined with national policies that put a low price on forest land to be cleared, decisively promote the destruction of mostly virgin forests all over Latin America. Cattle ranching implies definite destruction because "the forest is cleared away entirely, and on a scale that will not allow for recolonization by adjacent forest if the pasture land is abandoned." (Myers, 1982, p. 751).

The country with the largest forest reserves in the world but at the same time that with the largest absolute annual loss of tree cover is Brazil. The Brazilian case has become the most well-known, as the on-going vast forest destruction has caused considerable controversy at the international level and brought Brazil under attack from ecological movements and political parties all over the world.

The foundation for large-scale destruction of Brazil's vast Amazon forests was laid in the mid-1960s when this area began to be considered a strategic resource for Brazil's economic growth and an unexploited resource for the country's well-being.

Massive fiscal incentives from the government channelled investment into the region, and the large-scale clearing of the moist forest became privately attractive due to public financial support (Binswanger, 1987). Among the tax- and credit-privileged Amazon development projects, the conversion of forest land into pastures for extensive beef production has been of major importance for many years.

Slash-and-burn practices of migrant farmers have their part in forest destruction in Brazil, too. Land clearing by settlers following the establishment of new infrastructure as well as government-initiated settlement schemes play an increasing role in this context.

Cattle ranching is a dominant factor in the forest destruction of other Latin American countries, namely in Mexico, Colombia, Peru, Bolivia, and a number of Central American states. It has increased tremendously over the last three decades due to growing domestic and foreign demand for beef. The ranching areas expand almost exclusively at the expense of primary forests and will soon be the most important cause of forest destruction in Latin America.

In the small Central American states, the particularly strong dependence on agricultural exports as the main earner of foreign exchange puts the most severe pressure on the forest reserves. Since 1950, the region's forest lands have declined by almost 40 percent (World Resources Institute, 1989). Forest destruction is closely related to the production and export of beef (Shane, 1986). Over the last two decades, cattle breeding for that purpose has significantly increased in all Central American countries, particularly in Costa Rica, Guatemala, and Honduras. By far the largest importer of beef from these countries is the USA.

Costa Rica is an example of the situation. As in other parts of Central America, export-oriented cattle rearing was among the first causes of extensive land clearing (Ellenberg, 1989). It was followed by sugar and tobacco cultivation, which led to locally intensive land use. Nevertheless, forest destruction remained comparatively moderate in the whole region until the early 19th century due to low population densities and a very weak transport infrastructure. Major deforestation in Costa Rica occurred with the first coffee boom around 1830. Over the past decade, cattle ranches have been the major cause of forest destruction in the country. In 1950, pastures accounted for one eighth of the country's land surface, by 1975 for one third, and by 1980 for around 40 percent (Myers, 1982). In the 1980s, the clearing rate for ranching purposes has been about 500 km$^2$ per year.

The rate of forest destruction in Costa Rica is still accelerating; today it is one of the fastest in the world. An area of about 600 km$^2$ is lost each year, and almost 90 percent of the cleared area is afterwards employed as pasture (Ellenberg, 1989). The pastures can be used only very extensively, and many are only used for 5–8 years. Almost all the wood decomposes near where it was cut. An insignificant share of the cut trees reaches the sawmills or gets used as fuelwood or for producing charcoal.

A similar pattern as that described for Costa Rica has more or less been followed by other Central American countries. Extensive beef production in the region is privately profitable because the ranch owners usually buy the forest land they intend to transform into pastures for next to nothing. A higher price for land converted into pasture should provide a powerful incentive for introducing production systems with a higher stockage per ha and more sustainable ranching on the same area. Technical possibilities for that purpose (e.g., the employment of improved animal breeds, improved pasture management, etc.) are readily available. However, as long as Latin America's large landowners receive important financial support from the government for clearing new areas of primary forest, and moreover receive forest land almost free, there is little incentive to maintain the soil stocking capacity but rather to practise "shifting ranching." The situation is sometimes complicated by the fact that the owners of large estates often have considerable political power.

In Latin America, the reason for fire clearing by land-hungry peasants does not appear to be an absolute shortage of land but rather the lack and delay of land reform, which is a major social and political issue in almost all of these countries.

## Southeast Asia

Although the Southeast Asian countries have undergone a rapid expansion of their non-agricultural sectors over recent decades, one of their major problems has remained the satisfaction of the pronounced land hunger of their fast-growing population. In addition, the increase in food production can only partially keep pace with population growth. The industrial sector has not grown sufficiently to provide enough employment significantly to decrease the share of the population that depends on agricultural production for its livelihood. Therefore, "the need for more land remains a dominating political problem and the opening of new agricultural areas a main goal of regional and national development. This means conflicts in land use, especially an ever-growing pressure on forest reserves" (Uhlig, 1984, p. 8). The timber exploitation activities of settlers and shifting cultivators are considered the main immediate cause of most of the forest destruction. Southeast Asia is by far the most important region exporting tropical hardwood, with more than 80 percent of the internationally traded volume, almost exclusively from Malaysia and Indonesia (FAO, 1989; and Brünig, 1989). Although only few trees (often less than 20) are taken from a given area in the course of one logging operation, a number of surveys conducted in the region show that logging on average leaves one- to two-thirds of the remaining trees damaged beyond recovery (Myers, 1982), which makes the forests easy for migrant cultivators to clear. During the 1970s, farmers cleared at least 85,000 km$^2$ per year (Chandrasekharan, 1979), the greatest absolute loss of tree cover due to subsistence farming activities in any tropical region.

In the past, vast areas of forest land have been cleared for establishing plantations. Their products are still a major earner of foreign exchange for Southeast Asian countries, and therefore their future expansion seems more probable than any reduction.

Government-initiated settlement schemes are important in the region. They have gained priority in Indonesia, Malaysia, and Thailand and are held responsible for forest destruction in that they tend to proceed without sufficient financial and extension support.

Indonesia still possesses the region's largest reserves of closed forest. Around 150 million ha (about three quarters of the country's total land area) are still covered by woodland. This forest area must be considered as highly endangered, because it is reputed to be the world's most valuable reserve of tropical rain forest (Repetto, 1988), and the standard of living of the fast-growing population is low. In fact, Indonesia's annual deforestation is by far the highest in the region, with an estimated 700,000 ha per year, but most of it is caused by land clearance for farming (Repetto, 1988).

In Malaysia, too, deforestation proceeds particularly fast. Although the country is the world's most important exporter of tropical timber, a much greater share of its vast annual loss of tree cover is lost through agricultural activities. During the 1980s, the country's forests retreated by an annual rate of 1.2 percent or 250,000 ha per year (FAO, 1981). The country's high rate of timber exploitation for export coincides with an unfavourable land rights system; any citizen can establish a personal right to use a piece of land just by clearing and farming it. Hence, forest lands that are opened up and partly degraded by timber exploitation are rapidly cleared from the remaining tree cover and taken into possession by settlers, a process which is promoted by widespread rural poverty (Segal, 1983).

In the Philippines, too, the exploitation of the forest reserves has constituted a most welcome source of foreign exchange for many years. Combined with logging operations, shifting cultivation has contributed decisively to the destruction of the country's tree cover. Swidden (shifting cultivation) systems, which over long centuries were sustainable, have recently led to extensive forest destruction following increasing population growth and shortening fallow periods that have led to decreasing soil fertility. The deforestation has wrought the most negative consequences on the agricultural production as it has left upper watersheds unprotected and destabilized river flows.

Traditional agroforestry systems that would allow a sustained cultivation of the same soil are in retreat in Southeast Asia. This is considered to be mainly a consequence of increasing land shortage and of the competition with production systems in which speculative crops such as coffee and cloves are more essential (Mary and Michon, 1987).

# Conclusions

The extensive destruction and degradation of tropical forests is decisively promoted by agricultural policies in the form of producer incentives, massive public expenditures, and fiscal advantages for agricultural projects with questionable economic results. Governments that are in principle committed to observe the long-term interests of society, and hence to preserve the country's resources, give in to day-to-day social, economic, and political pressures. The forests are in this context treated as a free resource and serve as a short-term outlet for social and economic problems. If the presently prevailing trends continue, most tropical countries, have little prospect of possessing significant forest areas in the next century.

Although most governments of tropical countries seem to be aware of the threats posed by increasing forest destruction, they do little to modify those policies that contribute to accelerating the process. Major government parameters that could have a positive influence on forest conservation via developments in the agricultural sector are changes in tax and trade régimes, price incentives, and changes in land-tenure legislation. The policy changes needed to diminish the rates of deforestation and save forests at risk differ considerably from one region to another.

The African continent seems to be caught in a vicious circle of increasing population pressure, lack of non-agricultural employment opportunities, shortage of foreign exchange, and unclear land rights. This situation leaves little hope for saving the forests other than by financial support by richer nations, which might make sustainable agricultural production systems profitable to the individual farmer.

In the Latin American context, the removal of major fiscal advantages for large-scale forest clearing could contribute considerably to slowing down the rate of forest destruction. More complicated is the problem of forest destruction by shifting cultivators who use practices that they consider as necessary for their survival but which are extremely damaging to the forests. Here, appropriate agricultural reforms that would have to include measures of land redistribution seem necessary but might be very difficult to push through on a political level.

In Southeast Asia, the growth of non-agricultural sectors might take pressure off the forests as governments and individuals become less dependent on drawing on the forest land as a resource for creating income.

Policy changes require more than the availability of economic and technical solutions, however. Until now, the tropical forests have been used (or misused) for solving emerging political and social conflicts and for escaping from economic and fiscal pressure. Necessary changes in agricultural policies were postponed at the expense of the (seemingly free) forest reserves. How far the current processes of tropical forest destruction by agricultural activities are replaced by more sustainable methods of production on already-cultivated land is a political choice. Any move towards sustainable systems implies higher private costs that the individual farmer will not incur deliberately. An important step in introducing improved systems is to put a realistic implicit or explicit price on the forest land. Another step might be subsidies that make such systems privately profitable; depending on the situation, these would have to be paid by national or foreign donors who both have a vital interest in the long-term conservation of the tropical forests.

The mistaken policies of governments in tropical countries and the lack of financial support obviously contribute to tropical forest destruction, but the policies of governments in non-tropical industrialized countries, which hamper the development of non-agricultural branches in the economies of those countries, also play a role. Protectionism also plays an important part in this situation. These aspects should not be forgotten in the present context. If the process of deforestation cannot be stopped, irreversible damage to the environment that concerns the world at large will be the unavoidable consequence.

## Notes

[1] Universität Hohenheim.

[2] The tropical regions are defined as the regions with an average monthly temperature above 18° over the whole year. Land is defined as having a "forest cover" if it is "with trees whose crowns cover more than 20 percent of the area, and which are not used primarily for purposes other than forestry, whether reserved forest or not. This area includes also temporarily unstocked areas; i.e., forests in which trees have been temporarily removed by cutting or burning to such an extent that less than 20 percent of the area is covered by tree crowns. It excludes areas deforested by shifting cultivation and other wooded areas such as savannah and open woodland" (FAO, 1976, p. 4). This corresponds to the term "closed forest."

[3] Deforestation is defined as "the temporary or permanent removal of forest cover whether for agricultural or other purposes" (Grainger, 1983, p. 389). The difficulties of the practical methods that are applied for estimating deforestation and the problems associated with the reliability of the figures published about the rate of deforestation in various regions are described by Grainger (1983 and 1984).

[4] Agricultural land "comprises arable lands, orchards, vineyards, meadows, pasture, other grassland, agricultural land producing concurrent tree crops, and lands under shifting cultivation which are part of a recognized fallow rotation" (FAO, 1976, p. 4).

## References

Adelhelm, R., et al., Standortgerechte Landwirtschaft—Ansätze in der technischen Zusammenarbeit, in von Blanckenburg, P., and de Haen, H. (Eds.), Bevölkerungsentwicklung, Agrarstruktur und ländlicher Raum, Schriften der Gesellschaft für Wirtschafts und Sozialwissenschaften des Landbaues No. 22, Landwirtschaftsverlag GmbH, Münster-Hiltrup, Germany, 1986, pp. 363–389.

Binswanger, H., *Fiscal and Legal Incentives with Environmental Effects on the Brazilian Amazon*, World Bank, Washington, D.C., USA, 1987.

Brünig, E.F., "Internationaler Tropenholzhandel und Waldvernichtung in den Tropen," in Bähr, J., Corves, C., and Noodt, W. (Eds.), *Die Bedrohung tropischer Wälder*, Wissenschaftsverlag Vauk, Kiel, Germany, 1989.

Chandrasekharan, C. (Ed.), "Shifting Cultivation," *Forest News*, Vol. 2, No. 2, 1979, pp. 1–25.

Denevan, W.M., *The Role of Geographical Research in Latin America*, Conference of Latin Americanist Geographers, Muncie, Indiana, USA, 1978.

Deutscher Bundestag, *Zweiter Bericht der Enquête-Kommission, "Vorsorge zum Schutz der Erdatmosphäre," zum Thema Schutz der tropischen Wälder*, Bonn, Germany, 1990.

Dressler, J., "Standortgerechter Landbau (SGL) im tropischen Bergland, Situation und Entwicklungsmöglichkeiten landwirtschaftlicher Kleinbetriebe in Rwanda," dissertation, Universität Hohenheim, Germany, 1984.

Ellenberg, H., "Auswirkungen von Umweltfaktoren und Nutzungsweisen auf das Artengefüge und die Regeneration tropischer Regenwälder," *Entwicklung und Ländlicher Raum*, Vol. 3, No. 85, 1985, pp. 6–12.

Ellenberg, H., "Ursachen und Konsequenzen der Waldzerstörung in Costa Rica," in Bähr, J., Corves, C., and Noodt, W. (Eds.), *Die Bedrohung tropischer Wälder*, Wissenschaftsverlag Vauk, Kiel, Germany, 1989, pp. 31–46.

Esser, J., "Warum sind tropische Wälder Schutzwürdig?," in Bähr, J., Corves, C., and Noodt, W. (Eds.), *Die Bedrohung tropischer Wälder*, Wissenschaftsverlag Vauk, Kiel, Germany, 1989, pp. 17–30.

FAO (Food and Agriculture Organization of the United Nations), *Forest Resources in the Asia and Far East Region*, Rome, Italy, 1976.

FAO (Food and Agriculture Organization of the United Nations), *Map on the Fuelwood Situation in Developing Countries*, Rome, Italy, 1981.

FAO (Food and Agriculture Organization of the United Nations), *Yearbook of Forest Products*, Rome, Italy, 1989.

Grainger, A., "Improving the Monitoring of Deforestation in the Humid Tropics," in Sutton, S.L., Whitmore, T.C., and Chadwick, A.C. (Eds.), *Tropical Rain Forest: Ecology and Management*, Blackwell Scientific Publications, Oxford, UK, 1983, pp. 387–395.

Grainger, A., "Quantifying Changes in Forest Cover in the Humid Tropics: Overcoming Current Limitations," *Journal of World Forest Resource Management*, Vol. 1, No. 1, 1984, pp. 3–63.

Haffner, W., "Tropische Gebirge: Ökologie und Agrarwirtschaft," *Beiträge zur entwicklungsforschung Reihe I*, Giessen, Germany, 1982.

Kio, P.R.O., "Management Potentials of the Tropical High Forest with Special Reference to Nigeria," in Sutton, S.L., Whitmore, T.C., and Chadwick, A.C. (Eds.), *Tropical Rain Forest: Ecology and Management*, Blackwell Scientific Publications, Oxford, UK, 1983, pp. 445–455.

Khan, M.S., and Knight, M.D., *Fund-Supported Adjustment Programs and Economic Growth*, Occasional Paper No. 41, International Monetary Fund, Washington, D.C., USA, 1985.

Knerr, B., "Evaluierung des ländlichen Integrierten Entwicklungsprojekts Quinara, Guinea Bissau," Soziokulturelle und ökologische Rahmenbedingungen, Gutachten im Auftrag von Bundeslandwirtschaftsministerium/GTZ, Hohenheim, Germany, 1989.

Knerr, B., "Agricultural Policies and Deforestation in Sub-Saharan Africa," in Venzi, L. (Ed.), *The Environment and Agricultural Resource Management*, Proceedings of the XXIV Seminar of the European Association of Agricultural Economists, Viterbo, Italy, 1991.

Ley, A., "La Logique Foncière de l'État depuis la Colonisation: l'Expérience Ivoirienne," in Le Bris (Ed.), *Enjeux Fonciers en Afrique Noire*, Paris, France, 1982, pp. 135–141.

Mary, F., and Michon, G., "When Agroforests Drive Back Natural Forests: A Socio-Economic Analysis of a Rice-Agroforest System in Sumatra," *Agroforestry Systems*, Vol. 5, No. 1, 1987, pp. 27–55.

Myers, N., "Depletion of Tropical Moist Forests: A Comparative Review of Rates and Causes in the Three Main Regions," *Acta Amazonica*, Vol. 12, No. 2, 1982, pp. 745–758.

Myers, N., "Deforestation Rates in Tropical Forests and Their Climatic Implications," Friends of the Earth, London, UK, 1989.

Repetto, R., *The Forest for the Trees? Government Policies and the Misuse of Forest Resources*, World Resource Institute, Washington, D.C., USA, 1988.

Segal, J., "A Fragile Prosperity," *Far Eastern Economic Review*, 14 April 1983.

Shane, D.R., *Hoofprints on the Forest: Cattle Ranching and the Destruction of Latin America's Tropical Forests*, Institute for the Study of Human Values, Philadelphia, Pa., USA, 1986.

Sutlive, V.H., Altschuler, N., and Zamura, M.D., "Where Have All the Forests Gone?," Department of Anthropology, College of William and Mary, Williamsburg, Va., USA, 1981.

Uhlig, H. (Ed.), *Spontaneous and Planned Settlement in Southeast Asia*, Hamburg, Germany, 1984.

World Bank, "Memorandum and Recommendation of the International Bank for Reconstruction and Development to the Executive Directors on a Proposed Loan of $150 million equivalent to the Republic of Cameroon for a Structural Adjustment Program," Report No. P–5079–CM), May 16, 1989.

World Resources Institute, *World Resources Report, 1988–89*, 1989.

Zulu, J.B., and Nsouli, M., *Adjustment Programs in Africa: The Recent Experience*, Occasional Paper No. 34, International Monetary Fund, Washington, D.C., USA, 1985.

---

## Discussion Opening—*Kyrre Rickertsen* (Agricultural University of Norway)

Since World War II, deforestation has shifted from the temperate zone to the tropics. For example, during 1950–83, the area of forest and woodland fell by 38 percent in Central America and 24 percent in Africa. Simple projections indicate future deforestation that would reduce the tropical forest area by 10–20 percent by the year 2020.

Knerr's survey of the literature provides a detailed description of these large losses of tropical forests and some of the severe consequences of the losses. Furthermore, the negative effects on the tropical forests of agricultural policy as well as interactions of various government policies with agriculture are described. Finally, Knerr's conclusions appear to be reasonable, and she emphasizes that the solutions for better management of tropical forests may differ from region to region.

As a non-expert on tropical forestry, I will not go very much into the specific details in the paper but rather draw attention to two principles that I feel are not sufficiently emphasized.

First, I think it is uncertain that deforestation is always a negative form of land use. One sensible economic criterion for efficient forest management is to achieve the maximum net present value from all the forests' various possible uses over the long run. This implies that total benefits and total costs, discounted at an appropriate interest rate, for the various possible uses must be estimated. Then, the land should be devoted to the use—forest, agriculture, or other—that yields the greatest economic benefits.

Second, several of the benefits of the tropical forests are global. For example, the tropical forests contribute material for plant breeders and the pharmaceutical industry worldwide and may affect the global climate. If Third-World countries are expected to take these global benefits into consideration when they decide how to use their tropical forests, they should be compensated by the rest of the world when global considerations change the ranking of the alternative uses of the forests (e.g., "debt for nature" swaps).

*[Other discussion of this paper and the author's reply appear on page 329.]*

# Sustainable Land Use and Sustainable Development: Critical Issues

*Karen Liu and Chung-Chi Lu*[1]

**Abstract:** Sustainable agriculture has emerged as a key issue in agricultural development and natural resource management because of widespread and growing concern about the seriousness of degradation of the world's natural resource base and ever-increasing pressures on these resources from continuing rapid population growth. This paper examines the changes in land use and the problem of tropical deforestation affecting the world's land resource base for sustainable agricultural development. Global land-use changes have been slow in the last decade. However, changes in land-use patterns have been significant in many developing countries, especially the conversion of forest land into agricultural land to meet increasing demand for food and fibre. It is estimated that over 11 million ha of tropical forests are cleared every year in Latin America, Asia, Africa, and the Pacific Islands. The most serious consequence of tropical deforestation is soil erosion, which may greatly increase after forests are cleared. It is also apparent that tropical deforestation is contributing to global climate change. Improved agricultural sustainability, combined with policies to protect natural resources, is urgently needed to support the rapidly increasing population in developing countries.

## Introduction

Achieving agricultural sustainability in developing countries has become one of the most important issues in agricultural development in recent years because of growing concerns about rapid population growth and the accelerating decline of economically viable arable land. This paper examines the changes in land use and the rate of tropical deforestation affecting the conditions of the world's land resource base for sustainable agricultural development. The pressures of population growth and its adverse effect on land resource and food demand are discussed. Possible actions and policy interventions for a sustainable land resource base are suggested.

The natural resources of a country are its most valuable endowment. However, only recently have scientists become fully aware that growing population pressures have led to the narrowing of the resource base throughout the world. The observed result is that human-induced degradation of land resources has worsened largely due to the inadvertent, inappropriate use of technological innovations. It is estimated that each year 6 million ha of dryland are removed from production by mismanagement and 11 million ha of forest land are degraded by clearing. There are three major ways in which this is occurring: slash and burn farming in tropical areas, increased land conversion in temperate climates for settled agriculture, and high rates of inappropriate timber harvesting. Such a vast and growing degradation of the world's natural resource base has led to an increasing awareness of the importance of having a sound, sustainable natural resource base or environment.

Sustainable agriculture has emerged as a key focus in agricultural development and natural resource management because of growing concerns about the seriousness of degradation of the world's natural resource base and increasing pressures on those resources from rapid population growth. Because of the Green Revolution, there have been major successes in South and Southeast Asia in the past 20 years in increasing crop yields by the introduction of new high-yielding crop varieties and the use of inorganic fertilizers and pesticides. However, there is still a general perception that agriculture in developing countries is not able to respond to current needs for increased food and fibre production due to continuing degradation of natural resources—soil erosion, deforestation, desertification, and water contamination.

# World Population Growth

The world's population continues to grow rapidly, driven by very high growth rates in many developing countries. During 1950–89, global population more than doubled from 2,500 million to 5,200 million people, and another 1,000 million people will be added before the end of the next decade. Although some developed countries' population growth rates have fallen in the last decade, many developing countries are still growing rapidly because a large portion of their populations is in its reproductive years and contraceptive techniques are not widely used; e.g., India. Table 1 presents data on current and projected world population and its labour force. It highlights the relatively high population growth rates in developing countries. For example, Africa has the world's highest annual rates of population growth (3 percent overall, with 21 countries above 3 percent), followed by South America (above 2 percent in most countries). Asia, the most populous continent, also has a rate of increase (1.85 percent) that is slightly higher than the world average.

The increasing trend in world population growth is projected to continue well into the foreseeable future. According to recent projections by the United Nations, world population will increase by 3,200 million persons during 1990–2025. Of this growth, more than 90 percent will occur in the developing regions of Africa, Asia, and Latin America. Because of this ominous trend, these regions' share of total world population will increase from 67 percent in 1950 to 84 percent in 2025.

The rapid and continuous population growth and very large numerical increases in world population have several potentially adverse consequences. One major impact is heightened food demand. Another significant impact is the reduced sustainability of natural resources and resulting degraded quality of the environment.

In terms of world food production trends, cereal production has increased over 70 percent since 1965. It has also more than doubled in developing countries, with Asia leading as a result of the Green Revolution. However, the rate of increase in cereal production has slowed significantly since 1983, indicating that the major gains have already peaked. In terms of per-capita food production, China and the other centrally planned economies of Asia have shown the largest gains. Food production in other Asian nations is only slightly ahead of the population curve. In Latin America and the countries of the Near East, increases in food production appear to be staying only about even with population growth. By contrast, Africa is still continuing a long-term decline in per-capita food production that began in the early 1970s, with no reversal of this adverse trend in sight. Between the 1976–78 average and the 1986–88 average, per capita food production declined by 8 percent. Because of continuing rapid population growth in African countries and slow or non-existent agricultural productivity increases, it is becoming increasingly difficult to maintain agricultural development while at the same time providing adequate food supplies at reasonable prices. Therefore, one of the most important factors for sustainable agricultural development is the ability to balance the needs of larger populations in developing countries with the rate of increase in productivity of their natural resources.

# Global Land-Use Changes

Until recently, the history of humankind could be regarded as a process of appropriate natural resource use. However, the trend has shifted more recently to one of over-exploitation. The land base can be classified into urban, crop, range, pasture, forest, wetlands, and other categories. At the global level, shifts in land-use patterns have been gradual in the last decade. However, local or regional changes in land-use patterns have been very dramatic, especially in terms of the accelerating rate of conversion of forest land into agricultural land in an effort to meet the rapidly increasing demand for food and fibre in the developing countries.

Table 1—Size and Growth of Population, Labour Force, Share of Urban Population

| | Population (millions) | | | Average Annual Population Change (percent) | | | Urban Population as a Percentage of Total | | | Total Labour Force (millions) | Percentage of Labour Force in | | | | | |
| --- | --- | --- | --- | --- | --- | --- | --- | --- | --- | --- | --- | --- | --- | --- | --- | --- |
| | | | | | | | | | | | Agriculture | | Industry | | Service | |
| | 1960 | 1990 | 2025 | 1965–70 | 1975–80 | 1985–90 | 1960 | 1975 | 1990 | 1985 | 1960 | 1990 | 1960 | 1990 | 1960 | 1990 |
| Africa | 281 | 648 | 1,581 | 2.63 | 2.95 | 3.00 | 18 | 25 | 35 | 214 | 78 | 69 | 8 | 12 | 14 | 19 |
| N. and C. America | 270 | 427 | 595 | 1.64 | 1.47 | 1.28 | 63 | 67 | 71 | 176 | 18 | 12 | 32 | 29 | 50 | 58 |
| S. America | 147 | 297 | 498 | 2.47 | 2.27 | 2.08 | 52 | 65 | 76 | 94 | 44 | 29 | 22 | 26 | 33 | 45 |
| Asia | 1,667 | 3,109 | 4,890 | 2.44 | 1.86 | 1.85 | 22 | 25 | 30 | 1,299 | 75 | 66 | 10 | 15 | 15 | 19 |
| Europe | 425 | 498 | 512 | .67 | .45 | .23 | 61 | 69 | 73 | 226 | 28 | 14 | 39 | 39 | 33 | 47 |
| USSR | 214 | 288 | 352 | 1.01 | .82 | .78 | 49 | 60 | 68 | 143 | 42 | 20 | 29 | 39 | 29 | 41 |
| Oceania | 16 | 27 | 39 | 1.97 | 1.51 | 1.44 | 66 | 72 | 71 | 11 | 27 | 20 | 32 | 28 | 41 | 52 |
| World | 3,019 | 5,292 | 8,467 | 2.06 | 1.74 | 1.73 | 34 | 39 | 43 | 2,164 | 60 | 51 | 18 | 21 | 22 | 28 |

Table 2—World Land Area and Use, 1985–87, and Percent Change since 1975–77

| | Total Area | | Percent Change since 1975–77 |
| --- | --- | --- | --- |
| | Million ha | Distribution (percent) | |
| Cropland | 1,473 | 11 | 2.7 |
| Permanent pasture | 3,215 | 25 | -0.2 |
| Forest and woodland | 4,074 | 31 | -2.1 |
| Other land | 4,314 | 33 | 1.3 |
| World | 13,076 | 100 | |

Historically, increased food demand was met by enlarging the cultivated area. Although increasing yields have become more important since 1950, the cropland area continues to expand. Between 1964–66 and 1982–84, total cropland increased by 8.9 percent globally. The growth in total cropland between 1975–77 and 1985–89 has subsequently slowed, rising only 2.7 percent. There was an even greater percentage increase in many countries of Africa, Asia, Latin America, and Oceania, partly because of increasing population pressures (Table 2). However, the per-capita land area for agriculture has been declining due to the continuous growth of population. At the world level, without a major expansion of arable land, the world average of 0.28 ha of cropland per capita is expected to decline to 0.17 ha by the year 2025 under current population projections. In Asia, cropland per capita would decline to only 0.09 ha. For range and pasture land, the amount of land devoted globally to permanent pasture remained relatively stable (0.2-percent decline) between 1975–77 and 1985–89. However, intensity of rangeland use in many developing countries has increased as a result of rapid growth in human and livestock populations.

The increase in cropland has come at the expense of rangeland, forests, wetlands, and other areas that are both economically important and ecologically fragile. Urbanization and industrial development are the two major threats to the loss of existing cropland. Losses of agricultural land to other uses are most pronounced in the densely populated, rapidly industrializing countries of East Asia, including China, Japan, South Korea, and Taiwan, where nonfarm uses claim roughly 500,000 ha of cropland each year.

## Effects of Tropical Deforestation and Soil Erosion on Agriculture

Nearly all forms of global environmental degradation are adversely affecting agricultural production. The two most significant forms of land resource degradation include tropical deforestation and soil erosion. Since 1950, a period when population has been growing rapidly in developing countries, deforestation has been concentrated in the tropics.

Forests cover approximately 31 percent of the earth's land surface and are vital to both human activities and biological processes (Table 3). About half of these forests are in developing countries in the tropics. Today, the highest rates of land-use changes are also occurring in the tropics, where human populations are rapidly increasing.

Table 3—Distribution of Cropland and Forest Land among Major World Regions, 1985–87

|  | Total Land Area | | Cropland | | Forest and Woodland | |
|---|---|---|---|---|---|---|
|  | Million ha | Share (percent) | Million ha | Share (percent) | Million ha | Share (percent) |
| Africa | 2,964 | 22.7 | 185 | 12.6 | 689 | 16.9 |
| N. and C. America | 2,138 | 16.4 | 274 | 18.6 | 685 | 16.8 |
| S. America | 1,753 | 13.4 | 141 | 9.6 | 905 | 22.2 |
| Asia | 2,679 | 20.5 | 451 | 30.6 | 540 | 13.3 |
| Europe | 473 | 3.6 | 140 | 9.5 | 157 | 3.9 |
| USSR | 2,227 | 17.0 | 232 | 15.8 | 943 | 23.1 |
| Oceania | 843 | 6.5 | 50 | 3.4 | 156 | 3.8 |
| World | 13,077 | 100.0 | 1,473 | 100.0 | 4,074 | 100.0 |

Excessive tropical deforestation has become an urgent, global environmental issue. Based on a 1980 FAO assessment of tropical forestry research, it is estimated that over 11 million ha of tropical forests are cleared every year in Latin America, Asia, Africa, and the Pacific islands (FAO, 1988). Several, more recent, studies show that levels of deforestation have recently accelerated in Brazil, Costa Rica, India, Myanmar, the Philippines, and Vietnam. Forest clearing also increased sharply in Cameroon, Indonesia, and Thailand. Table 4 shows deforestation estimates for closed tropical forests in selected countries. Principal direct causes of tropical forest losses are their conversion and use for agriculture, fuelwood gathering by the rural poor, and poorly managed commercial logging. On a worldwide scale, the ever-expanding quest for food sources is the principal contributor. Other actions include the continued spread of shifting agriculture, clearing of closed forests for pasture, and heavy grazing pressures in the open forests of semi-arid regions.

Table 4—Deforestation Estimates for Closed Tropical Forests, Selected Countries

| Country | FAO Estimates (1981–85) | | More Recent Estimates | | |
|---|---|---|---|---|---|
| | 1000 ha | Annual Rate of Change (percent) | 1000 ha | Annual Rate of Change (percent) | Period |
| Brazil | 1,480 | –0.4 | 8,000 | –2.2 | 1987 |
| Cameroon | 80 | –0.4 | 100 | –0.6 | 1976–86 |
| Costa Rica | 65 | –4.0 | 124 | –7.6 | 1977–83 |
| India | 147 | –0.3 | 1,500 | –4.1 | 1975–82 |
| Indonesia | 600 | –0.5 | 900 | –0.8 | 1979–84 |
| Myanmar | 105 | –0.3 | 677 | –2.1 | 1975–81 |
| Philippines | 92 | –1.0 | 143 | –1.5 | 1981–88 |
| Thailand | 379 | –2.4 | 397 | –2.5 | 1978–85 |
| Vietnam | 65 | –0.7 | 173 | –2.0 | 1976–81 |

According to Brown (1990), soil erosion is slowly undermining the productivity of an estimated one-third of the world's cropland. Current estimates of annual losses of soil due to erosion are 25,000 million t globally. Soil erosion in Africa and South America is occurring at annual rates of about 7 t/ha due mainly to intensive cultivation and cropping. However, in the more fragile soils of developing tropical countries, the rapid rate of deforestation is the major factor contributing to soil erosion. Furthermore, soil erosion not only affects soil fertility but also causes degradation of aquatic resources through siltation.

## Agenda for a Sustainable Resource Base

A massive international effort is needed in order to protect soil, conserve water, and restore the productivity of degraded land. To bring the world's population into balance with its resources and environment, the relationship between agricultural and population dynamics of developing countries must be brought back into equilibrium. This is fundamental for sustainable agricultural development. Two aspects of population growth affect natural resource and environmental issues: increased numbers of persons and higher per-capita income. While increased population numbers present one kind of environmental impact, higher per-capita income is of even more concern regarding the adverse effects on worldwide environmental change. According to recent UN population projections, world population will increase by 3,200 million people during the 1990–2025 period. Most of this growth will take place in the developing countries. If this increase in population is realized, we must accelerate our educational programmes and stress economic development on a worldwide basis. The challenge for sustainable development is not only to meet the needs for the increasing size of population but also to meet the needs for a better quality of life. Joint cooperative efforts to provide education, family planning, population assistance, and economic development by national governments, international organizations, and the non-governmental organizations are essential to solving the world's population problem.

Comprehensive study of the soils, vegetation, climate, and the other physical resources of the country is also vital for land-use planning. The natural resource base must be surveyed and fully understood prior to or during the design stage to ensure that agricultural development projects are compatible with characteristics, properties, and capacities of the natural resource base to support sustainable agriculture. Resource mapping and geographical information systems as well as agroecological zoning are key ingredients for land-use planning for developing countries.

Development of policies and programmes on land resource conservation and environmental improvement is also vital to sustainable land use. Government authorities must reform policies that adversely affect farmers' use of land and their choice of technology, and must eliminate incentives that lead to forest destruction or growing of inappropriate crops on vulnerable lands. There is great need to build conservation and environmental improvement "incentives" into the economic and political system. Inappropriate or poorly enforced land-tenure systems can discourage farmers from conserving natural resources and investing in future productivity. At present, many countries do not have adequate laws to protect forests and rangelands from indiscriminate exploitation. For example, obtaining the clear title to land is often a problem in Brazil. In order to achieve sustainable land use, the constraints that threaten it must be alleviated.

**Adoption of improved land use practices.** Soil preservation is the most important resource for ensuring sustainability; loss of topsoil through erosion and a reduction in soil fertility by not replacing nutrients both turn a renewable resource into a nonrenewable one. Poor soil and water management in rainfed agriculture can cause severe land degradation. Positive steps that can be taken include adoption of conservation practices, shifting cultivation methods back to those traditionally practised in order to maintain soil fertility, expanded use of better technology and improved education, and employing low-input production practices, such as integrated pest management and best management practices. The proper balance of these measures can facilitate and maintain land resources.

**Actions for sustainable forestry development.** Because of the scale and adverse implications of continuing tropical deforestation, international organizations including the FAO, World Resources Institute, World Bank, and UN Development Programme surveyed forest conditions in the tropics. They formulated the "Tropical Forestry Action Plan," calling for an international commitment to reverse tropical deforestation. This plan includes work in five areas: forestry in land use, forest-based industries, fuelwood, conservation, and use of appropriate institutions to reach the people. The USA is also taking an active part in the worldwide effort to reverse tropical deforestation through the tropical forestry programme of the US Department of Agriculture's Forest Service.

Education and public awareness is one mechanism for changing direction on land resource management. It is important to build a public consciousness and awareness of the need for careful management of the resource base and then translate that consciousness into active policies and regulations. The educational process must be targeted to reach the public, particularly in developing countries, about the importance of a sound natural resource management programme for sustainable agriculture.

# Conclusion

Sustainable agriculture for developing countries has become a crucial issue in agricultural development and resource management. Agriculture is an extremely important engine of economic growth in developing countries. It provides not only food, fibre, and fuel, but also employment and income, as well as the raw materials, capital, and additional resources necessary for the development of other sectors. However, the present and future capabilities of these countries to provide adequate livelihood and food security are threatened by rapid population growth and degraded natural resources. Improved agricultural sustainability, combined with policies to protect natural resources, is therefore needed in order to support the rapidly growing population in developing countries.

## Note

[1]US Department of Agriculture and US Agency for International Development, respectively.

## References

Brown, L.R., *State of the World 1990*, N.W. Norton, New York, N.Y., USA, 1990.

FAO (Food and Agriculture Organization of the United Nations), *An Interim Report on the State of the Forest Resources in the Developing Countries*, Rome, Italy, 1988.

---

**Discussion Opening**—*Catherine Halbrendt* (University of Delaware)

This paper by Liu and Lu addresses a very pressing situation; i.e., sustainable development in less-developed countries through appropriate land use. To generate discussion, I will organize my comments in two parts: (1) the strengths and weaknesses of the major sections of the paper, and (2) additional issues concerning sustainable development not mentioned in the paper.

The global land use trends shown by regions and countries were very illustrative. The authors showed, by regions, respective accelerating rates of conversion of forest land into agricultural land to meet increasing food demand driven by population growth, especially in Asia and South America. However, this section of the paper fails comprehensively to identify the causes of the relative accelerating rates of land use by country upon which policies are based. More descriptive statistics should have been presented to infer comparatively why, for example, India's rate of land conversion is higher than Indonesia's although they are both in the Asian continent. Quantifying hypothesized linkages such as accelerating population growth, low productivity per capita, or policy incentives to rates of conversion by country are necessary to formulate effective policy reforms.

More in-depth analysis by agricultural enterprise by country would have enhanced the paper. More literature reviews on the effects by agricultural enterprise by country would have provided substantial insights into the differential impacts of deforestation. A summary quantifying the effects of economic incentives, technological adoption rates on productivity changes, and soil erosion by agricultural enterprise and country would definitely help formulate more prototype policy programmes useful in promoting a global sustainable resource base.

The section on formulating a global agenda for a sustainable resource base was very comprehensive. However, I would have liked the authors to focus more on discussing unique recommendations targeted towards sustainable development differentiated by a country's stage of development and resource endowments, as they undoubtedly require different policy recommendations.

The paper is useful as it addresses an important issue and proposes a useful agenda for decision makers, but, for practical purposes, it needs to be more focused and the analysis more comprehensive, such as including other very necessary elements essential for sustainable development; i.e., the socio-political sustainability aspects. Effectiveness of any sustainable land-use programme is highly dependent on whether social and economic policies established concurrently are politically sustainable.

*[Other discussion of this paper and the authors' reply appear on the following page.]*

**General Discussion**—*Daniel Pick, Rapporteur* (US Department of Agriculture)

One discussant felt that demographic factors behind deforestation were given much attention while demand-driven factors were ignored. He further commented that in dealing with natural forests, applying economic efficiency criteria is very complicated.

Knerr, in reply, agreed that the use of forest land and deforestation is not always negative. Yet, the negative effects are shifted to future generations who will have to pay for it. She further agreed that other nations that have vested interests in retaining the forests may have to pay for doing so. She warned that, for some countries, establishment of property rights will not be a good solution to deforestation, as small owners will continue to exploit the land, disregarding future generations.

A further point raised was that soil erosion is more than just an outcome of overcultivation, and no mention was made of water's contribution to the problem.

Lu, in reply, acknowledged that more research on specific countries is needed as well as research on the effects of deforestation on soil erosion. He further acknowledged that political and social stability were not included in their analysis.

Participants in the discussion included N.I. Isaksson (Swedish University of Agricultural Sciences), G.T. Jones (University of Oxford), F.G. Mack (Inter-American Development Bank), and N. Shanmugaratnam (Agricultural University of Norway).

# Political Economy Trade Negotiations: An Empirical Game Theory Analysis

## *Louis Mahé and Terry Roe*[1]

**Abstract:** The paper questions why agricultural trade compromise between the USA and EC is so difficult, whether a compensatory scheme be found that is both politically feasible and resource saving, and whether liberalizing policies by selected OECD countries will ease a trade compromise. These questions are addressed in a political economy context since, if the influence of special interests is ignored, trade compromises that both save resources and are politically feasible are unlikely to be searched for or found. The analysis entails the estimation of political preference weights, game theory, and a partial equilibrium world trade model based on 1988 data. The general answers are: the most influential special-interest groups face economic losses that, when coupled with their influence, tend to prevent a broad-based trade compromise given the current set of policy instruments; partial trade liberalization can occur if instruments are decoupled from production incentives, but free trade does not result; and partial liberalization by the rest of the OECD greatly increases the feasibility for the USA and EC to compromise. These results illustrate that interdependence in world trade has reached the point where bilateral action alone is unlikely to lead to real liberalization.

## Introduction

The economic history of EC–US conflicts over agricultural policy has been long and rich in events. Clearly, the difficulty of obtaining an agricultural trade agreement during the Uruguay Round of GATT negotiations suggests that economic efficiency is not the only factor motivating the negotiations between these antagonists. Another motivation is the balance of political pressures exerted by special interest groups on the governments in these countries. This balance typically protects producers relative to consumers and taxpayers in general. Hence, the search for a trade compromise confronts this political balance in each country.

If the political influence of special interests is ignored, then a search for a set of trade compromises that saves resources and is politically feasible is unlikely to be successful. In other words, the conventional neoclassical trade analysis where alternative trade compromises are based on net social welfare gains in each country is almost surely inconsistent with the balance of political power in these countries. Thus, the traditional type of analysis is of limited usefulness since trade compromises that are politically feasible cannot be identified from the larger set of compromises that merely save resources. Moreover, the typical compensatory payment scheme that compensates losers based on taxing some of the rewards from gainers from liberalization is misleading because this scheme assumes that one dollar of compensation to a group with less political power is equal to a dollar's worth of compensation to a group that has more political power.

While the USA and EC tend to be the major antagonists in the Uruguay Round, other OECD countries are also major participants in world agricultural markets. For many, their agricultural economies are adversely affected by the US and EC policies. And, as a group, their agricultural policies also affect world market prices and hence the budget costs of policies in the USA and EC. Hence, acceptable trade compromises between the USA and EC in the Uruguay Round may be dependent on policy changes in other OECD countries.

The general focus of this paper is on the political economy of the US and EC negotiations under GATT and the possible influence that the other OECD countries may have on the politically feasible set of possible trade compromises. More specifically, game theory, coupled with political economy and a partial equilibrium world trade model, is used to address the following questions: why a trade compromise between the USA and EC is so difficult when the net social gains from liberalization appear to be potentially large; whether a compensatory scheme can be found that can potentially lead to trade compromises that are politically feasible and resource saving; whether trade liberalization in other OECD countries would facilitate a trade compromise between the USA and EC, or, stated another way, in the context of political economy, how dependent acceptable US and EC trade compromises are on trade liberalization by other OECD countries.

## Conceptual Framework

Rausser and Freebairn (1974), Sarris and Freebairn (1983), and others (e.g., Tyers, 1989; and Riethmuller and Roe, 1986) have modelled the influence of special interest groups as the unconstrained maximization of a weighted, additive social welfare function over producer welfare, consumer welfare, and taxpayers. This paper adopts this approach and, for the remainder of the paper, refers to this type of a social welfare function as a political payoff function (PPF).

The PPF function for the USA or EC is:

$$(1) \quad V^i = \Sigma_j \lambda_j^i \pi_j^i (P_f^i, Z_f^i) + \lambda_c^i U^i (P_c^i, Z_c^i) + B^i (P_f^i, P_c^i, Z_j^i, Z_c^i, P_w)$$

where $i$ denotes the country, $-i$ denotes the other country, and the $\lambda_j^i$ is the political influence weight of the $j$th interest group (which is synonymous with the $j$th commodity) in the $i$th country. The $\lambda_c^i$ denotes the weight for consumers in the $i$th country. The weight associated with the government's budget $B^i$ is taken to be the numeraire.[2] The vector of domestic prices $P_f^i$, $P_c^i$ appearing in the $j$th interest group's profit function $\pi_j^i$ and the utility function $U^i$ of consumers depend on domestic policy instruments that affect domestic price (e.g., target prices, tariffs, and consumption tax), while the vectors $Z_j^i$ and $Z_c^i$ reflect those instruments that have indirect effects such as land set-aside, food stamps, and so on. Since markets are assumed to clear, the government's budget depends on domestic prices and a vector of world market prices $P_w$. The interdependency in world trade comes about through the effect of the other country's policy on world markets. Hence, $P_w$ is a function of countries' policy instruments.

Following Olson (1965), the interpretation of the PPF is that agricultural producers band together in lobbies to achieve through the government what they could not achieve in the market. However, the policies that they promote impinge on the welfare of other groups who lobby to counteract the agricultural lobby. Hence, in the PPF, a group's welfare weight $\lambda_j$ reflects the relative political influence wielded by the group in the determination of policies. The ratio of any two weights (e.g., $\lambda_j / \lambda_c$) reflects the amount of income loss to consumers per dollar gain to the $j$th producer that would leave the government indifferent, all else constant. Hence, if $\lambda_j > \lambda_c$, then the government is indifferent between a policy that transfers one European Currency Unit (ECU)[3] to producers for $\lambda_j / \lambda_c$ and a one-ECU loss to consumers.

The procedure for estimating these weights is based on an updated version of the world trade model that was initially developed by Mahé, Tavera, and Trochet (1988) and used to study the bilateral harmonization of US and EC agricultural policies (Mahé and Tavera, 1988). The model resembles the SWOPSIM model developed by the US Department of Agriculture (Roningen, 1986), except that it is designed to account for the actual policy instruments employed by countries. Consequently, it can account for both vertical and horizontal market interventions, including production and import quotas. It is a static partial equilibrium trade model that specifies production and demand elasticities for the USA, the EC, the region of Japan, the Nordic countries, Austria, Switzerland, Canada, Australia, and New Zealand, and, as an aggregate, the rest of the world. The model identifies eight commodities: wheat and coarse grains (grain), oilseed cakes, vegetable oil, feed-grain substitutes (including millings and other vegetable by-products, maize gluten feed, cassava, and citrus-pulp), beef, pigmeat and poultry, milk and milk products, and sugar. The model uses a set of production and demand elasticities similar to SWOPSIM. In a similar way to the procedure for initializing computable general equilibrium (CGE) models, the model is initialized to data for the base year 1988 so that its solution reproduces the base year data exactly. Simulation results are interpreted relative to the base year data.

## Estimation of Political Preference Weights

To characterize the economic game between the USA and EC, it is necessary closely to approximate the actual US and EC policy instruments. Accordingly, the instruments embedded in the model include, for the USA, deficiency payments, land set-aside, an export enhancement programme, import quotas for sugar, price supports for dairy, and, for beef, a tariff-linked import quota. For the EC, instruments are the variable levy (which fixes consumer grain prices), co-responsibility payments (which allow producer prices to depart from consumer prices), consumer prices equal to world prices for oilseed cakes and feed-grain substitutes (by previous GATT agreement), oilseed producer prices supported by a subsidy, and milk, sugar, beef, pigmeat, and poultry supported by a variable levy system. A production quota on milk was also implemented.

The next step is to assume that governments in the USA and EC choose these policy instruments as though they sought to optimize the PPF, Equation (1). Then, assuming differentiability, it is possible to use the first order conditions from this presumed optimization process to solve for the weights in (1); i.e., for the $\lambda_j^i$. The trade model is thus used, in effect, to derive these numerical derivatives. The result is a set of first-order conditions for each country; in matrix terms:

(2)  $\Delta V^i = \Sigma_j \lambda_j^i \Delta \pi_j^i (P_f^i, Z_j^i) + \lambda_c^i \Delta U^i (P_c^i, Z_c^i) + \Delta B^i (P_f^i, P_c^i, Z_j^i, Z_c^i, P_w) = 0$

(3)  $\Delta V^{-i} = \Sigma_j \lambda^{-i}_j \Delta \pi^{-i}_j (P^{-i}_f, Z^{-i}_j) + \lambda^{-i}_c \Delta U^{-i} (P^{-i}_c, Z^{-i}_c) + \Delta B^{-i} (P^{-i}_f, P^{-i}_c, Z^{-i}_j, Z^{-i}_c, P_w) = 0$

where, $i$ = the USA, $-i$ = the EC, and $\Delta V$, $\Delta \pi$, $\Delta U$, and $\Delta B$ are vectors of numerical derivatives obtained for small changes in the policy instruments of each country. Numerical estimates of the weights in (1) are then derived from:

(4)  $[\lambda^i] = [\Delta \pi_j^i (P_f^i, Z_j^i) \, \Delta U^i (P_c^i, Z_c^i)]^{-1} - [\Delta B^i (P_f^i, P_c^i, Z_j^i, Z_c^i, P_w)]$

(5)  $[\lambda^{-i}] = [\Delta \pi^{-i}_j (P^{-i}_f, Z^{-i}_j) \, \Delta U^{-i} (P^{-i}_c, Z^{-i}_c)]^{-1} - [\Delta B^{-i} (P^{-i}_f, P^{-i}_c, Z^{-i}_j, Z^{-i}_c, P_w)]$

The results from these computations appear in Table 1. Notice that those sectors that are the most protected in each country receive the highest weight. At the higher end of the scale, this includes producers of sugar and dairy products in both countries, while consumers and producers of pigmeat and poultry appear at the lower end of the scale. Clearly, the interpretation of these weights as revealing the political influence of the various groups in determining US and EC agricultural policy in 1988 must be conjectural. Hence, it is preferable to ask what these weights imply about the US and EC trade policy negotiations and whether the implications seem consistent with other information available about these negotiations. Their literal interpretation was mentioned above. For instance, the ratio $\lambda_{sugar} / \lambda_c$ for the USA is $1.90. This suggests that the US government was

Table 1—Policy-Goal Function Weights and Their Ranking by Interest Group for the USA and EC, Based on 1986

|  | USA | | EC | |
|---|---|---|---|---|
|  | Rank | Weight $(\lambda_{US})$ | Rank | Weight $(\lambda_{EC})$ |
| Sugar | 1 | 1.56 | 1 | 1.57 |
| Dairy products | 2 | 1.29 | 2 | 1.46 |
| Feed | 3 | 1.23 | 4 | 1.32 |
| Grain | 4 | 1.15 | 3 | 1.34 |
| Budget | 5 | 1.00 | 6 | 1.00 |
| Beef | 6 | 0.92 | 4 | 1.32 |
| Consumers | 7 | 0.87 | 8 | 0.83 |
| Pigmeat and poultry | 8 | 0.85 | 7 | 0.95 |

willing, at the margin, to give up $1.90 in consumer loss for every dollar it transferred to sugar producers in 1988. Of course, dividing $1.90 by the US population of sugar consumers and dividing $1.00 by the number of US sugar producers indicates that the marginal rate of substitution, on a per-capita basis, is small for the individual consumer relative to the producer gain. It is well known that distortions from government intervention tend to occur whenever the gain from intervention substantially benefits a few at a small individual cost to many.

## A One-Period Noncooperative Game

The procedure is to solve the world trade model for various trade liberalization scenarios for both the USA and EC. The net social gains obtained from the model for each of the above-mentioned special interest groups are then substituted into Equation (1) and a payoff matrix is formed. This procedure is consistent with game theory where the premise is that governments seek a treaty action space that makes at least one government no worse off than the *status quo* policies. These results are reported in the first panel of Table 2 for three alternative trade liberalizing scenarios. The scenarios are:

USA:
>   *sq*—The *status quo* of 1988.
>   *ber*—Ban on producer subsidies and export subsidies for all commodities except beef, sugar, and dairy products. Self-sufficiency in dairy products is allowed, while sugar and beef quotas remain at the *status quo*.
>   *pft*—A 30-percent decrease in the nominal protection coefficient for all commodities.
>   *ft*—Free trade in all commodities.

EC:
>   *sq*—The *status quo* of 1988.
>   *ber*—Ban on export restitution; *ad valorem* tariffs are used to attain self-sufficiency in grains, beef, pigmeat, poultry, dairy products, and sugar; price differentials, in percent, between producers and consumers remain at the *status quo*; the farm price of oilseed cakes is unchanged.
>   *pft*—A 30-percent decrease in the nominal protection coefficient for all commodities.
>   *ft*—Free trade in all commodities.

### Economic Results

In general, liberalization causes large increases in the world prices of grains, beef, sugar, and dairy products, decreases in the prices of oilseed cakes and feed-grain substitutes, and smaller changes in the prices of pigmeat and poultry. Three factors drive these results: crop production shifts in the USA from grains to oilseeds; feed input substitution occurs in the EC; and the EC substitutes grains for oilseed cakes and feed-grain substitutes. EC beef, dairy, pigmeat, and poultry producers also have lower feed input demand as the feed sector contracts.

As is well known, the EC variable levy system transfers income to producers from consumers and the budget. Hence, EC liberalization gives rise to large consumer gains that range from 6,400 million ECUs for *pft* to 24,400 million ECUs for free trade. Budget savings are also large but always smaller than the consumer surplus gains. Furthermore, most EC budget savings are realized under *ber* since most budget outlays are from export restitutions.

In the US case, most income transfers to producers occur through the budget, except for dairy products, beef, and sugar policies. Hence, consumer surplus gains in the USA range from only 2,200 million ECUs under *pft* to 7,510 million ECUs under free trade when sugar and dairy products are liberalized. In contrast, the budget savings range from 5,600 million

ECUs under *pft* to 16,540 million ECUs under free trade. Consequently, the greatest marginal budget saving occurs from *sq* to *pft* when deficiency payments on grains are removed.

The welfare effects of bilateral liberalization are dominated by own-country effects; i.e., liberalization in the USA does have welfare consequences in the EC and *vice-versa*, but they are always small compared to effects of any unilateral liberalization. For example, the budget savings in the USA from *ft* is at least 16,000 million ECUs, but the greatest change in budget savings to the USA from an EC liberalization is only 200 million ECUs from *sq* to *ber*. Those results motivate the political economy results reported next; i.e., that, with the exception of the results reported in Figure 1, neither the USA nor the EC is willing to choose policies that "pay off" the other country to induce it to liberalize.

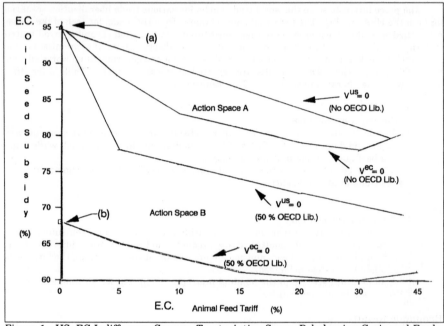

Figure 1—US–EC Indifference Curves: Treaty Action Space, Rebalancing Grain and Feed

## Political Economy Results: Game One

As mentioned, economic efficiency is not the only criterion motivating the negotiations. Hence, the economic results discussed above need to consider that influences of special interests vary as suggested in Table 1. Translating the welfare payoffs from the economic results into Equation (1) to obtain the political implications of trade liberalization gives rise to the political economy game depicted in Table 2.

By inspection, the *status quo* is the unique Nash equilibrium; it is a strongly dominant action for the USA and EC. Note also that when the USA plays the *status quo*, it always gains from EC liberalization, and *vice-versa*.[4] Hence, it appears in the self interest of each to encourage the other to liberalize while maintaining its own *status quo*. Moreover, there is no bilateral liberalization from which the EC gains; it always loses. But the USA gains if it pursues *pft* or *ber* and the EC pursues *pft* or *ft*. The EC would not be interested in these options, since it loses in each of these mutual liberalizations. Finally, it is irrational for the USA to propose *ft, ft* because it experiences a loss without decoupled payments.

Table 2—Policy-Goal Function Values for
Alternative US and EC Trade Liberalization Strategies and Decoupled Payments (1988)

| Panel one | Game One:  Using 1986 Action Space | | | | | | | |
|---|---|---|---|---|---|---|---|---|
| | sq | | pft | | ber | | ft | |
| sq | 0, | 0 | 182, | −292 | 412, | −1699 | 697, | −5407 |
| pft | −112, | 251 | 138, | −56 | −457, | −1722 | 1272, | −5551 |
| ber | −560, | 517 | −598, | 480 | −234, | −1554 | 233, | −4691 |
| ft | −2075, | 1020 | −2024, | 1154 | −1472, | −1433 | −877, | −4409 |

| Panel two | Game Two:  Using Decoupled Payment | | | | | | | |
|---|---|---|---|---|---|---|---|---|
| | sq | | pft′ | | ber′ | | ft′ | |
| sq | 0, | 0 | 182, | 2208 | 412, | 2057 | 697, | 16 |
| pft | 1102, | 251 | 1179, | 2298 | 1271, | 2061 | 3140, | −47 |
| ber | 2216, | 517 | 2212, | 2431 | 2484, | 2242 | 2968, | 640 |
| ft | 1559, | 1020 | 1600, | 2571 | 2099, | 2255 | 2600, | 868 |

| Panel three | Game Three:  Effects of OECD Liberalization | | | | | | | |
|---|---|---|---|---|---|---|---|---|
| | sq[a] | | pft | | ber | | ft | |
| sq | 294, | 551 | 476, | 285 | 681, | −1657 | 971, | −5085 |
| pft | 312, | 794 | 542, | 420 | 1018, | −1737 | 1796, | −5356 |
| ber | −376, | 1272 | −249, | 1071 | 15, | −1505 | 445, | −4360 |
| ft | −1704, | 1757 | −1599, | 1634 | −984, | −1419 | −444, | −4156 |

Note: Shaded areas denote solution to game. [a]sq, sq is positive since both the USA and EC benefit from OECD liberalization.

Notice that these results are remarkably consistent with the US and EC negotiating positions. The USA wishes to pursue trade liberalization provided the EC liberalizes. In this case, the USA is politically better off, and, at the same time, resources are saved (i.e., net social economic gains are positive). But it is not in the political interest of the USA to pursue trade liberalization on its own, nor to seek full liberalization. In this case, the economic results suggest that the USA would save resources by unilateral liberalization; but, in political terms, it loses (Col. 1, Table 2).

Suppose the budget savings from the various trade-liberalizing scenarios were reallocated to the losers from liberalization. Of course, in reality these decoupled payments would be in terms of maintaining the environment, helping farmers to adjust to lower farm prices, and so on. However, the payments are not allocated in the traditional compensatory way. Instead, payments are made to the loser with the highest political weight first, and then to the next most influential loser, and so on, until either all losers are fully compensated or all the budget savings from liberalization are exhausted. This distributional rule maximizes the PPF, given that the total transfer is no larger than the budget savings from trade liberalization and that no one is over-compensated.

**Political Economy Results:  Game Two**

The (′) appended in the payoff matrix of Game Two (Table 2) to an action reflects the addition of the transfer to each of Game One's liberalizations. Inspecting Game Two, only pft′, ft′ is not a treaty action. The payoffs of ft′, ft′ are consistent with the initial US proposal of free trade, and ft′, ft′ is the symmetric liberalization (liberalizations down the diagonal) that gives the USA the greatest payoff. This would suggest that the US proposal for free trade with decoupled payments is the treaty action that benefits it most. Subsequent US proposals have reduced the US payoff while offering others more, which is what one would expect in barter.

The Nash equilibrium of Game Two is *ber'*, *pft'*, because *pft'* is a dominant strategy for the EC and because *ber'* is a best response of the USA to all EC actions except *ft'*. The *status quo* is no longer the dominant strategy of the USA and EC, because the introduction of compensation allows the USA and EC to transfer income more efficiently between the budget and producers. It is in their own interests partly to liberalize regardless of the other's action.

Free trade with compensation is not the dominant strategy, because the budget savings are not sufficient fully to compensate the losers. This occurs because the compensatory scheme allows transfers only from the budget (and not from consumers) to producers. For the EC, the equilibrium occurs if it exports (if it continues to pay smaller restitutions) and if it subsidizes oilseed production. When the EC is autarchic, no more restitution savings can be obtained, but it remains politically expedient to transfer income from consumers to producers through higher domestic prices. For the USA, the savings result largely from reductions in grain support prices. Consumer support of dairy, beef, and sugar producers is still politically desirable. Thus, freer trade results. Free trade does not.

## OECD Effect on an EC–US Trade Compromise

The rest of the OECD is assumed to decrease its level of trade protection by 50 percent. The analysis of Game One is then repeated. These results appear in Panel three of Table 2. An important component of the EC's negotiating position is its interest in trading US and EC cuts in the support price of grain and oilseed cakes for EC tariffs on oilseed cakes and feed-grain substitutes. Whether these trade-offs can lead to a treaty that leaves both countries no worse off than the *status quo* is determined by the extent to which their respective policy indifference curves, based on Equation (1), "overlap." These results appear in Figure 1.

Let us consider Panel three, Table 2. When the rest of the OECD pursues partial liberalization (50 percent), world prices rise, particularly for grains, as import demand grows in Japan and the Nordic countries while exports from Canada and Australia decline somewhat. Because world prices rise, trade liberalization by the USA and EC is made easier since farm prices in the USA and EC do not fall as far as they did in Game One, and budget savings are greater. The solution to this game is also a dominant Nash strategy for both countries; the solution, in the absence of compensation, is *pft-sq*. Hence, it is in the interests of the USA and the EC to induce other OECD countries to pursue trade liberalization because, in so doing, the USA and EC can accomplish a treaty that would not, in the absence of our compensation scheme, be acceptable. In other words, the question of a treaty is really a multilateral problem, not just a bilateral US–EC issue. Interdependencies in world agricultural trade have reached the point where bilateral action is unlikely to lead to real trade liberalization.

Now, let us consider the second subcomponent, the trade-off between US and EC cuts in grains for EC tariffs on feed imports. In Figure 1, Action Space A denotes the treaty action space when the OECD and USA pursue the *status quo* and the EC harmonizes. The top line contains the smallest harmonizations the USA will accept, as measured by $V^{US} = 0$. The bottom line contains the largest harmonizations the EC will accept, as measured by $V^{EC} = 0$.

Next, the same analysis is repeated assuming that the rest of the OECD liberalizes its grain and oilseed protection by 50 percent. The top and bottom policy indifference curves for Action Space B have the same interpretations as in Action Space A. The US indifference curve for B is obtained by US liberalizations in grain and feed, holding the EC at the *status quo*, until the USA reaches its *status quo* value before OECD liberalization, point (a). At this point, $V^{US} = 0$. The remainder of the curve is then generated for the loci of points shown in Figure 1. At point (b), the EC liberalizes its grains and oilseed cakes subsidies until $V^{EC} = 0$. The remaining loci of points are obtained holding $V^{EC} = 0$.

Coordinate points in Space B yield non-negative values for $V^{US}$ and $V^{EC}$. Hence, OECD liberalization and the US–EC response to it increase the area of treaty action space and increase the overall reductions that could be obtained from the EC, relative to the smaller Action Space A; i.e., even larger possibilities exist for a US–EC compromise in these selected commodities. While the political economy and economic gains (net social welfare) are larger than in the case of Space B, they are still small relative to Panel two and three of Table 2.

## Notes

[1]Institut National de la Recherche Agronomique, France, and University of Minnesota, respectively.
[2]Since the maximum of $V^i$ is unchanged with a linear transformation, only relative rather than absolute weights are relevant.
[3]Monetary units in the empirical model are in ECUs.
[4]Gains (losses) refer to an increase (decrease) in the value of the PPF for the respective country unless otherwise indicated.

## References

Mahé, L.P., and Tavera, T., "Bilateral Harmonization of EC and US Agricultural Policies," *European Review of Agricultural Economics*, Vol. 15, 1988, pp. 327–348.

Mahé, L., Tavera, C., and Trochet, T., "An Analysis of Interaction between EC and US Policies with a Simplified World Trade Model: MISS," Background Paper for the Report to the Commission of the European Communities on Disharmonies in EC and US Agricultural Policies, Institut National de la Recherche Agronomique, Rennes, France, 1988.

Olson, M., *The Logic of Collective Action*, Harvard University Press, Cambridge, Mass., USA, 1965.

Rausser, G.C., and Freebairn, J., "Estimation of Policy Preference Functions: An Application to US Beef Import Quotas," *Review of Economics and Statistics*, Vol. 56, 1974, pp. 437–449.

Riethmuller, P., and Roe, T., "Government Intervention in Commodity Markets: The Case of Japanese Rice and Wheat Policy," *Journal of Policy Modeling*, Vol. 8, 1986, pp. 327–349.

Roningen, V.O., "A Static World Policy Simulation (SWOPSIM) Modeling Framework," Staff Report No. AGES–960625, Economic Research Service, US Department of Agriculture, Washington, D.C., USA, 1986.

Sarris, A.H., and Freebairn, J., "Endogenous Price Policies and International Wheat Prices," *American Journal of Agricultural Economics*, Vol. 65, 1983, pp. 214–224.

Tyers, R., *Implicit Policy Preferences and the Assessment of Negotiable Trade Policy Reforms*, Centre for International Economic Studies, University of Adelaide, Australia, 1989.

---

**Discussion Opening**—*Jacques Loyat* (Ministère de l'Agriculture et de la Forêt, France)

Mahé and Roe present an interesting contribution to the debate on EC–US conflicts over agricultural policies. The interest originates in the analysis of political power under GATT negotiations. I will discuss briefly the theoretical basis of the paper and its empirical approach.

There are two major streams of new thinking in trade theory. The first incorporates imperfect competition, by relaxing traditional assumptions of constant returns to scale, homogeneous products, and competitive markets. The second assimilates the theory of public choice. Some assume that governments possess a societal welfare function; others assume that governments respond to interest group pressures and set policies accordingly. In the present paper, the authors explicitly assume imperfect competition for international agricultural trade through interest groups. Here agents are price takers and economic rigidities are represented by the elasticities of the model; only governments are assumed to have strategic power.

The game is a one-period non-cooperative game between two governments. First, the model being a static comparative one, it is necessarily a one-period game. Second, even if preplay communication between players is possible, play is non-cooperative if commitments are not enforceable. This last assumption could be questioned for the EC and USA under GATT agreements.

But a more important aspect is surely the nature of the equilibrium that is realized; i.e., a Nash equilibrium. Even if we know how to define a Nash equilibrium, we do not really know how it is realized. A Nash equilibrium may or may not exist under a pure strategy game; all the more with a one-period game. The estimation of the political preference weights of the political goal functions is made through the unique 1988 equilibrium. It is therefore assumed that the 1988 equilibrium was a Nash equilibrium, from the moment when the governments were engaged in a severe negotiation for policy change. There are good reasons to think that the 1988 equilibrium was not a Nash equilibrium. The preference weights are fixed whatever the Nash equilibrium may be, but I would rather think that when the protection levels change, the weights themselves will change.

Four games are played by each government. The authors do not explain the choice of the different policies, and they do not discuss the feasibility of the scenarios generated by the MISS model. I observe that there is no symmetry between EC and US games. Let me take the case of *ber* policies, which give the Nash equilibrium in Game Two. For the USA, beef, sugar, and dairy products are unchanged. Second, there is a ban on producer subsidies, and export subsidies for all other products. For the EC, there is ban on export restitutions, and the objective is to reach self-sufficiency in grains, beef, pigmeat, poultry, dairy products, and sugar. The *ber* scenarios are evidently not politically equivalent for the EC and USA. Everything happens as if export strategies were playing no role for the EC government, while it is implicitly one of the main objectives of the US government.

By way of conclusion, this paper is interesting because it throws trade under imperfect competition into some relief. The empirical conclusions may not be sufficiently robust to give normative advice. A further step in this imperfect competition approach would be the discussion of other strategies by governments and traders and an analysis of international market structures. This could throw a new light on trade negotiations.

*[Other discussion of this paper appears on page 352.]*

# Domestic Policy Interdependence:
# Analysis of Dairy Policies in the USA and EC

*Mary A. Marchant, Steven A. Neff, and Alex F. McCalla*[1]

**Abstract:** This paper compares domestic dairy policies in the USA and EC and examines the impact of these policies on each dairy industry to explore domestic policy interdependence. The EC and USA have similar goals of improving farm income and use similar price support policies to achieve these goals. These policies have encouraged overproduction, generated surpluses and government stocks, and resulted in large government expenditures. Both have followed a mixed surplus disposal strategy with one key difference— the EC has used export subsidies to dispose of part of its surplus on the world market. In the mid-1980s, both the USA and EC took strong action to confront these problems. The EC's use of marketing quotas and commitment to reducing its stockpiles ultimately affected the world market as its exports fell in 1989. As a result, US stocks virtually disappeared, US prices increased substantially, and US surplus disposal programmes ended. By recognizing the policy linkage of EC domestic policy through the international market on US stockpiles, policy makers can choose policy instruments more precisely.

## Introduction

This paper compares US and EC domestic dairy policies and examines the impact of these policies on their respective dairy industries so as to explore domestic policy interdependence. By recognizing policy linkages, policy makers can choose policy instruments more precisely, resulting in reduced price and budget variability, increased stability of domestic and international markets, and increased stability of government stockpiles and of the programmes that use government stocks (e.g., domestic and international donations).

Since 1988, significant changes have occurred in both the US and EC dairy industries that indicate the existence of policy linkages. Both the USA and EC use price support programmes to support manufactured dairy products (butter, cheese, and milk powder) directly and support fluid milk prices indirectly. Both reinforce domestic support programmes with restrictive border policies. The USA restricts the amount of imports primarily by imposing a quota, supplemented with small tariffs. The EC imposes a variable import levy, which generates government revenues. In addition, the EC uses export subsidies, making EC dairy products competitive on the world market. The USA implicitly subsidizes exports through its international donation programmes.

US and EC domestic prices have been two to three times greater than the world price. These domestic prices have encouraged overproduction and generated surpluses that were purchased by each government, stored initially in government stockpiles, and disposed of using a variety of methods. The USA has used donations primarily to dispose of its surplus manufactured dairy products.[2]

In addition to donations, the EC also used export subsidies to sell surplus manufactured dairy products on the world market at world prices. The use of export subsidies as a method of surplus disposal for both EC butter and milk powder increased dramatically during 1985–88 due to the EC's extraordinary appropriation of funds to reduce its stocks.[3]

Supporting the dairy industry has been costly and variable, with record-high government expenditures by both the USA and EC in the 1980s. For example, US dairy programme costs during the 1980s ranged from $700 million in 1988/89 to $2,700 million in 1982/83, compared to an EC range of 3,300 million ECUs in 1982 to a maximum of 6,000 million ECUs in 1987.[4] The budget problem was magnified in the EC where dairy policy was draining the Common Agricultural Policy (CAP) budget.

Both the USA and EC took strong domestic action in the mid-1980s. In 1984, the EC instituted a marketing quota, reinforced with a super-levy penalty for its violation. In 1987, the quotas were further reduced and the EC made an extraordinary appropriation of funds to dispose of its mountain of butter and milk powder stocks. Milk powder stocks fell markedly to 14,000 t in 1988 from 847,000 t in 1986. Butter stocks were reduced to 223,000 t in 1988 from 1.3 Mt in 1986. This policy ultimately affected the world market since the EC had used the export market for surplus disposal.

Concurrent with EC policy setting, the USA sought to reduce its surplus through strong action in the 1985 Farm Bill by instituting (1) the Dairy Termination Program (also known as the Whole Herd Buy-Out Program), which paid dairy farmers to leave the industry for five years; and (2) the supply-demand adjuster (trigger mechanism), which for the first time related changes in the support price to government stock levels. Due to large government stockpiles upon passage of the 1985 Farm Bill, the US support price fell, and continued to fall from $12.60 per cwt (100 lbs) in 1985 to $10.10 per cwt in 1990 (annual average fat test). As a result of the Dairy Termination Program, dairy farmers left the industry. Both of these events caused US government stockpiles to decline.

Both US and EC policies of the mid-1980s affected the world market. Due to the EC's effective marketing quota and subsequent reduction in exports, world supply of manufactured dairy products, particularly milk powder, diminished. As a result of the decrease in world market supply, US stockpiles were drawn down, beginning in 1985, as stocks were placed on the world market. The export market then became a feasible outlet for disposal of US surplus stocks. The Minnesota-Wisconsin price (base price for manufactured dairy products) rose substantially, from $11.48 per cwt in 1985 to $14.93 per cwt in December 1989 (3.5 percent milkfat). Donations from government stockpiles came to an abrupt end, since surplus manufactured dairy products no longer existed. From the beginning of 1990, little, if any, US government stocks existed for cheese and milk powder, although there were still some butter stocks. The point here is that EC dairy policy affected the world market, which in turn affected the US market. Policy makers must recognize the international environment in which domestic policy is set (McCalla and Josling, 1985).

## Government Stock Accumulation

Examination of government stock levels provides insight into the overproduction problem since governments must first purchase surplus manufactured dairy products and then determine the appropriate surplus disposal method. Related to government stocks is the price support level. In general, high support prices encourage overproduction, generating surpluses that are purchased by the government and stockpiled. Milk powder has been the dominant US surplus manufactured dairy product since the mid 1970s. Commodity Credit Corporation (CCC) purchases peaked in 1983, corresponding to high support prices and large surpluses of manufactured dairy product. CCC stock levels for all manufactured dairy products peaked in 1984. As of 1989, CCC stock levels for butter, cheese, and milk powder decreased 63 percent, 95 percent, and over 99 percent from their 1984 peak levels. In the EC, both butter and milk powder have been dominant surplus manufactured dairy products. Stocks peaked in 1986 for both products. As of 1988, EC stock levels for butter and milk powder decreased 83 percent and 98 percent from their 1986 peak levels.

## Surplus Disposal Policies and Methods

Once stocks are accumulated, this surplus can be: donated as food aid via domestic and international donations; stored in domestic stockpiles; controlled via production controls, buy-out schemes, or a reduction in the price support level; and sold domestically or on the world market. No matter which method is chosen, taxpayers incur a cost for surplus disposal.

The USA and EC manage surpluses generated from domestic policies differently. The overproduction problem is more severe in the EC than the USA. When comparing government stock levels of manufactured dairy products to fluid milk production (on a milk equivalent basis), EC stocks peaked in 1986 and equalled 37 percent of production while US stocks peaked in 1984 and equalled 24 percent of production. Consequently, EC dairy policy has evolved into taking strong action to reduce surpluses, culminating with a marketing quota. The EC uses a variety of strategies to dispose of surplus manufactured dairy products, including domestic donations, subsidized consumption, export subsidies, and stock accumulation. The strongest policy measures have occurred in the supply control area. EC policies to curb production and reduce stocks have included: slaughter premiums and beef conversion programmes, which diverted cows from the dairy sector to the cattle sector (similar to the US Dairy Termination Program); and a producer co-responsibility levy used to finance sales promotion and school milk subsidy programmes. These EC policies failed to control the ever-growing surplus and ever-increasing stocks. Production dramatically outpaced consumption, even with expansion in exports. In the early 1980s, EC production increased 2.1 percent per year, while consumption increased only 0.2 percent per year. By 1983, total manufactured dairy product stocks reached a record peak, totalling 19 percent of EC milk production. The future looked rather bleak, with surpluses expected to continue growing. Storage costs, export subsidies, and price supports were draining the CAP budget. As a result, in 1984 the EC ministers agreed on new reforms using a five-year milk marketing quota system that imposed a "superlevy" penalty for over-base production and froze the target price.

The USA has also developed many different surplus disposal policies, including: domestic and international donation programmes (e.g., P.L. 480, food stamps, school lunch, and the special distribution programme of the 1981 Farm Bill, which directed the US Department of Agriculture to distribute Commodity Credit Corporation (CCC) commodities to the needy[5]); subsidized exports (e.g., the export enhancement programme and the dairy export incentive programme for the purpose of making US exports more competitive on the world market;[6] international marketing programmes (e.g., the targeted export assistance programme); and stockpiling.

With regard to specific surplus disposal methods for manufactured dairy products in the USA, for butter, the primary surplus disposal mechanism has been domestic donations. When large surpluses accumulated, surplus disposal methods also included exports and stock accumulation. Dominant EC butter surplus disposal methods include exports, export subsidies, and subsidized domestic consumption (e.g., Christmas butter sales).

In recent years, the EC has both increased its butter stocks and subsidized international sales, particularly to the former USSR. For cheese, which is supported in the USA, but to a much lesser extent in the EC, domestic donations and stockpiling have been the dominant US surplus disposal methods.[7]

Milk powder has been the leading surplus manufactured dairy product in both the USA and EC. In the USA, milk powder surplus disposal has consisted of stock accumulation, particularly during 1974–84, and export disposal, primarily international food aid. In the EC, the dominant surplus disposal method has been subsidized domestic consumption, with the majority used as animal feed. In summary, the USA has disposed of surplus processed dairy products using domestic and international donations along with stockpiling, whereas the EC has used domestic consumption and export subsidies. Both have thus used some form of domestic disposal.

Supply control is the most direct method to control surpluses. The USA has used voluntary supply control programmes, including the Dairy and Tobacco Act of 1983, whereby participating dairy farmers agreed to reduce herd size in order to receive a diversion payment financed by the dairy industry, and the Dairy Termination Program (Whole Herd Buy-Out Program), whereby participating dairy farmers left the dairy industry for five years and had to dispose of their herds. In addition to the Dairy Termination Program, the 1985 Farm Bill created a supply-demand adjuster (trigger mechanism) that linked the support price to projected CCC net removals. During 1988–90, if forecast CCC removals were greater than 5,000 million lbs (milk equivalent), then the support price decreased by 50 cents per cwt. If net removals were estimated to be less than or equal to 2,500 million lbs (milk equivalent), then the price support increased 50 cents per cwt. Since enactment, the support price has dropped 20 percent from the original 1985 level of $12.60 to the 1990 level of $10.10 per cwt of grade B milk, testing 3.67 percent butterfat. The USA has thus also attempted to control supply via support price reductions. The EC has taken even stronger action with its marketing quota.

## Dairy Programme Cost Comparisons

In the EC, the dairy support programme has been the largest expenditure in the CAP budget, 21.3 percent in 1988. Dairy expenditures have increased dramatically since the EC switched from being a net importer to a net exporter of manufactured dairy products in 1972. Government expenditures increased 18 percent per year during 1975–83, primarily due to the large dairy surplus storage costs and the cost of export subsidies corresponding to the expansion of export markets and the fall in the value of the dollar through exchange rates. The nominal cost of each of these two items increased 50 percent in the mid-1970s. The EC expenditures for agricultural support increased from $15,700 million in 1985 to $32,500 million in 1989.

In the USA, annual expenditures for the Dairy Price Support Program averaged $325 million during 1953–73. Costs fluctuated in the 1970s corresponding to the variability in milk production. In the 1980s, high support prices encouraged overproduction, which generated surpluses and resulted in large budget outlays associated with CCC purchases of this surplus. US government expenditures peaked in 1984, reaching nearly $2,800 million, which appears relatively small compared to EC expenditures.

## Trade Comparisons

Due to the perishability of fluid milk, only manufactured dairy products are traded. Once again, the magnitude of EC exports exceeds that of the USA. Milk powder is the dominant US export, while cheese is the dominant import—mainly specialty cheeses from the EC. For butter, US exports have fluctuated widely since 1950, while imports have been almost nonexistent during that period. In regards to US cheese trade, imports dominate. USDA exports are minimal. Virtually no imports exist for milk powder. In regards to US exports of milk powder, the CCC follows a mixed strategy, whereby approximately 75 percent of exports were international donations and the remaining 25 percent were international sales during 1975–85. In the early 1980s, exports of all manufactured dairy products increased, with milk powder being both the dominant surplus manufactured dairy product and also the dominant export manufactured dairy product from CCC stocks. Since 1985, both US exports and CCC stocks of all manufactured dairy products have declined dramatically and, from the beginning of 1990, have been virtually nonexistent.

The EC is the world's largest producer and exporter of dairy products with a 50 percent market share of world dairy trade, where dairy exports equal 13 percent of total EC agricultural exports. Prior to obtaining self-sufficiency in 1972, the EC was a net importer of dairy products. Export subsidies make EC products competitive on the international market but have been a costly budget item for the CAP.

Butter exports at the world market price using subsidies have dominated, peaking in 1980, declining until 1985, and increasing during 1985–88. Food aid has been a minor, yet fairly constant, outlet for exports. Prior to 1984, butter exports at reduced prices were nonexistent. Since then, the EC has sold reduced-priced butter to the former USSR, providing an even greater subsidy. Butter imports have been fairly constant.

For milk powder, exports at the world price using export subsidies follow a similar trend as that of butter—peaking (initially) in 1980 and dominating as the primary outlet for exports. Milk powder exports at the world price using subsidies increased dramatically during 1985–88, surpassing the previous 1980 peak. This dramatic increase in the use of export subsidies for both butter and milk powder is the result of the EC's commitment to reduce stocks. Special schemes for milk powder, consisting of sales to developing countries at reduced prices, only occurred prior to 1980. Since 1988, milk powder as food aid has maintained a relatively constant share of the export market. Unlike butter, milk powder imports are virtually nonexistent.

## Conclusions

It can be concluded that the USA and EC have a similar goal of increasing farm income and use similar price support policies to achieve that goal. These policies have encouraged overproduction, generated surpluses and government stocks, and resulted in large government expenditures. These impacts have been more severe in the EC. Both have followed a mixed surplus disposal strategy, with one key difference—the EC has used export subsidies to dispose of part of its surplus on the world market. In the mid-1980s, both the USA and EC took strong action to confront these problems. The EC's marketing quota and commitment to reducing its stockpiles affected the world market, resulting in decreasing EC exports. As a result, US stocks virtually disappeared, US prices for manufactured dairy products increased substantially, and US surplus disposal programmes (e.g., donations) ended. Thus, it appears that EC dairy policy affected the world market, which in turn affected the US market. Policy makers must recognize the international environment in which domestic policy is set. By recognizing the policy linkage of EC domestic policy, through the international market, to US stockpiles, policy makers can choose policy instruments more precisely.

## Notes

[1] University of Kentucky, US Department of Agriculture, and University of California (Davis), respectively.

[2] During 1955–89, domestic donations were the dominant disposal method for butter and cheese relative to international donations for milk powder.

[3] During 1974–83, the dominant disposal method for EC butter was the world market using export subsidies; during 1984–86, a combination of domestic and international strategies was used. For EC milk powder during 1974–86, the dominant disposal method was subsidized domestic consumption, with a primary outlet being animal feed for calves.

[4] An undetermined amount of this variability is due to fluctuations in international prices.

[5] The programme was extended under the temporary emergency food assistance programme and further extended under the 1985 Farm Bill.

[6] The dairy export incentive programme enabled US exports to meet the prevailing world price, using export subsidies in the form of dairy products from CCC stockpiles.

[7] See Marchant (1989) for detailed analysis of US and EC surplus disposal methods based on data from the US Department of Agriculture and the Commission of the European Communities.

## References

Marchant, M.A., "Political Economic Analysis of Dairy Policies in the United States," Ph.D. thesis, University of California, Davis, Calif., USA, 1989.

McCalla, A.F., and Josling, T.E., *Agricultural Policies and World Markets*, Macmillan Publishing Company, New York, N.Y., USA, 1985.

# Discussion Opening—*Lydia Zepeda* (University of Wisconsin)

Marchant, Neff, and McCalla trace the link between increases in US dairy prices and responses to EC policies (and the 1988 drought). These price incentives have led to increased US milk production, with the result that prices at the beginning of 1990 were near an all time high, and fell by record amounts by the end of 1990.

Proposals by US policy makers to the current crisis in the dairy industry exemplify the lack of linkage pointed out by the authors. Policy proposals focus on interregional differences within the USA and on increasing support prices, without addressing the interaction with world markets. In particular, changes in the Eastern Bloc, a major importer of dairy products, are ignored.

While Marchant, Neff, and McCalla focus on the links between the USA and EC, impacts on other economies are profound. Subsidies and donations of milk products reduce the cost of importing milk products by Third-World countries, but also create disincentives to their own producers. Instability in free market prices also creates incentives for self-sufficiency in milk production by Third-World economies. Food aid is ineffective if it destabilizes prices and supplies. In addition, smaller countries that export dairy products are hit hard by EC and US subsidies and fluctuating world prices.

In order to recognize the linkages between the dairy policies of the USA and EC, the policy goals of the two economies need to be examined. If the goal is to enhance farm incomes, a less destabilizing method than price supports or quotas might be found. Price supports will continue to be costly and result in overproduction, which affects world milk product prices. Price supports and quotas also create inequities within each system; large producers tend to benefit the most.

The need to recognize and address the links in policy and the objective of the policies has never been greater. The 1985 US policies, designed to enhance farm incomes, ignored the linkages and resulted in lower farm prices. In addition, the mounting costs of agricultural programmes needs to be dealt with. We can predict with high probability what will happen if support prices in the USA increase or if the EC continues export subsidies. We also know that buyout programmes and small support price cuts do little to counteract the continuing increase in the milk supply due to technological advances.

For policies to be effective, a clear goal must be defined and the context in which the policies exist needs to be recognized. The question is not whether we have the tools to do the analysis but whether policy makers will listen.

*[Other discussion of this paper appears on page 352.]*

# EC-1992: Implications for the Agrimonetary System and Commodity Markets

*Walter H. Gardiner and Timothy E. Josling*[1]

**Abstract:** The system of exchange rates, border taxes, and subsidies—the agrimonetary system—is incompatible with the goals of EC-1992. This system has led to a breakdown in the concept of common pricing and to distortions in trade patterns and has created large administrative costs to the EC. Elimination of the agrimonetary system would reduce EC prices, production, net exports, and budget expenditures and increase consumption. Major changes to the system will probably be reserved for a date beyond 1992.

## Introduction

Rapidly growing structural surpluses in a number of key commodity sectors, combined with rising budget outlays, sluggish world markets, and increased international tensions paved the way for some fundamental changes in agricultural policies in the 1990s. The importance of restructuring agricultural policies around the world was high on the agenda of the Uruguay Round of trade talks in the GATT. In addition, the 12 members of the EC have embarked on an ambitious and historic programme to eliminate national borders between their countries by the end of 1992. The goal of "Europe 1992," as the programme is known, is to achieve a true common market as envisaged by the EC's founders nearly 33 years ago. Until now, physical, technical, and fiscal barriers have prevented the EC from achieving greater economic efficiency.

The 1992 programme, which began in 1985, will continue until the end of 1992. All barriers that impede the free movement of goods, services, people, and capital among member countries are scheduled to be eliminated by 1992. The result will be a powerful trading bloc, whose sheer market size (in population) will exceed that of the USA. The new unified EC will have 320 million consumers with a purchasing power of $4,000,000,000,000 (Cecchini, 1988).

The EC's agrimonetary system refers to the mechanisms for fixing certain monetary sums—prices, subsidies, levies (taxes), and budget accounts—of the Common Agricultural Policy (CAP). Monetary matters have had a significant influence on the development of the CAP and on the EC's internal and external trade in agricultural products. Reluctance of EC member countries to allow exchange rate changes to be fully transmitted into their agricultural sectors led to the creation of a separate set of exchange rates for agriculture known as "green rates." These special exchange rates are used to convert policy prices denominated in European currency units (ECUs), the EC's monetary denominator, into each country's national currency. These special exchange rates have led to a breakdown in the concept of common pricing to a system of border taxes and subsidies (monetary compensatory amounts or MCAs) that distort trade patterns and create large administrative costs to the EC.

The EC's agrimonetary system is inconsistent with the goal of eliminating all internal barriers to trade by 1992. MCAs are currently collected at customs posts along the borders but are scheduled to disappear at the end of 1992. It would be unfeasible to maintain customs posts after 1992 for the sole purpose of collecting MCAs on agricultural products.

## Alternatives to the Agrimonetary System

The present system cannot be operated without border posts, and its possible replacement with a system of direct payments would be politically difficult. The only alternative would seem to be to abandon price differentiation. To abandon the policy of differential prices would imply increased uncertainty for domestic producers, whose support price would change with each change in the ECU rate for local currency.

Whether or not the EC moves towards monetary union and succeeds in removing border posts, the issue remains of how to handle the "switchover coefficient." The switchover coefficient is the premium placed on the ECU used for agricultural purposes in 1984 in order to avoid creating more positive MCAs. At the end of 1990, the switchover coefficient stood at 13.7 percent, an indication of the inflationary tendency of the "green ECU" system. Removal of the switchover coefficient would immediately drop the price of most agricultural products by about 14 percent. This could be compensated by an increase in ECU prices by the same amount. But such an ECU price rise might be taken by the trading partners of the EC as locking in the hidden price increases due to green-ECU appreciation. As the EC has made considerable play of its policy to hold down ECU prices and even agreed in April 1989 not to increase them during the GATT negotiations, such an action would be unpopular. Removal of the switchover coefficient could also be accomplished by changing green rates, as the same national prices could be ensured by offsetting the lower value of the agricultural ECU. This would have the disadvantage of re-creating positive MCAs for strong-currency countries (i.e., exposing the hidden positive MCAs in the present system). Once again, those positive MCAs would have to be reduced over time to lead back to common prices. In spite of the difficulties, it is probable that the EC would like to remove the switchover coefficient and return to a "regular" ECU for agricultural price purposes.

## Monetary Union and the Elimination of MCAs

In view of the many ways in which the present MCA system could be modified, it may be premature to speculate on the chosen method. Instead, one can put bounds on the outcome and discuss the likely implications of particular choices. In the next section, two such boundaries are explored. A "base"-case scenario presumes that the existing MCA system is retained, preserving the price relationships that exist at the moment. Rather than moving towards monetary union, national inflation rates keep their historical spread and exchange rates adjust accordingly. Though this scenario represents a *status quo* situation, it also sheds some light on the outcome if alternative policies substitute completely for the removal of MCAs, at least as far as the farm sector is concerned.

The second case considered is that of the removal, after 1992, of the MCA system and the switchover coefficient. This is accomplished in stages and completed by 1996. The assumptions are made that monetary union is also reached over the same time period and that inflation rates converge. The impact on agriculture from the harmonization of prices is such that this eventuality is unlikely; some form of transition or compensation seems more likely. But it is useful to establish the bounds of the problem. And a system of uniform prices, with ECU prices translated for all commodities at market exchange rates, would seem to be the only fully satisfactory solution to the problems of the green-money system.

## Adjustment to the EC's Agrimonetary System

The impact of changes in the agrimonetary system will vary from country to country in the EC. To analyse these impacts, one needs a way of quantifying the implications of policy change by country and commodity. The estimates given below were developed using CAPFRAME, a series of national models for the agricultural sector of EC countries in use in the US Department of Agriculture (Josling, 1990). CAPFRAME allows for a consistent series of projections of prices, market balance, financial flows, and policy impacts for each of 11 member states (with Belgium and Luxembourg treated as one economy). Commodities include wheat, barley, maize, beef, and dairy products, and the projections extend annually to the year 2000.

Two simulations were run using CAPFRAME: a base run that preserved the present MCA system and assumed no further shift towards monetary union, and a European Monetary

Union (EMU) run that assumed a movement from 1992–96 towards locked currency values, common inflation, and the dismantling of the MCA system. It is reasonable to assume that the actual outcome will fall somewhere in between these two extremes.

The results indicate that removal of the MCA system will have a negative impact on farmers in strong-currency countries. Gainers include consumers in those countries, farmers in countries with very weak currencies, and taxpayers in the EC as a whole. The extent of these changes is detailed below.

## Price Effects

The agrimonetary system governs the level and spread of support prices in various countries and for various commodities. These prices will be affected by the removal of this system in three ways: the abandonment of the switchover coefficient (i.e., the use of an unadjusted ECU), the removal of MCAs relative to that monetary ECU, and the removal of green rate divergences among products. Table 1 shows the percentage change in producer price in local currency for five commodities: wheat, barley, maize, beef, and milk. For the EC as a whole, cereal prices could be lower by between 9.6 percent (maize) and 7.9 percent (wheat), beef prices could be 10 percent lower, and milk prices could be reduced by 8.6 percent.

The price impact is felt most by farmers in strong-currency countries: Netherlands, Germany, Belgium-Luxembourg, and Denmark. Price falls for cereals, beef, and dairy products of the order of 10–15 percent in these countries can be attributed to the removal of the MCA system, including the switchover coefficient. A second group of countries is affected less, experiencing prices lower by 5–10 percent as a result of the removal of the system. This group includes Spain, France, Italy, and the Irish Republic. At the other extreme, UK farmers enjoy higher prices with the removal of the MCA system. The negative MCAs predicted in the base-case scenario are larger than the benefits from the artificially high ECU value implied by the switchover coefficient. UK prices are higher by between 4 and 7 percent in the EMU case scenario. Greek farmers also gain, but the benefit varies by commodity with little change implied by changes in the cereals régime.

## Market Balance Effects

The impact on production levels (Table 1) follows from the anticipated price changes. Cereal production goes down in all countries except the UK and Greece, with the impact on individual cereals being governed by the relevant cross elasticities. Wheat production in the Netherlands appears to be most vulnerable to the price changes, down 11 percent relative to the base-case scenario. Wheat production could increase by 5 percent in the UK, offsetting the decline in other countries. For the EC as a whole, production is estimated to be lower by less than 1 percent for cereals. More significant decreases are anticipated in beef, with an overall decrease in production of about 5 percent in the EMU/no-MCA case. In the dairy sector, removal of MCAs under these circumstances could reduce milk output by about 5 percent in most countries, with a 3-percent increase in the UK, leading to a decrease of 2.4 percent for the EC as a whole.

Consumption generally increases with the lower real prices expected from the removal of the MCA system (Table 1). This is particularly noticeable for beef, where increases of 6–11 percent are indicated for the strong-currency countries, and of 5–6 percent for the EC as a whole. Consumption of dairy products also increases, but by a lesser extent, from 3–4 percent relative to the base case in the strong-currency countries and just over 2 percent for the EC. Cereal consumption changes are less clear cut. Food use increases marginally (as price elasticities of demand are low in this sector), but use for feed is drawn in two different directions. Lower livestock prices tend to cut feed use, but lower cereal prices encourage the substitution of cereal for non-cereal feed. Thus, maize consumption increases by up to 7 percent in the strong-currency group, while barley use for feed declines in the same countries. Overall use of wheat is stable; that of maize increases by 1.2 percent, and overall barley use could decline by 1.8 percent.

Table 1—Impact of European Monetary Union, 1996

| | France | UK | Neth. | Bel./Lux. | Germany | Italy | Ireland | Spain | Portugal | Greece | Denmark | EC-12 |
|---|---|---|---|---|---|---|---|---|---|---|---|---|
| On Net Producer Prices (percent of base case) | | | | | | | | | | | | |
| Wheat | -9.7 | 7.8 | -15.3 | -12.2 | -13.3 | -8.1 | -8.1 | -8.7 | -9.7 | -0.4 | -10.9 | -7.9 |
| Barley | -9.7 | 7.8 | -15.3 | -12.2 | -13.3 | -8.1 | -8.1 | -8.7 | -9.7 | -0.4 | -10.9 | -8.6 |
| Maize | -9.7 | 7.8 | -15.3 | -12.2 | -13.1 | -8.1 | -8.1 | -8.7 | -9.7 | -0.4 | -10.9 | -9.6 |
| Beef | -11.4 | 3.6 | -14.6 | -12.2 | -13.0 | -8.6 | -9.9 | -9.7 | -9.7 | 15.3 | -10.9 | -9.9 |
| Milk | -9.7 | 7.0 | -14.6 | -12.2 | -13.0 | -8.6 | -8.1 | -9.7 | -9.7 | 15.3 | -10.9 | -2.1 |
| On Production (percent of base case) | | | | | | | | | | | | |
| Wheat | -0.5 | 5.0 | -11.0 | -0.9 | -4.1 | -1.4 | -3.9 | 0.3 | -0.7 | 0.0 | -0.6 | -0.2 |
| Barley | -1.2 | 1.7 | 0.0 | -2.2 | -1.1 | -0.3 | -2.5 | -0.7 | -0.5 | 0.0 | -1.1 | -0.6 |
| Maize | -0.5 | 0.0 | -0.8 | -0.6 | -0.7 | -0.5 | -0.8 | -4.5 | -0.5 | 0.0 | 0.0 | -0.8 |
| Beef | -4.2 | 0.9 | -5.5 | -7.8 | -6.0 | -6.8 | -9.6 | -2.7 | -2.1 | 4.3 | -3.2 | -4.7 |
| Milk | -2.3 | 3.0 | -4.7 | -5.1 | -4.7 | -1.5 | -1.9 | -2.9 | -3.0 | 17.1 | -4.8 | -2.4 |
| On Consumption (percent of base case) | | | | | | | | | | | | |
| Wheat | -0.1 | 0.2 | 3.4 | 0.7 | -0.7 | 0.7 | -1.1 | 0.6 | 0.1 | 0.2 | 0.9 | 0.2 |
| Barley | -1.9 | -0.2 | -1.0 | -1.5 | -3.6 | -2.7 | -3.4 | -0.8 | -0.2 | 6.4 | -2.9 | -1.8 |
| Maize | 0.0 | 0.1 | 3.2 | 1.2 | 0.2 | 0.2 | -1.3 | 4.3 | 0.0 | 1.6 | 1.5 | 1.2 |
| Beef | 8.3 | -1.8 | 11.3 | 9.2 | 9.5 | 6.3 | 6.0 | 5.5 | 5.7 | -7.7 | 6.6 | 5.6 |
| Butter | 2.4 | -0.8 | 4.0 | 2.6 | 3.4 | 1.9 | 1.3 | 1.8 | 1.5 | -2.9 | 1.2 | 2.3 |
| On Net Trade (1,000 t from base case) | | | | | | | | | | | | |
| Wheat | -171 | 800 | -189 | -27 | -418 | -176 | -11 | -14 | -5 | -3 | -44 | -259 |
| Barley | -16 | 173 | 10 | 4 | 264 | 64 | 3 | 20 | 0 | -48 | 47 | 521 |
| Maize | -72 | -1 | -61 | -13 | -16 | -50 | 1 | -380 | -4 | -25 | -1 | -620 |
| Beef | -216 | 34 | -59 | -48 | -257 | -179 | -36 | -49 | -12 | 26 | -13 | -810 |
| Butter | -23 | 5 | -16 | -5 | -35 | -4 | -3 | -1 | -1 | 1 | -4 | -84 |
| Net Economic Benefits of Policies due to EMU (million 1982 ECUs) | | | | | | | | | | | | |
| Wheat | -307 | 31 | 12 | 4 | -32 | 41 | 6 | 15 | 6 | -3 | -29 | -144 |
| Barley | -53 | 12 | 15 | 16 | 67 | 13 | 0 | 68 | 1 | -3 | -18 | 199 |
| Maize | -82 | -7 | 30 | 15 | 18 | 13 | 1 | 51 | 6 | -8 | 1 | 164 |
| Beef | 69 | 33 | -57 | 5 | -82 | 394 | -59 | 118 | 16 | -33 | -27 | 402 |
| Milk | 66 | -75 | -85 | 45 | 194 | 240 | -54 | 109 | 11 | 23 | -53 | 372 |

Lower production and stronger demand affects net trade (Table 1). The EC wheat export volume is estimated to decline marginally as a result of increased imports into the Netherlands, Belgium-Luxembourg, and Italy, counteracted by greater surpluses in the UK. France's net exports of wheat decline by 171,000 t and total EC net exports drop 259,000 t, less than 1 percent. In the case of barley, the level of net exports from the EC could rise, led by increases in exports by Germany and the UK, as feed consumption falls. This is more than offset by the increased maize imports into the EC–12 (620,000 t), up 54 percent as a result of the dismantling of the MCA system. The increase in imports is noticeable in the Netherlands and Spain, while French exports decline. Beef imports also increase (810,000 t) for the EC as a whole as a response to EMU and lower prices implied by the removal of MCAs. Such imports could climb by over 80 percent, with increased sales into France, Italy, Spain, and Portugal. Lower exports from Denmark, the Netherlands, Germany, and the Irish Republic contribute to this outcome. In the case of dairy products, higher imports of butter into Germany, Belgium-Luxembourg, Spain, and Portugal, coupled with fewer exports from the Netherlands, the Irish Republic, and Denmark, reinforce the trend for the EC–12 to reduce net exports to the world market.

### Financial Implications

The economic implications of the EMU/no-MCA scenario are summarized here. The combination of lower prices and lower production gives a reduction in farm receipts for all commodities. At the EC level, this drop is greatest for beef, milk, and maize. But this masks the differences among countries. The other side of the coin is that consumers could find expenditures on farm commodities reduced by EMU and the removal of MCAs. Expenditures on wheat for both food and feed decline moderately, with somewhat larger decreases in spending on barley and maize for feed. Consumer expenditures on beef decline by 4 percent and on dairy products by about 8 percent.

The financial aspects of the change in trade volume are reflected in the reduction in export earnings and increase in import expenditure. Export earnings for wheat decline by 9 percent and spending on imports of maize rises by 46 percent. The EC could spend over 70 percent more on beef imports: extra spending on butter imports and less earnings from skim milk powder exports could also be expected. This will also lead to a decrease in the expenditure on export subsidies. Budget spending could be cut considerably for wheat (161 million ECUs, at 1982 prices) and barley (63 million ECUs). Together with smaller changes in levy revenue and export subsidies on other commodities, the saving to the financial cost of farm programmes is estimated at 276 million ECUs.

In terms of net economic benefits and costs of the elimination of the EC's agrimonetary system, EC farmers lose in real income to the advantage of cereal users and taxpayers. Net changes in farmer benefits, consumer cost (including costs to feed compounders through higher ingredient prices), and the net cost to EC taxpayers are presented in Table 1. The beef sector reaps the largest net benefits (402 million ECUs in 1996), with Italy and Spain enjoying the largest increases, while Germany and the Irish Republic realize the largest net losses. At the other extreme, the EC's wheat sector actually show a net loss of 144 million ECUs, as producer losses offset gains to consumers and budget savings. Most of this is due to adjustments in France, the EC's largest wheat producer and exporter.

# Conclusions

The EC agrimonetary system, which began as a simple mechanism for converting agricultural prices, subsidies, and levies from a common accounting unit to the national currencies of the member countries, has evolved into complex web of rules and regulations that has created market distortions, not only among countries but also among commodity sectors within countries. The use of agricultural (green) exchange rates that no longer reflect market exchange rate changes has led to a breakdown in one of the EC's fundamental goals, common pricing. It has also led to a system of border taxes and subsidies (MCAs) that distort trade flows and to large administrative costs to both businesses and government to implement the system.

Elimination of the agrimonetary system with its green rates and border taxes/subsidies by 1992 will prove to be an extremely difficult task. The price differences that currently exist among countries as a result of the green rate system imply significant adjustments. A decision to harmonize at the highest price level would imply price increases for all countries. Harmonization at less than the highest level would imply price reductions for strong-currency countries such as Germany and the Netherlands and price increases for others. Price reductions will be strongly resisted and will likely require some form of compensation.

While the move to a common currency called for in the plan for full economic and monetary union would eliminate the agrimonetary system and its distortions, the political barriers to achieving full integration are formidable and will prevent it from occurring until well after 1992. The higher-than-expected cost of German economic and monetary union, which began on 2 July 1990, has caused some EC member countries that were pushing for a fast-track approach to EC economic and monetary union, particularly Germany, to shift their position more in line with the British go-slow approach. A more reasonable and likely solution to the agrimonetary dilemma in the near term is a tightening of the current arrangements, including a faster alignment of green rates with market rates, a gradual elimination of green rate differentials between commodities, and direct payments or tax credits made by national governments in place of MCAs.

## Note

[1] US Department of Agriculture and Stanford University, respectively.

## References

Cecchini, P., *The Benefits of a Single Market*, Commission of the European Communities, Luxembourg, 1988.

Josling, T.E., "CAPFRAME: Framework for Evaluation of the European Community Common Agricultural Policy (Software Documentation)," unpublished, 1990.

---

**Discussion Opening**—*Colin Brown* (University of Queensland)

The EC agrimonetary system has long been a source of concern. In discussing the paper by Gardiner and Josling, which examines the impacts of a possible demise in the agrimonetary system as the EC moves towards greater monetary and political union, several issues are worth raising.

Does the removal of border posts beginning in 1993 necessarily imply that the MCAs can no longer operate? Different VAT rates will also exist after 1992, with VAT settlement simply transferred from the border to the individual firm level. A system of MCAs paralleling or indeed tied to these VAT settlements could well operate.

Do the results in the paper adequately describe the main impacts on the various interest groups affected by changes to the agrimonetary arrangements? For instance, would inclusion of oilseeds and Mediterranean products in the analysis alter the impression from Table 1 of relatively large benefits to the southern member countries from a removal of the agrimonetary arrangements? Other inclusions such as sugar and intensive livestock, however, may exert the opposite effect. Furthermore, given the heterogeneity of EC agriculture, the farm-level effects and thus the reactions to the agrimonetary changes are not necessarily reflected in the commodity price changes of Table 1. Large cereal farmers in the Paris basin can be expected to react adversely to the lower effective cereal prices. However, the responses of their Danish counterparts will be tempered by the lower feed costs for the substantial intensive livestock production that characterizes the large cereal farms in Denmark.

What can be inferred from an analysis that compares the complete removal of the agrimonetary arrangements against the *status quo*? The results presented in the paper are based on the removal of MCAs relative to the monetary ECU, the removal of green rate divergences among commodities, and the abandonment of the switchover coefficient. But the first of these agrimonetary changes needs to be distinguished from the latter two, as in the context of this paper it reflects more the effects of a convergence towards monetary union rather than any fundamental changes to agrimonetary arrangements. Conversely, the remaining part of the results attributable to the commodity differentials and the switchover coefficient assumes no policy adjustment or compensatory measures. The results are qualified in the paper as being an upper bound of likely outcomes from a removal of the agrimonetary arrangements. For any noticeable effect to arise, however, implies that the agrimonetary system provides an institutionalized and relatively obscure method of support. Clearly there may have been elements of this in the past, given the complexity of the system. But with the much greater scrutiny of the agrimonetary arrangements in recent times, it is difficult to envisage situations in which offsetting compensation would not be sought by those groups adversely affected. Thus, whether the results reported in the paper are realized will depend on the more general issue of the resolve of the EC to lower agricultural commodity support.

Where MCAs may differ from other support measures is the tendency by EC agricultural ministers to use them as bargaining tools in the annual price rounds of the EC Council. Removal of the extent of this political manoeuvring may well alter the negotiated level of support. Another tangible effect of the agrimonetary system is the substantial administration costs alluded to in the paper. The extent to which the costs of administering these complex arrangements has eroded the relative benefits the measures are intended to provide is a topic worthy of further discussion.

*[Other discussion of this paper appears on the following page.]*

**General Discussion**—*Geraldo S.A.C. Barros, Rapporteur* (Universidade de São Paulo)

Discussion on the first paper by Mahé and Roe centred on the model used for the analysis. The possibility that equations in the model were overidentified given that the number of policy instruments exceeded the number of weights for particular groups was raised and explained in terms of the nature of the instruments considered. The relationship among weights for various groups across countries was also raised. It was also suggested that converting the model to a dynamic model may greatly enhance insights from the model.

The paper by Marchant, Neff, and McCalla raised some lively discussion on the direction of the link between dairy policies in the USA and EC and whether other participants in world dairy trade need to be considered. A further important omission from the analysis identified in the discussions was the need to consider cross-commodity linkages in both the USA and the EC. The strategic position of dairy policies in US agricultural policies *vis-à-vis* the dairy support arrangements in the EC, which are just one element of the CAP, was also raised.

Discussion on the paper by Gardiner and Josling was constrained by the absence of the authors. However, a comment was made that the removal of the switchover coefficient may be a non-issue given that it is only an administrative or accounting device.

Participants in the discussion included J. Beghin (North Carolina State University), N. Devisch (Belgian Farm Organization), T. Haniotis (Commission of the EC), J. Kola (Agricultural Economics Research Institute, Finland), and K. Thomson (University of Aberdeen).

# Environment and Sustainable Agriculture

**Sustainable Agriculture and Innovative Technologies**, Ariel Dinar *(University of California and US Department of Agriculture)*

Agriculture is a significant contributor to contamination of soil and water, as well as to other environmental contamination. Some suggest that agricultural pollution can be reduced by lowering the level of inputs used when choosing agricultural practices. However, this may result in food shortages. This paper suggests a different approach by demonstrating the use of two technologies (mobilized orchards and aeroponics). These technologies, which achieve sustainability through a controlled growing environment, use super-intensive inputs, minimize agricultural pollution, dramatically increase yield and quality, and provide more flexibility in supply. However, these technologies are sophisticated and require massive capital investment. Included in this paper is a simple economic analysis that demonstrates the economic superiority of these technologies over conventional agricultural practices, despite the heavy capital investment.

**Dynamic Systems and Limit Cycles for Modelling Sustainable Agriculture and Cooperation**, L.P. Apedaile *(University of Alberta)*

Concern about sustainable agriculture stems from the growing realization that deficiencies in meeting the social, economic, and ecospheric purposes of agriculture may jeopardize its role in feeding future generations. The problem arises because of the complexity of the agricultural system, which is difficult to model using the strong causality principle so successfully applied to other disciplines. Almost 20 years ago, Samuelson addressed this issue with modifications to the Lotka Voltera "predator-pray" model. More recently, Mandelbrot's discovery of fractal geometry and independent work on the persistence and stability behaviour of nonlinear dynamic systems have generated new hope for modelling the holism of complex systems. This paper examines these developments in the context of sustainable agriculture and the role of cooperative processes. Sustainability emerges as a matter of seeking flexibility and solving problems at the boundaries of systems rather than seeking the correct trajectory or arriving at an equilibrium. The conclusions are that sustainable agriculture is a purpose-related concept, that the domain of attraction about an equilibrium is more important than the equilibrium itself, and that the bifurcation and adjoining of sets of trajectories of system variables at system boundaries is at the centre of development processes for sustainable agriculture and cooperation.

**Analysis of Carrying Capacity for Sustainable Development**, Amar S. Guleria *(Himachal Pradesh University)*

"Carrying capacity" indicates the number of persons and animals to be supported per unit of land under constant and dynamic technologies without causing degradation of resources/environment. An attempt is made in this paper to estimate the carrying capacity of selected watersheds in Himachal Pradesh (India). The limits of carrying capacity in the study area varied from 190 to 271 persons per sq km. At the same time, around 20 percent of the cattle population was found to be surplus in the study area. It is emphasized that the concept of carrying capacity is more relevant in ecological and environmental studies to ensure sustained economic development. Carrying capacity can be used to determine the boundaries of feasible sustainable development/resource use. Some of the other features of a resource base that are useful in this context are exhaustibility, potential depletability, multifunctionality, stability, private versus public ownership, and whether the resource base is a closed or open system.

## Range Management and Rehabilitation through Stochastic Control and Markov Chains, K.M. Wang *(Western Australian Department of Agriculture)*

Rangeland ecology is in essence a dynamic system operating in a stochastic environment and involving very significant intertemporal effects. Combining stochastic control theory with Markov chains, this paper presents an integrated approach to range management and rehabilitation, which allows the optimal policy to be determined explicitly. Also, the concept of ecological sustainability and economic viability can be analysed quantitatively. The optimal grazing policy associated with water ponding treatment was found unable to rehabilitate all degraded range states. In those situations where there is still reasonable vegetation despite moderate degradation, rehabilitation through the optimal grazing policy is possible. In extreme situations where there has been severe depletion of the vegetation coupled with advanced soil erosion, reclamation is unlikely to succeed. Thus, a sustainable range management system should emphasize prevention of degradation through grazing management in the early stages. Optimal grazing strategies conserve young and old desirable perennial seedlings and exert increasing grazing pressure on the higher level of forage biomass and adult plant populations. Rotational grazing systems apply when range is not resilient to grazing. In the case of a resilient range, set stocking is the preferred strategy. Total destocking is optimal when range is severely defoliated or degraded.

## World Pig Production: Tradeoffs for Sustainable Agriculture, Z. Chen, R. Lent, and R. Saint-Louis *(Université Laval)*

Pig production presents a special case in the analysis of sustainable agriculture. An attempt is made to interrelate pig production systems, pigmeat consumption, trade, and externalities. Various production techniques among major pigmeat-producing countries of the world can be classified into a finite number of categories according to certain aspects of production: the low versus high feed cost method, free pasture versus total confinement with various models of low-to-high-investment housing in between, specialized versus integrated systems, etc. Production can generate both positive and negative externalities. In general, however, large-scale production is associated with negative externalities, although not exclusively with industrialized countries. Trade patterns reveal a potential way of "exporting" negative externalities from one country to another. As traditional barriers to trade decrease, more obscure barriers, such as asymmetric environmental regulation across countries, will become crucial variables. The issues involved in pig production externalities and regulation may be conveniently summarized in a Coasian framework.

## Implications of Constrained Factor Mobility for Sustainable Economic Development in China's Pastoral Areas, Greg Williamson *(University of Queensland)* and Ron McIver *(South Australia College of Advanced Education)*

A basic tenet of economics is that factors of production should be transferred to those areas where their marginal value product is highest. Any restrictions on such transfers limit the efficiency with which factors can be used and simultaneously increase the pressure on fixed factors such as land. For China, one of the consequences of this pressure has been the extensive degradation and desertification of large areas of its pastoral regions. This paper considers the impact of constrained factor mobility on the sustainable economic development of the pastoral regions of China and their prospects for future development given the above environmental complications.

## Biotechnology Development in Sri Lanka: Some Socioeconomic Implications, H.M.G. Herath *(University of Peradeniya)*

Although the use of biotechnology is a recent phenomenon in Sri Lanka, the trends observed indicate that undesirable effects can be generated. It is seen that most biotechnology work is concentrated in propagation techniques such as tissue culture. There are also ominous signs that the important food crops such as rice are being ignored. Present research appears to negate the benefits of donor-funded projects by simply facilitating the production of plant materials by wealthy entrepreneurs. The adoption of biotechnology products in Sri Lanka also appears contentious.

**Use of Chemical Fertilizer for Taro at the Subsistence Farm Level in the Lowlands of Papua New Guinea**, R.D. Ghodake and W.L. Akus *(Bubia Agricultural Research Centre)*

Taro *(Colocasia esculenta)* is grown as a semi-subsistence crop in Papua New Guinea, generally on newly cleared forest lands under the shifting cultivation method. The crop is very sensitive to soil fertility conditions. Recently, taro yield has started to decline in some localized areas due to increased intensity of cultivation and shorter fallows as a result of rising population pressure on land. Thus, there appears to be scope for the use of chemical fertilizers. However, inorganic fertilizers are rarely used on semi-subsistence farms for a bulky and staple crop like taro. This paper presents the results of statistical and economic analysis of the data obtained from two fertilizer experiments on taro conducted in a farmer's field in the lowlands of Papua New Guinea. The results, which consider technical response, marketing, farmer perceptions, field level losses, and risk factors, suggest high potential for the use of nitrogenous fertilizer for taro even under conditions of semi-subsistence. Phosphorus does not appear to be a limiting factor for taro production, while the effect of potassium on taro yield is not conclusive. These results further point to the need for more location-specific on-farm trials that involve wider levels and range of nutrients. Emphasis is implicitly placed on the need for the government to design and implement policies that will minimize risk and uncertainty related to the marketing of taro.

**Sustainable Land Use and Development in a Changing South Africa**, J.M. Erskine *(University of Natal)*

A focus on the consequences of future land reform in the less-developed areas of South Africa shows how remodelling land ownership and use may lead to both land degradation and a redistribution of resources to the disadvantage of the poor. The results of well-intentioned land reform and of equally well-meaning efforts to provide the infrastructure for greater economic development may thus be deleterious both for the land and for many of the people it supports. This conclusion concerning how the link between physical deterioration and socioeconomic differentiation operates in practice emphasizes the need for the evolution of new policies for managing the land simultaneously with the process of land reform; this will inevitably involve fundamental social changes, which in turn will require improved education and information distribution systems.

**Economic and Environmental Trade-Offs in Tropical Agriculture: The Case of Highland Milk Production in Costa Rica**, Katherine Griffith and Lydia Zepeda *(University of Wisconsin)*

Milk production is important economically and in terms of achieving nutritional goals in Costa Rica. However, cattle are responsible for much of Costa Rica's deforestation and erosion problems. Thus, a linear programming model is constructed to examine the environmental and economic trade-offs of milk production decisions in highland Costa Rica. Primary and secondary data are used to simulate three farm sizes. Seasonal and year-round production are examined, as well as land use regulation and introduction of lucerne as a forage crop to the region. The primary results indicate that credit and subsidies are the most useful tools to improve environmental sustainability and that research on high protein forages and labour-saving devices have the greatest potential for intensifying milk production.

**Trade-Offs between Deforestation and Productivity Gains: Implications for Sustainability**, Carlos Arnade *(US Department of Agriculture)* and Jorge Torres-Zorrilla *(Instituto Interamericano de Cooperación para la Agricultura)*

Previous research on agricultural sustainability focuses on externalities and social prices, leaving aside estimates of how long an agricultural economy can sustain current growth with existing technology. This study looks at trends in land use, productivity, population and economic growth and relates them to sustainability. The model defines a "food target equation" expressing that per-capita food demand equals per-capita supply, and an "environmental target equation" expressing that per-capita forest land cannot fall beyond a minimum. From these, agricultural land growth rate, deforestation rate, and period of sustainable growth are derived. Application to Brazil illustrates trade-offs between agricultural land expansion and productivity growth. Brazilian agricultural growth implies deforestation: per-capita forest land declined from 7 to 4 ha/person. Alternative simulations are summarized by per-capita forest land projections. At "best," low rates of population growth and high rates of economic growth and agricultural land increase violate the environmental target in 29 years. At "worst," high population growth rates and no economic growth or agricultural land increase violate the environmental target in 14 years. No scenario is sustainable, and, at best, Brazil has 29 years to reach 2 ha forest/person. Further research should consider relative prices, income growth, and non-neutral trade.

**Sustainable Agricultural Planning in the Texas Southern High Plains**, Eduardo Segarra *(Texas Tech University)* and John R. Abernathy and J. Wayne Keeling *(Texas Agricultural Experiment Station)*

Current sustainable agricultural planning efforts in the Texas Southern High Plains are reviewed and their economic and environmental significance addressed. In particular, stochastic dominance with respect to a function is used to rank irrigated and dryland cotton cropping systems. Results indicate that the cropping systems analysed seem to be a viable alternative to current cotton production practices in the area in terms of both their increased profitability and their environmental impacts.

**Environmental Approaches to Farm Support Policy in the EC**, T.N. Jenkins *(University College of Wales)*

The possibility of including environmental considerations in continued public support for EC farming is explored in this paper. Farming is an indispensable part of the management of the natural environment, and integrated agricultural and environmental policies are accordingly essential. The paper examines how policy might evolve if such integration is to find practical expression and, given the inefficiencies of the current farm support system, explores to what extent policy can support a farming industry acceptable in environmental terms without necessarily increasing the financial burden of support. It is shown that the integration of agricultural and environmental policy can potentially get the CAP off the hook of increasingly insupportable agricultural price support expenditure. A shift from product price support to environmental management payment to farmers is, at no extra financial cost to taxpayers, capable of simultaneously reducing agricultural output, maintaining farm incomes and farm populations, reducing food prices, and providing unequivocal environmental benefits such as reduced pollution, improved wildlife habitats and landscapes, and improved food quality.

**A Global Analysis of Energy Prices and Agriculture**, Steve Martinez, Brad McDonald, Miranda Otradovsky, and James V. Stout *(US Department of Agriculture)*

A multi-region computable general equilibrium model was used to assess the long-run impacts of higher energy prices on agricultural production, prices, and trade. An increase in the price of energy enters farmers' cost functions both through direct energy use and through the indirect influence of energy prices on intermediate inputs, especially fertilizer. The multi-region feature of the model permits inclusion of the effects of energy price shocks on other regions' economies and assessment of price changes in a global context. Because farming is highly energy intensive, agricultural output falls more than output in the manufacturing and services sectors of each region of the model. Real returns to farmland, a good indicator of farm welfare, fall in each of the four regions. The US land price declines by 3.5 percent, a drop comparable to that resulting from a 20-percent multilateral agricultural policy liberalization in a similar model.

**Improved Technologies for Latin America's New Economic Reality: Rice-Pasture Systems for the Acid Savannas,** Luis R. Sanint and Libardo Rivas *(Centro Internacional de Agricultura Tropical)* and Carlos O. Seré *(IDRC)*

This paper looks at economic trends and policies over the last 30 years in Latin America that led to exploitation of the continent's tropical savannas and rainforests and the ensuing environmental degradation and deforestation. Current preoccupation with these issues has resulted in increased demand for technological solutions in the agricultural sector that rely less on area expansion than in the past and in enhanced productivity while contributing to a sustainable agricultural or land-use system. Crop and legume-based pasture systems and, in particular, the rice-pasture system developed at CIAT for Latin America's acid savannas, are an example of such a technology. Several agronomic and economic advantages of the system are presented.

**Sustainable Agricultural Development: Zimbabwe's Smallholder Experience,** *1980–90,* Felix M. Masanzu *(Agricultural Marketing Authority, Zimbabwe)* and Brian D'Silva *(US Department of Agriculture)*

Zimbabwe's smallholder strategy since independence in 1980 has enabled the country to achieve its major objectives of food security, wider income distribution, rural development opportunities, and diversification of crops grown. Since 1980, the government has, through a combination of infrastructural development, input delivery facilities, agricultural credit, and pricing policies, provided the necessary incentives for smallholders to increase both their participation and agricultural production. Smallholders now deliver over 50 percent of all maize to the Grain Marketing Board, over 50 percent of all cotton to the Cotton Marketing Board, and over 90 percent of all sunflower. This has enabled Zimbabwe to achieve a high level of food security and near self-sufficiency in oilseeds. The costs to the economy can be seen in increasing deficits of the marketing boards. An analysis of these deficits, however, shows that they could be reduced if synchronization of producer and consumer prices were undertaken, efficiency in operation of the marketing boards enhanced, and greater autonomy given to the marketing boards. Zimbabwe has currently embarked on a structural adjustment programme; the successful implementation of this programme would enhance the sustainability of Zimbabwe's smallholder strategy well into the 21st century.

**Alternative Environmental Policies and Input Substitution in the Corn Belt: Single and Multi-Output Nonparametric Approaches,** Dan Primont *(Southern Illinois University)* and Richard Nehring and Agapi Somwaru *(US Department of Agriculture)*

This study investigates the impact on input substitution and profitability of production in maize farming of various agricultural chemical regulatory policies. The research evaluates the economic impact of: (1) banning atrazines, (2) banning triazines, and (3) restricting nitrogen use to 135 lbs per acre. Data on maize production from USDA's Farm Costs and Returns Survey are used to estimate a nonparametric model of profit maximization. Frontier specification of the profit function provides a reasonable benchmark from which to compare alternative policies. The results indicate that the triazine ban causes the greatest decrease in farm profitability.

### The Easily Forgotten Half of China: Towards Sustainable Economic Development in the Pastoral Region, Zhang Cungen, Zhou Li, and Xu Ying *(Chinese Academy of Agricultural Sciences)*

The pastoral region of China accounts for 52 percent of the land area of the country. In many respects it is the most strategically important half of China. Yet it attracts little attention both within China and in a global context. There have been great changes in the pastoral region since 1949. As in the rest of China, the pace of change accelerated in the 1980s. While significant economic development has occurred, grassland-based animal husbandry remains the major industry. Major socio-cultural, economic, ecological, and institutional constraints exist that will further restrict sustainable development in this half of China. This paper identifies these constraints and provides some policy suggestions that could help to achieve further sustainable development in the pastoral region of China.

### Risk-Reducing Input in US Agricultural Production, Richard Nehring *(US Department of Agriculture)* and Utpal Vasavada *(Université Laval)*

This study investigates the effect of using inputs on both the level and riskiness of agricultural production. Data on maize and wheat production from USDA's Farm Costs and Returns Survey are used to estimate production functions for these crops. The Just-Pope method is used to study how the use of pesticides, and of nitrogen fertilizer and pesticides, while increasing the level of production, also contributes to diminished riskiness of agricultural production. The impact of risk on input demand and implications for environmental policy are evaluated.

## International Trade and Policy

### Agricultural Trade and Regional Agreements, E. Wesley F. Peterson *(University of Nebraska)* and Fred Boadu *(Texas A&M University)*

The difficulties experienced in reaching agreement on rules for international agricultural trade have led some to suggest that the GATT multilateral approach is no longer appropriate. Regional economic organizations such as customs unions have been proposed as alternatives to the current system. In this paper, the implications for international agricultural trade of moving towards greater reliance on customs unions are analysed using the theory of clubs. This analysis highlights some of the problems with this alternative and suggests that it may not represent an improvement over the current system.

### Integration of EFTA Countries with the EC: How Compatible is Agriculture?, Lars Brink *(Agriculture Canada)*

EFTA countries are exploring various forms of integration with the EC. If integration were to include agriculture, differences in level and structure of agricultural policy support and protection (market price support, direct payments, import barriers, and export assistance) would require adjustment in EFTA policies. EFTA countries are very diverse in their agricultural trade with the EC and with other countries. The composition of support (market price support or direct payments) in Austria and Sweden most closely resembles that in the EC. Finland, Norway, and Switzerland provide a larger share of support as direct payments, and the unit PSE tends to be much larger than in the EC. Sweden is the only country using variable import levies to the same extent as the EC. Other countries rely much more on quantitative restrictions and state trading. In contrast to the EC, where export assistance is government funded, EFTA export assistance is at least partially funded by producers (or consumers). Agricultural policy in Sweden and Austria is most like EC policy, while Norway and Switzerland would require the most significant changes in level and structure of support and protection.

**Price Integration in the EC Sheepmeat Market: An Application of the Holmes-Hutton Test,** Daniel V. Gordon *(Norwegian School of Economics and Business Administration)*, Jill E. Hobbs *(Scottish Agricultural College)*, and William A. Kerr *(University of Calgary)*

Common markets such as the EC are, in part, organized to reduce barriers to the international movement of commodities. The benefits of trade liberalization will arise from specialization. This has two necessary conditions: market integration and transfer of resources. This paper provides a test for market integration. The integration of three markets is examined—the Paris market for French sheepmeat, the Paris market for British sheepmeat, and the British market for sheepmeat. While these markets exhibit a degree of integration, there are considerable lags, particularly between the Paris market for British sheepmeat and the British market. This suggests that the potential benefits to trade liberalization within the EC will not be fully realized.

**The Supply of Agricultural Products: Norwegian Agriculture and the EC,** Kyrre Rickertsen *(Agricultural University of Norway)*

The supply of 11 agricultural products is estimated on the basis of Norwegian data. A household utility maximization model is used to identify supply-determining variables. Following recent developments in econometrics, a large array of tests is applied to check the validity of the estimated equations. Short-run supply elasticities are calculated for the various products. The own-price supply elasticities are in the range 0.04–0.78. Furthermore, the estimated supply equations together with demand relations are used in a simulation model to investigate longer run effects on supply. A scenario based on Norwegian membership of the EC is simulated. The area under grains increases slightly in the year 2001 as compared with 1987–89. The (domestic) supply of milk, beef, pigmeat, carrots, and tomatoes is reduced in the simulations. The reductions are in the 5–45 percent interval. The supply of sheepmeat, eggs, and strawberries is increased. However, these results are uncertain, and 95-percent prediction intervals are presented.

**A North America Trading Bloc: The Relationship of CUSTA to MUSTA,** Carol Goodloe and John Link *(US Department of Agriculture)*

The vast differences in the economies of Canada and Mexico render much of the Canada-US Trade Agreement (CUSTA) inappropriate for a Mexican-US Trade Agreement (MUSTA), especially in agriculture. Canada is less worried than the USA about direct employment and trade competition from Mexico and more concerned about maintaining its preferred status in the large US market and displacement of its exports to the USA by Mexican exports. The differences between US-Mexico and US-Canada trade will dictate a MUSTA that differs from the CUSTA, although the CUSTA could serve as a model in areas such as tariff reduction and technical regulations. The potential for "unfair" competition as a result of differential treatment of the USA's northern and southern borders argues for eventual harmonization of at least some aspects of the CUSTA with a MUSTA to avoid trade diversion and reduce bureaucratic problems.

**Agriculture in a US-Mexico Free Trade Agreement**, Mary Burfisher *(US Department of Agriculture)*

This paper analyses current tariff and non-tariff barriers to US-Mexican agricultural trade. Its objective is to identify the types of barriers that present the most important impediments to trade and the commodities in which protection is highest. The paper then considers the potential for liberalizing bilateral agricultural trade, given the nature of current trade barriers and the treatment of agriculture in other free trade agreements, particularly the 1988 US-Canadian Free Trade Agreement, which is expected to provide a model for a US-Mexican Free Trade Agreement.

**Interaction between the New Hungarian Agricultural Policy and International Trade: Closer to a New Europe?**, László Kárpáti and András Nábrádi *(Agricultural University of Debrecen)*

Hungary was traditionally one of the food suppliers of Western Europe, a tradition stopped by World War II. The changing political situation in Eastern Europe and Hungary have made it possible to renew this tradition in order to meet the requirements of an integrated Europe, especially the Common Agricultural Policy of the EC. Hungarian agricultural policy has to be modified in parallel with the transition of the country's economic environment. To study this question, four different political scenarios were elaborated and possible effects on export structure and allocation of exported quantities to different groups of countries were also studied. Projection of the results shows that agriculture can remain a prosperous branch of the national economy of Hungary in the future.

**Domestic Rent-Seeking Effects of the International Coffee Agreement: An Indonesian Case Study**, Mary Bohman *(University of British Columbia)*, Lovell Jarvis *(University of California)*, and Richard Barichello *(University of British Columbia)*

The implementation of International Coffee Agreement (ICA) export quotas, by creating quota rents in coffee-exporting countries, encourages directly unproductive behaviour by domestic groups seeking to capture such rents. Such activity causes economic waste and alters income distribution. Farmers and the government in the exporting country, whom the developed country importers intend to be the beneficiaries of the ICA, are more likely to lose, while coffee-exporting firms and government officials may gain. Although quotas are known to create opportunities for rent seeking, such effects have not been examined in the context of international commodity agreements. This paper is concerned with the effects of the ICA, but there are many parallels with other commodity agreements. The paper uses a model of policy in exporting countries to show how the imposition of ICA quotas creates potential rents within coffee-exporting countries and analyses how domestic coffee policy determines the actual size and distribution of these rents among competing domestic groups. The model is applied to Indonesia.

**Optimal Import Quotas in Oligopolistic Processed Food Markets**, Steve McCorriston *(University of Exeter)* and Ian M. Sheldon *(Ohio State University)*

Recent developments in international trade theory suggest that, where markets are imperfectly competitive, there may be a justification for protectionism. This has come to be widely known as strategic trade policy. There has, however, been little consideration of the applicability of these arguments to agricultural trade. Since trade in processed food products is the most dynamic sector of agricultural trade and given that these markets are typically characterized (to varying degrees) by imperfect competition, it may nevertheless be the case that these recent theoretical developments may have some relevance for the use of protectionism in processed food markets. This paper, therefore, considers the use of import restrictions as a means of increasing national welfare, with the US cheese processing sector chosen as a case study. Using a computable partial-equilibrium model, the results suggest that the US government could increase national welfare by using import quotas to protect its cheese-processing sector; however, compared with the quota régime currently in use, the optimal import restrictions are more liberal in terms of the level of imports permitted.

**Industrial Organization as a Determinant of International Competitiveness in Food**, Dennis R. Henderson *(Ohio State University)* and Stuart D. Frank *(US Department of Agriculture)*

International trade is increasingly being viewed in the context of imperfect competition. This is particularly appropriate for food and other processed agricultural products as most food processing and manufacturing industries are oligopolistic. Industrial organization theory demonstrates a negative relationship between concentration of market power and domestic market performance. One theme emerging from the integration of industrial organization and international trade theories is that seller concentration is also negatively related to international market performance. This theme is tested and validated for US food manufacturing industries.

**The Influence of Intrinsic Attributes on the Loss of US Market Share in the Japanese Oilseed Market**, Joyce Cacho, Eluned Jones, Daniel B. Taylor, and David Kenyon *(Virginia Polytechnic Institute and State University)*

This paper uses a mathematical optimization model of the Japanese oilseeds and products market to analyse the impact of the current US soyabean marketing policy that excludes protein and oil intrinsic attributes from grades and standards on the US market share of Japanese oilseed imports. Intrinsic attribute data on competitor oilseeds, Brazilian soyabeans and Canadian rapeseed (canola), are available to market participants. The results show that a relationship exists between the foreign material content, the protein/oil composite, and both level and market share of US soyabean imports. Producers respond to price incentives provided by the market. Within the soyabean market structure, price signals have not encouraged producers to differentiate soyabean varieties by oil or protein production potential, but rather on the basis of yield. The 1990 US Farm Bill includes a "Grain Quality Incentives Act," which explicitly recognizes the importance of associating grades and standards with end-use value of the commodity. The results presented in this study indicate that potential exists whereby the USA could regain share in the Japanese market through identification of the threshold attribute levels.

**Export Supply and Import Demand Elasticities in the Japanese Textile Industry: A Production Theory Approach**, Timothy A. Park *(University of Nebraska)* and Daniel H. Pick *(US Department of Agriculture)*

The close link between Japanese industrial policy and domestic industries facing declining international competitiveness is clearly illustrated by developments in the Japanese textile industry. A production theory approach for modelling trade in intermediate goods is applied to derive export supply and import demand functions for the Japanese textile industry. Measures of biased technical change and the estimated elasticities are linked to structural adjustments in the Japanese textile industry. Technical change is biased towards domestic textiles and away from cotton imports, consistent with a pattern of domestic coordination and partnerships linking small-scale textile firms. As export markets contracted, textile firms became more price responsive to the Japanese domestic textile market. Although the price responsiveness of labour demand has increased over time, wage reductions played a relatively minor role in employment adjustment in the textile sector. The cross-elasticities showed that changes in imported cotton prices and in domestic textile prices have the largest impacts on labour demand in the textile industry and that these effects have increased over time.

**Government Grain Storage in a Market-Oriented World Economy: Acquisition and Dispersal Rules for Market Stability**, Robert D. Reinsel *(US Department of Agriculture)*

Under market-oriented conditions, grain price increases have led to calls for market intervention by import-dependent countries. Past experience shows that requests for international buffer stocks have increased during periods of high prices. During periods of low prices, major exporters have attempted to institute market-sharing schemes through international grain agreements with the hope of raising prices. Neither the cartel nor the buffer stock system has been able to provide price stability. Cartels were implemented but never really functioned, and international buffer stocks were never implemented. A yield-triggered stock acquisition and dispersal programme could buffer the market from weather-induced yield shocks, which are the noneconomic force in the market. Storage and dispersal of yield deviation from expected yield would allow true supply and demand shifts to become more readily discernible. Long-run supply and demand forces would provide appropriate signals to the market. If implemented unilaterally by both exporters and importers, price stability would be accomplished with low levels of stocks.

**An Econometric Analysis of World Wheat Trade: An Application of a Gravity Model**, Won W. Koo *(North Dakota State University)* and David Karemera *(Wayne State University)*

A commodity-specific reduced-form gravity model is derived from a partial equilibrium model of world trade. The model is then applied to the world wheat market to analyse factors affecting wheat trade flows. The study reveals that all independent variables, including production capacities, income, import and export unit value indexes, and trade policies, used in wheat trade play an important role in determining wheat trade flows. Long-term agreements achieved the highest performance by significantly enhancing international wheat trade. Credit sales also contribute to increased wheat movement. However, export enhancement and export refund programmes do not significantly increase wheat trade flows. Protectionist policies of supporting domestic prices greatly impair wheat trade.

**Accounting for Export Market Performance of Major Wheat Exporters: A Constant Market Share Analysis**, Michele Veeman, Terrence Veeman, and Xiao-Yuan Dong *(University of Alberta)*

In this paper, two versions of a constant market share (CMS) model are applied to analyse wheat export market performance of major exporters. Based on the analysis, an assessment is made of the uses and problems of CMS models for agricultural trade. These models are based on the assumption that a country's ability to increase market share reflects an increase in its competitiveness, after accounting for changes in the structural orientation of its exports in terms of its reliance on relatively fast- or slow-growing markets and products. The structural component of focus in the simpler model version is the regional orientation of wheat exports. The expanded model version also focuses on export concentration in different classes of wheat. The latter extension involved some limitations. It introduced ambiguities into the calculations of structural components. Econometric testing of annual changes in the calculated competitive effects of the two-model versions consistently indicated stronger performance of the results from the simpler model in explaining changes in export market growth. Although the CMS model does not identify sources of changes in competitiveness, it focuses attention on export growth that arises from regional concentration in exports and on exporters' relative performance.

**Domestic Policy Reforms and the International Audience: Canada's Agricultural Policy Proposals**, Theodore M. Horbulyk *(University of Calgary)* and Michael P. Kidd *(University of Tasmania)*

This paper provides a brief review of the principal criteria used to determine whether domestic agricultural policies are consistent with the general principle of liberalization of international trade in agriculture. The focus is on one aspect of Canada's recent agricultural policy proposals, those concerned with "financial farm support programmes" (the so-called "safety nets"). From a theoretical point of view, the proposed combination of domestic policies may induce output effects that are in clear contravention of the spirit of trade liberalization. Output effects may arise, for example, due to either sector-specific income transfers or the lack of precision with which such schemes can be administered.

**Effects of Macroeconomic and Trade Policy Changes on the Chinese Grain Market: A General Equilibrium Analysis**, Gil Rodriguez and H. Don B.H. Gunasekera *(Australian Bureau of Agricultural and Resource Economics)* and Will Martin *(World Bank)*

Excess demand in major sectors of the economy, a high level of government support to heavy manufacturing industries, and implicit taxation of agriculture continue to persist in China despite the economic reforms undertaken since 1978. Given this background, the focus in the paper is on the implications for the Chinese grain market of a reduction in domestic absorption and changes in some key trade policies. To analyse these implications, a general equilibrium model of the Chinese economy is used. The results indicate that China's ability to earn the foreign exchange required for grain imports could be increased by policy reforms such as the reduction of implicit export taxes imposed on key farm products like rice. Furthermore, compared with the trade policy reforms, macroeconomic policy changes, such as change in absorption, are likely to have greater impacts on China's grain sector and the overall economy.

**Effects of Export Promotion on Import Demand for US Cotton in the Pacific Rim**, Hosanna Solomon and Henry W. Kinnucan *(Auburn University)*

Government-subsidized export promotion of US cotton in the Pacific Rim is evaluated using an extended Armington trade model. Results suggest that US export promotion programmes over the 1965–85 study period had a significant impact on cotton trade flows in four of the six countries examined (Japan, South Korea, Hong Kong, and the Philippines). Marginal returns are estimated to range from US$11 in the Philippines to US$171 in Hong Kong, suggesting that a reallocation of funds or increased funding level could enhance the economic effectiveness of the promotion effort for US taxpayers, affected industry participants, and foreign third-party cooperators. The market share and product differentiation emphasis, coupled with the satisfactory statistical results obtained in this study, suggest that the Armington model is a useful methodology for examining the economic impacts of export promotion of agricultural commodities.

**Policies to Increase Agricultural Exports in Chile, Indonesia, and Egypt**, John B. Parker *(US Department of Agriculture)*

This paper shows recent drastic changes in exports in Chile, Indonesia, and Egypt. Chile, the "star" of agricultural exporters from South America, offers modern banking policies and a favourable climate for foreign investment. Agricultural exports increased dramatically during 1980–90. For example, exports of table grapes and apples are overwhelming other exporters such as the USA or Italy. The new horticultural exports, such as plums, cherries, pears, etc., are also increasing. For Indonesia, striking gains from about 30 new agricultural exports, such as live animals, confectionery, cakes, and soft drinks, helped offset the decline in exports of rubber and coffee. Exports of such tropical items as tea and spices are also increasing. Indonesia's total agricultural exports are now worth more than $3,000 million annually. Further gains are expected because of great progress with newer commodities like canned pineapple and cassava. Exports of horticultural items from Egypt increased, with oranges reaching $140 million. Cotton exports declined from the $500 million of the early 1980s to $123 million in 1990. In 1990, however, textile exports totalled $1,200 million, about double the mid-1980s' level. During 1980–82, Egypt's total agricultural exports averaged $780 million. Today Egypt promotes food crops like maize and rice. Egypt has also resisted changes in foreign investment or banking that could have helped cotton exports.

## Trade Policies and Poverty Alleviation: A Taxonomic Study, Habibullah Khan *(National University of Singapore)*

The paper provides an empirical investigation of the relationship between poverty alleviation and trade policies. The former is measured in terms of certain basic needs indicators and the latter is proxied by a recent World Bank classification of 41 developing countries. The Wroclaw taxonomic method is used to generate a number of composite indices for 1975 and 1985. The results indicate that as a country moves from strongly inward-oriented to strongly outward-oriented policies, the poverty situation improves gradually.

## Economic Adjustment and Agricultural Exports in Latin America, Yony Sampaio *(Universidade Federal de Pernambuco)*

Increased external indebtedness in Latin America generated a crisis at the beginning of the 1980s that required external and internal economic adjustment. As a consequence, on the external front, the currency was devalued in order to increase exports, and, internally, incomes and wages were depressed. The effect of economic adjustment on agricultural exports was positive, but, although acting simultaneously, wage decreases seem to have been much more influential, imposing high social costs in the process of adjustment. But the analysis shows that wage increases can follow from export increases, meaning that, if the adjustment is successful, the initial social cost can be compensated by income and wage growth in the medium term.

## Latin American Debt, Debt Forgiveness, and the Impact on Agricultural Trade, Daniel Pick and Barry Krissoff *(US Department of Agriculture)*

Financial conditions seriously deteriorated in many Latin American countries in the 1980s. A decade earlier, the Latin American countries borrowed significantly from foreign commercial banks, obtaining large capital inflows to purchase imported goods, including agricultural products. While borrowing continued in the 1980s, many of the Latin American countries suffered from a financial exodus. Net capital outflows have affected the ability of Latin American countries to import agricultural products, but there is considerable diversity in opinion regarding the magnitude of the impact. This estimation and simulation analysis indicates that a 50-percent debt forgiveness of four heavily indebted Latin American countries—Brazil, Mexico, Chile, and Venezuela—increases agricultural imports modestly, by approximately $400 million per year. Of that total, the USA and EC expand exports by $225 million, mostly in grains and meats. Two reasons may explain the modest increase in trade. First, agricultural imports of developing countries tend to be basic foodstuffs, which are less likely to be squeezed in times of financial stress. Second, as foreign exchange became less available to developing countries in the early- to mid-1980s, US credit guarantee programmes expanded and substituted for commercial borrowing guaranteed by the governments of developing countries.

## Development, Trade Policies, and Income Distribution: An Analysis of Brazil's Performance and Policy Options, Jackelyn Lundy (University of California)

This work attempts to provide some insight into the relationships among economic development, trade, and income distribution. The case of Brazil is used to analyse whether trade policy options can simultaneously promote economic growth and greater equity in income distribution. Specifically, a computable general equilibrium model was used to run three border policy experiments and to test for their effects on different income groups. The first experiment reduced tariffs on wage goods and increased tariffs on luxury goods. In contrast, the second experiment reduced tariffs on luxury goods and raised tariffs on wage goods. The third experiment followed a more traditional trade liberalization scenario by uniformly reducing tariffs for all goods. Even though the results of the experiments were mixed, they were revealing in terms of the amount of economic influence needed in order effectively to institute an economic policy and to overcome macroeconomic adjustments.

## Comparative Assessment of Competitiveness of Commercial Wheat Farm Operations in Canada and the USA, 1985–89, Shankar Narayanan (Agriculture Canada) and Michael E. Salassi (US Department of Agriculture)

Comparison of performance of typical large commercial wheat farms from the USA (major grain exporter) and Canada (medium-sized grain exporter) simulated over the 1985–89 trade-war period shows that in terms of their (export) competitiveness, the US wheat farm had relatively higher productivity and direct government payments resulting in higher total revenue, and the Canadian farm had relatively lower land costs and higher farm gate prices. The US farm, however, had lower non-land costs and higher flexibility to adjust to changing market conditions, although it was strained to service debt fully out of farm income. Overall, competitive performance remains very similar in both the farms. This implies that market effects of a further decline in grain prices and/or strengthening of own currency would enhance the burden of government payments relatively more for the Canadian farm, due to lack of scope for further improvements in cost or production efficiency. On the other hand, enhanced grain prices through global trade reform would benefit farm incomes as well as save government expenditures similarly in the two countries.

## Structural Adjustment of the Japanese Textile Industry and Impact on Trade, Fawzi A. Taha (US Department of Agriculture)

The Japanese textile industry grew rapidly until the early 1970s, but since then it has had to resort to structural adjustment due to rising production costs and keen competition. Japan's competitiveness in textiles has weakened, as labour has become increasingly expensive. Structural changes within Japan's textile industry over the past 20 years are a good example of the implication of the Heckscher-Ohlin theory of comparative advantage and international trade. Japanese textile policies are of particular relevance to other countries, industries, and commodities. Textile mills shifted from labour-intensive to capital-intensive methods of production, which consequently affected the commodity composition of trade and fostered a smooth transition from imports of raw fibres to value-added textile (yarn, fabrics, and ultimately clothing and finished goods). For example, the share of raw fibre in Japan's total textile imports fell from 82 percent in 1969 to 22 percent in 1989, while the share of yarn and woven fabric increased from 11 percent to 19 percent. The import share of clothing and other textile finished goods greatly increased from 6 percent to 59 percent during the same period. Adjustments of the Japanese textile industry included domestic specialization in the production of top-quality textiles, diversification to non-textile products, and the transfer of capital and machinery to low-labour-cost countries.

# Agricultural and Regional Development

**Tourism and Development: The Forgotten Issues,** Joseph Thuo Karugia *(University of Nairobi)*

Many Third World countries view tourism as a panacea for the many problems constraining their development. The benefits of tourism include its contribution to foreign exchange earnings, generation of employment and income, improvement of economic structures, and encouragement of entrepreneurial activity. However, there has been a tendency to promote tourism without taking into consideration its negative economic, social, and environmental impacts. This paper attempts to bring to the fore some of the issues that have received minimal attention when the role of tourism in Third World development is being assessed. The economic costs of tourism include investment in infrastructure for tourists, promotion, importation of goods and services for tourists, opportunity costs, and economic dependence. Social impacts include the effect of tourists on the moral conduct of host communities, the demonstration effect, and the effect of tourists on the health of host communities. The consequences of social and environmental impacts can seriously negate any benefits that might accrue from tourism. All these costs must be taken into account when promoting tourism as a vehicle for achieving development.

**High-Value Agricultural Products from Developing Countries,** Mary Burfisher and Margaret Missiaen *(US Department of Agriculture)*

This study analyses exports of high-value and processed agricultural products (HVPs) from 35 less-developed countries (LDCs) during 1970–87, and the roles of supply and demand factors in accounting for their export performance. HVPs account for 83 percent of LDC agricultural exports. Processed exports have undergone the most rapid growth and accounted for almost half of LDC high-value exports in 1986–87. Developed countries remain the major HVP export market for LDCs. LDCs' share of this market fell during 1970–87, although their share of developed country imports of highly processed HVPs increased. Domestic policies favourable to exports distinguished successful from unsuccessful LDC exporters. Several of the most competitive LDC exporters also adopted policies that promoted exports of specific HVPs. However, favourable domestic conditions were not sufficient; strong developed country demand was also necessary for successful LDC HVP export performance.

**Problems of Perception in Rural Development Programmes: Some Socioeconomic Implications for Smallholder Agriculture,** Diego Roldan *(Universidad del Valle)*

Most of the problems encountered in the process of implementation of integrated rural development programmes seem to have much to do with the so-called "problems of perception" regarding whether government researchers and extensionists perceive the needs of peasant producers and what the latter think those needs actually are. As a result of this conflict, what happens in practice is that, on the one hand, researchers and extensionists try to impose certain modern packages on peasants and, on the other hand, peasant producers are, to a great extent, not able to adopt them since they do not address their real needs. The problem seems to be partly one of the ideological factors underlying the practice of knowledge and training, which make technical officials view peasants as people who know little about modern techniques and therefore as people who have nothing to contribute or suggest in regard to farm production techniques.

**An Analysis of North Korea's Economic Development with Special Reference to Agriculture**, David W. Culver and Chinkook Lee *(US Department of Agriculture)*

The paper examines North Korea's record of economic development with special reference to agriculture. There was some initial success in grain production under the centrally directed economy and the *Juche* method and *Chungsan-ri* spirit of farming. However, too much emphasis on grain production has resulted in continuing shortages of other agricultural products, such as livestock and livestock products. Meanwhile, the early success of grain production now appears to have stalled. The economy as a whole is supply constrained, and effective demand is not recognized, a common failing of a planned economy.

**Impact of Rural Industry on Agricultural Development in China**, Zhang Cheng-liang *(Beijing Agricultural Engineering University)*

This paper analyses the impact of China's rural industry on agriculture with a tentative model illustrated by case study data. The model centres on the pattern of transfer of agricultural surplus labour to rural industry. Rural industry has a positive effect on agricultural labour productivity and agricultural transformation and a negative effect on agricultural land area, land productivity, and agricultural output, if labour transfer occurs. The two-sided impact has implications for policy making.

**Taiwan's Agricultural Development at the Crossroads**, Sophia Wu Huang *(US Department of Agriculture)*

Taiwan's rapid economic development has changed its agriculture dramatically. The traditional agricultural products—rice, sugar, and potatoes—are giving way to higher value, land-saving products. Taiwan increasingly depends on importing coarse grains, soyabeans, cotton, hides, and food products, while agricultural exports are no longer contributing much to foreign exchange earnings. There has been a significant increase in part-time farm households and a decrease in the intensity of land use. Surging production costs, a shrinking number and ageing farm workers, lack of an efficient scale of farming, and concern for pollution all bode ill for Taiwan's agriculture. Faced with a dwindling sector and increasing pressure on trade liberalization, Taiwan's agricultural development is indeed at the crossroads. The recently approved 6–year (1992–97) farm programmes should help to ease the transition of the agricultural sector to a smaller but more efficient and market-oriented farming operation. Given Taiwan's intention to join the GATT, its agricultural markets may open up further in order to conform with the GATT rules on agricultural trade.

**Structural Adjustment and Agricultural Development in Taiwan**, Chaur Shyan Lee *(National Chung-Hsing University)*

Structural adjustment is one of the most important ways of facing problems and challenges in agriculture. Problems of agricultural structure in Taiwan are: average farm size is very small, part-time farming predominates, age and poor training limit farmers, and there are wide regional disparities in rural development. In order to overcome these structuralal problems, the following remedial measures should be taken: establish a nucleus farming system for enlarging the size of farms of full-time farmers, which can help to solve problems of labour shortage in part-time farming; renew and retrain farmers through creating special aids to young farmers, enact a pension scheme for old farmers to encourage them to retire early, and strengthen vocational training for farmers; reinforce development in less-favoured areas through integrated regional programmes aimed at balancing regional development in rural areas; and introduce socio-structural programmes with direct income aids for the poorest farmers. Any policy on agricultural structure should be coordinated with general economic policy and regional development policy.

## On the Contribution of Public Investment in Research and Extension to Agricultural Output in Taiwan, Jhi-tzeng Shih *(National Taiwan University)* and Tsu-tan Fu *(Academia Sinica)*

The objective of this paper is to estimate the contribution of public investment in research and extension to Taiwan agricultural development since the Second World War (1954–81). The well-behaved inverted-V and polynomial distributed lag regression models are used. The empirical results estimated from both models are shown to be consistent. The optimal lag length identified by the $\bar{R}^2$ method for the inverted-V model is 13 years, whereas that identified by Schwarz-Bayesian criteria for the polynomial model is 12 years. The mean lag length for the true lag effect of public investment in education on agricultural output is around 6–7 years. The shape of the lag distribution identified follows an inverted-V or second-degree polynomial scheme. That means that the lag response of public investment in research and extension on Taiwan agriculture will last for a period of 12–13 years. The distribution of the lag effect will show an increasing trend in the early years and a decreasing trend after year 6 or 7. The highest lag effect occurred in year 6 or 7. The marginal internal rate of return to public investment in research and extension was about 65–69 percent during 1954–81.

## Agricultural Growth in the South Pacific Island Economies: Constraints and Key Issues, K.L. Sharma *(Bureau of Transport and Communications Economics, Australia)*

After examining the trends in agricultural growth in the islands of the South Pacific during 1980–87, this paper attempts to identify major constraints on and key issues in the performance of agriculture. Specific policies on productivity improvement, infrastructural development, and diversification of production and exports need to be instituted to achieve sustained agricultural growth in the Pacific Region.

## Agricultural Structure as a Determinant of Technology Choice in Bangladesh, Koichi Fujita *(National Research Institute of Agricultural Economics, Japan)*

One of the most notable features of agricultural structure in South Asian countries is that the skewed land distribution is not corrected through the tenancy market. The relatively privileged, often owning some land, occupy a larger share of rented-in land. This paper analyses the effects of agricultural structure on productivity and/or technology choice using multiple regression methods and 1983/84 Agricultural Census data for Bangladesh. Land quality variables are incorporated to be controlled. Small farmers use their limited land much more intensively than large farmers. Additional cropping of unremunerative crops such as *aus* rice is largely attributed to this. An inverse relationship between farm size and land productivity is observed under both traditional and modern technology. Tenancy itself could not be regarded as a constraint to technical progress, which refutes the theory of semi-feudalism in eastern India, including present-day Bangladesh. However, the inverse relationship is persistent even among owner-cum-tenants. The agricultural structure as a whole thus suppresses productivity. Land reform in terms of strengthening the land ceiling promises increased production.

## An Interregional Equilibrium Model to Evaluate the Impact of Agricultural Policies and Development Projects on the Agricultural Sector of Sri Lanka, Giovanni Vergani and Peter Rieder *(ETH Zürich)* and Cyril Bogahawatte *(University of Peradeniya)*

The objective of this study is to develop a suitable instrument for analysing the agricultural sectors of developing countries. This involves both the representation of the interdependencies that rule an economic sector and the evaluation of the reaction of the different market participants following the implementation of either agricultural policy measures or development projects. The methodology presented in this study relies on mathematical programming within which a linear programming approach was chosen, relying on a spatial equilibrium model. The model reproduces features of the Sri Lankan agricultural sector in order to be able to analyse the role played by each market interdependency. The simulation of different agro-political scenarios allows an impact analysis of exogenous interventions on the sector.

## Contract Farming and Sugar Cane Outgrower Schemes in Kenya, L. Maurice Awiti (University of Nairobi)

Contract farming and outgrower schemes in Kenya were studied to understand how they contribute to and limit increased agricultural growth and rural development in general. There is much concern in sub-Saharan Africa about the inability of the agricultural sector to maintain sufficient growth rates to match the demand of the rapidly growing population. There is a need to design alternative institutional arrangements so as to enhance the effectiveness of policy reform and technological innovation in promoting agricultural growth. This paper presents the results of a case study of sugar cane smallholder outgrower schemes in Kenya based on primary and secondary data. The basic description, performance, and impact as well as prospects for these schemes are presented.

## Structural Adjustment Policies and Agricultural Production in a Developing Economy: The Case of Tanzania in the 1980s and Prospects, A.K. Kashuliza and J.R. Rugambisa (Sokoine University of Agriculture)

This paper reviews the structural adjustment programmes implemented by Tanzania in the 1980s to reverse the declining trend in the economy. Selective policies with a direct bearing on agriculture such as producer prices and exchange rate adjustments implemented in the context of the Economic Recovery Programme (ERP) are described. This is followed by an assessment of the performance of the economy and in particular the agricultural sector in the face of ERP implementation. The paper concludes by suggesting short-, medium-, and long-term strategies for developing and sustaining agriculture and other sectors of Tanzania's economy.

## Application of Stochastic Dominance Criteria to Evaluate Bean Production Strategies in Central Province, Zambia, Mesfin Bezuneh (Clark Atlanta University)

Beans are the most important "relish" crop in the farming systems of Serenje District in Central Province, Zambia. Both leaves and dried beans are major food sources for home consumption, and dried beans have a commercial value in the system. Given this dual role, increasing bean yields would improve family nutrition as well as economic returns to capital and labour. However, low levels of production have been recorded over the years due to the use of local bean varieties, low fertility, and inadequate pest control. As a result, on-farm research on beans was carried out for four years to identify bean varieties and management strategies that would result in higher yields and economic returns. In this paper, results of the four years' on-farm research were analysed using stochastic dominance efficiency criteria in order to determine the most risk-efficient production management strategies. The results indicated that the Brazilian bean variety carioca, when used in combination with fertilizer and insecticide, performed best for the traditional and small-scale farmers in Serenje District who are usually highly averse to risk.

## Shifting Affirmative Action in South African Agriculture: Problems and Considerations, C. McKenzie and M. Lyster (Development Bank of Southern Africa)

The South African agricultural sector has been characterized by excessive state intervention in the form of affirmative action in favour of a limited number of European commercial farmers. In the context of the changing political economy, there is a real need to shift affirmative action to meet the needs of those denied support in the past. A number of mechanisms have been identified in order to make the agricultural sector more equitable. The implications of these are discussed in the context of their practical implementation with respect to available resources and recent experience in the South African commercial farming sector.

### Post-Harvest Rural Infrastructure Combination for Sustained Food Flow in Nigeria: The Case of the Kwara Agricultural Development Programme, J.C. Umeh *(University of Agriculture, Nigeria)*

The World Bank-assisted Agricultural Development Programme in rural Nigeria has led to an accelerated productivity growth in food supply. However, these gains are rapidly eroded soon after harvest as a result of poor post-harvest activities. With the aid of a principal component analysis model, this paper examines Nigerian efforts in forestalling the food-flow programme. Indices of food flow are constructed for two major post-harvest rural infrastructures—storage facilities and the rural road network. A third sample, which serves as a control and is thus not deficient in any of the post-harvest rural infrastructures, is also used. Each of the two rural infrastructures influence the rate of food flow, although storage facilities, in relative terms, are of greater importance. A combination of the two infrastructures will accelerate, on a sustained basis, the rate of food flow in Nigeria. Various policy implications of the results are indicated.

### Sources of Growth in Brazilian Agriculture Revisited: The Crop Sector, Léo da Rocha Ferreira *(Instituto de Pesquisa Econômica Aplicada and Universidade do Estado de Rio de Janeiro)*

This paper describes some of the results of a study of changes in the pattern of growth of Brazilian agriculture in the 1980s and its relationship with public investment in the sector. A modified version of a shift-share model was used to analyse the sources of agricultural growth. The partial results obtained show that productivity (yield) gains were of greater importance as a source of growth in the 1980s than the expansion of the cultivated area found in previous decades. The possible relationship between this change in the pattern of growth and the increase in public investments in agriculture is discussed.

### Stabilization Policies and Agriculture in Brazil, 1986–90, Gervasio Castro de Rezende *(Instituto de Pesquisa Econômica Aplicada)*

This paper shows that the increase in inflationary instability in Brazil after 1986 and the policies adopted by the government since then to fight hyperinflation caused the terms of trade of agriculture to become much more unstable at the same time that the government—pressed by a growing fiscal crisis—withdrew the support it had been providing to agriculture. Under these circumstances of increased risk, agricultural investment and output fell drastically in 1990 and were expected to remain at low levels in 1991. Turning to the current stabilization strategy, the paper points to its depressing effects on agriculture and stresses the need for policies specifically designed to support agriculture if the latter is to recover and contribute to the success of the stabilization programme itself.

### Intersectoral Capital Transfers: The Role of Relative Prices, Luis Romano *(Instituto de Agricultura, Colombia)*

Several theories of economic development indicate that one of the basic functions of the agricultural sector in developing countries is to generate an economic surplus to provide capital resources for industrial development. While there are different mechanisms for transferring these resources from one sector to another, one of the most important is the price system. The main objective of this paper is to provide an analytical and empirical framework to estimate the amount and direction of the intersectoral capital transfer through relative prices and to calculate terms of trade among sectors. The concepts of implicit transfer, implicit deflator, terms of trade, and index numbers are used. A significant net transfer from agriculture to other sectors was found for the 1970–88 period in Colombia, indicating that the agricultural sector has been an important source of capital for the country's economic development. Also, the industrial sector transferred capital resources to the service sector over the final time period, which is surprising but fits with trends in financial and government activities in Colombia in recent years.

**Argentina: Possibilities and Restrictions of Agricultural Policy in an Adjustment Policies Framework**, Edith S. de Obschatko *(Instituto Interamericano de Cooperación para la Agricultura)*

Argentina is at a critical point in its economic development. GNP is similar to that of 1970, per-capita income has seriously decreased, infrastructure is deteriorating, and external debt is seven or eight times the value of annual exports. In the current world context, Argentina needs to develop an export strategy to re-start economic development based on the production and processing of agricultural, fishery, and forestry products. The macroeconomic adjustment framework is favourable for the development of the export-oriented agro-industrial sector because of non-intervention policies. However, the scarcity of public funds reduces the possibility of addressing the three principal constraints on sustained growth: technological gap, agricultural protectionism practices, and dangers to natural resources. In order to achieve sustained growth of the agricultural and agro-industrial sector, a new approach to agricultural policy is needed. This approach is not based on market interventions, but relies principally on four features: state modernization, increased coordination among macroeconomic policy and sectoral policies, greater coordination among national and provincial governments, and increased interaction and cooperation among the public and private sectors to enhance their respective technical capacities and resource availabilities.

**Marketing and the Rural Poor**, John C. Abbott

The numbers of poor people of the world are increasing. Most live in the rural areas of developing countries. The role of marketing in alleviating such poverty is examined through the projects of the International Fund for Agricultural Development, which has taken some path-breaking initiatives.

# Farm Structure, Land Use, and Rural Problems

**Segmentation of the Life of the Farm Household: Trends and Actual Situation in Japan**, Kiyomitu Kudo *(Chugoku National Agricultural Experiment Station)*, Tokuya Kawate *(National Agricultural Research Centre)*, Keiichi Murano *(National Institute of Sericultural and Entomological Science)*, and Kiyoko Arai *(National Agricultural Research Centre)*

During the pre-war years (before 1945), farming households in Japan preserved their strong unity under the leadership of their patriarchs. Today, however, there are a number of signs that the traditional patriarchy is rapidly being replaced by a new type of household managed by husband and wife rather than by the patriarchy. Based on a questionnaire in Ibaraki Prefecture, this report studies the actual life of farm households in recent years from the viewpoint of housework, family budget management, transfer of family budget management rights, decision making, living space, leisure time, and attitude towards living with parents. The study reveals that the family relationship in farm households is changing from one of vertical cooperation among different generations to one of lateral links among different family units within farm households.

## An Economic Analysis of Family Size Decision Making with Reference to the Developing Areas of South Africa, C.D. Fairlamb and W.L. Nieuwoudt *(University of Natal)*

South Africa's annual population growth rate in the traditional sector is 2.9 percent. This study identifies economic factors affecting family size choice to provide policy makers with a strategy for reducing fertility. A neoclassical utility framework was used to analyse linkages between family size decisions and socioeconomic variables. Household utility for "child services" and "standard of living" was maximized subject to the resource constraints of time, labour, and income. A demand curve for children was specified within a simultaneous model of family decision making. A stratified sampling technique was used to collect household data from Ulundi and Ubombo in KwaZulu; 175 women in three equal occupational strata were interviewed. The simultaneous model was estimated by two-stage least squares regression analysis. Dummy dependent variables were estimated by probit analysis. Child education, women's opportunity cost of time, and formal market participation were negatively related to fertility, reflecting substitution from numbers of children (time intensive) to fewer, more educated children (less time intensive) as opportunity costs rise. Child labour was positively related to fertility. Strategies to reduce population growth rates should therefore include improvements in women's education and employment opportunities to raise their time costs and time-saving devices to reduce demand for child labour.

## Poverty among US Rural Husband-Wife Households: An Econometric Analysis, Wallace E. Huffman *(Iowa State University)*

The paper presents a human-capital-based household model of a household's cash income and poverty status. The data are for US farm and rural nonfarm households included in the annual Current Population Surveys of 1978–82. The empirical analysis sorts out the relative importance of human capital and family size, local labour market conditions, and local cost of living and amenity factors on income and probability of a household being in a state of poverty. For farm and rural nonfarm households, the human capital and family size variables are shown to be significant determinants of household income and poverty status. Local economic conditions are shown to matter significantly for income and poverty status of rural nonfarm, but not farm, households. Although married males and females have quite different frequencies of participating directly in income-generating activities, the marginal effects of men's and women's schooling on household cash income and probability of poverty are very similar.

## Ordering of Multiple Objectives of Small Farmers in Bendel State, Nigeria, O. Dicta Akatugba *(Bendel State University)* and M. Upton and T. Rehman *(University of Reading)*

Consideration is given to the line of thought that farmers have multiple objectives and thus behave in such a way as to attain satisfactory levels among the different objectives rather than the traditional maximization of a single objective. The procedure for ordering the farming objectives of farmers on the basis of a sample of 150 small farmers in Bendel State, Nigeria, uses both a numerical rating scale and a paired comparison. Results support the fact that even though profit is an important objective for small farmers, it is neither the only nor the highest priority consideration in their decision making.

## Allocation of Activities by the Farm Household: Organization of the Farm Firm, Gregori M. Chiesa R. *(Università di Udine)*

The paper analyses the organization of the farm firm as an example of productive activities inside or outside the farm. The analysis involved a concise survey of some of the main usable interpretative models in approaching the problem, a comparison of the casual variables used by each, and a quantitative assessment of the heuristic capacity and compatibility of these interpretative models. The models adopted are the microeconomic "farm household" model (conveniently integrated with an interpretation of industrial vertical disintegration processes) and a simple description of the productive process through the "stocks and flows" model. The models are complete alternatives in construction, open to some common interpretations (allocation outside the farm is negatively correlated to farm size) and some differing interpretations (allocation outside the farm is linked to minimization of production costs or reduction of factor inactivity), and useful in interpreting particular phenomena such as the substitution effect of manual for mechanized labour and the correlations between structural features of the farm and phases in the family life-cycle (the first model), that part-time farms engage in work outside agriculture to render farm activity compatible with employment outside agriculture and that farms externalize or internalize activity to set requirements for manual work in terms of time (the second model). A quantitative analysis applied to a sample of approximately 300,000 farms led to results in line with this set of theories, confirming how alternative theoretical models may lead to complementary explanations of empirical phenomena.

**A Comparison of Cost Efficiency of Part-Time and Full-Time Cash Grain Farming in the USA**, Gerald W. Whittaker, Mary C. Ahearn, and Hisham S. El-Osta *(US Department of Agriculture)*

Individuals associated with farming are more likely than the average individual in the USA to be engaged in multiple job-holding. Concerns have been expressed that part-time farming may not be healthy for US agriculture because of production inefficiency. This paper analyses the cost efficiency of a sample of cash grain farms from USDA's Farm Costs and Returns Survey which represents 133,463 farms. Two-thirds of these farms participated in off-farm economic activity. A non-parametric "best practice" frontier is established following the method of Färe *et al.*, and the average cost efficiency of two groups, those farm operators who do not participate in off-farm income generating activities and those who do participate, is compared. A statistical test of the means leads to the conclusion that there is no difference in cost efficiency between part-time and full-time cash grain farms in the USA.

**Economic and Financial Planning in Farming: A Case Study**, Manuel Cabanes and Francisco Amador *(Universidad de Córdoba)*

Several research projects indicate that decision making in farming is characterized by the presence of multiple objectives and by a non-maximizing attitude on the part of the decision maker as far as the proposed objectives are concerned, adopting instead an attitude of "satisfying." In general, farm planning objectives, such as gross profit, level of employment, etc., are usually included, but the inclusion of a set of coordinated economic and financial objectives is not so frequent, and these are the ones that really have an important part to play when the decision maker prepares an action plan for the following year. The objective of this paper is to show in a real case study how multi-criteria techniques can help to define suitable short-term economic and financial policies.

**Substitution between Labour and Capital in South African Commercial Grain Production**, Johan van Zyl and Helmke Sartorius von Bach *(University of Pretoria)*

This paper is particularly concerned with the analysis of trends concerning farm labour-capital substitution in five major commercial grain producing areas of South Africa. Discussions involve labour and machinery usage in quantitative terms, capital-labour substitution, and the policy relevance of the findings. Only trends and annual changes are considered. The identified trends and other characteristics of the agricultural labour market have definite implications. Capital is relatively scarce in South Africa, while there is an abundance of unskilled and semi-skilled labour. Capital should thus be used with a great deal of discretion to maximize income and work creation opportunities. Policy makers should therefore review certain policy aspects that impair job creation opportunities in agriculture and that have resulted in distorted prices of production factors relative to their scarcity.

**New Human Resource Structures in the Hungarian Farm Production System**, Katalin Daubner *(Budapest University of Economic Sciences)*

Industrial-type farm production systems in Hungary are operating as a form of inter-farm cooperation where one of the large farms at the centre performs coordination, research, technical development, information, and several service activities. In the process of production, the member farms join the system with their specific branches and technological procedures. The human resource endowment of these centres is illustrated on the basis of information about education, technical training, practical experience, and land-labour ratios. Recent features on the partner farms are the disappearance of "dictatorial" management approaches, organiza-tional decentralization in various forms, and the reappearance of large-scale cooperatives.

**Land Use and Potential for Agriculture in Kenya**, Cameron Short *(Carleton University)* and Kangethe W. Gitu *(Ministry of Planning and National Development, Kenya)*

This paper describes a method used to create a national agricultural land data base for Kenya combining land use by land potential class. Information from a detailed Geographic Information System (GIS) data base was extrapolated to the whole country with a less detailed national GIS data base. At the national level, the overall area cultivated could be increased by 32, 35, and 101 percent for high-, medium- and low-potential land, respectively.

**Improved Production in Traditional Agriculture with Specific Reference to a Land Market**, W.L. Nieuwoudt *(University of Natal)*

Despite population pressure, land is underutilized in KwaZulu, South Africa. A market failure situation arises due to inadequate property rights. Using data from 564 farmers, a discriminant analysis shows that the following factors were positively associated with surplus production of food: inputs purchased on credit, renting of land, savings, and wage remittances. A variable with a high standardized coefficient was that surplus farmers tend to rent additional land. Creating an opportunity cost in land by promoting renting arrangements will benefit both owners of land and users of land. Farmers are presently reluctant to rent their land due to the perception that land rights are jeopardized. Policies should focus on reducing the transaction costs in renting arrangements through legal protection. Further, a modest tax on the rent could be paid to chiefs to obtain their support. Promotion of such a rental market in land has been recently proposed elsewhere, in the former USSR and in Mexico. In a further analysis, principal components were used to cluster the socioeconomic variables. Different household types were identified, implying that policies may affect different households differently.

**An Empirical Analysis of the Relationship between Personal Distribution of Non-Labour Resources and Income and Geographical Redistribution of Population: The Case of Rural-to-Urban Migration in Iran**, Mohamadreza Arsalanbod *(Urmia University)*

The purpose of this paper is to analyse empirically the relationship between personal distribution of non-labour resources and income within rural areas and population redistribution from rural to urban areas in Iran. Using multiple regression analysis and a number of variables as proxies for explanatory variables normally used in theoretical models of rural-urban migration, it was found that distribution of non-labour resources and income within the rural areas has a strong influence on rural-to-urban migration. The ratio of the difference between the population born in the rural area and the population that resided in the rural area over the population that resided in the rural area was used as a measure of rural-urban population redistribution. The variance of farmland divided by average farmland is used as a proxy for the distribution of non-labour resources and income within the rural area. The correlation between the dependent variable and the distribution of agricultural land was 0.7695 and significant at the 0.001 level. The $F$ value for this variable was also significant at the 1-percent level.

**Perception of Regional Leaders on Agricultural Problems and Revitalization: A Cognitive Map Analysis of Tono-shi, Iwate Prefecture, Japan**, Toshiyuki Monma *(Tohoku National Agricultural Experimental Station)*

This study demonstrates the usefulness of cognitive map analysis applied to the clarification of regional agricultural problems and future development planning in Tono-shi, Iwate Prefecture. Regional leaders clearly recognize that the factors that are important for the revitalization of regional agriculture and the rural community include reorganizing the present small-scale rice production system and promoting further rice farm enlargement by further land mobilization and contract rice operation, which will eventually lower the cost of rice production. For the effective use of labour freed by land mobilization and contract rice operation, promotion of such farm products as vegetables, beef cattle, and forest products is important. Production systems need to be established that are low cost but also promote high-value products adapted to diversified consumer demands. The promotion of forestry is perceived as an important measure for increasing farmer's income as well as environmental conservation.

**The New Village Economics**, Peter H. Calkins *(Université Laval)* and Michel Benoît-Cattin *(CIRAD, France)*

This paper explores the potential strengths of using village-level economics as a complement and corrective to household and macro-level studies of the economy. Village geographical areas and analytical methods are inventoried. Gaps in current thematic coverage and analytical tools are identified. Under-utilized tools, with mathematical programming examples from Africa and Asia, are illustrated.

**Income Distribution in US Agriculture**, Hisham S. El-Osta and Mary C. Ahearn *(US Department of Agriculture)*

When the size distribution of personal income is examined for US farm operator households, the households within each income quartile break into two distinctive groups based on farm assets. Using data from USDA's 1988 Farm Costs and Returns Survey, it is found that more farm operator households with farm assets of less than $500,000 across all quartiles of the income distribution, in comparison to those with farm assets of $500,000 or more, have sales of $40,000 or less. In general, these households earn less income from farming, receive less in government payments, are more likely to be organized as sole proprietors, and are more likely to specialize in beef, pigs, and sheep. These households, with the exception of those in the middle income quartiles, also earn less income from off-farm sources. In terms of human capital, operators of these households and their spouses have less education.

**The Economic Consequences of a Changing Rural Economy: Implications for a US Rural Policy**, Ralph D. Christy *(Louisiana State University)* and James T. Bonnen *(Michigan State University)*

The process of transforming agrarian societies into industrial service-based economies poses economic and social consequences for rural areas and their residents. The purpose of this paper is to provide a pragmatic conceptual understanding of public policy imperatives for the US rural economy. It considers problems confronted by rural economies in the USA and outlines public policies that are needed in response to these emerging problems. The paper uses a framework that suggests that the basic characteristics of an economy give rise to critical problems, and these problems require policy responses. The relationship between characteristics and problems, between problems and policy, and, in turn, between policy and characteristics, are dynamic ones. This framework is further extended to include the "theory of change," which seeks to identify the fundamental forces that influence the economic process: institutions, people, technology, and physical resources. In designing a development strategy for a rural community, region, state, or nation, all four of these prime movers must be examined for their roles and included in some balanced mix in the policies that constitute the development plan and strategy.

### Public Choice and Agricultural Reform in South Africa, Konrad Hagedorn *(Universität Hannover)*, Nick Vink *(Development Bank of Southern Africa)*, and Johan van Zyl *(University of Pretoria)*

This paper uses arguments from public choice theory to explain the relative efficiency of farmers in generating political influence. The success of farmers, however, also creates an opportunity for free riding by sub-groups. In South Africa, a sub-group within the group of European farmers has captured most of the benefits of farm policy, to the detriment of all other farmers and all consumers and taxpayers.

### Farm Structural Changes as Transition to Eastern European Market Economies: Ownership and Prospects in Hungary, Katalin Sebestyén *(Agricultural Economics Research Institute)* and Ferenc Fekete *(Budapest University of Economic Sciences)*

The paper is divided into the following main sections: changes in agricultural land ownership and land use patterns, economic reconstruction and agricultural policy objectives, and market orientation and social stratification. The major lessons drawn from Hungary's history in the past five decades, particularly since the 1956 revolution, are sketched. Some characteristics of the recent reform of farm structures and ownership are also analysed. From the beginning of the transition period marked by the political events of 1988/89, land ownership relations have become of central interest. The paper views the recent economic crisis and expected conditions of capital supply and labour motivation. While the secondary and tertiary economic (employment) sectors grew at a considerable rate, the productive and export capacities of Hungarian agriculture declined. In the past, relatively high incomes were not accompanied by appropriate rates of savings. In the transition from an autocratic system characterized by overcentralized planning and bureaucratic coordination to a democracy with several parties and to a market economy of mixed (including some private) ownership, Hungarian agriculture may serve as a model for and stabilizer of the overall developing political system.

### Transition Towards a Market-Oriented Agriculture, Csaba Forgács *(Budapest University of Economic Sciences)*

The paper deals with the major questions of the transition period from centrally planned towards market-oriented agriculture in Hungary. The roots and the basic structure of the present agricultural performance, dominated by large-scale farming, are illustrated. The most important questions of the transition period are highlighted. Special attention is given to agricultural performance between the early 1960s and the mid-1980s. The framework of the New Agricultural Policy (NAP) is discussed in terms of the requirements of a market-oriented system, reforming land ownership, the privatization process, and the transformation of large farms into more efficient ones. The paper also covers the problem of restructuring both agricultural performance and the organization of the sector. Conclusions concentrate on how to introduce the NAP in practice and the major features it should include to develop a new environment-oriented agriculture that can also compete on the world market.

**The Shift to a Deflationary Era and the End of the Road to Farm Expansion: Implications of the Farm Crisis in the 1980s**, Shinnosuke Tama *(Hirosaki University)*

The aim is to indicate from an historical perspective that the 1980s marked a turning point for Japanese agriculture. After the passage of the Agricultural Basic Law in 1961 many support measures for farm expansion were introduced and facilitated by the inflation of the 1960s–1970s. Farm expansion did not take place in Japan, however, except in Hokkaido. On the contrary, an increase in part-time farming became the general trend, since the Japanese *Ie* (family) system prevented farmers from transferring property. The shift of the world economy from inflation to deflation in the 1980s dramatically changed the circumstances surrounding Japanese agriculture. Hokkaido, the major producing region of Japanese agriculture, faced a farming crisis similar to that in the USA because of scaled-down support measures. Moreover, part-time farming based upon the *Ie* system proved to be the more sustainable system during the deflationary era.

**Evidence on Adjustment and Agriculture: Directions of Change for Developing Regions**, Manuel Vanegas Senior *(Makerere University)*

Many developing countries, under structural adjustment, have now implemented policy reforms. The majority of them have aimed at reducing distortions in the agricultural sector, redressing financial imbalances, and liberalizing their trade régimes. Many have succeeded to some extent in the last two of these, but others have failed, and progress in agriculture and pricing policy has been limited. Real producer prices paid to producers for food and export crops continue to be eroded by inflation. In Africa, price and marketing controls continue to have an adverse effect on the performance and needs of small farmers. Excluding Argentina and Brazil, Latin America needs to import about $2,000 million of foodstuffs to feed its people, and inappropriate trade régimes and extremely large fiscal deficits are hindering the efficient use of its productive potential. In the long run, supply responses to change in prices are strong. The question of how to deal with the problems of the short run remains unsolved. Malawi shows a short-run elasticity of 0.09, compared with 0.63 in Chile. Building human capital, strengthening institutions, and infrastructural improvement can be more of an incentive to agricultural output than prices.

**Searching for the Old Strategy: Polish Peasant Farming in the 1990s**, Zbigniew Kowalski *(Academy of Technology and Agriculture, Poland)*

In mid-1990, when the market-oriented price system for agriculture was discussed in Poland, peasant farmers were among the most ardent supporters of the change. Now, their enthusiasm has abated abruptly and they want state intervention in the market. The newly introduced system faced farmers with problems they never encountered under Communist rule. Although they had been anxious during the previous 45 years about the long-term existence of family farming under collectivization, they were, in fact, quite secure. Farm bankruptcy was something unknown to Polish farmers. The durability of family farms was even confirmed by the Constitution Act. After the Communist system had been dismantled, anxiety about the future turned, paradoxically, into reality. The paper discusses the most vital problems faced by Polish farmers during the present period of rebuilding the market mechanism in the economic environment shaped by the previous system of central planning.

# Price Analysis and Technology

**General Equilibrium Approach: Lessons from Theory,** Jacques Loyat *(Ministère de l'Agriculture et de la Forêt, France)*

Applied general equilibrium (AGE) models are part of a growing literature on international trade policy analysis. Insofar as models are used in a normative way, this poster seeks to summarize the limits of such models. A general competitive economy is a Walrasian one, where prices are market signals; an equilibrium is established through agents' preferred actions. The temporary equilibrium hypothesis achieves a good level of simplification for applied models, but it is not Pareto efficient. Incomplete information and uncertainty do not conflict with general equilibrium, but they make applied models more complicated. The limit of the AGE approach is reached with non-Walrasian equilibrium. Equilibrium with rationing generalizes the notion of Walrasian equilibrium by allowing markets not to clear. The Hicksian fixed price model provides a different method of analysis that links micro- and macroeconomics and where prices are taken outside the model. The conjectural analysis introduces interaction among agents, the effects of which are subject to conjecture. These are important extensions of the understanding of the economy, insofar as models cannot disregard history. AGE models give a simplified representation of an idealized exchange economy, where generally no comparison between prediction and experience is done.

**Direct Comparison of General Equilibrium and Partial Equilibrium Models in Agriculture,** James V. Stout *(US Department of Agriculture)*

Both partial equilibrium and general equilibrium models are used for the study of agricultural policy liberalization. Partial equilibrium models can include detailed descriptions of agricultural markets and agricultural policies but are criticized for not considering intersectoral effects. General equilibrium models capture the intersectoral effects but often lack the agricultural market detail of the partial equilibrium models. The model described in this paper represents an attempt to reconcile partial equilibrium and general equilibrium modelling techniques by building a general equilibrium model based on the same agricultural sector information as in the US Department of Agriculture's partial equilibrium SWOPSIM model. The model is able to produce agricultural sector supply and demand response consistent with partial equilibrium elasticity estimates because it uses a generalized form of CES function called a "constant-difference elasticity" function to represent producer behaviour towards production (output) decisions and consumer behaviour towards consumption decisions. Results for the general equilibrium model's agricultural sector are not significantly different from the results for the partial equilibrium model on which it is based. The general equilibrium model shows how changes in non-agricultural supply and demand take place after agricultural policy liberalization, and it allows welfare changes to be easily and unambiguously calculated in terms of either equivalent or compensating variation.

**Factor Markets and Agricultural Policy Shocks,** Bradley J. MacDonald *(US Department of Agriculture)*

Nearly all existing computable general equilibrium (CGE) models use quite restrictive functional forms for the specification of technology, i.e., Cobb-Douglas or CES. While some authors have used flexible functional forms, the global stability properties for equilibrium analysis of these forms can be problematic. This paper applies recent theoretical work on the nonseparable constant elasticity of substitution (NCES) functional form to the specification of technology in the Japanese component of an existing four-region CGE model. The results of a unilateral agricultural policy liberalization by Japan under the NCES form are compared to earlier results using the standard CES form. The use of improved functional forms in this example is of great importance for the analysis of the effects of liberalization on factor markets and of modest importance for the effects on agricultural output and prices.

**Effects of Japanese Monetary Policy on Agricultural Income**, Tada Minoru *(Shikoku National Agricultural Experiment Station)*

Japanese agriculture will be strongly influenced by macroeconomic factors such as foreign exchange rates or interest rates as well as agricultural policies, since its markets will be tightly linked to the world markets after the introduction of market liberalization for beef and oranges. The effects are analysed of Japanese monetary policy (i.e., official discount rate and money supply) on agricultural income through the fluctuations of the yen-dollar exchange rate and prices of imported agricultural products. Equations composed of a yen-dollar exchange rate determination, a money supply-inflation multiplier, the relationship between exchange rate and agricultural product prices, and production and income functions of rice and non-rice sectors are estimated. These estimated results show that a 1-percent increase in the rice price leads to a 1.74-percent increase in the income of the sector and that a 1-percent increase in the money supply growth rate and a 1-percent decline in the official discount rate lead to increases of 0.32 and 0.03 percent, respectively, in the income of the non-rice sector. A 1-percent increase in the rice price and a 5.4 percent increase in the money supply growth rate thus seem to have equivalent effects on agricultural income.

**Exchange Rate Appreciation: The Case of Taiwan's Grain Prices**, Cameron S. Thraen *(Ohio State University)*, Ain-Ding Liaw *(Council of Agriculture, Taiwan, China)*, and Donald W. Larson *(Ohio State University)*

This study investigates the macroeconomic linkage between exchange rates and the Taiwan grain sector. The appreciation of Taiwan's currency against the US dollar during the 1980s may decrease domestic prices and increase imports of maize, sorghum, wheat, and soyabeans. The exchange rate impact on Taiwan's prices, imports, domestic consumption, and production of grains is analysed by developing a simultaneous equation model that incorporates the exchange rate, cross-price, income, and target price variables. Elasticities are estimated from both the structural trade model and its reduced form. Taiwan's grain imports, with respect to the world price in real domestic currency, are elastic for maize, sorghum, and soyabeans and inelastic for wheat. Taiwan's domestic target price policy has no significant effect on grain imports. The empirical results of the reduced form model suggest that Taiwan's exchange rate appreciation has decreased domestic prices for maize, sorghum, soyabeans, and wheat. The exchange rate elasticities are positive and inelastic for all grains. The elasticity of grain prices in Taiwan with respect to real income is positive for maize and negative for the other grains.

**The Production Response of the Firm to the Elimination of Commodity Support Programmes in the USA**, Phil Johnson and Eduardo Segarra *(Texas Tech University)*

The production response of a farm in the Texas High Plains to the elimination of support programmes for certain agricultural commodities in the USA is estimated using duality theory. Linear programming methods were used to derive optimal profit-maximizing combinations of enterprises with current farm programmes and without farm programmes for a grain/cotton farm. An indirect profit function was estimated using ordinary least squares methods. Duality theory was used to derive the farm supply functions for each commodity with and without farm programmes. The results indicate that a short-run increase in the supply of each commodity due to shifts of the supply curves would be expected.

**Fiscal Reform and the Chinese Wool Industry**, Du Yintang *(Chinese Academy of Social Sciences)*, Gregory J. Williamson *(University of Queensland)*, and Ross G. Drynan *(University of Sydney)*

Reforms aimed at improving allocative efficiency and promoting the modernization and growth of the Chinese economy were set in motion in 1979. Fiscal reforms giving all levels of government incentives and responsibility for their own budgets were introduced to encourage greater efficiency. This contributed to a phase of investment in regional wool processing facilities by lower levels of government in those regions where there is little other industry and thus little opportunity for collecting tax. Further, after the central government relaxed national wool market policies, provincial and county governments had varying incentives to introduce their own wool market controls, with those dependent on the wool industry for revenue often rigidly controlling the market.

**The Supply of and Demand for Malaysian Palm Oil**, Karsheng Au and Milton S. Boyd *(University of Manitoba)*

This paper investigates the supply response, demand for, and stocks of Malaysian palm oil using an econometric model. Both the long-run price elasticities of supply and the price elasticities of demand for Malaysian palm oil are quite elastic. The relatively high elasticity of export demand suggests that a large increase in quantity supplied would increase export earnings for Malaysia because the palm oil price would only decrease slightly in relation to the increased quantity. Restrictions on production to raise prices and revenue would be likely to fail because the increased palm oil price would not offset the revenue lost, due to the lower quantity exported. But since the population elasticity of export demand is much higher than the income elasticity, it appears that most of the increase in export demand for palm oil in the future will be through population increases, rather than increases in income.

**A Political Economic Analysis of Philippine Sugar Pricing**, Rigoberto Adolfo Lopez *(University of Connecticut)*

The focus is on the motivation for government intervention in the Philippine sugar market. Regression analysis was performed using real consumer prices and the ratio of domestic/world sugar prices as dependent variables and a set of commodity and political market variables as regressors. The results support the notion that the government has responded to distributional concerns as well as political economic issues beyond the sugar market in the post-war period. The goals and weights placed on various societal groups changed markedly over the industrialization period (1945–61), the Marcos' years of political upheaval (1974–85), and the recent period under President Aquino (1986–92). The US quota, in particular, has played a central role.

**Livestock Pricing Policies and their Effects in Five Sub-Saharan African Countries**, Timothy O. Williams *(International Livestock Centre for Africa)*

Livestock pricing policies in many developing countries are often instituted without a good appreciation of the consequences of such policies for allocative efficiency, output, and trade. This paper evaluates, in a comparative cross-country context, the objectives and instruments of livestock pricing policy in five sub-Saharan African countries—Côte d'Ivoire, Mali, Nigeria, Sudan, and Zimbabwe—during the 1970–86 period. It assesses the extent to which pricing policy objectives have been attained and also estimates the effects of price interventions on output, consumption, trade, and government revenues. The empirical results indicate that, in comparison with real border prices, a certain degree of success was achieved in stabilizing real domestic producer prices. The results also show that since the early 1980s, there has been a gradual shift away from taxation of producers. However, consumers still appear to gain as much as producers in three of the study countries, with negative consequences for foreign exchange earnings and government revenues. The analysis reveals the importance of domestic inflation and exchange rates as key variables for livestock pricing policies and highlights the need to address the macroeconomic imbalances that cause exchange rate distortions and high domestic inflation at the same time that direct price distortions are being tackled.

**Prices of Livestock Products in Japan and other Countries**, Michio Sugiyama and Katsuyuki Oguri *(Gitu University)*

The price of livestock products (except eggs), especially beef, pigmeat, and chicken, has been high in Japan and higher than in the countries that export to Japan. In order to analyse the reasons for these high prices and clarify the marketing process, comparisons were made between countries in terms of prices at different stages of the marketing chain. Beef prices in Japan are 2.8 times those in the USA, and the prices of calves, live cattle, carcasses, and portioned and sliced meats are higher than in the USA. This is because the marketing channels are complicated and the cattle and meat merchants are numerous and scattered. Also, the marketing form or style changes with the marketing level; live animal, carcasses, and portioned and sliced meat. Integration of the broiler industry has advanced in Japan, but cutting and further processing have not been fully integrated. The reason why the processing cost of meat has been higher in Japan than in Thailand is the high price of land and labour. Pigmeat and egg prices are also analysed. Marketing and distribution costs could be reduced by enlarging the scale of production and integrating marketing and processing.

**Production Efficiency of Multiproduct Milk Plants in Taiwan**, Ming-Ming Wu and Tsorng-Chyi Hwang *(National Chung-Hsing University)*

This research applies the multiproduct translog cost function to analyse the efficiency of dairy plants producing milk for drinking, including fresh flavoured, preserved, and fermented milks, in Taiwan. The sample points are monthly data for 168 firms for 1988 and 1989. Dairy plants have an inelastic demand for domestic raw milk with respect to price. A small increase in raw milk supply may largely depress the milk price. Enlarging the difference between market and government fixed prices may increase the use of imported milk powders as substitutes for domestic raw milk. Providing that the short-run supply elasticity is smaller than that of demand, most risks of the decreased price should be borne by dairy farmers. The demand for research funds is elastic with respect to prices. Since research and development is beneficial to the performance of material, manufacturing, and management inputs, it is suggested that dairy plants invest in research and development and increase expenditure on it. Manufacturing costs and wages have a high degree of substitution. If the wage rate increases, dairy plants should increase capital inputs to improve automation. Plants already producing all dairy products are not recommended to increase outputs at the same time. Plants producing preserved milk may increase inputs to seize scale economies. Plants producing fresh flavoured and preserved milks should increase the production of fermented milk. There is no need for plants producing fresh milk only or plants producing all products except fresh milk to add the production of new products.

**Impacts of Alternative Government Policies on Farm Investments**, Doo Bong Han *(Korea Rural Economics Institute)* and John B. Penson, Jr. *(Texas A&M University)*

The focus of this study is modelling investment in durable farm inputs under alternative expectations patterns in the US farm sector. The neoclassical investment model is chosen over other approaches because of the direct linkages that can be established between farmers' investment and government policies, including macroeconomic policy and farm policy. The estimated equations for the alternative expectations patterns are endogenized into the AG-GEM model to examine the extent to which each hypothesis filters the impact of two topics of current interest: no chemical use and high deficits. *Ex ante* simulation results suggest that farmers' expectations may also depend on the durability of an asset or a commodity. Shorter lived assets (equipment) are more sensitive to current information-based expectations than longer lived assets (structures).

## Investment Behaviour of Farmers: An Empirical Assessment of Alternative Expectations Specifications, Geert Thijssen *(Wageningen Agricultural University)*

Demand functions for capital were estimated using an incomplete panel of Dutch dairy farms. Three forms of expectations formation about the path of future variables were used: static expectations, non-static expectations, and rational expectations. The models based on static and non-static expectations fit the data well. The own-price elasticity of the output is 0.10 in the short term, 0.21 in the intermediate term, and 0.84 in the long term for the model based on static expectations of prices and fixed factors. For the model based on non-static price expectations, the results are approximately the same. The results with respect to the rational expectations model are not consistent with the received theory. It is possible that this rejection of the rational expectation hypothesis is due to the instrumental variable estimation technique used.

## Cross-Country Comparisons of Technical Progress and Efficiency Gains in EC Agriculture, Eldon Ball and Hyunok Lee *(US Department of Agriculture)*

Not all countries are identical in their productivity performance. Their relative productivity efficiency differs. Productivity growth of the most efficient, or "best practice," countries requires new technical innovations. Those countries not performing at full efficiency, however, can potentially improve productivity by increasing efficiency with existing technology. The emerging EC–1992 will probably lead to increased flows of resources and technologies among member countries. Thus, the new economic environment will help countries improve efficiency levels. Evaluating a country's relative performance provides information on the potential efficiency gain of that country. Identifying the "best practice" technology, however, provides information on how to reallocate resources to capture those efficiency gains.

## Technology Adoption and Productivity: A Simultaneity Problem, Lydia Zepeda *(University of Wisconsin)*

An adoption model is estimated to determine factors affecting the adoption of a record-keeping system (DHIA) by California dairy farmers. Since productivity both influences technology adoption and is influenced by the adoption, a single-equation adoption model contains simultaneity bias. Thus, productivity and DHIA adoption are estimated as a mixed system of equations with continuous and discrete endogenous variables. Generalized probit does not rid the system of simultaneity bias, so two-stage least squares, as suggested by Heckman, is used. When the results are compared with biased single-equation estimates, the implications are quite different. System estimates indicate that DHIA does indeed improve productivity, while productivity has no effect on adoption of DHIA. The biased single-equation estimates indicate that education and industry involvement enhance DHIA adoption, while productivity diminishes it.

## Technology Adoption on Third-World Small Farms, Ganesh P. Rauniyar and Frank M. Goode *(Pennsylvania State University)*

The study examines the interrelationships among five technological adoption practices and identifies packages of practices. The socioeconomic determinants of these five individual practices are then compared with the determinants of adoption of a corresponding package of practices to determine whether explanatory variables explaining single practices are significantly different from the package model. Finally, conclusions are drawn from the package of practices models. The results suggest that five practices are summarized by two distinct but independent packages, referred to as the "basic" and the "advanced technology"

382

packages. Adoption of the basic technology package is explained by household educational level, revenue from sale of maize, proximity to the output market, amount of credit, number of cattle owned, and per-ha family labour units available on the farm. The number of household members working off the farm, the level of education of the male heads of household, and less variability in rainfall are important for "advanced technology" package adoption. Furthermore, the variables explaining adoption of individual practices are different from the variables explaining adoption of a package of practices. The study concludes that inferences would differ depending upon the nature of interrelationships among technological practices. If the practices are interrelated, as in this study, then the policies based on individual practice models will be misleading.

**The Impact of Tractorization on Employment in Agriculture Using the Translog Cost Function**, Farman Ali *(University of Agriculture, Pakistan)* and Ashok Parikh *(University of East Anglia)*

The objectives of this study are to study the impact of tractorization on employment and input use in the Northwest Frontier Province of Pakistan (NWFP) using farm-level survey data for 98 farms. The study established that human labour and tractors are substitutes for one another. Similarly, draught power and tractors are substitutes. These relationships are expected in NWFP agriculture. The demand for fertilizers is price elastic, and, if a subsidy scheme were operated, it would significantly increase this demand. On the other hand, the demand for human labour is price inelastic, due to the existence of surplus family labour. The policy of taxing tractor use can lead to greater use of both human labour and draught power. If there were a choice between taxing tractor services versus a subsidy on human labour, the former would be preferred if the objective was to increase labour employment.

**Application of Risk Analysis in the Adoption of New Maize Technology in Zaïre**, Glenn C.W. Ames *(University of Georgia)*, Donald W. Reid *(University of South Alabama)*, and Li-Fang Hsiou *(University of Georgia)*

A quadratic programming model was used to analyse the adoption of new maize technology, subject to minimum food security requirements and agricultural policy constraints, in the Kasai Oriental Region of Zaïre. Net returns to four levels of maize technology for the primary and secondary rainy seasons were evaluated in combination with staple food crops for four cropping systems with and without mandatory cotton production. The results indicate that cropping systems that include new maize technology were more risk efficient than those with local maize varieties. Minimum food security requirements were met. The analysis indicates that mandatory cotton production was not risk efficient at the prevailing price and yield levels in the Kasai-Oriental farming system. Mandatory cotton production not only lowered the level of profitability but also increased the variability of returns for a given level of expected returns. Zaïre and other African countries must analyse policy constraints to agricultural growth and evaluate on-farm technology. To stimulate the adoption process, the technology package needs to be evaluated from a risk efficiency perspective.

**Method for Assessing and Verifying Efficiency of Farm Management in Vegetable Growing**, Masayuki Yoshida *(Chiba University)*

The aim is to prove that discriminant analysis can be applied to assessment of management efficiency in vegetable growing and that statistical methods can be used effectively to perform an assessment by variety of crop. Discriminant analysis is found to be effective for analysis of a single production activity, and the method is considered superior to the direct comparison method in that it makes analysis of factors possible. Discriminant and standardized scores are useful to discover analytically defects in farm management and to make an assessment of these defects; in order to make a diagnosis by variety, regression analysis and other statistical methods are useful to analyse such determining factors of significant management indexes as shipment price and yield per 10 ares. It was also found that, in regression analysis, calculated residuals can be used for the diagnosis.

## Market Factors, Government Policies, and Adoption of New Technology by Small Farmers in Honduras, Miguel A. Lopes *(Centro Internacional de Mejoramiento de Maiz y Trigo)*

This paper reports the findings of a recent study on the potential effects of new soil erosion control and sorghum technologies on the income and productivity of small farmers in Honduras and explores the relationship between potential adoption rates and government agricultural policies. A discrete stochastic programming model, based on the results of a socioeconomic survey of farmers in the region, was used for the analysis. Results indicate that the soil conservation technologies are profitable and enable more intensive and sustainable hillside farming systems. When resource constraints and key policy variables are relaxed, improved sorghum cultivars combined with fertilizer and insecticide would allow further productivity gains once the land is improved. Income gains of 70 percent, risk reduction by 15 percent, and doubled expected grain production can be achieved when both technologies are introduced and the cereal price collapse is avoided. Suggested alternatives to moderate the price reductions expected from increased cereal production include storage bins for surplus sorghum and increased small animal production using surplus grain for feed. Policy initiatives that encourage these activities and offer greater access to official credit at more favourable conditions would facilitate adoption of the technologies.

## The CAP Faces European Unification and the GATT, David R. Kelch and Mary Lisa Madell *(US Department of Agriculture)*

The unification of Europe through the EC harmonization programme, German unification, economic and monetary union, and political union will affect the production, consumption, and net trade of food and agricultural products in the EC. The Common Agricultural Policy (CAP) of the EC is also under pressure from its agricultural trading partners in the Uruguay Round of the GATT. The EC's harmonization programme threatens the agrimonetary system, which is the price-setting mechanism of the CAP and its most distorting aspect. German unification will add more productive capacity to the CAP in products that are already in surplus. The GATT negotiations should result in lower CAP prices and reductions in other forms of production-enhancing support. The adoption of market-oriented agricultural policies in Eastern Europe will add to the oversupply of temperate zone products in Europe. All these pressures should lead to lower CAP prices and less farm support.

## Estimation of Wheat Production Response Functions by Province for China: Implications for Trade, Shwu-Eng H. Webb *(US Department of Agriculture)* and Catherine Halbrendt *(University of Delaware)*

This paper uses crop production acreage and price data by province over the 1979–88 period to estimate provincial wheat production functions for China. The study shows that wheat production during the 1980s in China was responsive to changes in crop prices. Peasants in China still have very limited power in allocating land in response to prices. Within this constraint, the economic incentive in terms of crop prices does provide a mechanism to allocate variable inputs and, thus, affect yield and production. This paper further analyses the impact of continuing reform in agricultural production on provincial wheat production. The effects of crop prices on wheat self-sufficiency are discussed.

**An Application of Composite Forecasting to Predict the Impact of EC Milk Quotas on Milk Supply at the Farm Level**, Franco Rosa *(Università Cattòlica di Piacenza)*

Milk quotas introduced in the EC since 1984 have induced reactions from milk producers due to changes in market prospects and expected revenues in the short to medium term. An attempt is made to simulate, using alternative forecasting techniques, the future supply of milk in Italy and to evaluate farmer reaction to quotas. Statistical tests showed that composite forecasting using adaptive and minimum variance weights was able to predict milk supply more accurately than other forecasting methods (single and composite). The forecasts obtained predicted the supply pattern quite closely and confirmed the decline in supply and the overreaction of producers to the market uncertainty caused by the quota restriction.

**The Growing Demand for Food Quality: Implications for Agricultural and Trade Policy**, Harald von Witzke *(University of Minnesota)* and Ian M. Sheldon *(Ohio State University)*

The growing demand for many food quality components is usually expressed in the form of a growing demand for food quality standards. Such standards represent public goods and thus involve market failure. The general focus of this paper is on both the nature of this market failure and its central implications for agricultural and trade policy. If competitive markets do not generate the necessary standard-assuring mechanisms, the nature of such a market failure needs to be understood. By setting out a simple model of contractual enforcement, it is possible to show that asymmetric information on food quality components is sufficient for the quality enforcement mechanism not to work in the case of food safety. Given that public institutions will tend to set nationally divergent food quality standards, many will act as barriers to trade. Therefore, political coalitions between consumers and agricultural producers are likely to gain in importance, which will add a new dimension to attempts at international agricultural and trade policy coordination.

**On the Importance of Including Microeconomic Information in General Equilibrium Models**, Federico Perali *(University of Wisconsin)*

The main objective of the study is to evaluate the advantages for policy making of including microeconomic information within general equilibrium models. What is often neglected is the richness of information contained in Sah and Stiglitz's description of the economics of price scissors, both in regard to producer and consumer choices and to government behaviour in choosing the rules to reform the terms of trade. In particular, input, land, labour, and consumption allocation decisions are not modelled by taking into account the theory of the farm household and government behaviour in the process of price formation. Following the analytical structure exposed in Sah and Stiglitz, the present study proposes some modifications of classical structuralist general equilibrium models to incorporate such micro-information and discusses some policy implications.

# Food Consumption, Marketing, and Credit

**Food Consumption Patterns of Asian Countries**, Michio Kanai *(National Research Institute of Agricultural Economics, Japan)*

This paper seeks to classify 27 Asian countries by their food consumption patterns using a cluster analysis method. Five possible groups, Asia high- and middle-income, West Asia, Southeast and East Asia low-income, South Asia low-income, and isolated countries, were found. The food consumption patterns of countries belonging to the Asia high- and middle-income group change more rapidly over time and seem to become relatively similar.

**Changes of Diet Patterns in Japan, Korea, and Taiwan: Application of Principal Component Analysis to Food Balance Sheets**, Yasuhiko Yuize and Hiromichi Inaba *(Chiba University)* and Youko Miura *(Chiba Keizai University)*

Economic development tends to change diet patterns, which has an effect on the food industry. The significance of this change is assessed by applying principal component analysis to the food balance sheets of Japan, Korea, and Taiwan during 1965–87. The Japanese experience is analysed in detail, so that changes in the three countries can be compared on the basis of the Japanese case. Finally, changes in the diet patterns in the three countries are classified by plotting the scores of the second and third components with those of the first component. Three main components in the three countries can explain almost the all changes in food patterns. The first component, the trend or structural change factor, suggests that Japan and Korea belong to the same group, while Taiwan is different. The third component, the stability or nutrition adjustment factor, suggests that Japan and Taiwan belong to the same group, with Korea the exception.

**Estimating Consumer Purchase Intentions for Organic Produce**, Chung L. Huang and Sukant Misra *(University of Georgia)*

The study postulated a sequential probit model to estimate consumer purchase intentions for organic produce based on data collected from a mail survey of Georgia residents. The first equation estimates the probability that a respondent would prefer to buy organic produce. Results suggested the profile for potential organic buyers is consumers who have fixed attitudes on use of chemical pesticides on fresh produce, want produce to be tested and certified as residue-free, tend to believe that organic produce offers better nutritional value, are younger than 30 years or older than 60 years, and have annual household income of less than $20,000. The second equation estimated the probability that a potential organic produce consumer would buy organic produce with apparent sensory defects. Results suggested that consumers who would accept low sensory quality are likely to be white, have a higher educational level, and have a larger family. The study concludes that consumers have found organic produce attractive mainly because of concerns about food safety associated with pesticide residues. The most important factors for the marketing potential of organic produce are testing and certification for freedom of chemical residues, good sensory qualities, and a reasonable price that is competitive with conventionally grown produce.

**Optimal Marketing Decisions under Imperfect Competition with Price Uncertainty: A Programming Model with Stochastic Demand Functions**, Teruaki Nanseki *(National Agricultural Research Centre, Japan)*

This paper presents a new optimal marketing planning model where stochastic demand functions, transport costs, and various marketing constraints are incorporated. The model is formulated by quadratic programming with stochastic linear coefficients. This model is applied to problems of marketing pimentos in Miyazaki Prefecture, Japan. The risk averse and risk neutral optimal solutions are obtained and compared with actual quantity marketed. The results show that the risk averse solution is closer to the actual marketing quantity than the risk neutral solution, which is equivalent to that of an ordinary quadratic programming model with deterministic revenue functions. This implies that risk plays an important role in the marketing planning model under imperfect competition as well as in the model under perfect competition.

**Branded Product Licencing: An Alternative International Marketing Strategy for Food and Beverages**, Ian M. Sheldon and Dennis R. Henderson *(Ohio State University)*

Empirical evidence shows international licencing of the production and marketing of branded food and related products to be an important aspect of the globalization of the food industry. Empirical evidence on the extent of international licencing is presented. Recent theoretical literature on licencing has dealt only with the licencing of process technologies. The paper therefore considers the possible motives for branded product licencing using a simple game-theoretic structure. The results suggest that imperfect competition in overseas markets and imperfect information may be important determinants of international product licencing. For a licensor, product licencing can be treated as a substitute for either exporting or direct foreign investment or as part of a long-term strategy for overseas market development. For a licencee, licencing may represent a lower cost method of product line extension and/or a means of discouraging market entry by a foreign competitor.

### Farmers' Selling Decisions for Vegetables in Taiwan: The Case of Cabbage, Tsorng-Chyi Hwang and Ming-Ming Wu *(National Chung-Hsing University)*

Four vegetable marketing channels are proposed for cabbage farmers in Taiwan to sell their products. The aim is to find ways of improving cooperative marketing for vegetables. Eighty-one observations are estimated using the seemingly unrelated regression technique. The main findings are that small, full-time cabbage farmers tend to prefer direct marketing but those involved in processing prefer local dealers. Cabbage farmers do not trust grading standards and would like to have unstable prices. For other marketing channels, the nearby wholesale market is preferred by experienced farmers and those with labour and cash shortages. Cooperative marketing is preferred by growers, those with labour and cash shortages, and those near collection points.

### The Competitive Structure among Japanese Wholesale Fruit and Vegetable Markets and Consumer Welfare, Fujishima Hiroji *(Chugoku National Agricultural Experiment Station)*

Currently, 22 Mt of fresh produce (fruit and vegetables), or three times as much as rice, is marketed every year in Japan. Nearly 90 percent of this produce is distributed through wholesale markets. The objective of this report is to explain variations in the competitive structure among wholesale markets for fresh produce in the last two decades and the increase in consumer welfare due to this variation. Competition among the large, central wholesale markets plays the dominant role, although it was weak in the past. An increase in the distribution areas for the central wholesale markets and distribution to supermarkets has increased competition. A high level of competition among the central wholesale markets has enabled consumers to purchase any item for a long period throughout the year, more items than before, and each item at almost the same price in any part of the country.

### The Reorganization and Future of the Rice Control System in Japan, Tokuzo Mishima *(Hokkaido University)*

In 1942, the Japanese government introduced the Food Control System *(Shokuryo Kanri Seido)* in order to administer the production and consumption of food during wartime. This system evolved from the end of the war (1945) until the early 1960s as a system to protect the interests of both the rice-producing farmers and rice consumers. As a result of overproduction and a decrease in demand for rice since the late 1960s, the system was reorganized by weakening its former controlling function and, at the same time, strengthening its marketing function. However, it is apparent that the liberalization of domestic distribution will inevitably lead to the participation of large firms in the rice market and liberalization of rice imports. Since rice is a staple food and a key crop, it is essential to introduce necessary reform to the rice control system and also to ensure its functioning in order to maintain self-sufficiency in rice.

**Scale and Scope Economies of Japanese Agricultural Cooperatives: Multiproduct Cost Function Approach**, Tamotsu Kawamura *(Iwate University)*

This paper clarifies the present state of scale and scope economies in Japanese multipurpose agricultural cooperatives using multiproduct production theory. Their translog multiproduct cost functions are first estimated, and measurements of scale and scope economies are then derived from the estimated cost functions. The main conclusions are that Japanese agricultural cooperatives have multiproduct scale economies, product-specific scale economies are observed only in credit cooperatives but these also have scope economies, and there is cost complementarity between credit and purchasing cooperatives, while other combinations are not effective for cost saving.

**Eroding the Remnants of Apartheid in Namibia: Structural Adjustment in the Beef Industry**, Diethelm Metzger *(farmer, Namibia)* and Helmke Sartorius von Bach and Johan van Zyl *(University of Pretoria)*

The paper considers the regional influence of prices and access to markets on beef cattle numbers in Namibia, given the present inequalities, using econometric analysis of time-series data. This is particularly relevant because Namibia is in a process of structural adjustment following its recent independence and in view of the beef sector's importance in the Namibian economy. The results accentuate the role of access to markets in beef production in Namibia. In cases where access is severely restricted due to lack of infrastructure like processing facilities and adequate transport, for example in the communal regions, beef producers do not act on price incentives or react to climatological and ecological variables. On the other hand, beef producers with limited access to markets, mainly due to high transport costs, do react to environmental changes, but not to price incentives. Only producers with easy access to markets react to both environmental changes and price incentives. The major conclusion is that the present production and marketing structure in Namibia with respect to beef is probably non-optimal. The results highlight the need for an overall policy that accounts for all related industries, producers, consumers, and other relevant factors simultaneously.

**The Marketing of Live Small Ruminants and Meat in Highland Balochistan**, Khalid Mahomood *(Arid Zone Research Institute, Pakistan)* and Abelardo Rodriguez *(International Centre for Agricultural Research in the Dry Areas)*

The study identified livestock/meat marketing practices in highland Balochistan, Pakistan, through interviews with producers, village dealers, wholesalers, commission agents, butchers, and consumers. The majority of small ruminants are produced under transhumant/ nomadic pastoralist systems. Producers showed a lack of knowledge regarding market forces and quality of livestock, thus limiting their ability to increase their income. However, they incorporate liveweight in their perception of livestock price per unit of weight. The average liveweight of sheep and goats was 24.1 kg/head with an estimated farm-gate price of Rs 496/ head. The price paid by consumers was Rs 716/head. The services of intermediaries in the marketing chain represent 31 percent of the price paid by the consumers. Meat grading is absent, but there is government regulation of the retail price. Consumers thus have no mechanisms to convey their degree of dissatisfaction to producers through intermediaries in the marketing chain. Most of these services could be improved for the benefit of consumers and producers: the overall volume of the market could be higher, meat quality could be more uniform, and some marketing costs could be reduced. Extension efforts to improve producer market awareness, however, will face the risk minimizing strategy of the highland Balochistan pastoralists.

## Inappropriate Rural Banking Policy Assumptions in a Developing Economy: Evidence from a Discriminant Analysis, Aja Okorie *(University of Agriculture, Nigeria)*

Ten years after the establishment of the Rural Banking Scheme in Nigeria, there are clear indications that the problems and issues that led to it are still prevalent. This study questions the validity of the central assumption of the scheme, that increasing the physical proximity of banks to the rural people enhances rural savings mobilization and increases the flow of funds to the rural sector. Rural residents were surveyed, and discriminant analysis of the data showed that four variables were significant in discriminating between rural bank users and non-users: household income, years of formal education, gender of respondent, and the awareness of the existence of the rural bank branch. The proximity of the bank to the respondent's residence was not a significant determining variable. These findings have important implications for rural bank designers and implementors in Nigeria and other developing countries. They suggest that the current excessive emphasis on physical distance as a critical factor in rural bank development in Nigeria should give way to a more comprehensive strategy, which would incorporate the four critical variables identified in this study.

## Loan Repayment in Rural Financial Markets: A Multinomial Logit Analysis?, Nelson Aguilera-Alfred and Claudio Gonzalez-Vega *(Ohio State University)*

This paper analyses the repayment performance of loans disbursed by a typical developing country specialized lender, the Agricultural Development Bank of the Dominican Republic. It shows that loans in default are just one dimension of the repayment problems faced by specialized lenders and that rescheduling and payment of arrears should not be ignored. It also shows that, by following through time the status of loans disbursed in a particular period, the factors determining repayment performance may be better identified. It also shows how the results of multinomial logit analysis can be used by lenders to analyse the various types of potential repayment problems that they encounter.

## Agricultural Loan Recovery Strategies in a Developing Economy: A Case Study of Imo State, Nigeria, Andrew C. Iheanacho *(University of Maiduguri)* and Aja Okorie *(University of Agriculture, Nigeria)*

Recovery of loans is critical to long-run viability of agricultural credit systems in most developing countries. There exists little understanding of the strategies and considerations lending agencies adopt in dealing with the problem. This paper thus provides some empirical evidence on loan recovery strategies of agricultural lending agencies in Imo State, Nigeria. It shows that both formal and informal lending agencies strive to deal with the defaulters in what they perceive as the most cost-effective way. Adoption of any strategy depends on a complex interaction of factors on the demand and supply sides of credit. These suggest that the lender, the borrower, and socio-political factors must be taken into consideration in planning effective loan recovery strategies in developing countries.

## Applying Contestability Theory to Rural Informal Credit Markets: What Do We Gain?, Emmanuel F. Esguerra, Geetha Nagarajan, and Richard L. Meyer *(Ohio State University)*

The proposition of the theory of market contestability is applied to analyse whether certain identifiable features of the rural credit markets lend support to the contestable market hypothesis. Data of the type typically available from field surveys are used to argue that contestability of the rural informal credit market in the Philippines, given the methodological and data limitations, is empirically difficult to prove. The paper further argues that the effort to find evidence of market contestability is not necessary to argue against government regulation of informal financial markets.

## Are Land Banks Feasible as Market-Driven Substitutes or Complements of Land Reform? Some Insights for Central America, John Strasma *(University of Wisconsin)*

With land reforms in Central America completed or politically unviable, land banks could enable the landless to buy land parcels in the market. Pilot programmes are not replicable, but this paper suggests that land banks could be solvent, could transfer a significant amount of land, and could be viable without continuous government or external subsidy. Downpayments would be indispensable, and these banks would not provide costly social services. Buyers could resell the land freely to other *campesinos*, who would assume their land debt. Payments should be indexed to product prices where inflation is a problem. Sellers will have to carry much of the paper transaction as land contracts, increasing the land price to provide what they deem an acceptable return. Loan delinquency was eliminated successfully in a private Salvadoran land sale programme and in a Dominican Republic rice production credit programme. The former uses renting with option to buy; non-payers can be evicted quickly for non-payment of rent. In the Dominican Republic example, informal, temporary foreclosure by peers is used as the enforcement mechanism. Thus, creative methods can eliminate loan delinquency, a major threat to a land bank. The real key is to eliminate paternalism, allowing unsuccessful farmers to sell freely to others, as in any other business.

## Informal Finance through Land Pawning Contracts: Evidence from the Philippines, Geetha Nagarajan *(Ohio State University)*, Cristina C. David *(International Rice Research Institute)*, and Richard L. Meyer *(Ohio State University)*

Land pawning contracts in which the pawner temporarily transfers his land cultivation rights to the pawnee in return for a loan with an agreement to redeem it on loan repayment have increased in importance in Philippine rice-growing villages. This paper uses cross-sectional data from farm households in five heterogenous production environments to analyse the determinants of the choice of pawning contracts. The analysis shows that land pawning is an informal credit instrument used by small farmers to obtain large loans to finance productive investments, such as non-farm employment, where the returns to investment are high. An econometric model was developed to examine the factors affecting the choice of pawning contracts and the observed loan size. The results suggest that farm households with poorer quality land, smaller farm sizes, and lower physical and human assets pawn out land, while wealthier farm households with larger farm sizes and greater physical and human assets pawn in. The observed loan size is explained by reputation of pawners and rice cropping intensity in the region.

## Agricultural Budget for Land Improvement: Economic Significance and Issues, Junichi Shimizu *(National Research Institute of Agricultural Economics, Japan)*

Since World War II, investment in land improvement, along with expenditure on the food control system, has had a major share in public expenditure on agriculture in Japan. The primary economic function of land improvement is its use as a means of executing fiscal policy to control total demand in a Keynesian fashion. Its secondary function is its use in adjusting resource allocation. The post-war agrarian reform in Japan has resulted in a system of land ownership in a which a large number of farmers own small parcels of dispersed land. Today, when the issue has become one of cost reduction, government intervention to develop farmland has significance as supplying public goods. In fact, the emphasis of land improvement investment has shifted from irrigation and drainage to farmland partition adjustment. A future issue is likely to be comprehensive development of social overhead capital in rural communities.

## Socioeconomic Impact of Credit on Rural Indian Households:  A Gender Analysis
Elavia Behroz Hosi *(University of Baroda)*, Kathleen Cloud *(University of Illinois)*, and Alam Zafar and Chavan Hema *(University of Baroda)*

This paper presents a methodology for gender analysis of levels and effects of credit at the household level in Indian rural households.  A regression function is used for estimating the impact of credit on the income and the value of production gained by men and women borrowers from activity/asset finance using data collected by a primary survey.  The major policy implication of the analysis is that credit given in small amounts has little effect on poverty eradication, particularly in the case of women.  The impact of credit on incremental income and work days was higher in the case of men than for women.  However, the opposite was true in the case of value of production.  Application of the proposed methodology to the experience of other Third World countries will provide insight and data for translating research activities into effective programmes for enhanced women's participation in rural credit.

## Wheat Policy Reform in Ecuador:  Analysis of the Distributional and Nutritional Effects of the Wheat Import Subsidy, Andres Guarderas *(Banco del Pacífico, Ecuador)* and David R. Lee *(Cornell University)*

Wheat consumption has risen sharply in developing countries in recent years, partly due to widespread food subsidies that have often encouraged the consumption of wheat-based products at the expense of "traditional" food staples.  These food subsidy policies have often been key targets of structural adjustment and policy reform programmes instituted over the past decade.  This paper examines the household economic and nutritional effects of the removal of the wheat import subsidy in one such country, Ecuador, in 1989.  The analysis centres on the estimation of Tobit food demand equations for a sample of 9,518 urban households, divided into four geographic groups and four income quartiles.  The results show that the wheat import policy induced higher consumption of wheat-based products in Ecuador over the 1973–88 period, particularly among Sierran consumers.  High-income consumers benefited disproportionately from the subsidy, due to both their greater absolute demand for wheat-based products and their lower estimated cross-elasticity of demand for substitutes.  Caloric demand and nutritional effects are estimated to be minimal, however, due to the fact that the higher caloric demand elasticities among lower income groups are offset by lower absolute caloric availability.  Alternative, more efficient, subsidy policies are simulated and analysed.

## Marketing Problems of Rural Sources of Livelihood:  Can the Transaction Cost Approach Help in Dealing with Them?, Petri Ollila *(University of Helsinki)*

The paper sees the problems of rural sources of livelihood as market failures.  The prevailing marketing system is not capable of transmitting information leading to proper production decisions among rural entrepreneurs.  An attempt is made to experiment with the usefulness of the transaction cost approach (TRC) in dealing with these problems.  The paper concludes that TRC has potential but at its present stage lacks operational research methodology.  Combining TRC with the industrial organization approach is suggested.

## Food Self-Sufficiency, Fertilizer Use, and Access to Formal Credit:  A Test of Relationships on Small Farms in Ethiopia, William Grisley *(Centro Internacional de Agricultura Tropical)*, Wilfred Mwangi *(Centro Internacional de Mejoramiento de Maiz y Trigo)*, and Gethun Degu *(Institute of Agricultural Research, Ethiopia)*

The relationships between household food self-sufficiency and fertilizer use and between fertilizer use and credit obtained from an agricultural credit bank are investigated for a 1989 sample of 160 rural households living in the highland area of Areka in southern Ethiopia.  A logit model is used in estimation; 30 of the 160 farmers were found to be self-sufficient in food production. Obtaining bank credit increased the odds to 1 in 20 (in favour of fertilizer) that fertilizer would be used, and use of fertilizer increased the odds of being self-sufficient in food to 2 in 5. Household food self-sufficiency can thus be enhanced by strengthening the credit-fertilizer linkage. Households obtaining credit from informal sources did not tend to use fertilizer. The number of coffee trees was not significant in explaining food self-sufficiency. A probable reason is the production of food crops under coffee trees. This result can have important implications for other countries in eastern Africa as farmers are prohibited from intercropping coffee. Larger families decreased the odds of self-sufficiency, while farm size was not associated with self-sufficiency.

**Willingness to Pay for pST-Treated Pigmeat**, A. Elnagheeb, W.J. Florkowski, and C.L. Huang *(University of Georgia)* and C. Halbrendt *(University of Delaware)*

Progress in application of porcine somatotropin (pST) justified an assessment of consumer acceptance of pST-treated pigmeat. A survey of the metropolitan area of Atlanta, Georgia, USA, collected information about consumer attitudes towards lean pigmeat produced with biotechnologically developed pST. A qualitative dependent variable model was used to identify socioeconomic consumer characteristics influencing the willingness to pay for lean pigmeat. The model was modified to account for the selectivity bias of the sample data. Results indicate that frequent pigmeat consumers were willing to pay more for lean pigmeat produced using pST in contrast to respondents who frequently ate beef, were older, and had relatively high income. Probabilities associated with the willingness to pay a specific premium were calculated. In general, the average respondent was willing to pay an additional $0.18 per kg of lean pigmeat produced using pST.

# Author Index

## IAAE Occasional Papers Nos. 1–6